Papermaking Science and Technology

a series of 19 books
covering the latest
technology and
future trends

Book 15

Materials, Corrosion Prevention, and Maintenance

Series editors
Johan Gullichsen, Helsinki University of Technology
Hannu Paulapuro, Helsinki University of Technology

Book editors
Jari Aromaa, Helsinki University of Technology
Anja Klarin, ÅF-IPK AB

Series reviewer
Brian Attwood, St. Anne's Paper and Paperboard Developments, Ltd.

Book reviewer
Dave Bennett, Champion Corp.

Published in cooperation with the Finnish Paper Engineers' Association and TAPPI

Cover photo by Martin MacLeod

ISBN 952-5216-00-4 (the series)
ISBN 952-5216-15-2 (book 15)

Published by Fapet Oy
(Fapet Oy, PO BOX 146, FIN-00171 HELSINKI, FINLAND)

Copyright © 1999 by Fapet Oy. All rights reserved.

Printed by Gummerus Printing, Jyväskylä, Finland 1999

Printed on LumiMatt 100 g/m^2, Enso Fine Papers Oy, Imatra Mills

Certain figures in this publication have been reprinted by permission of TAPPI.

Foreword

Johan Gullichsen and Hannu Paulapuro

PAPERMAKING SCIENCE AND TECHNOLOGY

Papermaking is a vast, multidisciplinary technology that has expanded tremendously in recent years. Significant advances have been made in all areas of papermaking, including raw materials, production technology, process control and end products. The complexity of the processes, the scale of operation and production speeds leave little room for error or malfunction. Modern papermaking would not be possible without a proper command of a great variety of technologies, in particular advanced process control and diagnostic methods. Not only has the technology progressed and new technology emerged, but our understanding of the fundamentals of unit processes, raw materials and product properties has also deepened considerably. The variations in the industry's heterogeneous raw materials, and the sophistication of pulping and papermaking processes require a profound understanding of the mechanisms involved. Paper and board products are complex in structure and contain many different components. The requirements placed on the way these products perform are wide, varied and often conflicting. Those involved in product development will continue to need a profound understanding of the chemistry and physics of both raw materials and product structures.

Paper has played a vital role in the cultural development of mankind. It still has a key role in communication and is needed in many other areas of our society. There is no doubt that it will continue to have an important place in the future. Paper must, however, maintain its competitiveness through continuous product development in order to meet the ever-increasing demands on its performance. It must also be produced economically by environment-friendly processes with the minimum use of resources. To meet these challenges, everyone working in this field must seek solutions by applying the basic sciences of engineering and economics in an integrated, multidisciplinary way.

The Finnish Paper Engineers' Association has previously published textbooks and handbooks on pulping and papermaking. The last edition appeared in the early 80's. There is now a clear need for a new series of books. It was felt that the new series should provide more comprehensive coverage of all aspects of papermaking science and technology. Also, that it should meet the need for an academic-level textbook and at the same time serve as a handbook for production and management people working in this field. The result is this series of 19 volumes, which is also available as a CD-ROM.

When the decision was made to publish the series in English, it was natural to seek the assistance of an international organization in this field. TAPPI was the obvious partner as it is very active in publishing books and other educational material on pulping and papermaking. TAPPI immediately understood the significance of the suggested new series, and readily agreed to assist. As most of the contributors to the series are Finnish, TAPPI provided North American reviewers for each volume in the series. Mr. Brian Attwood was appointed overall reviewer for the series as a whole. His input is gratefully acknowledged. We thank TAPPI and its representatives for their valuable contribution throughout the project. Thanks are also due to all TAPPI-appointed reviewers, whose work has been invaluable in finalizing the text and in maintaining a high standard throughout the series.

A project like this could never have succeeded without contributors of the very highest standard. Their motivation, enthusiasm and the ability to produce the necessary material in a reasonable time has made our work both easy and enjoyable. We have also learned a lot in our "own field" by reading the excellent manuscripts for these books.

We also wish to thank FAPET (Finnish American Paper Engineers' Textbook), which is handling the entire project. We are especially obliged to Ms. Mari Barck, the project coordinator. Her devotion, patience and hard work have been instrumental in getting the project completed on schedule.

Finally, we wish to thank the following companies for their financial support:

A. Ahlstrom Corporation
Stora Enso Oyj
Kemira Oy
Metsä-Serla Corporation
Rauma Corporation
Raisio Chemicals Ltd
Tamfelt Corporation
UPM-Kymmene Corporation

We are confident that this series of books will find its way into the hands of numerous students, paper engineers, production and mill managers and even professors. For those who prefer the use of electronic media, the CD-ROM form will provide all that is contained in the printed version. We anticipate they will soon make paper copies of most of the material.

List of Contributors

Aromaa Jari, D.Sc (Tech), Senior Lecturer, Helsinki University of Technology

Klarin Anja, D.Sc (Tech), MBA, Senior Consultant, ÅF-IPK AB

Preface
Jari Aromaa
Anja Klarin

Preventive maintenance is becoming a key factor in the process industry to obtain maximum operating time and peak equipment performance with minimum cost. This requires maximum performance from the equipment using the assets and people available. Corrosion is a poorly understood phenomena causing production interruptions and profit loss. Corrosion and corrosion prevention are factors that directly influence process operation, economics, and safety. Plant shutdowns, product contamination, and loss of valuable products are problems no manager wants. Corrosion and corrosion prevention are primarily economic problems. Cost considerations usually determine material selection and application of corrosion prevention methods. Proper material selection can decrease investment costs, and proper corrosion prevention and maintenance procedures will increase profit. Safety and reliability of process equipment has primary importance when using hazardous materials or operating at high pressures and temperatures. Material selection must then use reliability rather than economics. For many structures and components, material selection uses mechanical and physical properties. Corrosion resistance has secondary importance. Corrosion resistance might be an important factor, but engineering properties also require consideration.

This book provides an overview of materials, environments, and their interaction. This interaction is corrosion. Methods of corrosion control, monitoring, maintenance, etc., provide a starting point for solving corrosion problems. This book is not a list of solutions to particular problems. Such a list would be impractical since the coverage would not be complete and current because materials and environments change constantly. The book also does not give detailed process descriptions. The reader should study relevant books in this series for such information. The examples given for corrosion forms, corrosive environments, and corrosion prevention are common in the pulp and paper industry. Managing corrosion is a problem of information distribution and management. Unfortunately, understanding why, where, and when corrosion damages happen is more difficult than understanding mechanical failures of machinery.

Maintenance has become an increasingly vital business function through its goal of assuring maximum equipment operating time at peak performance and lowest total costs. The modern approach of preventive maintenance includes routine care, regular inspections and planned repairs, and overhauls and reconditioning to assure the highest equipment service reliability. Corrosion prevention should be a major task in every preventive maintenance program.

This book has two major sources. The first part on materials and corrosion used metallurgy and corrosion courses from the Helsinki University of Technology, Department of Materials Science and Rock Engineering. The ideas in the corrosion portion come from classic textbooks such as those of U. R. Evans, N. D. Tomashov, M. Pourbaix, and H. H. Uhlig. The second part covering maintenance uses the MBA thesis of an author of this volume. The maintenance review uses literature published on maintenance since the middle 1950s. Interviews with pulp and paper producers and equipment vendors have been sporadic and therefore have received only brief mention. Similar to the section on corrosion problem solving, the authors have avoided specific product names and companies. Comprehensive and detailed information would be impossible to provide, and it would not be current.

The authors are indebted to Professor Seppo Yläsaari at Helsinki University of Technology who encouraged them to write this book.

Espoo
May 1999

Jari Aromaa   Anja Klarin

Table of Contents

1. Introduction 12
2. Construction materials 19
3. Electrochemistry of metallic corrosion and corrosion test methods 67
4. Corrosion forms 123
5. Corrosion in different environments 175
6. Corrosion prevention 249
7. Corrosion monitoring 331
8. Corrosion management 347
9. Concepts of maintenance in the pulp and paper industry 359
   Conversion factors 429
   Index 431

CHAPTER 1

Introduction

1   Corrosion and corrosion prevention 13
2   Maintenance 15

CHAPTER 1

Jari Aromaa and Anja Klarin

Introduction

Corrosion is the natural degradation process of most construction materials. The term has frequent association with metals, but it is also applies to plastics, concrete, wood, etc. Although most people are familiar with corrosion and corrosion prevention in their daily lives, they often ignore this phenomenon in large scale operations. Several studies of the cost of corrosion in developed countries give an estimate of 3%–4% of the gross national product. Another estimate is that using existing technology and knowledge could prevent one-third of this cost. This does not include hidden costs such as energy and raw material waste on a global scale or environmental considerations:

- The Report of the Committee on Corrosion and Protection by T.P. Hoar in 1971 concluded that the cost of corrosion in the United Kingdom was 3.5% of Gross National Product (GNP).

- Battelle and the Specialty Steel Industry of North America estimated in 1996 that the annual cost of metallic corrosion in the United States economy is approximately US$300 billion. The report estimates that about one-third of the cost of corrosion (US$100 billion) is avoidable and could be saved by broader application of corrosion resistant materials and application of proper anti-corrosive practice from design through maintenance. The estimates result from a partial update by Battelle scientists of the findings of a study conducted by Battelle and the National Institute of Standards and Technology titled "Economic Effects of Metallic Corrosion in the United States." The original work in 1978 used an estimate that in 1975 metallic corrosion cost the United States US$82 billion (4.9 percent of the GNP). Approximately $33 billion was avoidable by use of best practices available.

- In the Scandinavian pulp and paper industry, the cost of corrosion is approximately US$ 25–30 for a ton of pulp. Direct material costs are about 40% of these costs. Indirect costs such as loss of production or poor quality are more difficult to quantify.

Corrosion damages are not always avoidable. Sometimes it is not economical to avoid them. Corrosion problems are primarily economical problems. They require a balance between cost of corrosion and cost of prevention.

Corrosion science and engineering are interdisciplinary fields. The main topics are chemistry, electrochemistry, and material sciences. Corrosion engineers must understand manufacturing processes and know the environment in which they are operating. The operational environment can be a chemical process or a construction project. Engineers must know the internal workings and how various changes will influence it.

Introduction

Corrosion engineers must also have considerable common sense to decide which corrosion prevention measures will be most economical. One can calculate the profitability of a corrosion prevention investment using common methods. Because corrosion prevention is not an exact science, experience gained over time has more value here than in many other engineering fields. Success in corrosion prevention requires knowledge in many fields including the following:

- Equipment design and construction
- Material science
- Chemistry
- Mill operation
- Inspection.

1 Corrosion and corrosion prevention

Corrosion is a physicochemical interaction between a metal and its environment that results in unwanted changes in the properties of the metal. It may lead to impairment of the metal, a technical system, the environment, or a product. Modern corrosion theory uses two basic elements: mixed potential theory and formation of surface films. Most corrosion phenomena are electrochemical. Different metal and solution heterogeneities and geometric factors will cause potential differences leading to corrosion. Some factors are unavoidable, but proper manufacturing and installation procedures can remove others.

Corrosion is the result of a cathodic reduction reaction. Cathodic reactions consume electrons that release on the anode and transfer to the cathode. This corrosion cell is a closed electric circuit. The cell voltage is the driving force, and the current flowing in the cell obeys Ohm's law. Higher cell voltage and lower resistance values in the system give higher cell current. An anode or cathode can support more than one reaction, but the total anodic currents must equal total cathodic currents. This requirement leads to mixed potential theory. Corrosion processes begin with removal of atoms from the metal surface by reactions with the environment. These metal ions may react with the species in the electrolyte and form reaction products. Under appropriate conditions, these reaction products may form surface films on the metal. If the surface film protects the metal from further corrosion, the metal is in a passive state or the metal is passive.

Classification of corrosion forms uses the appearance of damage they cause on a metal. Visual examination can identify most corrosion forms, although the corrosion cause may remain unclear. No universally accepted list of corrosion forms exists since

CHAPTER 1

many are interrelated and the terminology has changed over time. The most fundamental classifications are uniform and localized corrosion. In uniform corrosion, most metal surfaces respond to the corrosive environment with a uniform weight loss or thinning. Uniform corrosion causes the greatest material loss on a tonnage basis, but localized corrosion causes more rapid failures. Localized corrosion is a problem with passive metals. The passive surface film protects most of the metal surface from uniform corrosion, but localized corrosion occurs in places where the passive film is weaker or the corrosive environment is more severe.

The number of variations in corrosive environments is enormous. For example, papermaking processes are very similar, but every mill has its own characteristics. The differences may be a few degrees in temperature or a few tenths in pH value. The operational parameters of a mill will often vary with production rate changes, introduction of new chemicals, or changes in composition of raw materials. Most corrosion phenomena happen in aqueous solutions. Aqueous environments range from thin moisture films to bulk solutions and include natural environments and chemicals. The corrosiveness or corrosivity of an aqueous solution depends on the amount of dissolved gases, salts, and organic compounds and the presence of microorganisms. The concentrations of dissolved species, pH value, hardness, conductivity, etc., describe aqueous solutions. The factors influencing a corrosive environment are numerous. They can have corrosion accelerating and corrosion retarding effects. The best solutions to corrosion problems often result from studying minor changes in solution chemistry.

The basic methods of corrosion control are sound design and selection of proper materials. Corrosion prevention begins before the drawing board stage. The principles for a sound structure considering corrosion are very simple but are sometimes difficult to achieve. The design of process vessels and equipment can be complex and must often comply with governmental regulations. The purpose of a sound design is to ensure a system lifetime with adequate safety margins without overestimating material thicknesses. The rate, extent, and type of corrosion damage that are tolerable in a structure vary depending on the specific application. Some basic rules for good general design and design details from the corrosion viewpoint are available. The most fundamental rule is to avoid heterogeneities, but small details to avoid in a sound design are numerous. Most engineering materials are thermodynamically unstable and they corrode. Many factors that will limit the number of possible materials influence material selection. Corrosion resistance is seldom the main factor in materials selection. Often only the mechanical properties have use to determine material suitability for a particular application. Alloying and manufacturing processes give a metal its physical and mechanical properties. The manufacturing and processing methods used to obtain the necessary product also influence metal properties. These are properties of the bulk metal. The corrosion resistance depends on a very thin surface film. The stability of the surface film determines corrosion performance. The importance of materials selection should not be underestimated. Optimum material selection is often the best way to solve corrosion problems particularly when changes in the process environment could cause product degradation.

Totally stopping corrosion is usually not practical. Decreasing corrosion rate to an acceptable level is a better goal. Several corrosion prevention methods exist. Economic factors will determine the most feasible method to solve the problem. Corrosion control problems extend beyond mere correction of a basic problem such as a leaking pipe. It encompasses all the procedures used to design, build, and operate a unit without excessively severe corrosion problems. Effective corrosion control therefore requires cooperation from many groups including management. Since about one-third of the cost of corrosion is avoidable and could be saved for example by application of best anti-corrosive practice, distribution of this information is important. Corrosion control and treatment are important to management because corrosion of equipment and structures has an extensive impact on operational and structural safety. Economy is another basic consideration, since corrosion will eventually weaken construction to the point where replacement or reinforcement is necessary. The tasks for a corrosion management group should include process and equipment design, material selection, manufacturing and construction, operation, maintenance, and economics. All organizations must develop and apply corrosion control procedures to satisfy particular requirements. Corrosion prevention is an important component of preventive and predictive maintenance programs.

2 Maintenance

Maintenance is a complex phenomenon covering labor and equipment. Its goal is to achieve maximum operating time with peak equipment performance at minimum cost. This means obtaining maximum performance from the equipment using the available people. How does this occur in practice at a pulp and paper mill? It depends primarily on the size and age of the mill, the degree of automation, and the time maintenance personnel must divert to capital projects. The size and productivity of the maintenance workforce, the level of preventive maintenance practiced, and the standard applied to defining a satisfactory degree of maintenance are important considerations in daily work at a pulp and paper mill. A comprehensive book on maintenance should contain all these subjects and discuss the different methods and ways to estimate the phenomena. This book tries to be comprehensive and can be a handbook for maintenance in the pulp and paper industry. The authors try to provide a general overview of maintenance parameters in the United States, Canadian, and Nordic pulp mills; hints for maintenance managers; new ideas for financial controllers on estimating equipment efficiency and personnel productivity; and advice for equipment vendors.

Interest in maintenance has grown considerably since the late 1980s. This growth is evident in the increasing number of conferences and papers devoted to the subject. The primary reasons for increased focus on maintenance are cost control and margin improvement. When the cost of producing a ton of pulp is nearly equal to the sales price for that ton, operating budgets require scrutiny to minimize all costs. Maintenance is a large portion of the operating budget at any mill. No single tool measures the effectiveness of maintenance. A centralized maintenance workforce system may be most appropriate at a certain time for an organization. At another time, decentralized maintenance

CHAPTER 1

workforce systems are best. This review discusses some maintenance analysis methods. Each mill should choose the most suitable method.

As pulp and paper mills become more "closed," i.e., the process loops undergo tightening to prevent unwanted air emissions or effluent discharge, the tasks required of maintenance and production people will change and grow dramatically. Maintenance may sometimes appear to be the "poor stepchild" of production. This antiquated view is giving way to realization that maintenance has a substance of its own. Maintenance is no longer merely a "cost of doing business." This book will concentrate on showing the parameters of significance regarding maintenance in the pulp and paper industry.

The modern approach to preventive maintenance includes routine care, regular inspections to detect faults before they progress to serious disruptions and repairs, and overhauls and reconditioning to assure the greatest equipment service reliability. Corrosion prevention and corrosion management as part of preventive maintenance deserve more attention than they usually receive. Corrosion should not be a problem of secondary importance that is delayed or forgotten. Using suitable tools of routine care, inspection, and planned repair work, mitigation of corrosion can produce large benefits. Including corrosion management in a preventive maintenance program is not a difficult task.

CHAPTER 2

Construction materials

1	**General**	**20**
1.1	Alloying	24
1.2	Manufacturing and processing	25
1.3	Heat treatment	26
1.4	Surface films	27
2	**Carbon and low alloy steels**	**28**
2.1	Mechanical properties	30
2.2	Corrosion resistance and applications	32
3	**Stainless steels**	**34**
3.1	Types of stainless steels	35
3.2	Corrosion resistance and applications	39
4	**Nickel alloys**	**42**
4.1	Corrosion resistance and applications	43
5	**Titanium**	**45**
5.1	Corrosion resistance and applications	46
6	**Welds**	**48**
6.1	Welding techniques	49
6.2	Welding problems	51
7	**Plastics and elastomers**	**54**
7.1	Thermoplastics	55
7.2	Thermosets	57
7.3	Elastomers	60
7.4	Corrosion resistance and applications	61
8	**Summary**	**63**
	References	65

CHAPTER 2

Jari Aromaa

Construction materials

Materials science forms the basis for engineering. The properties of materials define and limit the capabilities that a device or structure can have and the techniques that can be used for fabrication. Material properties depend on the material microstructure that depends on composition and processing. Many engineers often think first of mechanical properties such as strength, ductility, or impact resistance. Properties also include electrical and thermal conductivity, corrosion resistance, the ability of the material to be fabricated or coated, etc. For many materials, the formation of the microstructure begins with a combination of elements or mixing of components often in liquid form. Solidification causes different structures to form. Mechanical processing also controls the material properties and fabricates a desired shape. Materials may be fabricated in a variety of ways including casting, machining, forming, and joining. These operations also modify the final microstructure and affect the properties of the materials For the pulp and paper industry, the main metallic material groups are carbon steels, stainless steels, nickel alloys, and valve metals. This chapter provides an overview on metallurgy and manufacture of these alloys, their properties, and their use.

Many factors that limit the number of possible materials influence material selection. All engineering materials are subject to some form of corrosion, but corrosion resistance is seldom the main factor in material selection. Providing a list of suitable materials for all environmental conditions is impossible because minor variations in an environment may influence corrosion behavior. Figure 1 shows some properties that can affect the choice of a structural material. Corrosion resistance is only one factor.

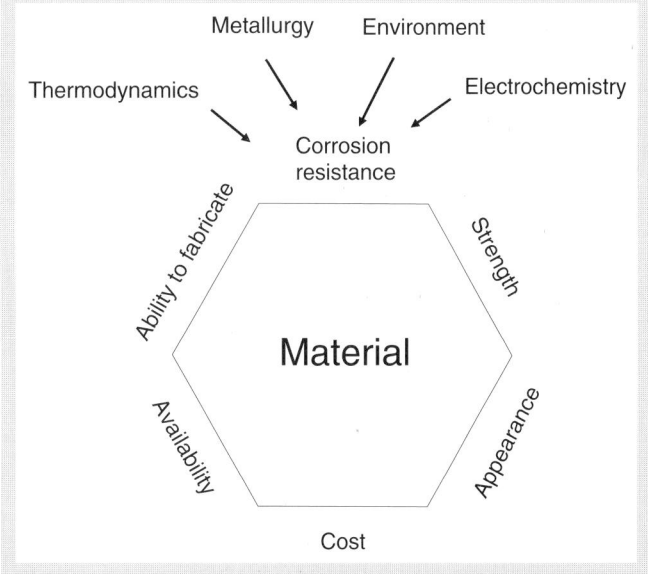

Figure 1. Factors influencing selection of a construction material.

Corrosion resistance itself is also subject to various factors including material properties and environmental factors. In engineering applications, mechanical behavior is more important than appearance. In architecture, appearance has utmost importance. For process industries, cost and corrosion resistance can be important factors. Fabrication or the ease of forming and joining is a consideration. Availability and cost will certainly be decisive factors.

Large structures often use carbon or low alloy steel because of its low price and adequate mechanical properties. The corrosion resistance of steel is poor. A paint coating or similar protective material controls the corrosion. Corrosion resistant alloys such as stainless steels and nickel-based alloys find use in the process industries where environments are very corrosive. A current trend places more stringent demands on pollution control, performance, and equipment lifetime. This will undoubtedly increase the use of corrosion resistant materials. Unless materials are corrosion resistant in their intended operational environments, their useful strength, toughness, thermal conductivity, etc., will have little use.

The main structural materials for the pulp and paper industry are carbon steels and stainless steels. Sometimes nickel alloys, titanium, and plastics have use. Although this chapter primarily discusses metallic materials and their manufacture, properties, and use, some information on plastics is also included. The identification of materials uses the American Iron and Steel Institute (AISI) types or the Unified Numbering System (UNS) number developed by the American Society of Testing and Materials (ASTM) and the Society of Automotive Engineers (SAE). References to other national standards are supported by AISI type, UNS number, or chemical analysis. Chemical analyses are in percent by weight unless otherwise stated.

1 General

In nature, metals usually exist as minerals, but some metals occur in their metallic state. The most common minerals are metal oxides, sulfides, or carbonates. If a mineral deposit has economic value, it is an ore. Metals are extracted from their respective ores, refined, and worked with considerable energy expenditure. Figure 2 shows the accumulation and release of excess energy during the life cycle of a metal. The excess energy stored in the metal has a tendency for release. Metals will react with their environment causing the metal to revert to its mineral state. The reactivity of a metal to the environment is a measure of corrosion resistance. A more reactive metal needs more energy to produce metal from its minerals. This usually equals a greater tendency to corrode.

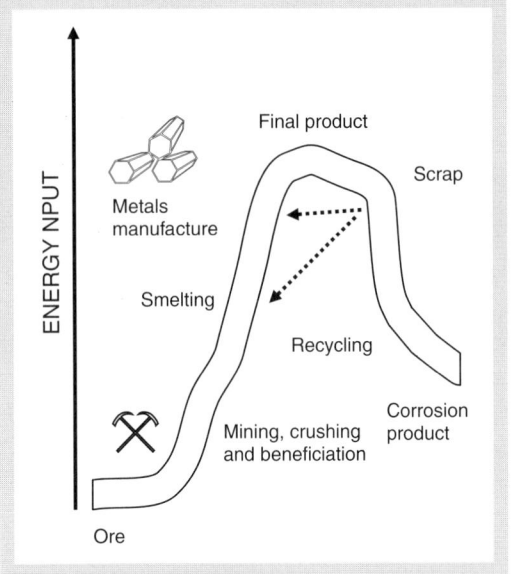

Figure 2. Accumulation and release of energy during metal manufacture and corrosion.

Manufacture of metals usually uses smelting of their respective mineral concentrates followed by purification and casting into some suitable form. Further refining or working steps are possible. All steps affect metal composition, microstructure, and properties. The major metallurgical factors that influence corrosion resistance are chemical composition, manufacturing method, and heat treatment, since these properties influence purity and microstructure.

Metals are crystalline solids. The atoms of a solidified metal are in constant vibratory motion. The amplitude of motion increases as temperature increases. When the temperature approaches the melting point, atoms dislocate from the crystal lattice and melting begins. During melting, the rigid crystal lattice breaks down. As a molten metal cools, the first crystals nucleate at some temperature below the melting point. This is called super cooling. The nuclei grow by incorporating atoms from the surrounding melt and connecting with other nuclei to form grains that subsequently connect with other grains. With slow cooling, a columnar and branched or dendritic structure forms. With rapid cooling, a very small equiaxial grain structure forms. When molten metal cools, it shrinks a few percent. Dissolved gases cause pores in the structure unless it is deoxidized or "killed." For example, deoxidization of steel uses addition of silicon, aluminum, or titanium in the melt. The selection of a deoxidization compound has some effect on material performance. Excess silicon can decrease corrosion resistance in pulping liquors, but sufficiently high aluminum content in unalloyed construction steels will eliminate the brittle fracture tendency resulting from hot working. The solidification process and heat treatment influence the grain size and phases in the microstructure. Compounds and structures that are stable only at high temperatures are possible at room temperature by using fast cooling.

At room temperature, the microstructure of a metal is a network of grains joined at grain boundaries. Inside a grain, the metal has a crystalline structure. The grain boundaries occur because the contracting grains have different crystallographic orientations. The crystal structure is usually face-centered cubic (fcc), body-centered cubic (bcc), or hexagonal. Aluminum, copper, nickel, γ-iron (iron at 910°C–1390°C), silver, and gold have fcc structure. Chromium, molybdenum, vanadium, tungsten, and α-iron (iron up to 910°C) have bcc structure. Magnesium, zinc, and titanium have hexagonal structure.

A metal structure is not a crystal lattice with atoms arranged in perfect order. Although most atoms are in such a state, a fraction of the atom sites that may be less than one in a million is not perfect. Usually, this imperfect arrangement determines the properties of the macroscopic specimen. The structure has lattice defects classified by their geometry as point defects and one-, two-, and three-dimensional defects:

- A point defect is a vacancy such as a missing atom or a foreign substitute or interstitial atom in the lattice.

- A dislocation is the most common one-dimensional or line lattice defect. In a dislocation, parallel crystal planes do not match perfectly. Extreme cases are edge dislocation and screw dislocation. An extra plane of atoms results in an edge dislocation, and a different alignment of neighboring planes is a screw dislocation. Most dislocations are mixtures of these.

- Stacking faults or two-dimensional dislocations and grain boundaries are the most common two-dimensional or plane defects.

CHAPTER 2

- Inclusions, precipitations, voids, and cracks and zones depleted or enriched in some alloying elements are three-dimensional or volume defects. Inclusions are particles of foreign material in a metallic matrix. The particles are usually compounds such as oxides, sulfides, or silicates but may be any substance that is foreign to and essentially insoluble in the matrix. Precipitates originate from the metallic matrix as constituents of a supersaturated solid solution.

Point defects and line defects are within the grain. Plane and volume defects occur primarily outside individual grains. Lattice defects have significant effects on physical, mechanical, and chemical properties of the metal. These effects can be beneficial or harmful. For example, dislocation movement gives metals their ductility, but voids and cracks decrease tensile strength.

Usually, only mechanical properties determine material suitability for a particular application. The fundamental difference of brittle and ductile fracture has utmost importance. Little or no plastic deformation precedes a brittle fracture. In a ductile fracture, metal deforms plastically near the fracture area. A brittle fracture requires very little energy and proceeds rapidly. A ductile fracture uses considerable energy in plastic deformation. A ductile fracture is a shear deformation with sliding action on crystal planes. The brittle fracture is a crystalline fracture proceeding along grain boundaries or cleavage fracture proceeding through the grains. A fracture in which most grains have failed by cleavage gives bright reflecting facets. Table 1 summarizes the differences in ductile and brittle fracture behavior.

Table 1. Differences in ductile and brittle fracture behavior.

	Ductile	Brittle
Mode	Shear	Cleavage
Movement	Sliding	Snapping apart
Plastic deformation	Yes	No
Occurrence	Gradual	Sudden
Fracture appearance	Dull and fibrous	Bright
Microscopic appearance	Dimpled, ruptured	Cleavage

Several factors influence ductile or brittle behavior. With almost all metals, ductile behavior is more likely at higher temperatures. Brittle behavior is more likely at lower ones. This behavior is particularly common with body-centered cubic metals such as most ferrous metals. From the materials selection viewpoint, the ductile-to-brittle transition temperature is an important material property. When temperature is lower, many materials change their fracture behavior from ductile to brittle at some temperature range. A lower temperature gives a more suitable material for low-temperature service.

Soft, relatively tough metals are usually more apt to show ductile behavior. Hard, strong metals are often brittle. Small size, thin section, and simple geometry of the structure promote ductile behavior. Large, thick sections are more prone to contain various discontinuities. If the structure has notches or internal discontinuities, these will cause stress concentrations. If the highly loaded areas cannot share the load with the remainder of the structure, ductile behavior will not happen and brittle fracture will be more likely. Low loading rates promote ductile behavior because the shear mode has time to distribute the load along crystal planes. Torsion loading will usually have ductile behavior, but tension or compression will cause brittle behavior. Table 2 provides a summary of these factors.

Table 2. Factors affecting ductile and brittle behavior.

	Ductile	Brittle
Temperature	High	Low
Strength of metal	Low	High
Size	Small, thin	Large, thick
Geometry	No stress raisers	Notches, etc.
Loading rate	Low	High
Loading type	Torsion	Tension or compression

Metal properties and the tests that measure them have several categories. Tensile strength or ultimate tensile strength is the ratio of maximum force sustained to original cross-sectional area. Yield strength is the maximum obtainable stress. Above this value, a material shows an increase in strain without a proportional increase in stress. Below the yield point, the material behaves elastically and no permanent deformation occurs with load application and removal. For metals with no clear yield point, an offset of 0.2% from proportionality of stress and strain has practical use. These properties are measured by slow application of stress to provide stress-strain curves. Figure 3 shows a general example.

Brittleness is the tendency of a material to fracture without plastic deformation. Ductility or toughness is the ability of a material to deform plastically without fracturing. Measurement of these

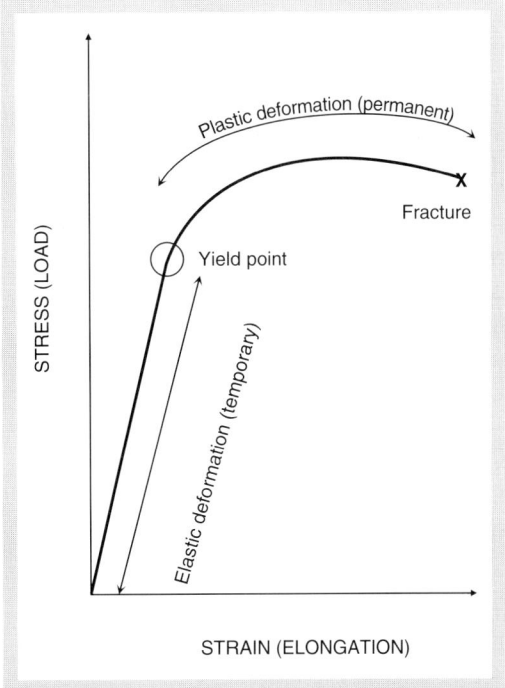

Figure 3. Typical stress-strain curve showing elastic and plastic deformation areas.

properties shows the tendency of a material toward brittle fracture or determines its ability to form. Several methods are available for ductility measurements. The most common is the Charpy V impact test. Factors that increase yield strength or tensile strength will decrease ductility. Only a decrease in grain size will have a beneficial effect on both properties. Hardness is a measure of the ability to withstand penetration. Toughness and hardness relate closely, but heavy local plastic deformation always influences hard-

CHAPTER 2

ness measurements. Fatigue resistance describes the ability to withstand cyclic or fluctuating stresses. Measurement uses repeated application of stresses below the tensile strength. Fracture toughness is a measure of resistance to crack growth and the ability of a metal to absorb energy and deform plastically before fracturing. Impact strength describes material behavior under conditions to promote fracture. Testing uses rapid loading with measurement of the absorbed energy when a single blow breaks the specimen. Creep resistance measurement uses an applied stress at high temperatures while the material deforms continuously.

1.1 Alloying

Alloying is the addition of one or several metallic elements to another metal. Its use provides a specific combination in properties of strength, hardness, wear, corrosion resistance, etc. The amount of a single alloying element varies from less than 1% to more than 40%. The base material, i.e., the matrix, can be less than 50%.

The microstructure of an alloy may contain a single phase or several phases that vary in composition, mechanical properties, and chemical properties. If the alloying elements have nearly equal atomic size and equal crystalline structure, they form a solid solution where alloy metal atoms replace base metal atoms or occupy positions between base metal atoms in the crystal lattice. These are substitutional and interstitial solid solutions, respectively. In steel, those alloying elements with less than 41% of the base metal atom size such as hydrogen, nitrogen, carbon, and boron take interstitial positions in the base metal crystal lattice. If the solubility is limited and phase transformations occur as temperature changes, the microstructure has two or more phases depending on alloy composition, solidification, and heat treatment.

The structural changes and properties of metal alloys vary with composition and temperature. A phase diagram or equilibrium diagram allows study of these variations. The horizontal axis is the alloy composition in weight percent or atomic percent, and the vertical axis is the temperature. Most phase diagrams are for binary alloys. An important diagram is the iron-carbon system. Figure 4 is a hypothetical binary phase diagram that also includes the resulting structures at various compositions.

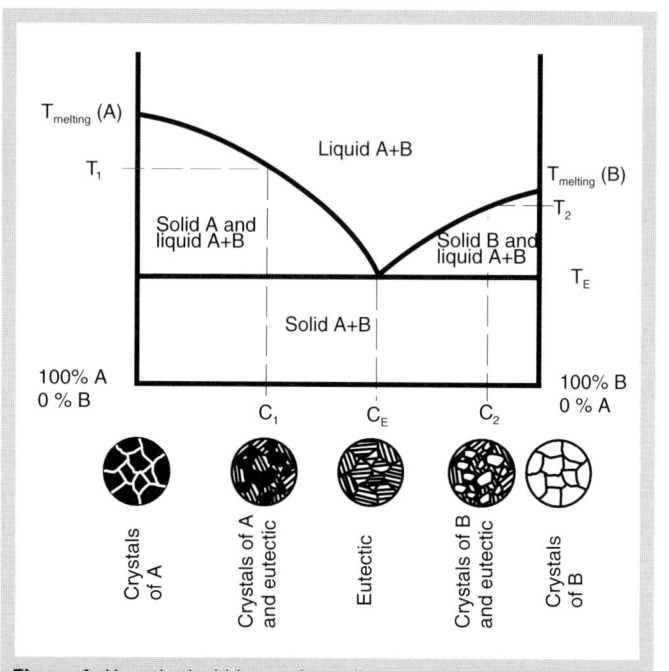

Figure 4. Hypothetical binary phase diagram with resulting structures at different compositions.

The material in Fig. 4 is a eutectic system, and the metals, A and B, have complete insolubility. Unlike a pure metal, alloys solidify over a temperature range resulting in a "pasty" (solid plus liquid) phase between liquid and solid phases. The diagram has four phase fields:

- Liquid (A+B)
- Solid A + liquid
- Solid B + liquid
- Solid (A+B).

The microstructure of alloys in such a system depends on composition. Pure metals have their own characteristic structures. At composition C_1, the first solid to deposit is metal A. With further cooling, deposition of A from the liquid continues. The composition of liquid travels along the phase boundary between solid A + liquid and liquid (A+B). At the eutectic temperature, T_E, the remaining liquid solidifies giving a duplex structure of small quantities of A and B side-by-side. This composition is the eutectic composition, and the phase is the eutectic. Solidification of alloy with composition C_2 results in crystals of metal B and eutectic. The solidification process is inherently a nonequilibrium process. Even when the composition and microstructure can be estimated from the phase diagram, large differences in concentration between various phases may result. During cooling and solidification, redistribution of the solute between liquid and solid phases often results in segregation in the final product.

1.2 Manufacturing and processing

The manufacturing and processing methods used to obtain the desired product also have an influence on metal properties. Processing controls microstructure, and microstructure controls properties. Metals may be wrought or cast. Wrought metal processing includes rolling, forging, extrusion, and drawing. The main effect on microstructure is elongation of the grain in the direction of the working process. Making thick sheets uses hot rolling, and making thin sheets uses cold rolling. Hot or cold rolling or extrusion produce rods and profiles. Drawing produces seamless tubes.

Casting includes static, continuous, pressure, and centrifugal processes. Static ingot castings have grain structures that are more equiaxial and usually coarser than corresponding wrought alloys. Pressure and centrifugal castings contain some preferred grain orientation such as dendritic grains near the metal mold surfaces. All castings can have porosity and impurities as slag or oxidation product inclusions. Good foundry practice can eliminate these defects. Figure 5 shows an example of grain growth and orientation during static casting. Three different zones usually occur in a cast alloy. The chill zone is near the mold wall. Here the grains are small and equiaxial. During pouring, the liquid contacts the cold mold and undergoes rapid cooling. Many solid nuclei form on the mold wall and begin to grow into the liquid metal. Very soon after the pouring, the temperature gradient at the mold wall decreases and the crystals in the chill zone begin to grow dendritically. The cooling rate is most rapid perpendicular to the

CHAPTER 2

wall. Dendritic growth is therefore fastest in this direction resulting in columnar grains. In the center of the part, cooling rate decreases sharply resulting in equiaxial but coarse grains randomly oriented. This three-zone structure is typical for most metal products, although sometimes the coarse central part does not exist.

In many cases, cast alloys are more susceptible to corrosion than their wrought counterparts. This is because the cast alloys can still be nonuniform in local composition or microstructure produced by the casting process. For example, regions with less than average content of protective alloying elements are more easily dissolved and some nonmetallic impurities may favor cathodic reactions increasing local dissolution rate. Cast alloys may also have slightly different composition to ease casting. This can sometimes affect corrosion performance. Solution annealing some cast materials for a reasonably long period is possible to homogenize the material and optimize corrosion resistance.

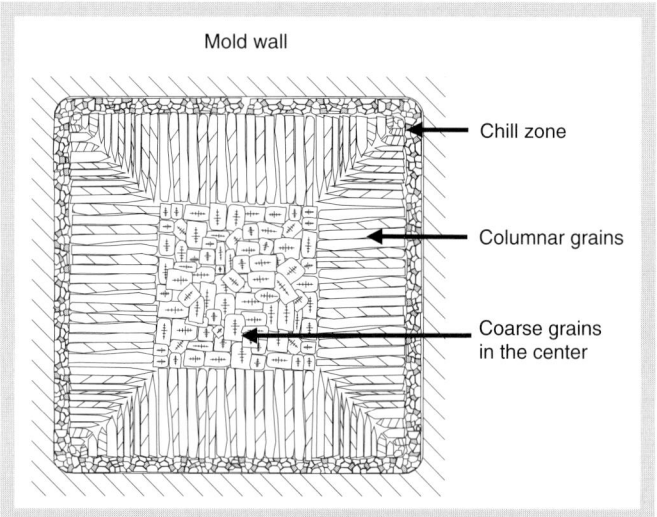

Figure 5. Microstructure of a cast ingot with different grain sizes and orientations.

Welds have in their microstructure some similarities to cast structures. The rapid thermal cycle causes a treatment that resembles quenching more than normal casting and cooling. During the usual fusion welding, base and filler materials are heated to their melting point and rapidly cooled. The original microstructure and properties of the metal change in the region near the weld. This region is the heat-affected zone (HAZ). The solidified weld metal diluted with material melted from the surrounding base metal has coarse columnar structure. Next to the weld is a zone of melted, solidified base metal that is not mixed with weld metal. Coarsely grained and finely grained zones that have only partially melted during the welding follow this. Porosity and slag defects are also possible in welds. The heating and cooling cycle during welding can cause high residual stresses unless it is stress relieved. Working of welds can refine the metal grain size, but most welds contain large, coarse, dendritic grains. Alloy segregation can also occur as the weld solidifies and cools. Heat treatment and selection of filler metal can improve the weld microstructure.

1.3 Heat treatment

The purpose of heat treatment is to produce the desired properties by controlled application of heat and rate of heating and cooling. The main objectives are to produce hardness and ductility. The common heat treatment methods are hot working (rolling, forging, etc.), annealing, and hardening. Hot working means deforming the metal plastically at a temperature and strain rate such that recrystallization occurs simultaneously with deformation. This avoids any strain hardening in contrast with cold working. Mild steels are hot worked at 870°C–1100°C. All working occurs before the steel cools below 870°C. The worked part cools freely in air. A low degree of deformation or finishing of deformation at high temperatures may cause grain growth and subsequent decrease in ductility. Normalizing can recover the ductility. Cold working usually but not necessarily occurs at room temperature. In cold working, the metal is plastically deformed under conditions of temperature and strain rate that induce strain hardening. Cold working occurs below the recrystallization temperature of the metal.

Annealing means heating to a suitable temperature, maintaining that temperature, and then cooling at a suitable rate to reduce hardness, facilitate cold working, produce a desired microstructure, etc. Annealing is a generic term that primarily refers to reducing hardness but also simultaneously to producing desired changes in other properties or in microstructure. When applied only for the relief of stress, the process is actually stress relieving or stress-relief annealing. Annealing is another term for normalizing, i.e., restoring increased grain size back to the original, soft annealing to produce suitable microstructure before hardening, etc.

Hardening can use various techniques. For steels, the primary hardening method is martensite hardening, although precipitation hardening also has use. Martensite is a term for microstructures formed by phase transformation without diffusion. An acicular pattern in the microstructure for ferrous and nonferrous alloys characterizes martensite. In alloys where the solute atoms occupy interstitial sites in the lattice such as carbon in iron, the structure is hard and highly strained. When the solute atoms occupy substitutional sites such as nickel in iron, the martensite is soft and ductile. For carbon steels, hardening by martensite transformation includes heating to produce an austenitic structure, quenching to transform the austenite to martensite, and tempering to decrease hardness and increase toughness. Quenching usually means rapid cooling of metals from a suitably elevated temperature. It uses immersion in water, oil, polymer solution, or salt. Tempering means reheating of hardened steel or hardened cast iron to some temperature below the eutectoid temperature for decreasing hardness and increasing toughness. The process sometimes also applies to normalized steel and nonferrous alloys. Tempering of steels occurs at about 200°C for at least one hour followed by slow cooling in air. Aging is a change in properties of a metal that generally occurs slowly at room temperature but more rapidly at higher temperatures. Aging has use for hardening after rapid cooling or cold working. The change in properties is often due to a phase change but never due to a change in chemical composition.

1.4 Surface films

Alloying and manufacturing processes give the metal its physical and mechanical properties. These are bulk metal properties. Corrosion resistance depends on a very thin surface film. Corrosion begins with removal of atoms from the metal surface by reactions with the environment. Metal ions may react with the species in the electrolyte and form reaction products. Under appropriate conditions, these reaction products may form surface films on the metal. The stability of the surface film determines corrosion performance. If the surface film protects the metal from further corrosion, the metal is in a passive state or the metal is passive. Typical passive metals are stainless steels, aluminum, and titanium. Suitable alloying such as chromium in stainless steels can enhance passivation.

Active metals will usually corrode uniformly with uniform weight loss or thinning. Passive metals will not corrode noticeably, but they will fail under specific environmental conditions at small, local sites. Typical localized corrosion forms are pitting, crevice corrosion, and stress corrosion cracking. Localized corrosion can lead to more rapid failure than uniform corrosion. Localized corrosion of passive metals is therefore more dangerous than uniform corrosion of active metals. Passivation is a complex phenomenon influenced by film formation, composition, and strength. All passive metals can become active upon damage or contamination of the passive film. A fresh surface that contains heterogeneities, scratches, notches, etc., is more susceptible to corrosion than a smooth surface with a reaction product layer. For example, a passivating treatment is used after cold rolling and pickling of stainless steels.

The formation of a protective surface film depends on the metal composition and environment. When the dissolution rate is sufficiently high and the environment is suitable, precipitation of a reaction product film becomes likely. Formation of a protective film is a kinetic phenomenon. The ability of a corrosion product film to protect the metal is the result of several complex phenomena. The primary factors are chemical composition and stability, thickness, adhesion, mechanical properties, and crystal structure including defect density of the film. Pure oxide films are often protective. The passive films are a mixture of crystalline and amorphous oxides, hydrated oxides, absorbed water molecules, etc.

Not all surface films are protective. During heat treatment, hot rolling, and forging for example, so-called mill scale forms on a steel surface. Mill scale is a 10–100 μm thick layer of iron oxides. FeO (wüstite) is next to the metal, Fe_3O_4 (magnetite) is an intermediate layer, and Fe_2O_3 (haematite) is an outer layer. A surface with mill scale has a blue-gray color. Mill scale is not a protective surface film. Under mechanical or differential thermal stresses, the surface cracks and spalls leaving bare areas of metal. Another nonprotective oxide film is a heat tint on stainless steel. Heat tint forms when the surface of a stainless steel weld is exposed to air before cooling. Mill scales and similar films on other metals, corrosion products, etc., require removal before surface treatments. Paint films applied over scales are useless since they will definitely spall.

2 Carbon and low alloy steels

Carbon and low alloy steels are the most important groups of structural materials. Steels are relatively inexpensive to manufacture in large quantities. The range of mechanical properties that is achievable by alloying and heat treatment is very extensive. Mechanical properties are the result of proper alloying and combined use of different strengthening processes.

The steelmaking process has two steps. First, iron ore is smelted with coke in a blast furnace to produce molten, carbon-rich pig iron. A converter changes pig iron to steel by burning away most carbon with oxygen. Simultaneously, some impurities are removed to justify the steel composition. Steel is then deoxidized and cast to blooms or billets that can be wrought to desired shapes. Casting can use molds or a continuous process. The blast furnace removes sulfur and silicon impurities, and the converter removes phosphorus and manganese. A typical carbon steel contains about 0.55%–0.60% Si, 1.5%–1.7% Mn, 0.06% P, and 0.06% S maximum. Sulfur and phosphorus cause poor welding and low impact strength. Silicon as SiO_2 decreases ability to machine and form. High levels of manganese decrease impact strength. Low silicon content such as below 0.01% by weight in a batch kraft digester may sometimes improve corrosion resistance. Many modern steels have lower impurity content than those acceptable several decades ago that still occur in specifications.

Pure iron has two crystal forms: body-centered cubic α-iron and face-centered cubic γ-iron. Figure 6 shows the iron-carbon binary phase diagram with predominant phases of α-ferrite, austenite (γ), δ-ferrite, and cementite that is an iron carbide (Fe_3C) compound. The α-ferrite or usually only ferrite is a solid solution that forms as carbon atoms occupy the interstitial sites of the face-centered cubic iron lattice. The maximum solubility of carbon is about 0.025% by weight. Ferrite is usually tough and easily workable both cold and hot. Ferrite has a clear yield point. It may be prone to cleavage fracture. This is the most common type of brittle fracture. Austenite is also a solid solution of iron and carbon, but the carbon atoms occupy interstitial sites of a body-centered cubic lattice. The maximum solubility of carbon is 2.06%. Although the austenite is not stable at room temperature, the structure may remain partially austenitic when it cools rapidly. Because of the high carbon content, austenite is harder than ferrite. It is also tough and

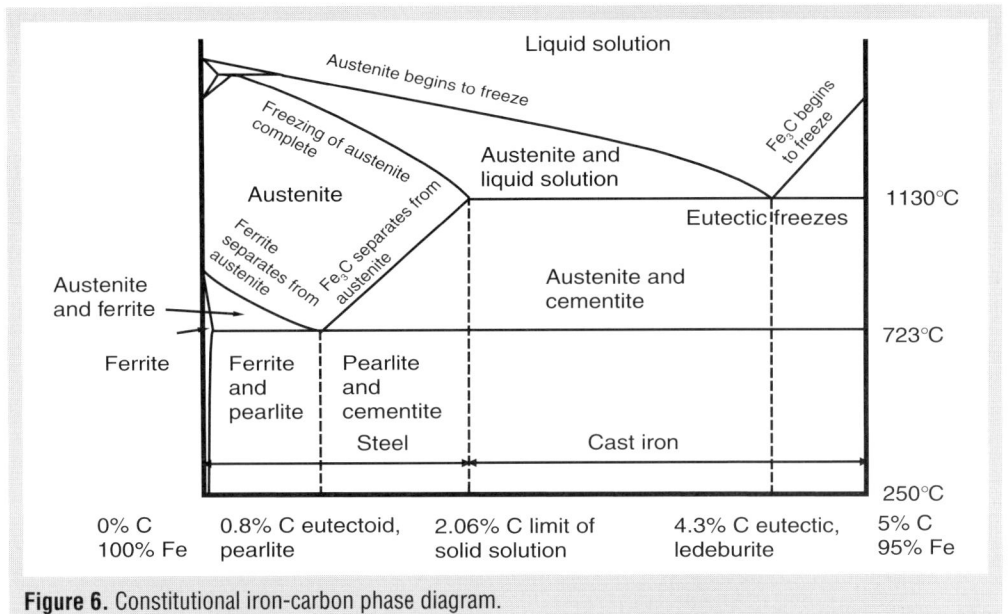

Figure 6. Constitutional iron-carbon phase diagram.

easily workable. Austenite has no yield point and is not prone to cleavage fracture. δ-ferrite has a face-centered cubic structure and exists only at high temperatures. If the carbon concentration is so high that it cannot dissolve in iron, the surplus carbon forms cementite. Cementite is an iron-carbon compound with chemical composition of Fe_3C. It is a metastable phase that decomposes at 1300°C–1900°C. It has no melting point. The cementite is a very hard and brittle compound.

Steel is often reheated and cooled several times to obtain the shape and desired microstructure necessary for specific mechanical properties. These heating and cooling processes have names that describe the specific purposes. The names relate to the temperatures in the iron-carbon phase diagram. The treatments are completely in the solid state. During annealing, steel is heated to immediately below the eutectoid temperature, A_1 = 723°C, to promote recrystallization and remove the effects of prior cold work and grain deformation. This allows further forging or rolling. Austenitizing heats the steel to the austenite phase region to dissolve more carbon. Since the temperature is near the austenite-ferrite phase boundary (A_3 temperature) to minimize grain growth, the ferrite and carbide structures formed on cooling are finely grained and have good mechanical properties. The cooling rate from the austenite region determines the microstructure that forms. Faster cooling usually results in finer structures and better mechanical properties. Quenching or rapid cooling causes austenite to transform into a tetragonal structure called martensite that is usually tempered later. Tempering is reheating of quenched steel below the A_1 temperature of 150°C–700°C to allow diffusion and precipitation of very fine carbides combining strength and toughness. The austenite that does not transform to martensite during quenching causes problems because it may decompose later producing brittle cementite layers that can cause cracking. Pearlite and bainite structures form during slower cooling so that ferrite and cementite form instead of martensite.

2.1 Mechanical properties

Carbon steels are basically iron and carbon alloys containing up to 2% carbon by weight. Alloys with higher carbon contents are cast irons. Carbon steels have different classifications. According to their use, they can be structural steels (UNS G and UNS K series with less than about 0.60% carbon) and tool steels (UNS T series with more than 0.60% carbon). The resulting mechanical properties depend primarily on the carbon content and strengthening methods that the carbon content makes possible. Low carbon steels or mild steels contain up to 0.30% carbon by weight. These steels are typical structural steels that also have use as castings. They have low strength but high ductility. Heat treatment cannot harden them except by surface hardening processes. Their uses are hot worked and especially hot rolled sheets, rods and profiles, machinery compounds, bolts and rivets, and cold-rolled thin sheets. Medium carbon steels contain 0.30% and 0.60% carbon. They have a balance of strength and ductility. Heat treatment can harden them, but the process only affects thin sections or the outer layers of thick parts. Medium carbon steels are suitable for machinery parts. High carbon steels contain more than 0.60% carbon by weight. They can be hardened and used for tools and wear resistant and high strength parts.

Structural carbon steels include unalloyed carbon steels with yield strength of 200–260 MPa, manganese alloyed steels with yield strength of 230–320 MPa that are being replaced by finely grained steels with yield strength of 260–450 MPa, and quenched and tempered steels with yield strength of 450–900 MPa. Pressure vessel steels have the same requirements as structural steels, but they must retain their toughness at low temperatures. The main requirements for carbon steels are strength and the ability to weld.

Low alloy steels were developed primarily to improve mechanical properties of carbon steels. Small amounts of alloying elements primarily influence the relationship between the cooling rate of the steel melt and the resulting microstructure. Hardness and strength will therefore be affected. Their corrosion resistance is usually the same as that of carbon steel, but their atmospheric corrosion resistance is often better. Low alloy steels contain up to 4%–5% by weight of various alloying elements. These are primarily chromium, nickel, copper, manganese, vanadium, and molybdenum. For example, high strength, low alloy (HSLA) steels with minimum yield strengths of 275 MPa are available for use as high strength construction materials. Because HSLA alloys are stronger, they have use in thinner sections. This makes them particularly attractive for components where weight reduction is important. Typically, HSLA steels are low carbon steels with up to 1.5% manganese strengthened by small additions of elements such as niobium, copper, vanadium, or titanium. They sometimes use special rolling and cooling techniques. Weathering steels are a class of low alloy steels developed to resist atmospheric corrosion. The main alloying element is copper in small amounts with smaller quantities of various other elements. When exposed to relatively uncontaminated atmospheres with cyclical wetting and drying, they form a brown protective rust film on their surface. With constant immersion, the protective film will not form. The main use is in construction requiring a special architectural effect or when painting is difficult or useless due to heavy mechanical wear. Heat resistant steels retain their mechanical properties at high temperatures, and they do not oxidize easily. Their alloy uses up to 9%–10% chromium, some molybdenum, and silicon to form a protective oxide layer. The 9% Cr steels may suffer from 475°C embrittlement upon prolonged exposure around this temperature due to their ferritic microstructure.

Carbon equivalent provides an estimate of the ability to weld. Equation 1 from the International Institute of Welding provides the calculation using the composition[1]. This equation is used when the carbon content is more than 0.18% or when the cooling rate after welding is more than 12 s from 800°C to 500°C. Other equations to calculate carbon equivalent are also available. If the carbon equivalent is 0.41 or less, ability to weld is usually good with all methods and without preheating.

$$C_{ekv} = C + \frac{Mn}{6} + \frac{Si}{6} + \frac{Cr + Mo + V}{5} + \frac{Ni + Cu}{15} \tag{1}$$

Strengthening provides the mechanical properties of steel. The following discussion describes the common strengthening methods by reference to simple iron-based systems. The basic mechanisms for strengthening iron alloys include solid solution strengthening by interstitial and substitutional atoms, refinement of grain size, dispersion strengthening, and work hardening. Solid solution additions or refinement of grain size strengthen the austenite or ferrite matrix. The behavior of interstitial atoms of carbon and nitrogen has prime importance. Amounts as low as 0.005% by weight carbon and nitrogen cause a sharp transition between elastic and plastic deformation under tensile stress. Austenite can form a solid solution with up to 2% by weight carbon that can be retained in solid solution by rapid quenching. The phase transformation leads to martensite. High carbon martensite is normally very hard but brittle resulting from interstitial solid solution hardening. Tempering of martensite gives better toughness and ductility while retaining some strength. Solid solution strengthening by substitutional atoms uses common alloying elements such as Mn, Si, Ni, and Mo. Their effect is additive. Addition of these alloying elements is usually for another reason so the strengthening is an additional benefit. Refinement of ferrite grain size gives higher yield strength. The increase is proportional to the square root of grain size. Grain sizes today of about 2–10 μm result from closely controlled rolling practices and micro alloying.

Controlling the dispersion of other phases in the microstructure can improve the strength of the matrix. The most common other phase is a carbide such as cementite in carbon steels. Other possible phases include nitrides and inter metallic compounds. Most dispersions produce strengthening with a decrease in toughness and ductility. This is dispersion strengthening. Work hardening that results from plastic deformation is particularly important for obtaining high strength rod and wire. The wear resistance of common stainless steels is also the result of work hardening. Work hardening can have negative effects. In some cases, common austenitic stainless steels work harden due to continuous vibration. This can make parts and structures brittle and cause cracking.

2.2 Corrosion resistance and applications

The corrosion resistance of carbon and low alloy steels is poor. Barrier coatings such as paints and rubber or plastic linings usually protect them. Hot-dip galvanizing often prevents atmospheric corrosion. Carbon steels corrode in most atmospheric environments when the relative humidity exceeds 60%. After a moisture film forms, the corrosion rate depends on time exposed to moisture, pH, and the presence of chlorides and sulfur dioxide. The corrosivity of the micro climate is more important than that of the macro climate. The composition and surface condition of the steel and the angle of exposure also influence the corrosion rate. Increases in carbon, manganese, and silicon content reduce the corrosion rate. Initially, the corrosion rate is high. It falls after the first year of exposure because of the formation of surface film.

Aqueous corrosion resistance of carbon and low alloy steels in process environments is usually poor. The alloying elements or residual impurities at normal levels have no significant effect on corrosion rates under immersed conditions, and the corrosion products are not protective. In fresh water, a typical corrosion rate is 0.05 mm/year, but uneven attack may penetrate much deeper and faster due to formation of loose, porous, and nonprotective corrosion film products. If product purity demands allow, steel also has use in applications with corrosion rates up to 0.2–0.5 mm/year because of its low cost assuming regular maintenance shutdowns. In such cases, the design considers a corrosion allowance. The equipment wall thickness is larger to ensure desired service life. Steel has use for organic chemicals except low molecular weight acids and neutral or alkaline aqueous solutions at moderate temperatures. It is a suitable material for storage of concentrated sulfuric acid and handling sodium hydroxide up to 50% and 55°C. Carbon steels are sensitive to stress corrosion cracking for example in concentrated sodium hydroxide solutions and in hydrogen sulfide environments.

Despite their poor corrosion resistance, the use of carbon and low alloy steels in the pulp and paper industry is common. Carbon steels have use in structures that are not in direct contact with corrosive environments and have protection with paint. In some cases, carbon steels are also suitable for corrosive environments especially in alkaline media. Large equipment such as a kraft pulp digester or recovery boiler often uses low alloy steel due to its low cost. Many noncritical parts in heavy equipment may also use steel. Carbon steel is suitable without coatings to reinforce concrete structures because of the concrete alkalinity. Various chemicals that penetrate into the concrete could cause dealkalization and corrosion of the steel. Protection of reinforcement bars with zinc, organic coatings, inhibitors, or cathodic protection is an option.

The inside of a kraft pulp digester is an alkaline environment where carbon steel should perform well. As long as the oxide layer has no damage, carbon steel will not corrode. Protective measures such as cladding and electrochemical protection are becoming more common due to some field failures. The damage was the result of forceful flow during charging and blowing or residual stresses. New digesters today rarely use unprotected carbon steel in contact with cooking liquors. In other pulping processes, carbon steel will usually corrode due to the acid environment, chlorides, or mechanical wear.

Pulp washing and evaporation systems have also been suitable applications for mild steels since weak black liquor and wash water are alkaline. In storage tanks, corrosion under pulp deposits has occurred. As chloride concentration increases and pH decreases due to closure, this might become a more serious problem. In closed systems using acidic bleach plant filtrates in pulp washing, carbon steels may corrode. In chemical recovery, carbon steel is suitable in green liquor piping and tanks, but waterline corrosion may be a problem. Carbon steel is also useful for handling green liquor to the clarifiers and causticizers. Since green liquor contains solid particles, it may cause erosion. White liquor was earlier considered mildly corrosive to carbon steel, but waterline corrosion and corrosion from sulfur compounds have resulted in replacement with stainless steel.

In bleach plant solutions, uncoated carbon steel will not survive. In the paper mill, carbon steel is suitable for structures, foundations, etc. Usually, carbon or low alloy steels are not suitable for direct contact with the stock because rusting will cause product discoloration. Paper machine roll journals and shafts can be carbon and low alloy steels. The main problem here is corrosion fatigue. The trend for use of carbon steel in papermaking is decreasing. Closure of the processes will raise the solution temperatures and increase concentrations of dissolved ions. This will always increase the corrosivity of the solution. In the future, very few places will remain where unprotected carbon steel contacts aqueous solutions.

A recovery boiler primarily uses carbon and low alloy steels. The severe corrosion of carbon steel boiler tubes first encountered in Finland in the beginning of the 1960s resulted in changes for tube material. Carbon steels corroded in the reducing smelt zone due to hydrogen sulfide in flue gases. The maximum operating temperature was limited to about 300°C. In the oxidizing combustion zone, the problem was hot corrosion due to carryover of black liquor particles and condensed salts. Stainless steels were at least ten times more resistant, but use of solid stainless steel would have been risky due to stress corrosion cracking on the steam side. This resulted in development of composite tubes with an outer surface of stainless steel. Old carbon steel tubes are often coated today with thermally sprayed stainless steel on the smelt side. Heat resistant ferrite steels also have use as tube materials. These steels are alloyed with 1%–3% or 10%–13% chromium and about 1% molybdenum. Austenitic chromium-nickel steels have better creep resistance, but they are more expensive, have higher thermal expansion, and are prone to localized corrosion.

3 Stainless steels

Stainless steel is an iron-based alloy with at least 11%–12% chromium to obtain a protective surface film. Modern stainless steels were developed in the beginning of the 20th century. They are the most commonly used alloys for corrosion resistant applications. Stainless steels are available in several types. Their primary classification uses the microstructure. The most important groups are austenitic, ferritic, and martensitic steels. Austenitic stainless steels are those where the austenite phase is dominated by stabilization of the austenite phase with nickel or manganese additions. Similarly, ferritic stainless steels are alloys where the ferrite phase is stable or made more predominant by alloying with chromium. Martensitic grades contain chromium but no nickel. Their higher carbon content allows hardening. Besides the above mentioned alloys, duplex steels with austenitic-ferritic structure and precipitation hardened stainless steels for high structural strength are available. Martensitic and precipitation hardening stainless steels primarily have use for mechanical strength since their corrosion resistance is lower than that of the other grades.

The early practice in stainless steel production was to melt and refine the steel in an arc furnace. The arc furnace or electric arc furnace is now solely a melting unit. Refining occurs in another vessel. The process steps are charging, melting, decarburizing, reducing, and finishing. The charge includes steel scrap, ferrochromium, and other alloying elements. The charge is melted and decarburized. Oxygen gas was previously injected in the molten metal to oxidize and remove carbon. Oxygen injection leads to oxidation of chromium that must be reduced for economical reasons. Chromium reduction uses a basic slag and a reducing agent. In the finishing stage, desulfurizing, alloying, and temperature adjustments occur. After deoxidation, the steel is ready for casting.

The problem with arc melting and oxygen decarburization was that oxygen blowing is effective for reducing carbon content only to 0.04% after which chromium losses become too high. To overcome these problems, an argon-oxygen decarburization (AOD) practice was invented in the 1970s. AOD is a component of a duplex process where metal is melted in a furnace and decarburization and further refining occur in another vessel. In the AOD process, a mixture of argon and oxygen is blown from the bottom of the vessel. This method allows reducing carbon below the allowable 0.03% maximum level of low carbon grades. The method also provides accurate control of alloying elements.

3.1 Types of stainless steels

Austenitic stainless steels are the most common group. They are iron-based alloys containing 16%–18% or more chromium and sufficient nickel to assure a fully austenitic structure. This is the most widely used corrosion resistant alloy group in the process industry. Only carbon steels have broader application. The steels containing chromium and nickel are the AISI 300 series. Those containing chromium, nickel, and manganese are the AISI 200 series. The reason for manganese alloying is replacement of the more expensive nickel as an austenite promoter. Annealed austenitic stainless steels typically have low strength but possess good ductility and toughness. Austenitic stainless steels can be hardened by cold working but not by heat treatment. In their annealed state, they are all nonmagnetic although some may become magnetic by cold working. They have excellent corrosion resistance and very good ability to form. Ferritic and martensitic stainless steels contain more than 12% chromium. They belong to the AISI 400 series and are magnetic. Ferritic grades contain a low amount of carbon. Heat treatment will not harden them. They have good ductility and corrosion resistance. Martensitic grades contain more carbon, and heat treatment will harden them. They have good ductility, but they are corrosion resistant only in mild environments. Most stainless steels, i.e., alloys with more than 50% iron, belong in the UNS S series.

The microstructure of stainless steel relates to the composition by using the Schaeffler diagram of Fig. 7. The diagram has use primarily for determination of residual ferrite in the welds, but it is also suitable for wrought and heat treated stainless steel grades. The microstructure depends on the relative amounts of austenite and ferrite stabilizers. The nickel equivalent and chromium equivalent form the axes of the diagram. Equations 2 and 3 provide their calculation. In the equations, the element symbols refer to the alloy composition in weight percent.

$$Ni_{equivalent} = Ni + 30 \cdot C + 30 \cdot N + 0.5 \cdot Mn \tag{2}$$

$$Cr_{equivalent} = Cr + Mo + 1.5 \cdot Si + 0.5 \cdot Nb \tag{3}$$

Cobalt and copper will also increase the nickel equivalent, and vanadium, aluminum, titanium, and tungsten will increase the chromium equivalent. To obtain the desired microstructure, the steel composition requires control within narrow ranges. The

CHAPTER 2

Figure 7. Schaeffler diagram[2].

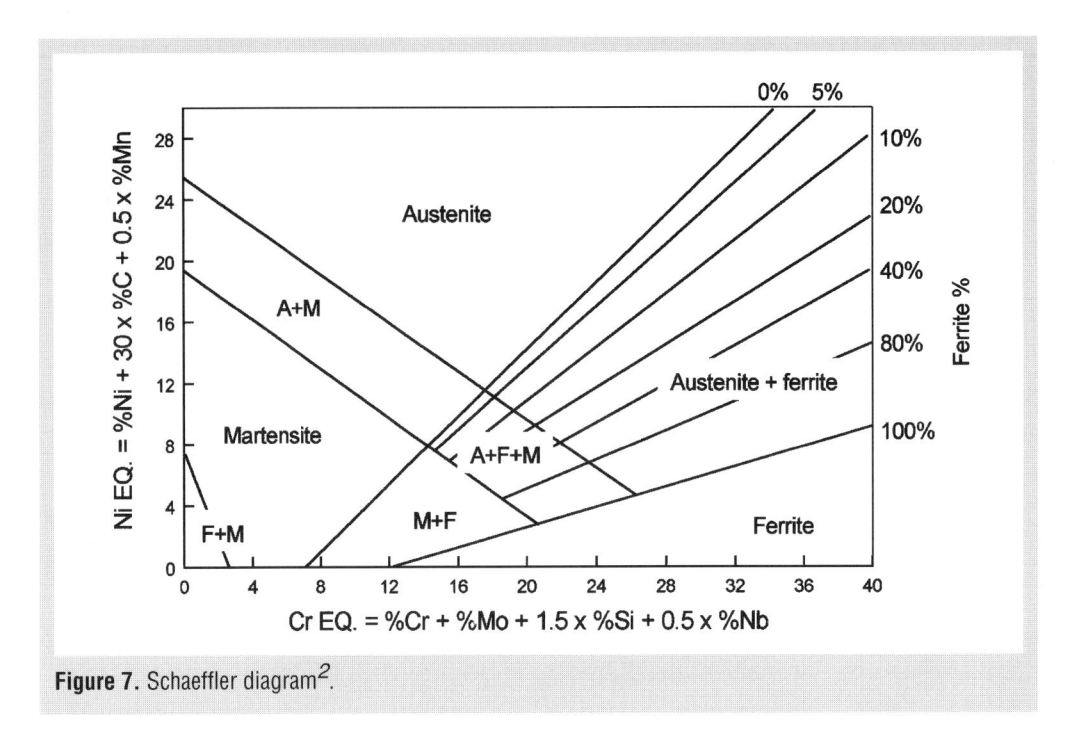

addition of some alloying metals to obtain desired properties usually requires other compositional changes to maintain the microstructure. For example, an increase in chromium and molybdenum content to improve corrosion resistance also requires additional nickel to maintain fully austenitic microstructure as Fig. 7 shows.

Figure 8 shows relationships of selected stainless steel types, alloying elements, and their effects. The basic stainless steel is the austenitic alloy with 18% chromium and 8% nickel (AISI Type 304 or UNS S30400). The 18/8 alloy is suitable for oxidizing conditions. To improve corrosion resistance in reducing conditions, 2%–3% molybdenum is added (AISI Type 316 and UNS S31600). A further development of Type 316 is Type 317 (UNS S31700) with 3%–4% molybdenum. The carbon content of these three alloys is 0.05% maximum. This can cause sensitization to intergranular corrosion during welding or heat treatment of 5–6 mm or thicker sheets. Because the thick sections cool slowly, the chromium carbides have more time to form. To reduce sensitization, the maximum carbon content decreases to 0.03% in the low carbon grades 304L, 316L, and 317L (UNS S30403, S31603, and S31703, respectively). Stabilized grades such as 321 (UNS S32100) or 347 (UNS S34700) also have use to avoid sensitization and intergranular corrosion of the weld. When the amount of alloying elements increases further, another group called super austenitic stainless steels results. They are often called 6% Mo alloys or UNS 31254-type alloys. These alloys may contain less than 50% iron. Further increase of alloying elements results in nickel alloys.

Ferritic steels are less common in the process industries. The basic, general purpose ferritic grade is AISI 430 (UNS S43000) with 16% Cr. AISI 409 (UNS S40900) has lower chromium content for less severe environments. AISI 409, AISI 439, and AISI 444 are stabilized to improve welding. AISI 444 also contains molybdenum to increase corrosion resistance in reducing conditions. This grade has the highest molybdenum content of the 400 series ferritic stainless steels. Ferritic stainless steels with even higher amounts of alloying elements are super ferritic steels. They became available in the 1970s. The super ferritic steels typically contain about 30% Cr and 4% Mo with some nickel. An example is 29–4–2 (Cr–Mo–Ni) for use in heat exchangers and power plant condenser tubing.

Duplex steels that contain austenite and ferrite grains combine the best properties of austenitic and ferritic grades. They combine strength and stress corrosion cracking resistance of the ferritic steels with the toughness and ability to weld of the austenitic steels. The first duplex stainless steels such as UNS S32900 did not contain nitrogen. During welding, they formed a continuous ferrite phase in the heat affected zone. This decreased toughness and corrosion resistance. The least alloyed duplex grade commonly used is UNS S32304 with 21.5%–24.5% Cr and 3%–5.5% Ni. Addition of chromium to increase corrosion resistance also requires an increase in nickel content. Molybdenum alloying increases resistance to reducing conditions. The basic duplex grade is S31803 manufactured by several companies under designation 2205. This grade contains 21%–23% Cr, 4.5%–6.5% Ni, and 2.5%–3.5% Mo. Molybdenum alloying often occurs with higher chromium content. This usually requires addition of nickel or nitrogen to maintain the balanced microstructure for grades UNS S31803 and S32750. Super duplex grades often contain more than 25% chromium and 6%–7% Ni. Their

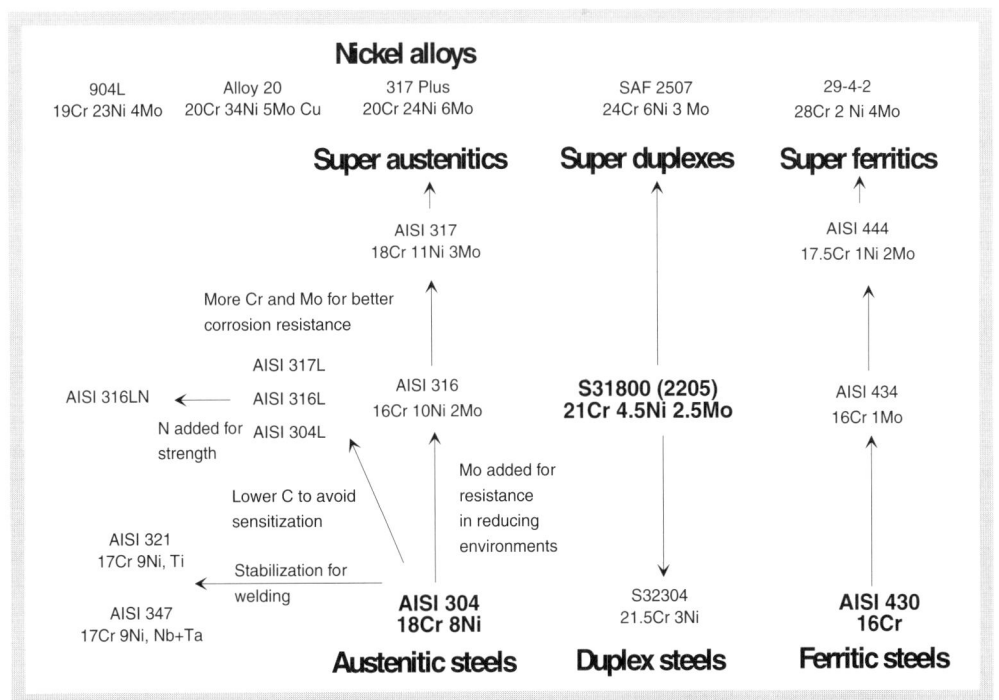

Figure 8. Relationships between selected stainless steel types including alloying elements and their effects with approximate compositions.

CHAPTER 2

molybdenum content is more than 3%. Like other stainless steel groups, the super grades are alloys with pitting resistance equivalent (PRE) above 40.

Austenitic grades have a good ability to form, strength in cold worked conditions, and ability to weld. Lower alloyed stainless steels such as S30400 and S31600 have yield strengths around 290 MPa in the annealed condition. Some higher alloyed stainless steels containing nitrogen possess higher yield strengths in the annealed condition. Cold working often increases strength especially in higher alloyed austenitic stainless steels. Austenitic grades have good corrosion resistance under oxidizing conditions. Alloying with molybdenum increases corrosion resistance under reducing conditions. These types are prone to pitting, crevice corrosion, and stress corrosion cracking especially in halide solutions. Welding and heat treatment may cause sensitization that leads to intergranular corrosion in oxidizing acids. Limiting the carbon content to 0.03% maximum, using titanium or niobium stabilized grades, and avoiding temperatures of 550°C–700°C can avoid sensitization. Austenitic steels have poor ability to machine, but alloying sulfur or selenium (AISI 303 and 303Se) can improve this. These grades have lower corrosion resistance, they are not suitable for forging, and they do not withstand abrasive wear.

Ferritic grades are less expensive than nickel containing austenitic grades but usually have lower ductility and toughness. They have good corrosion resistance under oxidizing conditions, but their corrosion resistance under reducing conditions is lower than that of austenitic grades. Molybdenum alloying can improve corrosion resistance under reducing conditions. The grades are only slightly susceptible to stress corrosion cracking. They have good ability to machine if chromium content is below 18%. Ferritic grades cannot be heat treated or welded as well as austenitic grades because they are prone to several embrittlement phenomena. Embrittlement due to grain growth is possible above 900°C. The so-called 475°C embrittlement occurs when heating the steel between 370°C and 540°C. The heating results in an increase of tensile strength and hardness but decrease of ductility and impact resistance. The σ-phase is a very brittle inter metallic compound, FeCr, whose precipitation causes σ-embrittlement if more than 14% chromium containing alloy is exposed for long times at 650°C–900°C. Ferritic grades are prone to brittle fracture at low temperatures, and they are not as creep resistant as austenitic grades.

Duplex stainless steels usually have 50%/50% microstructures, but the smaller phase volume is at least 30%. The microstructure is ferrite grains in austenite matrix or vice versa. The purpose of a ferritic/austenitic duplex microstructure is to combine the strength and stress corrosion cracking resistance of the ferritic steels with the toughness and ability to weld of the austenitic steels. Duplex stainless steels have been available since the 1930s but used only in unwelded constructions and castings. The problem with the original duplex grades was that they could lose the optimum balance of austenite and ferrite phases causing reduction in corrosion resistance and toughness. Nitrogen alloying and use of AOD technology solved this in the 1970s. Ferritic-austenitic grades have good corrosion resistance under oxidizing conditions, and they are not prone to stress corrosion cracking. Their strength is about twice that of austenitic grades. They may show grain growth embrittlement when welding thick sheets, and molybdenum alloyed grades may show σ-embrittlement. Precipitation of intermetallic phases resulting in decrease of corrosion resistance happens at 800°C–900°C for exposure times longer than 5–10 min, and 475°C embrittlement can happen at 370°C–540°C by heating for longer times. Duplex stainless steels fall into groups using their composition. Low alloyed grades do not contain molybdenum, but they have use to replace austenitic AISI 316 when improved stress corrosion cracking resistance is necessary. Medium alloyed grades contain 2%–3% molybdenum. Their corrosion resistance is comparable to high alloyed austenitic grades such as 904L. High alloyed grades and super duplex grades contain more than 25% chromium and usually 6%–7% Mo. They find use for special applications, and their corrosion resistance is comparable to super austenitic grades.

3.2 Corrosion resistance and applications

Stainless steels are not as intrinsically noble as gold and platinum. The corrosion resistance of stainless steels depends upon a passive surface film of chromium oxide that is self healing in many environments. A minimum level of 11%–12% chromium alloyed in the steel and adequate oxygen supply are necessary to maintain the surface film. If the film cannot repair itself after mechanical or chemical damage, the surface of steel will be active locally and very rapid corrosion may result. Stainless steels usually perform well in oxidizing environments but poorly in reducing environments. In very strongly oxidizing environments such as bleaching solutions, general transpassive corrosion of low alloyed stainless steels is possible. In strongly reducing environments, the protective passive film decomposes over the entire metal surface resulting in uniform corrosion. Different forms of localized corrosion caused by aggressive ions are more probable. Estimation of the effect of steel composition on corrosion resistance often uses chromium, molybdenum, and nitrogen content. The most popular index is the Pitting Resistance Equivalent (PRE) obtained from Eq. 4 where the element symbols refer to the alloy composition in weight percent.

$$PRE = Cr + 3.3 \cdot Mo + (12.8 - 30) \cdot N \tag{4}$$

A high PRE value is an indication of good corrosion performance. The PRE value usually has a linear relationship with the critical pitting temperature (CPT) or critical crevice temperature (CCT) above which corrosion starts in some tests. The PRE values are merely rough estimates of the corrosion resistance, and small differences are irrelevant in practice. The PRE value does not include other alloying elements that could have an influence. No equation expresses the effect of impurities on pitting corrosion, but nonmetallic inclusions like sulfides are initiation points for pitting. Higher pitting resistance is possible with lower levels of impurities present.

The purpose of chromium is to form a passive layer and provide resistance to uniform and localized corrosion. Higher chromium content increases corrosion resistance but may adversely influence mechanical properties, ability to weld, etc. Nickel promotes the austenitic microstructure and improves mechanical properties and fabrication. It improves resistance to stress corrosion cracking with more than 8%–10%. Complete resistance in most applications occurs at 30% Ni. Molybdenum with chromium increases corrosion resistance especially in reducing solutions. The maximum amount of molybdenum in austenitic stainless steels produced by common melting and decarburizing practices is 6%. With nitrogen alloying and special heat treatment, some companies produce alloys with more than 6% Mo. Carbon is the primary element providing strengthening of the steel. In high amounts, it can cause sensitization. Nitrogen alloying provides strength and increased resistance to pitting and crevice corrosion. Titanium and niobium avoid sensitization. Addition of 1%–2% copper improves corrosion resistance of some higher alloys. PRE also distinguishes various steel groups. For example, super stainless steels are those with PRE above 40. Note that the multipliers in PRE equations may be different in different sources. Table 3 shows compositions of selected stainless steels with calculated PRE values[3,4].

Table 3. Approximate compositions and properties of selected stainless steels[3, 4]. Single values are maximums unless otherwise specified. Yield strength and hardness are for annealed plate, sheet, or strip. PRE values are calculated with %Cr + 3.3 • %Mo + 16 • %N.

AISI or UNS number	COMMON NAME	TYPE	PRE	Cr	Ni	Mo	C Max	OTHER	Yield strength MPa	Hardness max.
AISI 304		austenitic	18–20	18–20	8–10.5		0.08		205	201 HB
AISI 304L		austenitic	18–20	18–20	8–12		0.03		170	201 HB
AISI 316		austenitic	23–28	16–18	10–14	2–3	0.08		205	217 HB
AISI 316L		austenitic	23–28	16–18	10–14	2–3	0.03		170	217 HB
AISI 317		austenitic	28–33	18–20	11–15	3–4	0.08		205	217 HB
AISI 317L		austenitic	28–33	18–20	11–15	3–4	0.03		205	217 HB
S31725	317LM	austenitic	31–36	18–20	13–17	4–5	0.03	Cu 0.75 max.	205	88 HRB
AISI 321		austenitic	17–19	17–19	9–12		0.08	(5·C) Ti min.	205	217 HB
AISI 347S		austenitic	17–19	17–19	9–13		0.08	(10·C) Nb min.	205	201 HB
31254	254 SMO	austenitic	42–45	19.5–20.5	17.5–18.5	6–6.5	0.02	0.5–1 Cu, 0.18–0.22 N	300	241 HB
AISI 409		ferritic	14–16	14–16	0.5		0.08	(6·C)–0.75 Ti	205	180 HB
AISI 430		ferritic	16–18	16–18			0.12		205	183 HB
AISI 439		ferritic	17–19	17–19	0.5		0.07	(12·C)–1.1 Ti, 0.15 Al	205	183 HB
AISI 444		ferritic	23–28	17.5–19.5	1	1.75–2.5	0.025	0.8 (Ti+Cb), 0.025 N	276	83 HRB
S32900	329	duplex	26–35	23–28	2.5–5	1–2	0.2		485	98 HRB
S31803	2205	duplex	21–23	21–23	4.5–6.5	2.5–3.5	0.03	0.08–0.20 N	450	293 HB
S32304	2304	duplex	22–28	21.5–24.5	3–5.5		0.03	0.05–0.6 Cu, 0.05–0.2 N	400	290 HB
S32550	Ferralium 255	duplex	32–44	24–27	4.5–6.5	2–4	0.04	1.5–2.5 Cu, 0.10–0.25 N	550	31.5 HRC
S32750	2507	duplex	38–47	24–26	6–8	3–5	0.03	0.24–0.32 N	550	310 HB
S32950	7–Mo Plus	duplex	32–43	26–29	3.5–5.2	1–2.5	0.03	0.15–0.35 N	485	293 HB

Stainless steels are sensitive to various fabrication defects that leave weak points on steel surfaces. Shipping of stainless steel products usually uses a pickled, ground, or cold rolled condition without oxides, free iron, or other foreign material on the surface. Oxides may increase the corrosion rate, and iron may initiate pitting. The surfaces should have protection during fabrication as much as possible with paper, film, etc. Carbon steel tools should not be used to avoid iron embedding. Grinding abrasives contaminated by iron from carbon steel or carbon steel wire brushes may also cause iron embedding. During fabrication and installation, stainless steels are frequently cleaned chemically to remove iron contamination especially. This usually uses nitric acid solution. It is misleadingly called "passivation." The correct term for the acid cleaning process is pickling. Pickling uses an acid mixture of oxidizing nitric acid and some reducing acid such as HF, H_2SO_4, or HCl. The reducing acid provides descaling, and nitric acid restores the passive film. Passivation in nitric acid sometimes uses sodium dichromate. All pieces require cleaning before passivation since dirt, grease, and cutting fluids will interfere with passivation. Solvent cleaning followed by alkaline cleaning and water rinse is a suitable procedure. Surface treatments range from a hot-rolled, annealed, and pickled surface with R_a = 6.0 µm to mechanical polishing for a mirror finish.

Stainless steels have had a successful history in pulp and paper manufacturing service since the 1920s. The two most common steels are austenitic types AISI 304/304L and 316/316L. Type 304 has use for inland paper mills and 316 for coastal mills where sea transportation of logs causes higher chloride levels. More corrosive environments require higher alloyed grades. Use of more than 3% Mo containing austenitic grades and duplex steels is increasing. Austenitic stainless steels that are often clad on carbon steel have had use for many years as construction materials for digesters and pulp storage towers to reduce maintenance costs from corrosion. In a kraft digester, AISI 304L or 316L for high chloride risks has use as linings. Overlay welding of a digester usually uses molybdenum containing grades. Another way to obtain a pressure rating for a digester is to use a solid, high strength duplex stainless steel to combine the advantages of low maintenance and reduced initial material cost. The strength of a duplex stainless steel allows reduction of wall thicknesses. A typical grade would be S31803 with 22% Cr and 3% Mo. Nitrogen alloyed types AISI 304LN and 316LN have also had use. Duplex steels have found use for example in heat exchanger tubing when stress corrosion cracking has been a problem. In white liquor, the increase in corrosion rate with increasing chloride content is much lower for duplex grades than for austenitic grades. The sulfite process is much more corrosive than other pulping processes requiring stainless steel use for most process equipment. Type 316L has had traditional use, but type 317L or 4.5% molybdenum steels (slightly more alloyed than 317L) have had use when corrosion resistance of type 316L is inadequate. Preparing sulfite cooking liquor may require UNS N08904 type alloys with 20Cr–25Ni–2Mo–2Cu due to sulfuric acid formation. In mechanical pulping, AISI 304L is often satisfactory, but 316L is necessary for higher temperatures. In semichemical and recovered paper pulping, types 304L and 316L have been satisfactory[5–7].

The higher temperatures and concentrations of aggressive species resulting from system closures will certainly increase use of stainless steels in brownstock washing and chemical recovery. The common types AISI 304 and 316 have been successful in most applications. An increase of chloride concentration may raise the risk of pitting and crevice corrosion and stress corrosion cracking. This may require use of more alloyed austenitic grades or duplex grades such as S31803 or S32304.

The solutions in a chemical recovery process may also be too corrosive for carbon steel. Type AISI 304/304L is probably sufficiently resistant even in the immediate future in green and white liquors. Higher temperatures or places where risk of solution concentration is possible require type 316/316L. In some cases, the passive film on stainless steel is sufficiently strong to withstand the erosion and fluctuating liquid levels that damage carbon steels[6, 7].

Bleaching processes are highly corrosive. Type 316L is the lowest grade with successful use, and the use of 317L has been common. Since recycling of process water will increase chloride content, use of more alloyed austenitic grades (4.5% or 6% molybdenum) and duplex grades has increased. Use of N08904 (20Cr–25Ni–2Mo–2Cu) and S31254 (20Cr–18Ni–6Mo) type alloys is common today since they will serve in chlorine dioxide environments if its concentration is not too high[5, 7]. Most new bleaching processes such as peroxide and ozone are assumed to be less corrosive than traditional bleaching processes such as chlorine or hypochlorite due to reduced chloride ion concentration. Many mills today like the option to change from one bleaching chemical and sequence to another. This makes material selection even more troublesome and increases use of high alloy grades.

High demands on product purity and continuous operation have increased use of stainless steel in paper mills. The primary types are AISI 304 and 316 or low carbon grades with a predominance of the latter. Even large cast stainless steel equipment has use for suction press rolls as an example. Martensitic stainless steels have use in rolls. Paper machine headboxes use pickled and passivated products. They can be electrolytically cleaned and polished to obtain maximum smoothness and eliminate possible initiation places for corrosion.

The most important high temperature application of stainless steels is in composite tubes of the recovery boiler. The materials used on the outer surfaces of these tubes have changed from an AISI 304 type to more highly alloyed grades with up to 25% Cr and 20% Ni.

4 Nickel alloys

Nickel and the nickel based alloys are corrosion resistant in many environments and temperatures. Some types are extremely corrosion resistant in certain media, but nickel alloys are usually more expensive than iron or copper based alloys or plastics. Nickel has use as a construction material, plating on a less noble metal, or cladding. Deposition of the plating may use an electrochemical process or an autocatalytic process when it is actually a Ni–P alloy. Most nickel alloys are available under trade names. References to the UNS numbering system or other systems have infrequent use.

Nickel alloys are a continuation in the development of stainless steels. The base metal is nickel rather than iron. The solubility of alloying elements in nickel is higher than in iron, and nickel is not susceptible to ductile-to-brittle transition with decreasing temperature. The primary alloying metals are essentially the same as in stainless steels, i.e., chromium and molybdenum. As conditions become more severe, consideration of upgrading from stainless steel to a nickel alloy is common. For best corrosion resistance, two alloy groups exist: alloys with corrosion behavior of nickel itself and alloys whose corrosion resistance results from chromium alloying as in stainless steels. The main alloy types in the first group are pure nickel, Ni–Cu, and Ni–Mo. In the second group, Ni–Cr, Ni–Cr–Mo, Ni–Fe–Mo, and Ni–Fe–Cr–Mo alloys with or without copper are available. Alloys with better corrosion resistance than common stainless steels are super austenitic stainless steels or medium nickel alloys. They have compositions between traditional austenitic stainless steels and nickel base alloys. Although their nickel content is below 50%, the UNS numbering system classifies some as nickel alloys since they contain more than 50% alloying elements. These nickel-chromium-iron alloys with generally more than 30% nickel behave as high nickel alloys. Most nickel alloys are solid solution alloys. The precipitation hardening alloys have only limited use in corrosion service. Super alloys are nickel alloys primarily intended for high temperature service. These alloys with high temperature strength are usually multiphase alloys with precipitation strengthening elements such as Al, Ti, and Nb.

Nickel alloys are manufactured by electric arc melting and refining argon oxygen decarburization (AOD) or vacuum oxygen decarburizing. Vacuum induction melting also has use. Nickel alloys have a good ability to weld. Different nickel alloys can be welded together or to stainless steels. Alloys are also machinable, but work hardening may become a problem. Solid solution nickel alloys (Cu–Ni and Cr–Ni–Mo with or without Fe and Cu) have common use in the annealed or annealed and cold worked condition. These alloys are usually not suitable for strengthening by heat treatment. Alloys in cold worked condition are usually not welded because the mechanical strength of the weld would be lower than that of the cold worked region. In the annealed state, these alloys can be welded. The yield strengths of solid solution alloys are about 200–1400 MPa. The maximum yield strength depends on alloy composition, cold work characteristics, and specified ductility. Table 4 shows compositions and properties of some nickel alloys[4, 8]. In nickel alloys, chromium improves resistance to oxidizing media and molybdenum to nonoxidizing media as in stainless steels. Iron has typical use to reduce costs, and it improves corrosion resistance in some cases. Small additions of copper and tungsten provide resistance to nonoxidizing media.

4.1 Corrosion resistance and applications

The main uses of nickel alloys are for alkali solutions, reducing conditions where stainless steels corrode, warm chloride solutions that cause stress corrosion in stainless steels, and strong chlorine and chlorine dioxide containing solutions causing pitting and crevice corrosion of stainless steels. The alloys of the first group, i.e., alloys without chromium, are suitable for alkaline and nonoxidizing environments. Commercially pure nickel resists most alkaline solutions excluding ammonium hydroxide. It is especially useful for concentrated sodium hydroxide and dry chlorine. Nickel-copper alloys perform well in nonoxidizing acids such as deaerated H_2SO_4, HCl, and H_3PO_4. Nickel-molybdenum alloys such as Alloy B and Alloy B-2 are even better in nonoxidizing acids, most organic acids, and reducing salt solutions. Oxidizing impurities or dissolved oxygen can cause rapid corrosion of Ni–Cu and Ni–Mo alloys. For example, Alloy B-2 is particularly suitable for equipment handling reducing chemicals such as hydrochloric acid. Premature failure may occur if using B-2 where iron or copper is present. Chromium-containing alloys are suitable for oxidizing environments. These alloys form a passive layer containing chromium oxide. As with stainless steels, a minimum chromium content is

CHAPTER 2

necessary. Alloying with molybdenum and some minor elements can improve the corrosion resistance. Many Ni–Cr and Ni–Cr–Fe alloys with more than 50% Ni have excellent high temperature properties. About 6% minimum Mo with high nickel and high chromium provides excellent resistance in oxidizing chloride solutions. Alloys like C-22, C-276, G-3, G-30, and 625 are typical examples.

Nickel alloys with more than 40% nickel are not susceptible to chloride-induced stress corrosion cracking. Nickel alloys are sensitive to stress corrosion cracking in some environments. Pure nickel is sensitive in mercury, sulfur, and sulfides. Nickel-molybdenum alloys (Alloy B types) are sensitive to HCl, although they are designed for reducing conditions. Chromium containing alloys may show stress corrosion cracking in hot NaOH solutions.

Nickel and nickel alloys have use in pulp and paper mills where conditions are the most corrosive such as the bleaching circuit and pollution control equipment. These environments have high temperature and they are acidic. They contain chloride, chlorine, chlorine compounds, oxygen, peroxides, etc. The oxidizing environment with chloride often requires use of chromium containing high molybdenum alloys such as 625 (N06625), C-276 (N10276), C-4 (N06455), C-22 (N06022), and a cast version of C-4 (C-4C alloy). C-276 and 625 have performed well in C and D stage bleaching equipment. Cases also exist where highly alloyed stainless steels actually perform better than nickel alloys due to their broader passive ranges. The 600 and 800 type alloys have had use for digester liquor heater tubing since the high nickel content prevents stress corrosion cracking (SCC) caused by chlorides. The super austenitic stainless steels have found wide use in the pulp and paper industry primarily in applications of high temperatures and chloride concentrations. Type 904L alloy (N80904) has use in bleaching processes[6, 7].

Table 4. Approximate compositions and properties of selected nickel alloys[4, 8]. Single values are maximums unless otherwise specified. Yield strength and hardness are for annealed plate, sheet, or strip.

UNS number	COMMERCIAL NAME	TYPE	Ni	Cr	Fe	Mo	C Max.	OTHER	Yield Strength MPa	Hardness HRB
N02200	Nickel 200	Pure nickel	99 min		0.4		0.15		100	65
N06600	Inconel 600	Ni–Cr	72 min	14–17	6–10		0.15	0.5 Cu	240	
N08020	Carpenter 20Cb3	Fe–Ni–Cr	32–38	19–21	bal.	2–3	0.07	3–4 Cu	241	95
N08800	Incoloy 800	Fe–Ni–Cr	30–35	19–23	bal.		0.10	0.5 Cu, 0.15–0.6 Al	205	
N10665	Hastelloy B2	Ni–Mo	bal.	1	2	26–30	0.02	1 Co	352	100
N06022	Hastelloy C22	Ni–Cr–Mo	bal.	20–22.5	2–6	12.5–14.5	0.015	2.5 Co, 2.5–3.5 W	310	100
N06455	Hastelloy C-4	Ni–Cr–Mo	bal.	14–18	3	14–17	0.015	2 Co		
N06625	Inconel 625	Ni–Cr–Mo	bal.	20–23	5	8–10	0.1	3.15–4.15 Nb	276–414	
N10276	Hastelloy C-276	Ni–Cr–Mo	bal.	14.5–16.5	4–7	15–17	0.02	2.5 Co, 3–4.5 W	283	100
N06030	Hastelloy G30	Ni–Cr–Fe–Mo	bal.	28–31.5	13–17	4–6	0.03	5 Co, 1–2.4 Cu		
N06985	Hastelloy G-3	Ni–Cr–Fe–Mo	bal.	21–23.5	18–21	6–8	0.015	5 Co, 1.5–2.5 Cu	241	100
N08026	Carpenter 20Mo-6	Ni–Cr–Fe–Mo	33–37	22–26	bal.	5–6.7	0.03	2–4 Cu	241	
N08825	Incoloy 825	Ni–Cr–Fe–Mo	38–46	19.5–23.5	bal	2.5–3.5	0.05	1.5–3 Cu, 0.6–1.2 Ti	241	
N08904	904L Alloy	Ni–Cr–Fe–Mo	23–28	19–23	bal.	4–5	0.02	1–2 Cu	220	

5 Titanium

Titanium is a lightweight metal with excellent corrosion resistance in many natural and industrial environments. Titanium is a common element, but it requires considerable energy during manufacture because it of its high reactivity. Its excellent corrosion resistance is due to a thin, adherent oxide film. Once considered expensive, titanium is often now economical when considering an entire life-cycle. Cost-effective use is possible when using its properties in the design rather than to substitute titanium for another metal. Commercially pure titanium and titanium alloys fall into three groups depending on their microstructure: α-titanium, α/β-titanium, and β-titanium. Pure titanium is α-titanium. It has a closely packed hexagonal structure. α-titanium is stable up to 882°C where it transforms to body-centered cubic β-titanium that remains up to the melting point of 1668°C. Alloying with aluminum stabilizes the α-phase, and alloying with iron, copper, chromium, molybdenum, nickel, and vanadium stabilizes the β-phase. The two most important impurities, oxygen and hydrogen, are α- and β-stabilizers, respectively. Commercially pure titanium and α-alloys have the lowest strengths but good corrosion resistance. β-alloys have use because of their high strength. Some corrosion resistant alloys have been developed for environments where pure titanium is not suitable. Titanium vessels, heat exchangers, tanks, agitators, coolers, and piping systems find use in aggressive solutions such as nitric acid, organic acids, chlorine dioxide, inhibited reducing acids, and hydrogen sulfide.

To produce titanium, the basic ore is converted to sponge in two distinct steps. First, the ore is mixed with coke or tar and converted to titanium tetrachloride. The crude titanium tetrachloride is a colorless liquid that is purified by continuous fractional distillation. Reaction with magnesium or sodium under an inert atmosphere gives metallic titanium sponge. Titanium is converted from sponge to ingot by blending crushed sponge with the desired alloying elements to ensure uniformity of composition and then pressing into briquets welded together to form an electrode. The electrode is melted in a consumable electrode vacuum arc furnace. To insure homogeneity of the final ingot, a second or sometimes third melting operation is necessary. Titanium alloys are also produced by electron beam or plasma melting and cast into ingots. Wrought products are produced by conventional metallurgical processing in air. Castings can also be produced under shielding gas. α-alloys are generally fabricated in the annealed condition. The ability to weld for unalloyed titanium and α-alloys is good, but proper inert gas shielding is necessary. α/β-alloys are usually fabricated at elevated temperatures followed by heat treatment. The cold forming possibilities of these alloys are limited. β-alloys may be fabricated using any of the techniques employed for α-alloys. Some β-alloys can be welded and may be heat treated to increase strength after welding. Welding must use an inert environment or an inert cover gas.

The classification of titanium and its alloys often uses ASTM standards, although many countries have national standards. In the ASTM standard, titanium alloys have numbers as Grades 1–12. In the UNS system, they belong to the R series with refractory alloys based on tantalum, zirconium, etc. Grades 1–4 are commercially pure titanium, and they differ in their oxygen content. Grade 1 has the lowest and Grade 4 has the highest oxygen level. Increasing oxygen content increases strength but reduces ability to form and ductility. Grade 2 has common use because it has the best combination of strength and ability to form and weld. Small alloying additions or impurities do not affect the corrosion resistance in normal environments. In ASTM Grades 7 and 11, addition of Pd (0.15%) improves resistance to reducing acids and crevice corrosion since Pd alloying improves passivation. Titanium Grade 12 contains 0.3% Mo and 0.8% Ni. It is a low cost alternative to Grades 7 and 11. It also has improved crevice corrosion resistance. Al and V alloying in titanium Grades 5 and 9 reduces resistance in strong reducing or oxidizing acids. The alloys normally used in corrosive environments are the α-alloys of commercially pure Ti, Ti-0.3% Mo-0.8% Ni, Ti-0.15% Pd, and the α/β-alloy Ti-6% Al-4% V. In most applications, commercially pure titanium is the most resistant alloy, but Mo-Ni and Pd alloys have better resistance to crevice corrosion. Table 5 shows compositions and properties for selected titanium alloys[9].

The α-alloys have yield strength of 170–480 MPa. Alloy composition rather than heat treatment usually gives variations. The α-alloys have use in the annealed or stress relieved condition. The α/β-alloys have yield strength from 860 to more than 1200 MPa. Strength can vary by alloy composition and heat treatment. Water quenching is necessary to obtain higher strength levels. The α/β-alloys have common use in the annealed or solution treated and aged condition. The β-alloys have yield strength from 800 to more than 1350 MPa. β-alloys can be aged to high strength, but then they lack ductility. Over 1250 MPa yield strength is possible through cold work and aging treatments. β-alloys are difficult to weld without embrittlement. Their common use is in the solution treated and aged condition.

5.1 Corrosion resistance and applications

The corrosion resistance of titanium depends on the formation of a surface oxide film that is primarily TiO_2. The film forms instantly with exposure of a fresh surface to air or moisture. A damaged oxide film can often repair itself if some traces of oxygen or moisture are present. The oxide film of titanium is thermally very stable, and most environments do not attack it. In common industrial environments, uniform corrosion, crevice corrosion, and hydrogen uptake are problems that require consideration. Pitting and stress corrosion cracking seldom occur in the chemical industry. Intergranular corrosion and corrosion fatigue are not significant. Since titanium and its alloys are highly reactive and passivating metals, fretting and wear are possible due to repeated passive film removal and reformation.

Corrosion resistance of titanium is superior to most alloys in oxidizing and mildly reducing media. The corrosion resistance of unalloyed Grades 1–4 are equal. Palladium alloyed Grades 7 and 11 are more resistant in a less oxidizing environment. Titanium is suitable for use with oxidizing mineral acids, mixed acids, alkaline solutions, inorganic salt solutions, and most organic acids and chemicals. Titanium cannot be used in concentrated (red fuming) nitric acid where it may react vigorously, hydrofluoric acid, dry chlorine gas, chlorinated compounds, and hot formic acid. Pickling and descaling of titanium with amine inhibited acid or boiling sulfamic acid have caused rapid corrosion.

Hydrogen pickup may cause embrittlement. Hydrogen can result from the corrosion process itself, and hydrogen evolution at titanium may be enhanced by galvanic coupling when titanium acts as a cathode. Titanium can be protected anodically when used in reducing media. Environments reported to cause SCC in titanium include acetic acid, chlorinated and fluorinated hydrocarbons, and silver and AgCl. These environments are not common in the pulp and paper industry. Alloys with high oxygen and aluminum levels are particularly susceptible to SCC. In α- and α/β-alloys, the characteristic aqueous stress corrosion cracking process is cleavage fracture of the α-phase. The α-phase is susceptible to SCC when oxygen content is above 0.2% by weight and aluminum content is above 5% by weight. Aluminum alloyed grades have seldom use in corrosion resistant applications.

The main applications of titanium in the pulp and paper industry are in bleach plants. Titanium is resistant to oxidizing conditions, and chlorides or acidity do not heavily influence it. Titanium has become a standard material for drum washers, diffusion bleach washers, pumps, piping systems, and heat exchangers especially for the equipment developed for chlorine dioxide bleaching systems. Titanium resists solutions of chlorites, hypochlorites, chlorates, perchlorates, and chlorine dioxide. Oxidizing impurities extend titanium's passivity to lower pH levels. A limiting factor of titanium alloy application in aqueous chloride solutions can be crevice corrosion in metal to metal and gasket to metal joints or under deposits. Titanium is generally highly resistant to alkaline media. Titanium is often the choice for alkaline media containing chlorides, oxidizing chloride species, or both. Even at higher temperatures, titanium resists pitting and stress corrosion cracking causing problems for stainless steels.

The most difficult problem to date has been use of titanium equipment for the alkaline peroxide stage. Existing equipment intended for the chlorine dioxide stage has frequent use. Titanium alloys shift from a passive to active state when pH, hydrogen peroxide concentration, and temperature are too high. Salts causing natural hardness in water such as Ca^{2+} and Mg^{2+}, silicate, and lignin can inhibit corrosion. Transition metal ions and complexing agents can accelerate corrosion. When using existing titanium equipment, the location of hydrogen peroxide feed is important to avoid local concentrated solutions.

Titanium alloys are susceptible to iron impurities. During manufacturing, iron chips or dust can adhere to titanium surfaces. Steel and iron tools are therefore not allowable for titanium processing. Chlorinated cutting or cooling fluids are also not suitable because they might cause stress corrosion cracking. Titanium is also prone to notch sensitive cracking. Titanium can usually be safely used with metals such as stainless steel, nickel alloys, and copper. One should nevertheless be careful with galvanic couplings involving titanium because of the risk of hydrogen evolution and embrittlement of the metal.

The properties and characteristics that are important to design engineers are corrosion resistance, erosion resistance, and high heat transfer efficiency. Titanium exhibits exceptional resistance to a broad range of acids, alkalis, natural waters, and industrial chemicals. Titanium has very good resistance to erosion, cavitation, or impingement attack. Titanium's high strength allows using thinner walls in equipment. The erosion and corrosion resistance of titanium permits higher operating velocities. Due to recycling of waste fluids and the need for greater equipment reliability and life span, titanium has become a common material. In the pulp and paper industry, titanium has use for washers, pumps, piping systems, and heat exchangers in the bleaching section. This is particularly true for equipment in chloride dioxide bleaching systems. One major application for titanium is heat transfer when the cooling medium is seawater, brackish water, or polluted water. Titanium condensers, shell and tube heat exchangers, and plate and frame heat exchangers find extensive use in power plants, refineries, and chemical plants. Reported failures due to corrosion on the cooling water side are very rare.

Table 5. Compositions and properties of selected titanium alloys[9]. Single values are maximums unless otherwise specified. Yield strength and hardness are for annealed plate, sheet, or strip.

UNS number	Common name	Type	Ti	O	Pd	Mo	Ni	Al	V	Yield Strength MPa
R50250	Grade 1	α	com. pure	0.18						170–310
R50400	Grade 2	α	com. pure	0.25						175–450
R50500	Grade 3	α	com. pure	0.35						380–550
R50700	Grade 4	α	com. pure	0.40						485–655
R52400	Grade 7 or Ti-Pd	α	bal.	0.25	0.12–0.25					275–450
R52250	Grade 11 or Ti-Pd	α	bal.	0.18	0.12–0.25					170–310
R53400	Grade 12	near-α	bal.	0.25		0.2–0.4	0.6–0.9			345
R56400	Grade 5 or Ti-6-4	α/β	bal.	0.20				5.5–6.75	3.5–4.5	830

6 Welds

Welding has extensive use in fabrication because it is an inexpensive and rapid technique. Ability to weld is the capacity of the metal to join satisfactorily on a process scale. The criteria are metallurgical compatibility of base metal and filler for a specific welding process, ability to produce mechanically sound joints, and serviceability under required mechanical conditions and corrosive environment. The ability to weld is often an evaluation of the susceptibility to various types of cracking. Several empirical equations and rules relating alloy composition or certain parameters to cracking susceptibility are available. For example, carbon content and impurities are very significant for welding ability. For carbon steels, the carbon equivalent should be less than 0.41 to ensure good welding[1]. When the carbon equivalent is 0.41–0.45, welding is usually good when using dry alkaline consumables. The carbon equivalent works best for moderate to high carbon levels. The carbon equivalent indicates microstructure in the heat affected zone. It is especially useful to indicate martensite formation that affects the hardness of the weld. For stainless steels, the carbon content should be less than 0.03% to avoid sensitization.

Welding is a metal joining process that involves local melting. The heating, melting, and solidification cycle affects metallurgical and compositional properties of the weld. The heating cycle is very rapid. Peak temperatures are above the melting temperature, and cooling rates are rapid. This causes different effects compared with conventional heat treatments. The weld metal should be matching or higher in alloy content than the base metal to achieve strength and corrosion resistance. Inert gases such as helium or argon have frequent use to prevent oxidation. Alloys that can passivate easily will also oxidize easily during welding if the shielding gas or flux is inadequate. Problems in welding typically arise in the base metal next to the weld where the heat from the welding operation can produce undesirable microstructure and properties. For example, excessive grain growth makes any material weak. Formation of martensite may make steel brittle. Controlling the amount of energy used in welding is the key to avoiding failures.

The weld consists of three principal regions: the weld metal, heat-affected zone (HAZ), and parent metal. Different metals are sometimes joined to create a dissimilar metal weld. The weld metal should be equal to or stronger than the weaker material and have better corrosion resistance than the materials being joined. Figure 9 shows the regions of a weld schematically. The weld solidifies in various regions. The composite region or the weld nugget consists of filler metal diluted with surrounding material. Next to the composite region are an unmixed zone that has completely melted and solidified, the weld interface or fusion line, and a partially melted zone of base material. At the fusion line, the peak temperature melts the base metal. The true heat-affected zone is beyond the partially melted area. It is subject to micro structural changes. Most notable is the coarsened structure in the middle of the HAZ surrounded by refined structure. In the heat-affected zone, carbide precipitation is possible. This may cause intergranular corrosion in the HAZ or knife-line attack near the fusion line.

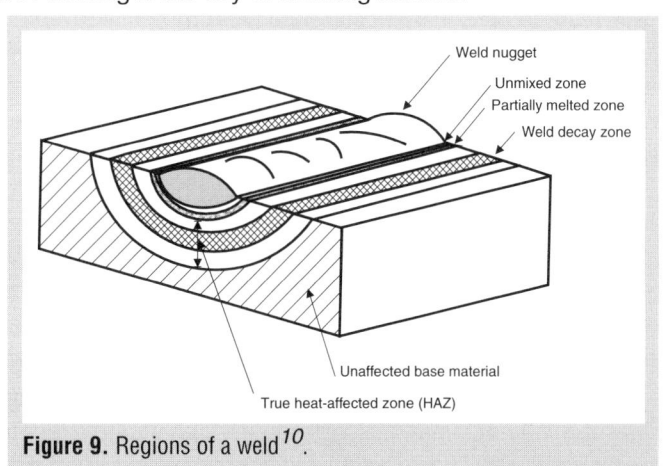

Figure 9. Regions of a weld [10].

6.1 Welding techniques

When welding using a fusion process, the edges of a component melt together to form weld metal. The fusion welding processes include arc, gas, power beam (laser or electron beam), and resistance welding processes. The European standard, EN 24063:1992, "Welding, brazing, soldering and braze welding of metals," groups the main welding processes as follows:

CHAPTER 2

- Arc welding
- Resistance welding
- Gas welding
- Forge welding
- Other welding processes
- Brazing, soldering, and braze welding.

Factors that will affect selection of a welding process include: material type, material shape (plate or tubular), quality and strength requirements, and degree of mechanization. The choice may be restricted by the cost of the process and availability of plant or workforce skill. In selecting a suitable process, one must also consider the type of application, such as portability of equipment, whether it can be used on site, and whether it is manual or mechanized.

The most important welding techniques for corrosion resistant alloys are different types of arc welding and resistance welding. Since these are the techniques common in construction of process equipment, this chapter does not discuss other methods. The most common arc welding processes are gas shielded metal arc welding (GSMAW) and shielded metal arc welding (SMAW). Gas metal arc welding (GMAW) is an American description of the gas shielded metal arc welding process. MIG welding means the use of an inert gas. Gas shielded tungsten arc welding (GTAW or TIG) uses an ac/dc arc burning between a nonconsumable electrode and the workpiece to melt the joint area.

In gas shielded metal arc welding, the heat is comes from burning an arc between the consumable electrode and the workpiece. The process consists of a dc arc burning between a thin bare metal wire electrode and the workpiece. The arc and weld area are enveloped in a protective gas shield. The wire electrode is fed from a spool, through a welding torch, and into the weld zone. The process is used on all thicknesses of steels, aluminum, nickel, stainless steels, etc.

Shielded metal arc welding is a term for the manual metal arc (MMA) welding process. In shielded metal arc welding, heat comes from a low voltage and high current metal arc. An ac/dc arc burns between a consumable electrode and the workpiece to melt the joint area. The consumable electrode is filler metal coated with flux coating. The arc and the weld pool are shielded by gases and slags that result from the decomposition of the electrode coating. The process has wide use in fabricating industries for construction and repair of plain carbon and low alloy steels. Stainless steel, nickel, Ni-Cr, Ni-Cr-Mo, and cast iron electrodes also have common use as do surfacing type electrodes for building up wear resistant surfaces.

In principle, submerged arc welding (SAW) is the same as shielded metal arc. It does use a separate slag feeding process and has a higher energy input rate. Similar to GSMAW welding, SAW involves formation of an arc between a continuously-fed bare wire electrode and the workpiece. The process uses a flux to generate protective gases and slag and to add alloying elements to the weld pool. An inert gas atmosphere such as in metal inert gas (MIG) welding may protect the molten metal. Before welding, a thin layer of flux powder is placed on the workpiece surface. The arc moves along the joint line. Since the flux layer completely covers the arc, heat loss is low and the operator cannot see the weld pool. SAW is ideally suitable for longitudinal and circumferential butt and fillet welds. Material thickness has virtually no restriction with suitable joint preparation. The most commonly welded materials are carbon-manganese steels, low alloy steels, and stainless steels. The process is also capable of welding some nonferrous materials. In electroslag welding, resistance heating of a consumable electrode and a molten conductive slag generate the heat. The energy input rate is higher than in arc welding[11].

In TIG welding, the area of the arc is enveloped in a protective gas shield that protects the weld pool and electrode from contamination while maintaining a stable arc. The filler wire is added by hand or feed rollers attached to the torch. TIG is applicable in all industrial sectors but is especially suitable for high quality welding. Because deposition rate can be quite low, MMA or MIG may be preferable for thicker material and for fill passes.

Oxyacetylene welding commonly called gas welding is a process that relies on the heat of a flame to melt the material being welded. Fusion can be autogenous or require addition of a filler material.

Fusion welding is a complex process with many variables. The welding method affects weld properties. In arc welding, an intense heat source moves over the workpiece. The heat flow in the direction of travel is small compared with that in the perpendicular direction. For the base material, the welding process is therefore a short heat pulse. In electroslag welding, the heat source is large and moves slowly to preheat the material ahead of the heat source. Heating rate influences grain growth. Stronger grain growth occurs with high energy processes such as electroslag welding. Cooling rate determines the final microstructure. Multirun welding can result in microstructure refinement, better toughness, and lower residual stresses compared with a single run weld[11].

Resistance welding is an economical and popular method for manufacturing thin sheet constructions especially for stainless steel. Electrical resistance welding is most suitable for repetitive production work when making many joints on the same metal of the same thickness. Common resistance welding methods are spot welding and seam welding. In spot welding, two electrodes hold the parts together. As current passes through the electrodes, heat obtained from resistance produces the weld. The weld nugget is created from molten workpiece surfaces under the electrodes. The electrode shapes may vary, but a clear distance always exists between two weld nuggets. In seam welding, the electrodes are circular. A series of overlapping spot welds forms the weld seam. The distortion with this method will be greater than in spot welding[12].

6.2 Welding problems

Welding often causes residual tensile stresses in the weld metal and compressive stresses in the surrounding base metal. Welding is a geometric discontinuity that concentrates any applied or residual stresses. This makes welds the weak point in constructions since all forms of environmentally induced cracking will preferentially occur in the weld. The skill of the welder and quality of materials and equipment significantly influence the occurrence of defects such as porosity, slag inclusions, convexity, and misalignment. Metallurgical changes during welding can often decrease corrosion resistance. The weld metal should therefore be higher in alloy content to ensure its nobility compared with base metal. Nonequilibrium cooling of the weld metal may lead to alloy segregation. When welding stainless steels or high strength alloys, the consumables must be free from moisture to avoid porosity and hydrogen penetration.

The problem with carbon and low alloy steel welding is martensite formation. These may be susceptible to hydrogen induced cracking. By keeping the carbon equivalent low, martensite formation is not as likely. Welding of hot rolled steel with yield strength less than 350 N/mm^2 and thickness up to 20–30 mm usually does not require special arrangements. If the yield strength is higher than 350 N/mm^2, tensile strength higher than 500 N/mm^2, or thickness higher than 25 mm, preheating, lower heat input, or post heat treatments may be necessary. Using the correct temperature is particularly important in field welding and repairs when the amount of molten metal is small compared with the surrounding metal. This may cause excessively rapid cooling resulting in martensite formation. When welding high strength steels, the heat input requires restriction to maintain impact strength requirements. The allowable heat input is lower with higher steel strength and lower required ductile-to-brittle transformation temperature. Common post weld heat treatments for constructions are stress relieving at 550°C–600°C and normalizing. Normalizing occurs for pressure vessel constructions at 900°C–920°C when the amount of cold work is more than 5%–10%.

Problems in austenitic stainless steel welding are cracking and loss of corrosion resistance. Weld metal pitting may result from welding with no filler or under-alloyed filler. Crevice corrosion of weld metal may result from micro fissures, lack of penetration in one-sided joints, or entrapped welding flux. Precipitation of carbides may occur in the heat-affected zone causing sensitization. Sensitization may result in intergranular corrosion and possibly intergranular stress corrosion cracking. The fusion line may corrode due to formation of an unmixed zone or carbide precipitation of Nb- or Ti-stabilized steels. Ferritic stainless steels are difficult to weld due to various embrittlement phenomena. Contamination from common atmospheric gases and carbon may cause intergranular corrosion. Duplex stainless steels require special procedures and consumables to maintain the balance between phases. High ferrite welds are brittle, and high austenite welds are susceptible to chloride induced localized corrosion. The consumables are typically higher in nickel content than the base metal[12].

Corrosion of stainless steel welds has undergone wide study. The main problems with stainless steel welds are weld decay, knife-line attack, and stress corrosion cracking. Weld decay is a result of precipitation of chromium carbides at the grain boundaries that occurs at 450°C–800°C. As a result of carbide precipitation, the stainless steel is sensitized. Since chromium that is essential for passive film forming is depleted at the grain boundaries, intergranular corrosion can occur. Sensitization does not affect mechanical properties unless the material corrodes. Knife-line attack sometimes occurs with stabilized steels. Then the fusion line is attacked. Knife-line attack has occurred in autogenous welds (no filler metal) of 6% Mo super austenitic steels in chlorine dioxide service. Stress corrosion cracking is the most dangerous failure mode of welds. The residual stresses caused by the welding process can be sufficiently high to cause SCC in a chloride bearing environment. Annealing to relieve the stresses is the only positive method to avoid stress corrosion cracking. Figure 10 shows schematically the corrosion sites of a stainless steel weld. During welding, some base metal will be heated to the carbide precipitation range. The location of the potential sensitized zone depends on the energy input, material thickness, and cooling rate.

Various smaller defects will also affect corrosion resistance of a weld and its surrounding area such as slag, weld metal splatters, or arc strikes if these are not removed. Arc strike is the location where the welding electrode has contacted the metal surface to melt a small volume of metal. Avoiding these strikes requires run-on and run-off tabs.

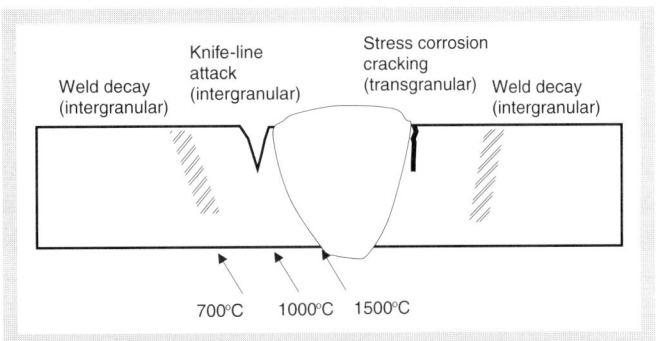

Figure 10. Corrosion sites in stainless steel welds showing typical peak temperatures during welding[13].

Welds in nickel alloys are also subject to sensitization and intergranular corrosion. Welding titanium structures is more problematic compared with other metals. Titanium does not react at normal temperatures, but above 400°C it absorbs oxygen and becomes brittle. Contamination will lead to brittle failures. Even a fingerprint can cause a failure. During welding, everything must be clean and done in an inert atmosphere to exclude all contaminants. If performed correctly, loss of strength is usually less than in the welding of steel. Welding by the gas tungsten arc welding process (TIG) using argon gas shielding has common use.

Welding of dissimilar metals can cause problems. When welding carbon steel and stainless steel, mixing in the melt pool can decrease corrosion resistance of the stainless steel. Using more alloyed filler metals can avoid this. Stainless steel has a higher thermal expansion coefficient. It will therefore expand and contract at a faster rate than carbon steel. Stainless steels conduct heat more slowly than carbon steel. This promotes sharper thermal gradients and possibly more warping. The weld zones may also remain longer in the carbide precipitating temperature range. To ensure mechanical properties and corrosion resistance, welding must use care. Attention to the method and filler metal selection, cleanliness, and good workmanship are necessary. Weld metal that matches the composition of a parent metal will usually have lower corrosion resistance due to segregation effects during solidification.

Welding heats the base metal causing oxide films known as heat tint on stainless steels to develop in the HAZ. Removal of the weld oxide layer after welding decreases impurity accumulation and pit initiation. Grinding, pickling, and passivation of welds in critical places are good practices.

CHAPTER 2

7 Plastics and elastomers

Plastics, composites, and other nonmetallic materials are useful for protecting carbon steel equipment, structures, and components. Structures can use polymers or composites in tanks, piping, pumps, and valves. Many small but important components such as seals, gaskets, and caulks are often polymers. The variety of polymers is enormous. To avoid confusion, the information about a polymer must include its generic class (polyester, epoxy, etc.), thermophysical behavior, and mechanical properties. The thermophysical behavior divides polymers into thermoplastics that soften and flow upon heating and thermosets that do not.

Thermoplastics soften with increasing temperature and return to their original hardness when cooled. Most are meltable, and they are recyclable. Thermoplastics usually have linear or branched structure consisting of flexible, entangled molecular chains. After polymerization of a thermoplastic, the finished polymer chains have no physical connections between them. Use of thermoplastics is only suitable to temperatures well below their softening point without some form of reinforcement.

Thermosetting materials harden on heating and retain their hardness when cooled. The final polymer is less temperature sensitive than thermoplastic materials. Thermosetting materials assume their final form with catalysis or heating. Thermosetting polymers crosslink by chemical bonds when cured making them unsuitable for recycling. A highly cross-linked, three-dimensional network structure exists throughout the part.

A composite is a combination of several materials. A nonmetallic composite is typically a chemically resistant polymer reinforced with a stronger material such as glass reinforced plastic (GRP) also known as fiber reinforced plastic (FRP). With proper polymer and additive selection, one can tailor corrosion resistance to match the environment. FRP composites have taken their place as an important construction material in the manufacturing industries. High strength to weight ratios, durability, and corrosion resistance make composites useful for many process applications.

Almost all thermoplastic materials are combinations of polymer resin systems with additives such as stabilizers, fillers, reinforcing fibers, etc., designed to provide specific properties. The common thermoplastic polymers in industrial service are fluorocarbons, polyolefins (polyethylene, polypropylene, etc.), and polyvinyl chlorides (PVC and PVDC). Thermosetting polymers usually have use as composites. Typical thermosetting polymers are epoxies, polyesters, furans, and phenolics. Most thermoset composite parts use polyester and epoxy resins. Polyester systems predominate in volume. Elastomers are organic materials with elastic properties comparable to rubber. They consist of linear polymer chains with light crosslinking. Stretching an elastomer causes the chains to untangle partially but not deform permanently like the thermoplastics. They are produced by vulcanization of linear polymers to form a network structure. Their uses are typically as seals and gaskets, flexible hoses, and coatings. Figure 11 gives a comparison of various types of polymer materials and their characteristics.

Construction materials

Thermoplastics
- Fully polymerized in raw state
- When softened or melted, can be formed. Cooling gives dimension and shape.
- Molecules are entangled but can be dissociated from one another without damaging them.

Fluorocarbons
Polyolefins
Polyvinyl chlorides

Thermosets
- Partially polymerized polymer and a catalyst.
- Polymerization or "curing" bonds molecules by crosslinking resulting in a stable structure.

Polyesters
Epoxies

Elastomers
- Linear polymer chains lightly crosslinked.
- Stretching an elastomer causes the chains to untangle partially but not deform permanently.
- Produced by vulcanization of linear polymers to form a network structure.

Natural and synthetic rubbers

Composite material
- Two or more materials dissimilar in form and properties.
- Clear boundaries between the components.
- Continuous matrix holds the reinforcement together, and discontinuous reinforcement provides the primary load-carrying capability.
- Resin-rich gel coat gives corrosion resistance.

Gel coat
Glass reinforced polyester layer

Figure 11. Polymer materials and their characteristics.

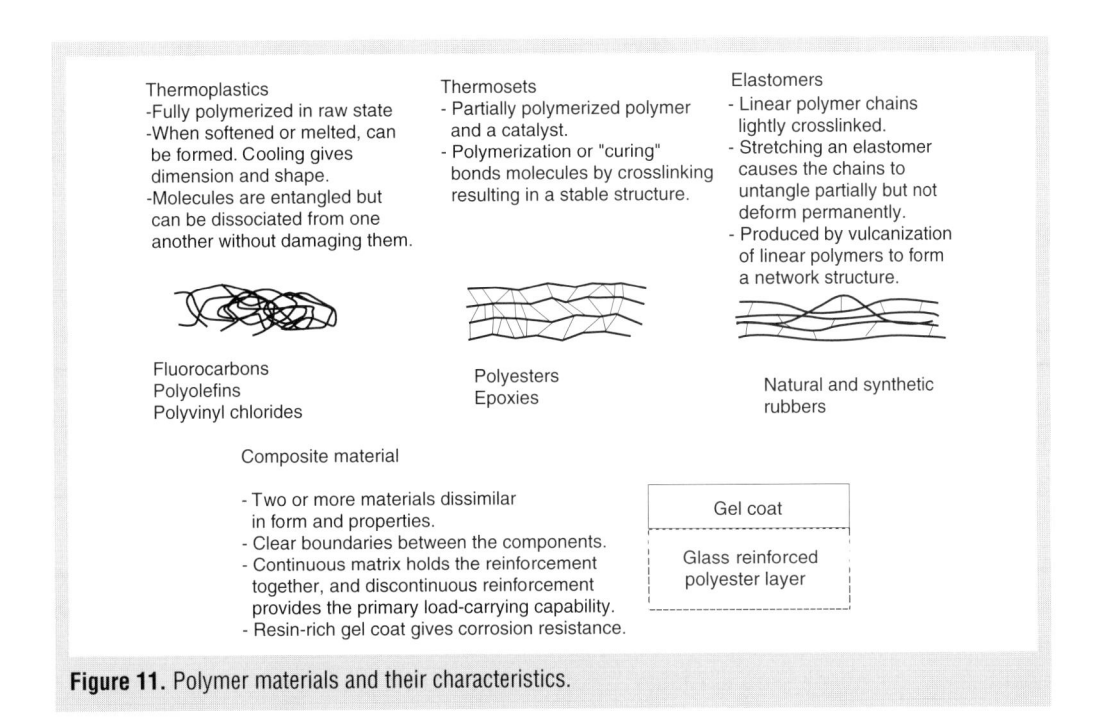

7.1 Thermoplastics

Fluorocarbons are a versatile and important group of thermoplastics used in the process industries. Most fluorocarbons can handle extremely corrosive environments, and they are seldom attacked by solvation. Some limitations exist for the use of fluorocarbons. The materials are porous. Adsorption of a component from a stream followed by reaction with another component or polymerization can lead to surface degradation and blistering. Overheating should be avoided, and thermal cycling can cause fatigue of the materials. Vacuum service requires careful design because collapse of pipe liners has occurred. The fluorocarbon materials are difficult to manufacture. Design and application are more limited than with other thermoplastic materials. The materials have use as impellers, mixers, and small containers. The most important use of fluorocarbons is lining for steel vessels. Lined pumps and valves are also available. Fluorocarbons are often expensive.

There are two general categories of fluorocarbon polymers. They are either fully or partially fluorinated. Fully fluorinated polymers include polytetrafluoroethylene (PTFE), fluorinated ethylene propylene (FEP), and perfluoroalkoxy (PFA). Examples of partially fluorinated polymers are ethylene-tetrafluoroethylene (ETFE), chlorine group containing polychlorotrifluoroethylene (PCTFE) and ethylene chlorotrifluoroethylene (ECTFE), and polyvinylidene fluoride (PVDF). The fluorocarbons are usually chemically resistant and are suitable for high temperatures. They will withstand adsorption of impurities. In general, the fully fluorinated polymers are more corrosion resistant and have higher operating temperatures than those with partial fluorination. All fully fluorinated polymers have nearly identical corrosion resistance but differ in maximum operating temperature. Partially fluorinated polymers differ in corrosion resistance and maximum temperature.

PTFE is the original, totally fluorinated thermoplastic material. It is weather resistant, resists all common solvents, is stable to high temperatures, has low friction coefficient, and is resistant in weak and strong acids and alkalies. PTFE is difficult to work and produce in usable forms. This restricts its use. It has use as a liquid sprayed coating, but these often remain thin and porous. PTFE has the highest temperature stability up to 290°C and is therefore resistant to aggressive environments at high temperatures. The main uses of PTFE are gaskets, filter material, valves, and piping. FEP has heat resistance of about 200°C but is more workable than PTFE. Its corrosion resistance is equal to PTFE. FEP has use as lining in pipe fittings and in smaller vessels. FEP coating uses a powder form with thicknesses up to 2 mm. ETFE copolymers are similar to FEP, and the corrosion resistance is equal. They are also available as sprayed powder coatings.

PFA has use in molding and extrusion applications requiring high temperature strength, chemical inertness, and ability to process as a melt. Many industrial components have linings of PFA powder coatings. PVDF or PVF2 has excellent resistance to aging, chemical attack, and abrasion. It has common use in piping systems, valves, storage tanks, heat exchangers, etc. PVDF coatings are liquids and are not sprayed at less than 100 m thickness. They are also porous and therefore not suitable to very corrosive environments. PVDF may degrade slightly in outdoor exposure and is attacked by fuming sulfuric acid.

Partial substitution of the fluorine group with a chlorine group will improve some polymer properties. PCTFE is stable up to 180°C and has slightly lower chemical resistance than the totally fluorinated materials. It undergoes slight attack by ketones, esters, and chlorinated and aromatic hydrocarbons but not acids or alkalies. The manufacturing properties are good, and the material has use as a coating and prefabricated liner for chemical applications and as o-rings, gaskets, and seals. ECTFE has excellent creep, wear, and strength properties and good chemical resistance. It is suitable for pipe and tank linings. Corrosion resistance is equal to ETFE. ECTFE coatings are available as thick powder coatings.

The use of polyolefins such as high and low density polyethylene (HDPE and LDPE), polypropylene (PP), and polybutylene (PB) is extensive in the chemical industry. These are the lowest cost polymers for piping, smaller structural parts, and containers. Low density polyethylene has extensive use for piping because of its ease of handling and fabrication on site. The chemical resistance of LDPE is good but not as good as high density polyethylene or polypropylene. LDPE may embrittle under outdoor exposure, and it does not resist oxidizing acids and hydrocarbons. The upper temperature limit for the material is 60°C. High density polyethylene and polypropylene are very similar in chemical applications. Their mechanical properties and chemical resistance are better than those of LDPE. Only strong oxidizing acids will attack these materials significantly. The better mechanical properties of these two products extend their use into larger shapes and liners and as solid containers. Polybutylenes are chemically resistant to most solvents, acids, alkalies, and inorganic salt solutions but strong oxidizing acids should be avoided. The materials are resistant to creep, abrasion, and stress cracking.

Polyphenylene oxide (PPO) has primary use in pump parts and other applications where impact strength and reasonable abrasion resistance are necessary. The chemical resistance, heat stability, and dimensional stability of the material are good, and the temperature limit is 120°C. Polyphenylene sulfide (PPS) has good chemical resistance excluding strong oxidizing acids, and the temperature range is -170°C– +190°C. Coatings prepared from PPS are available.

Polyvinyl chlorides are easily workable and can be heat or solvent welded or machined to adapt fittings. Two classes of primary PVC material are available: a flexible, normal grade and a rigid, high impact grade. The latter is normally used. The upper temperature limit is 65°C for the normal and 60°C for the high impact grade. PVC is useful as a piping material for a wide range of products found in industry. PVC coatings are available as liquids for trowel application. Piping, solid valves and pumps, smaller equipment, and structures may use polyvinyl dichloride or vinylidene chloride (PVDC). It also has use as a liner for vessels. PVDC has better chemical resistance and mechanical properties than PVC including heat stability up to 100°C. The chemical resistance of PVC is good in inorganic media, but strong acids may attack it slightly. Almost all solvents can attack PVC.

7.2 Thermosets

Thermosetting polymers have primary use as composite materials. The polymers usually have no distinct phases on a macroscopic scale. Composite materials consist of two or more materials on a macroscopic scale with phases that are dissimilar in form and properties. The basic concept of a composite is that its properties are superior to those of the independent constituent materials. Most composites contain matrix and reinforcement phases. The matrix is the less stiff and weaker continuous phase. The purpose of the matrix is to hold the reinforcement together and distribute mechanical loads. The reinforcement is a discontinuous phase that actually carries the load. Common thermoset resins are epoxies, polyesters, vinyl esters, and phenolics. The reinforcing fibers are usually glass or carbon fibers with a diameter of about 10 µm. Aramid fibers and polyethylene roving or mat also have use as reinforcement. The thermosetting polymer is usually liquid before processing. As it hardens, the formation of a network structure causes reinforcement. Corrosion resistance of a composite depends on the resistance of the matrix polymer. The outermost layer of a composite structure is usually pure resin to avoid absorption of solution through the fibers. In corrosion resistant applications, the thickness of this layer is often more than 1 mm.

The corrosion resistance of fiber reinforced plastics can be improved by using dual laminates, i.e., FRP with a thermoplastic corrosion liner. Dual laminate is a thermoplastic construction bonded to a fiber reinforced plastic structure as Fig. 12 shows. A typical laminate construction consists of the following:

CHAPTER 2

- Thermoplastic liner using as typical materials polyvinyl chloride (PVC), polypropylene (PP), polyvinylidene fluoride (PVDF), ethylene chlorotrifluoroethylene (ECTFE), ethylene tetrafuoroethylene (ETFE), fluorinated ethylenepropylene (FEP), and perfluoroalkoxy (PFA).
- Thermoset bonding layer that provides a mechanical and chemical bonding with the FRP layer.
- A layer of electrically conductive material (typically a carbon fiber veil) laminated immediately behind the thermoplastic. This is for spark testing or leak detection in the event of liner damage.
- A secondary corrosion barrier is laminated on top of the FRP layer using thermoset corrosion resistant resins and alternate layers of polyester or glass fabrics and chopped strand glass fiber mat. The barrier must be resin rich to insure corrosion resistance.
- The FRP structure consists of either a hand lay-up or filament winding construction.

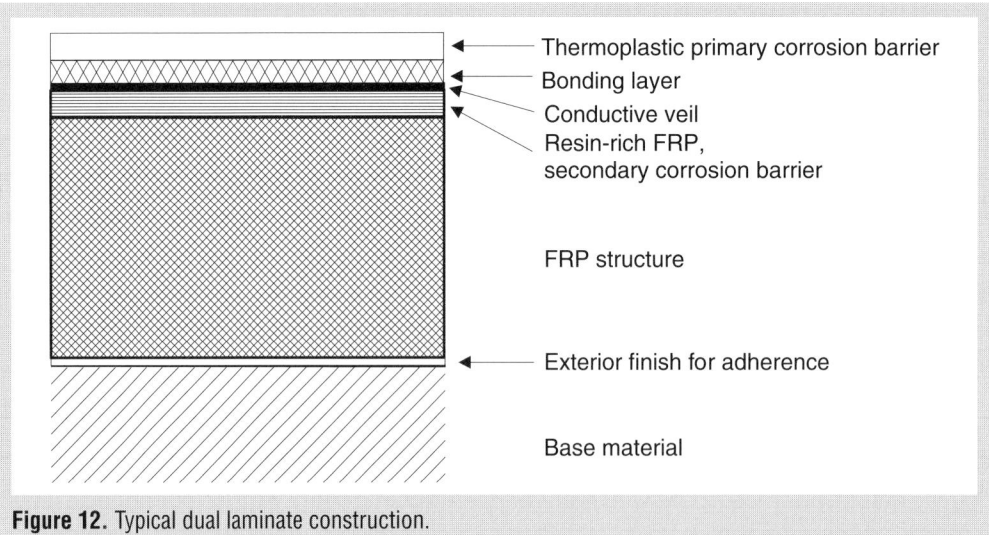

Figure 12. Typical dual laminate construction.

Successful application depends on matching the properties of a given corrosion-resistant plastic material to the application requirements. Resin selection has the single biggest impact on long-term performance of the equipment and can have an equally significant impact on cost. The first step in determining the suitability of FRP composite materials is characterization of the environment. Resistance to expected chemicals must be verified by published information or by testing. After verifying the material's compatibility with the chemicals, the following factors must be considered:

- Operating temperature
- Continuous immersion vs. splash and spill
- Effect of vibration, abrasion, and fretting
- Structural loads, compliance with building codes, and safety (fire) codes
- Design details such as treatment of cut ends to prevent absorption or galvanic action due to carbon fiber reinforcement.

Epoxies are the most common resins. Epoxies reinforced with glass fibers have very high strength and are very resistant to heat. They have use at low to medium temperatures and even above 100°C. Composites with epoxy resins give the best combination of corrosion resistance and mechanical properties. Chemical resistance of an epoxy resin is excellent in nonoxidizing and weak acids but not in strong acids. Alkali resistance is excellent in weak solutions. Epoxies are available as castings, extrusions, sheets, adhesives, and coatings. They have use as pipes, valves, pumps, small tanks and containers, linings, and protective coatings. Epoxy resin is typically cured with amine or anhydride curing agents to vary the thermal and chemical performance. Aromatic amines give optimal properties particularly in alkaline environments. The anhydride curing agents also have good chemical resistance and may be more suitable in acid environments.

Polyesters are inexpensive, but they have high shrinkage that results in a poor fiber and matrix bond. Polyester resins reinforced with glass fibers also have good strength and good chemical resistance except to alkalis. Some special materials using bisphenol-A epoxy are more alkali resistant. The temperature limit for polyesters is about 90°C. Polyester resins are used for similar but less severe services than vinyl ester resins.

Vinyl esters are used when toughness and fatigue resistance must be greater than those of polyesters. They also have better high-temperature and corrosion resistance than polyesters. Vinyl ester pipe grades handle most epoxy applications. They have superior corrosion resistance to strong acids, chlorine, and oxidizing solutions. Vinyl ester resins are most often specified in highly concentrated acids and oxidizing acid services. The term vinyl ester is a generic one to describe the chemical nature of the resin. Standard vinyl esters are synthesized from bisphenol-A epoxies.

Phenol-formaldehyde or phenolic resins are the oldest completely synthetic polymers. Phenolics have good high-temperature properties, but they require high curing pressures. The void content in a cured part is high[14]. Phenolic plastics filled with asbestos, carbon or graphite, and silica have low cost, good mechanical properties, and chemical resistance except to strong alkalis. Phenol-formaldehyde resins containing more phenol than formaldehyde (novolac resins) have use for acid environments. Furan (furfuryl alcohol) resins have much better alkali resistance than phenolic plastics. Furan resins are particularly suitable for solvents and mildly alkaline and nonoxidizing acid service.

CHAPTER 2

Most resins and some fibers absorb moisture. This increases weight and decreases strength and stiffness. The resin properties degrade since the use temperature approaches the glass transition temperature. UV light degrades the resin, and micro cracking weakens the composites and provides a path for moisture absorption.

7.3 Elastomers

Rubbers are the most typical elastomers. They consist of a basic elastomer that is natural or synthetic rubber and several additional components. The corrosion resistance depends on the basic elastomer. The additives are primarily fillers, softeners, and compounds for vulcanizing. Rubbers have use as soft or vulcanized materials. Vulcanizing is a process to harden rubber by producing sulfur bridges between molecule chains. Typical rubbers are natural rubber (NR) and its synthetic counterparts such as styrene butadiene (SBR), butyl rubbers (IIR), polychloroprene (CR), chlorosulfonated polythene (CSM), and nitrile rubber (NBR). General purpose types such as NR, SBR, and IIR are not resistant to oils, but the special grades such as CR, CSM, and NBR do have resistance to oils[15]. The most important properties of rubbers are elasticity, lack of stiffness, good abrasion resistance, and good chemical resistance. The primary applications for rubbers are sealing, electrical insulation, corrosion protection, and wear protection. In corrosion and wear resistant applications, rubbers primarily have use as linings. Soft rubbers are easy to install on clean surfaces. They are cut, applied, and air bubbles and solvents between rubber and substrate are removed. The applied layer is tested with high voltage and repaired if necessary before vulcanizing. In a coating plant, vulcanizing uses an autoclave at 130°C–150°C. In the field, lined vessels are closed and filled with water that is heated to 95°C–97°C with steam. If vulcanizing with water is not possible, self-vulcanizing rubber types must be used. After vulcanizing, the coating quality is rechecked with high voltage, and its hardness is measured.

Vulcanized natural rubber is more common than soft natural rubber. Vulcanized natural rubber resists salt solutions, dilute alkalis and acids, moist chlorine gas, and chlorine solutions. The highest operating temperature is 85°C. It is prone to cracking when cold and when subject to large and rapid temperature changes. Rapid temperature changes can also cause delamination. Soft natural rubber has good mechanical properties and almost equal corrosion resistance compared with vulcanized natural rubber. The highest operating temperature is 70°C–80°C. A thick layer of soft natural rubber gives efficient protection against wear. Styrene butadiene rubber is very resistant to dilute and concentrated caustic soda.

Butyl rubbers (IIR) are elastic materials with low permeability and good corrosion resistance. Their mechanical properties are poor. They are suitable for corrosion protection applications where hard rubbers are not suitable due to temperature variations. Butyl rubbers may be difficult to apply and process. Development of halogenated grades overcomes this. Butyl rubbers retain their elasticity at high temperatures. They have use up to 100°C–120°C[15]. They have good resistance to alkalis and slightly better resistance to acids than natural rubber. Butyl rubbers have wide use in transportation tanks and flue gas desulfurization.

Chloroprene (CR) has poor resistance to oils. It has low permeability and good wear resistance but limited chemical resistance. Typical uses are mildly corrosive environments that may contain organic compounds and seawater. Maximum operating temperatures are about 100°C. Soft chloroprene is resistant to dilute and concentrated caustic soda even at high temperatures.

Chlorosulfonated polythene (CSM) is resistant to acids, alkalis, and oils. Its resistance to hypochlorite makes it suitable for use in bleaching plants. The highest operating temperature is 120°C. CSM is difficult to process and therefore expensive. In certain cases, butyl rubber can substitute for CSM. Nitrile rubbers have use primarily in environments that contain large amounts of oil. Other rubber types swell due to absorption. Nitriles also resist dilute acids.

Different rubber types may fail due to several factors. Atmospheric oxygen can cause oxidation of natural rubber, styrene butadiene rubber, and nitrile rubber. The material usually becomes hard and brittle, but it may sometimes soften and lose adhesion. Ozone is a destructive compound for rubbers. The molecular chains of the rubber react with ozone causing cracks. Ozone can attack natural, styrene butadiene, isoprene, butadiene, and nitrile rubbers. Strongly oxidizing chemicals can attack natural rubber resulting in loss of adhesion, brittleness, or softening. Solutions containing active chlorine such as chlorine or chlorine dioxide in acid solution or hypochlorite in alkaline solution can cause damages as can other oxidizing solutions. Solution absorption causes swelling and eventually softening. Rubbers absorb organic solutions that have a similar chemical nature. Natural rubber absorbs hydrocarbons such as oils. Nitrile and chloroprene rubbers absorb esters, ketones, etc.

7.4 Corrosion resistance and applications

Plastics do not corrode in the electrochemical sense. Degradation of plastics has three major causes. An active species from the environment can absorb into the plastic causing swelling or internal reactions with the polymer. This normally causes softening and distortion but weight loss can also occur. Oxidation of the resin molecule can occur in oxidizing conditions with solutions containing active chlorine for example. This often causes hardening, embrittlement, and cracking. With certain resin components, polymerization of the resin can continue after curing resulting in hardening, shrinkage, and cracking of the material. Degradation of a plastic is not a surface effect like corrosion of metals. It occurs internally. Absorption of only a few ppm of an aggressive species from otherwise innocuous process streams can result in total loss of mechanical properties. Failure of thermosetting polymers usually results from the breakdown of chemical bonds or dissolution in particular solvents possibly enhanced by temperature, stress, impact, or wear. Blistering may occur with hot, small molecule solutions when molecules penetrate the plastic surface layer. Blistering is often only an aesthetic problem. Hardening results from removal of softener and additives leaving only the pure, hard base material. Cracking is usually the result of stresses and occurs in heat treated areas. Plastics often have excellent resistance to weak mineral acids, and inorganic salt solutions do not effect them. The use of plastics is limited to moderate temperatures and pressures.

CHAPTER 2

Plastics are less resistant to mechanical wear and have high thermal expansion rates. Thermoplastics usually have low strengths. The corrosion resistance of reinforced plastic equipment depends on the resin-rich layer (gel-coat) next to the process stream. The reinforced layers provide strength. Any exposed glass in a laminate can influence the chemical resistance of glass laminates. Diffusion of water between resin and fibers reduces strength of a laminate.

Major resin suppliers often give data to eliminate obviously incompatible polymers. Table 6 provides some information on polymer compatibilities. The best way to avoid failure is to test under worst case conditions. Temperature, stress, and time are critical factors. Test temperatures should not exceed those experienced during use. Stress levels up to the supplier recommended long-term stress limit should be tested. These are usually 10–20 MPa. If the material is chemically resistant at these stress levels, the risk of field problems is low. Any physical changes at elevated temperatures are likely to affect chemical resistance. For example, the glass transition temperature is the temperature where the polymer changes from a hard glassy material to a rubbery one. Above the glass transition temperature, mechanical strength decreases. The exposure time during testing should be sufficiently long. Short term exposure can produce microscopic stress cracks in the surface. These invisible cracks will not necessarily influence initial performance. As the part ages or undergoes exposure to environmental stresses, the cracks may propagate and cause failures.

Table 6. Chemical resistance of polymers where R = resistant, A = attacked, S = slight effect, and E = embrittles[16, 17].

Material	Weather resistance	Weak acid	Strong acid	Weak alkali	Strong alkali	Solvent
PTFE	R	R	R	R	R	R
FEP	R	R	R	R	R	R
ETFE	R	R	R	R	R	R
PFA	R	R	R	R	R	R
PVDF (PVF2)	S	R	A H_2SO_4	R	R	R
PCTFE	R	R	R	R	R	S
ECTFE	R	R	R	R	R	R
LDPE	E	R	A oxidizing	R	R	R
HDPE	E	R	R – A	R	R	R
PP	E	R	A oxidizing	R	R	R
Polybutylene	E	R	A oxidizing	R	R	R
PPO	R	R	R	R	R	R – A
PPS	R	R	A oxidizing	R	R	R
PVC	R	R	R – S	R	R	R – A
PVDC	R	R	R – S	R	R	R – A
Epoxy	R	R	A	R	R	R – S

In digesters or washing plants, reinforced polymers have use for handling fumes and flue gases. To recover heat from the flue gas, a scrubber must remove particles. Scrubber and gas ducts often use plastic. Plastics and polymer composites have common use in a bleach plant. Composites have been successful in process and effluent piping for chlorine containing liquids, recovery boiler stacks and scrubbers, chemical and filtrate tanks, ventilation ducts, etc. Vinyl ester based GRP has been successful in bleaching stages since it is resistant to chlorine, chlorine dioxide, and sodium hypochlorite. Pulp washer hoods in bleaching plants exposed to caustic and bleach chemical vapors have been used since the 1970s. Vinyl ester based composites have been used to handle corrosive gases at temperatures up to 170°C being emitted from a black liquor recovery furnace. Liquid handling equipment such as pumps and valves use fiberglass reinforced resins. For example, polysulfone resin provides chemical resistance to caustic and sodium hypochlorite.

In the pulp and paper industry, polyolefins are common. Bisphenol-A polyester laminates can handle corrosive vapors in chlorine dioxide bleaching. High density polyethylene is suitable for effluent lines, and polypropylene and polybutylene also have use for this application. Polypropylene resists 60% sulfuric acid, concentrated hydrochloric acid, and 50% sodium hydroxide up to about 65°C. Polypropylene and polyvinyl chloride can handle acidic and alkaline filtrates up to approximately 60°C. PVC cannot handle hypochlorite solutions above ambient temperature, but PP has use up to 50°C. PVC can handle most acids and alkalis up to 60°C but not 98% sulfuric acid[16].

The most common fluoroplastics are polyvinylidene fluoride and polytetrafluoroethylene. PVDF has use as steel pipe linings. Pump components may be coated with it or made from the solid reinforced polymer. PVDF can handle all acid and alkaline filtrates in a bleach plant, wet chlorine gas, chlorine dioxide, and 50% NaOH up to 50°C. It is unsuitable for concentrated sulfuric acid or hydrogen peroxide[16]. PTFE has had use in pump linings. It is chemically inert in bleaching environments and has frequent use as a reinforcing material.

The original uses for reinforced thermosetting plastics were those where corrosion resistance comparable to high alloyed stainless steel or titanium at reasonable cost was necessary. Reinforced plastics today have routine use for mildly corrosive environments. The common resins for these applications are novolac epoxies, vinyl esters, and bisphenol-A polyesters and furans. Phenolic resins cannot withstand corrosive chemicals. Epoxies have use for inexpensive piping. Furans are resistant to solvents. For corrosion resistant applications in bleach solutions, bisphenol-A polyester and vinyl ester resins have common use. The latter materials are becoming dominant because of better mechanical properties. The reinforcement is usually glass fiber. Composite materials are suitable for chlorination, chlorine dioxide, and alkaline extraction stage filtrates.

The main reason to use a dual laminate in the process industry is the superior chemical resistance of the thermoplastics compared to FRP only. Dual laminates have found use in numerous applications. Examples from the pulp and paper industry include piping and headers in chlorine drying towers and sodium hypochlorite tanks and chlorine dioxide generators and washer drum suction boxes in bleach plants.

CHAPTER 2

8 Summary

For the pulp and paper industry, the main metallic material groups are carbon steels, stainless steels, nickel alloys, and valve metals such as titanium. Under specific conditions, all these metals will corrode. Corrosion of metallic materials is not a problem unless it significantly affects production rate, product quality, safety, etc. Corrosion is a problem when the corroding equipment can not fulfill its function due to leaks, cracks, etc.

The corrosion resistance depends on many factors. These relate to material composition and microstructure, construction details such as welds, and the corrosive environment.

The basic construction material is carbon steel. If its corrosion resistance is not sufficiently high, then it is replaced by stainless steel, nickel alloy, or titanium. The corrosion resistance of stainless steels increases with an increasing amount of alloying elements, Cr, Ni, Mo, and nitrogen. Stainless steels are suitable for oxidizing environments when temperature and aggressive anion concentration are not too high. As conditions become more severe, consideration of upgrading from stainless steel to a nickel alloy is common. Titanium is suitable for highly oxidizing environments where temperature and chloride concentration are high.

Material selection is not limited to metallic materials. In many applications, polymers are a cost-effective solution. Polymer materials have use as construction materials and as coatings and linings. The corrosion resistance of reinforced plastic equipment depends on the resin-rich layer (gel-coat) next to the process stream. The variety of polymers is enormous. They often have resistance to weak mineral acids and inorganic salt solutions. Data to eliminate incompatible polymers is available. The use of plastics is limited to moderate temperatures and pressures. Plastics are less resistant to mechanical wear and have high thermal expansion rates.

References

1. Anon., The Metals Blue Book, Welding Filler Metals, Metals Data Book Series, vol. 3, CASTI and American Welding Society, Edmonton, 1995, Chap. 4.

2. Anon., The Metals Black Book, Ferrous Metals, 2nd edn., Metals Data Book Series, vol. 1, CASTI and American Welding Society, Edmonton, 1995, Chap. 6.

3. Anon., The Metals Black Book, Ferrous Metals, 2nd edn., Metals Data Book Series, vol. 1, CASTI and American Welding Society, Edmonton, 1995, Chap. 18.

4. Treseder, R. S., NACE Corrosion Engineer's Reference Book, NACE International, Houston, 1991, pp. 153–223.

5. Jonsson, K. -E., in Handbook of Stainless Steels (D. Peckner and I. M. Bernstein, Ed.), McGraw- Hill, New York, 1977, Chap. 43.

6. Garner, A., in ASM Metals Handbook, vol. 13, ASM International, Metals Park, 1987, pp. 1187–1220.

7. Anon., Stainless steels for pulp and paper manufacturing, Nickel Development Institute Report No. 9009, American Iron and Steel Institute, Toronto, 1982, pp. 5–47.

8. Anon., The Metals Red Book, Nonferrous Metals, Metals Data Book Series, vol. 2, CASTI and American Welding Society, Edmonton, 1995, Chap. 2.

9. Anon., The Metals Red Book, Nonferrous Metals, Metals Data Book Series, vol. 2, CASTI and American Welding Society, Edmonton, 1995, Chap. 3.

10. Streicher, M. A., in Forms of Corrosion – Recognition and Prevention (C.P. Dillon, Ed.), NACE International, Houston, 1982, Chap. 6.

11. Easterling. K., Physical Metallurgy of Welding, 2nd edn., Butterworth-Heinemann, Oxford, 1992, Chap. 1.

12. Anon., Welding of stainless steels and other joining methods, Nickel Development Institute report No. 9002, American Iron and Steel Institute, Toronto, 1993, pp. 15–43.

13. Jarman, R. A., in Corrosion (L. L. Shreir, R. A. Jarman, and G.T. Burstein, Ed.), 3rd edn., vol. 2, Butterworth-Heinemann, Oxford, 1994, Chap. 9.5.

14. Fontana, M. G., Corrosion Engineering, 3rd edn., McGraw-Hill, New York, 1987, Chap. 5.

15. Pitman, J. S., in Corrosion, (L. L. Shreir, R .A. Jarman, and G. T. Burstein, Ed.) 3rd edn., vol. 2, Butterworth-Heinemann, Oxford, 1994, Chap. 18.7.

16. Brydson, J. A., in Corrosion, (L. L. Shreir, R. A. Jarman, and G. T. Burstein, Ed.), 3rd edn., vol. 2, Butterworth-Heinemann, Oxford, 1994, Chap. 18.6.

17. Treseder, R. S., NACE Corrosion Engineer's Reference Book, NACE International, Houston, 1991, pp. 224–234.

CHAPTER 3

Electrochemistry of metallic corrosion and corrosion test methods

1	**Basic electrochemistry**	**67**
1.1	Electrode polarization	72
1.2	Electrode surface	77
2	**Corrosion cell**	**78**
3	**Passivation**	**82**
4	**Corrosion research methods**	**86**
4.1	Physical measurements	87
4.2	Electrochemical techniques	92
	4.2.1 Polarization curves	94
	4.2.2 Potentiostatic and galvanostatic tests	98
	4.2.3 Polarization resistance measurement	102
	4.2.4 Electrochemical impedance spectroscopy	103
	4.2.5 Calculating corrosion rates	106
4.3	Instrumentation	109
4.4	Sample preparation	115
5	**Field testing**	**118**
6	**Summary**	**119**
	References	121

CHAPTER 3

Jari Aromaa

Electrochemistry of metallic corrosion and corrosion test methods

This chapter will give an introduction to the electrochemistry of corrosion and explain some common electrochemical test methods. Corrosion is a physicochemical interaction between a metal and its environment that results in unwanted changes in the properties of the metal. It may lead to impairment of the metal, technical system, environment, or product. Corrosion is spontaneous destructive oxidation of metals. Corrosion is due to one or more corrosive agents or corrodents. Corrosive agents in contact with a given material influence reaction rates or stabilities of surface films. In a normal atmosphere, all metals except the noblest ones such as gold, platinum, and palladium corrode spontaneously. The corrosion system consists of one or more materials and those parts of the corrosive environment that influence corrosion.

Corrosion can be chemical or electrochemical. Chemical corrosion is direct oxidation of the metal. Electrochemical corrosion is oxidation in wet environments caused by charge transfer reactions involving at least one anodic and one cathodic reaction. Modern corrosion theory uses two basic elements: mixed potential theory and formation of surface films. In a corrosion situation, all the electrons released in one reaction must be theoretically consumed in another reaction. This results in a closed circuit and leads to mixed potential theory. The corrosion resistance of any material depends on its nobility and its ability to form stable surface films. Noble materials are less prone to react. Stable surface films will form a barrier between the metal and its environment to prevent further reactions. This is passivity.

1 Basic electrochemistry

When studying chemical reactions, one must study thermodynamics and kinetics of reactions. Thermodynamics is the probability of a reaction to happen. Kinetics is the reaction rate. Thermodynamics will tell whether corrosion reactions are thermodynamically possible or not. In the first case, a metal may corrode. To obtain a more definite answer, estimating reaction rates is necessary. In the second case, metal will not corrode and studies on reaction rates are not necessary.

In electrochemical reactions, chemical energy converts to electrical energy or vice versa, i.e., electrons are released or consumed. Electrochemical reactions usually occur at some surface. The reactions that release electrons are anodic reactions. Reac-

CHAPTER 3

tions that consume electrons are cathodic reactions. The surface on which anodic reactions occur is an anode. Cathodic reactions occur on a cathode. Anodes and cathodes are electrodes. An electrode is a point at which conduction changes from ionic to electronic and charge transfers over the interface by an electrochemical reaction. Anodic reactions are oxidation or dissolution reactions, and cathodic reactions are reduction or deposition reactions. Oxidation and reduction reactions are all redox reactions. A redox reaction is an electrochemical reaction in which both reduction and oxidation occur simultaneously. A reduction reaction consumes the electrons produced in an oxidation reaction. These are sometimes considered half reactions of a redox reaction. The combination of oxidation and reduction reactions also forms an electrochemical cell. The corresponding redox reaction is a cell reaction. If the reaction rates of the electrode material are very small, the reactions of the solution species determine the electrode potential. It is a solution redox potential. Otherwise, the electrode potential is the equilibrium or corrosion potential of the electrode material.

An electrochemical cell is galvanic, reversible, or electrolytic. A galvanic cell is one in which the cell reaction proceeds spontaneously to produce power. An electrolytic cell is one in which an external source drives the cell reaction. It consumes power. A reversible cell has no current flows and is always in a state of equilibrium. The cell reaction in a reversible cell is not spontaneous. In a reversible cell, an infinitesimal change in the cell potential can cause it to proceed in either direction. Every electrochemical reaction can spontaneously or with an external current source proceed in the anodic or cathodic direction. For example, energy can be stored in an automobile battery as chemical compounds by forcing the electrochemical reactions in one direction with an external current source. The energy stored in the battery can be released as electricity by spontaneous electrochemical reactions proceeding in the opposite direction as Fig. 1

Unloading
- spontaneous reaction

Anode   Cathode

Anode: $Pb + H_2SO_4 = PbSO_4 + 2 H^+ + 2 e^-$
Cathode: $PbO_2 + H_2SO_4 + 2 H^+ + 2 e^- = PbSO_4 + 2 H_2O$

Loading
- forced reaction

Anode   Cathode

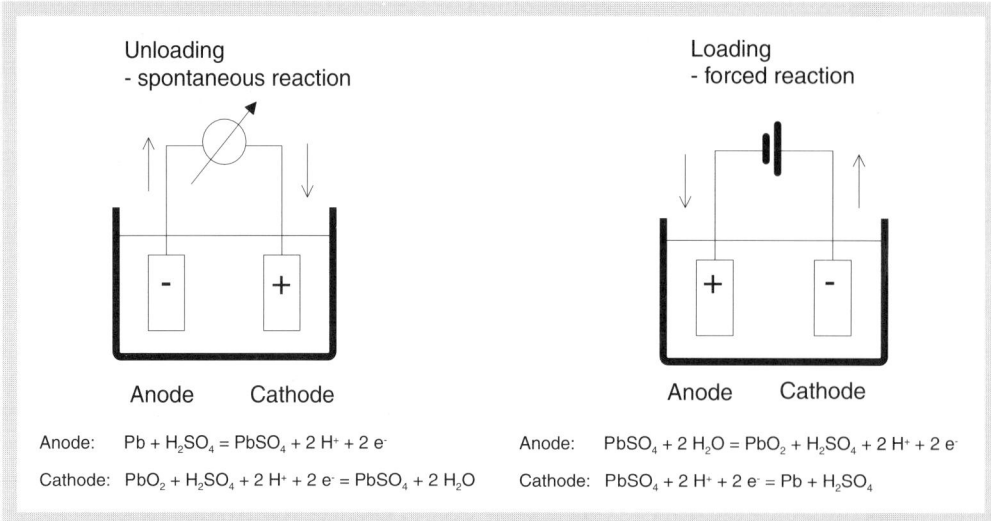

Anode: $PbSO_4 + 2 H_2O = PbO_2 + H_2SO_4 + 2 H^+ + 2 e^-$
Cathode: $PbSO_4 + 2 H^+ + 2 e^- = Pb + H_2SO_4$

Figure 1. Reversible nature of electrochemical reactions showing that energy stored in an automobile battery can release in spontaneous reactions.

shows. Note the change of signs of the electrodes. An international convention calls the electrode where oxidation occurs the anode. The electrode where reduction occurs is the cathode. Nothing is said whether the electrode is positive or negative. The sign of a given electrode depends on whether it converts chemical energy into electrical energy or vice versa. If chemical energy converts into electrical energy, the electrochemical reaction is spontaneous such as corrosion or a nonrechargeable battery. The anode is negative, and the cathode is positive. If electrical energy converts into chemical energy, the reactions require driving with external current such as electrolysis and electroplating. In this case, the anode is positive and the cathode is negative. Whether a cell is galvanic or electrolytic, electrons flow into cathodes for the reduction reaction. The electrode into which electrons flow is the most positive electrode of the cell. Convention writes and draws it on the right.

Every chemical or electrochemical reaction has an equilibrium state. In this state, the anodic and cathodic partial reactions of an electrochemical reaction have equal rates. The system is in a dynamic equilibrium state, and no net reaction occurs. For example, when a copper sheet is immersed in copper sulfate solution, the anodic dissolution rate of copper from sheet to solution in the equilibrium state equals the cathodic deposition rate from the solution to the surface of the sheet. Theoretically, one can calculate the equilibrium state of an electrochemical reaction from thermodynamic values. This is the standard electrode potential, E_0, or equilibrium potential of the electrochemical reaction. The standard electrode potential corresponds to a determined standard state of 1 atm, 25°C, activity of reactive species of 1 or ideal solution of 1.0 mol/L, and equilibrium potential to any other state.

Thermodynamic calculations allow estimations on the tendency of a reaction to occur. If the change of so-called Gibbs free energy, ΔG [kJ/mol], of the reaction has a negative value, the reaction can proceed in that direction. If the value of ΔG is positive, then the reaction cannot occur spontaneously. For more negative ΔG values, the reaction is more likely to occur spontaneously. Any chemical reaction can proceed in either direction. The values for ΔG for the forward and reverse reactions are numerically the same but opposite in sign, since $\Delta G = \Delta \Sigma G$ (products) - $\Delta \Sigma G$ (reactants). The value of the free energy change of a chemical reaction changes only in sign when the direction of the reaction reverses. If the anodic reaction has a negative ΔG value, the corresponding cathodic reaction has a positive value. If the equilibrium potential of an electrochemical reaction is low, it will more likely proceed in the anodic direction (negative ΔG) than in the cathodic direction (positive ΔG). If the equilibrium potential is high, the reaction will more likely proceed in the cathodic direction.

The electrochemical series tabulates standard electrode potentials. Some sources call the electrochemical series oxidation/reduction potentials, electromotive series, etc. The sign of the potential may also be opposite. The reference state of electrochemical series is the hydrogen evolution reaction, H^+/H_2 reaction. Its standard electrode potential has been universally assigned as 0 V. This electrode is the standard hydrogen electrode (SHE) against which all others are compared. For example, the standard electrode potential of the Fe/Fe^{2+} reaction is -0.440 V and that of Cu/Cu^{2+}

reaction is +0.337 V. The standard electrode potentials are calculated from Gibbs free energy values by Eq. 1 that is applicable only in the standard state mentioned above.

$$E_0 = \frac{\Delta G}{z \cdot F} \quad \text{anodic direction}$$

$$E_0 = \frac{-\Delta G}{z \cdot F} \quad \text{anodic direction} \tag{1}$$

At other conditions, the equilibrium potentials are calculated by the Nernst equation:

$$E = E_0 + \frac{R \cdot T}{z \cdot F} \cdot \ln \frac{[OX]}{[RED]} \tag{2}$$

where R   is   the molar gas constant = 8.3143 J/mol·K
       T   absolute temperature, K
       [OX] and [RED] refer to the concentrations of oxidized and reduced species, e.g., metal ion and solid metal.

If the reaction is cathodic, i.e., a reduction reaction, the reduced species are reaction products. The standard electrode potential, E_0, and the logarithm term are now added together.

Instead of concentrations, one should strictly use activities in thermodynamical calculations. Most basic calculations assume that the system is ideal and activity of a reacting species equals its concentration. This is usually not valid. The difference is activity coefficient, γ. The activity of a species is therefore its concentration multiplied by the activity coefficient. The following discussion uses activity and concentration. Concentration means "effective concentration." The activity coefficient will often change with concentration and temperature and can be greater or less than one. Temperature or ion concentration changes will change the equilibrium potential value. Temperatures lower than 25°C will result in higher values. Those higher than 25°C will give lower potential values. An increase in ion concentration will result in lower potential values. For gaseous components, the concentration is the component partial pressure in the surrounding atmosphere. When using a gaseous component as a bleaching agent (O_2, O_3, or Cl_2) for example, increase of the component partial pressure will increase equilibrium potential of the corresponding redox reaction and therefore the oxidizing power. Oxygen enriched air is a more powerful oxidizer than normal air. Higher ozone content in an O_3/O_2 mixture gives higher oxidizing power.

Le Chatelier's principle is important when studying equilibrium states. In a closed system, any chemical reaction will eventually reach an equilibrium state. Gibbs free energy describes the equilibrium constant of the general reaction, aA + bB = cC + dD according to the following:

Electrochemistry of metallic corrosion and corrosion test methods

$$\Delta G = -RT \cdot ln(k) \; ; \; k = \frac{[C]^c \cdot [D]^d}{[A]^a \cdot [B]^b} \tag{3}$$

More negative G values give higher values for the equilibrium constant, k. The equilibrium of the reaction will lie on the reaction side. The situation with $G = 0$ is often falsely assumed to mean that nothing happens. If the free energy change of the process is zero, the system is at equilibrium and can shift to another state in response to any external disturbances. According to Le Chatelier's principle, the equilibrium will shift in response to changes in temperature, pressure, or reagent concentration to counter the change. For the general reaction, adding reactants or removing reaction products will shift the equilibrium to the process side. For pressure increases, the system shifts to lower the total pressure of the system. If the reaction produces heat, then lowering the temperature will shift equilibrium to the product side. If the reaction requires heat, then raising the temperature will shift equilibrium to the product side.

In equilibrium state, the electrode is in standard electrode potential or in equilibrium potential and no net reaction occurs on the electrode. Dynamic equilibrium means that anodic and cathodic partial reactions have equal rates. This reaction rate is the exchange current. A higher exchange current means the electrochemical reaction starts more easily when the electrode potential changes. Measurement of the rate of an electrochemical reaction is electric current. Electrochemical reactions follow all normal chemical stoichiometric relations. They also follow certain stoichiometric rules related to electrical charge. These are Faraday's laws of electrolysis:

- The mass of an element reacted at an electrode is directly proportional to the amount of electrical charge passed through the electrode.
- If the same amount of electrical charge is passed through several electrodes, the mass of an element reacted at each will be directly proportional to the atomic mass of the element and the number of moles of electrons required to discharge one mole of the element.

Faraday's empirical laws relate to the number of electrons required to discharge one mole of an element. The charge carried by one mole of electrons is one Farad. One Farad equals 96485 coulombs [C], and one coulomb is equal to one ampere-second [As]. In general terms, the relationship between current and reacted mass is Faraday's Law:

$$\Delta m = \frac{I \cdot \Delta t \cdot M}{z \cdot F} \tag{4}$$

where Δm is mass, g
   I current, A
   Δt time, s
   M atomic weight, g/mol
   z valency of the metal (number of electrons)
   F Faraday's constant, 96485 C/mol.

CHAPTER 3

The ratio, M/zF [g/As], is the electrochemical equivalent of the metal. Since current and mass are directly proportional, corrosion rates can be calculated from current by differentiating Eq. 4 with respect to time.

1.1 Electrode polarization

When the potential of an electrode changes, it is no longer in the equilibrium state. Increasing the potential starts an anodic net reaction, and decreasing the potential starts a cathodic net reaction. These generate a current flowing through the electrode. Consequently, forcing current through the electrode will change its potential. The deviation from equilibrium potential is polarization, and the difference between actual electrode potential and its equilibrium potential is overpotential. The sign of overpotential and current are positive for anodic reactions and negative for cathodic reactions. The nature of polarization is such that movement of electrons from an anode or to a cathode is faster than the corresponding electrode reactions. If the reaction mass transfer rate of metal ions is slower in the anodic dissolution than the removal of electrons through the external circuit, surplus positive charge accumulates on the anode. The potential then shifts to the positive direction.

For every electrochemical reaction, increasing polarization increases the current and vice versa. If the polarization is low, i.e., the potential shift per unit increase in current density is small, no factors will retard the electrochemical reaction rate. If the polarization is high, the electrode potential must change considerably to obtain higher currents. Polarization has various meanings and interpretations depending on the system under study. For an electrochemical reaction, this is the difference between actual electrode potential and reaction equilibrium potential. Anodic polarization is the shift of anode potential to the positive direction, and cathodic polarization is the shift of cathode potential to the negative direction. In an electrochemical production system driven with an external current source, polarization is a harmful phenomenon. It will increase cell voltage and therefore production costs. In a spontaneous system, anodic reaction rates will equal cathodic reaction rates. In this kind of system, anode and cathode will polarize to adopt the same potential. In a battery, polarization is again harmful since it decreases the cell voltage. In a corroding system, high polarization is beneficial. A system that polarizes easily will not pass high currents even at high overpotentials. The reaction rates are therefore small. The polarizability of an electrode can be estimated with exchange current density, i_0. A high i_0 value in the order of 1 mA/cm² indicates a reaction that will not polarize significantly. The i_0 value of 10^{-3} mA/cm² indicates intermediate tendency to polarize. If the i_0 value is about 10^{-6} mA/cm², the electrode will polarize easily.

Electrode polarization is not a simple phenomenon. Electrochemical reactions are heterogenous reactions. They occur in several phases in subsequent steps. The reaction path usually includes the following steps:

- Transfer of reactive species from electrolyte to electrode surface
- Adsorption

- Charge transfer step at the surface
- Desorption
- Transfer of reaction products from the surface to the electrolyte.

Any step can be rate determining, i.e., the slowest one determines the total reaction rate. As current flows through the electrode, the electrode potential shifts from its equilibrium value, and the electrode polarizes. The resulting overpotential consists of several factors. The most important are activation, concentration, and resistance overpotentials. The activation overpotential results from the limited rate of a charge transfer step, concentration overpotential comes from the mass transfer step, and resistance overpotential is the result of ohmic resistance such as solution resistance. Depending on the nature of the slowest step, the reaction is activation, mass transfer, or resistance controlled.

The relationship between overpotential and current density of a single, activation controlled electrochemical reaction is the Butler-Volmer equation, Eq. 5. As the equation shows, the rates of anodic and cathodic partial reactions are exponentially dependent on the overpotential. The net current is the sum of the anodic and cathodic partial currents.

$$i = i^+ + i^- = i_0 \cdot \left[e^{\frac{\alpha \cdot z \cdot F}{R \cdot T} \cdot \eta} - e^{-\frac{(1-\alpha) \cdot z \cdot F}{R \cdot T} \cdot \eta} \right] \tag{5}$$

In Eq. 5, the factor α is the symmetry factor with values between 0 and 1. If the value of α is 0.5, then anodic and cathodic partial reactions have equal characteristics. If α is higher than 0.5, the rate of the anodic partial reaction increases faster with increasing overpotential than the rate of the cathodic reaction. At values $0 < \alpha < 0.5$, the rate of the cathodic partial reaction increases faster. Two approximations traditionally simplify the Butler-Volmer equation. The high overpotential (high current density) approximation uses the fact that one exponent term in Eq. 5 is small at sufficiently high overpotential. The overpotential and logarithm of current density therefore relate linearly. This assumption is valid if the absolute value of overpotential, $\eta \gg RT/zF$, e.g., $\eta \gg 25.7/z$ mV at 25°C. The high overpotential approximation of the Butler-Volmer equation for an anodic reaction becomes the following:

$$i = i_0 \cdot e^{\frac{\alpha \cdot z \cdot F \cdot \eta}{R \cdot T}} \tag{6}$$

After taking logarithms and rearranging, this gives

$$\eta = \frac{R \cdot T}{\alpha \cdot z \cdot F} \cdot 2.303 \cdot \log(i_0) + \frac{R \cdot T}{\alpha \cdot z \cdot F} \cdot 2.303 \cdot \log(i) \tag{7}$$

that simplifies to the following:

CHAPTER 3

$$\eta = a + b \cdot \log(i) \qquad (8)$$

The semi logarithmic presentation of Eq. 8 is the Tafel equation or Tafel plot. The factor $b = 2.303RT/zF$ is the Tafel slope. Theoretically, at 25°C the Tafel slope has values of 30 mV, 40 mV, 60 mV, or 120 mV for reaction mechanisms involving 4, 3, 2, and 1 electrons, respectively.

In the low overpotential approximation of $\eta \ll RT/zF$, the exponent terms are sufficiently small, e.g., $RT/F = 25.7$ mV at 25°C, that they can be approximated as $e^x = 1+x$. Near the corrosion potential, the overpotential and current density relate linearly. In the low overpotential approximation, the symmetry factor has no influence. The exchange current density determines the shape of the curve as follows:

$$i = i_0 \cdot \frac{z \cdot F}{R \cdot T} \cdot \eta \qquad (9)$$

In the case of a single activation controlled reaction, measurement of overpotential and current density points will give the value of the exchange current density as the slope of the (i, η) plot. Consequently, the factor $RT/i_0 zF$ has the units of resistance multiplied by area. It is the charge transfer resistance, R_{ct}. Note that the current density unit in $RT/i_0 zF$ must be A/m². Using mA/cm² will give R_{ct} values that are ten times too high. The linear polarization method also uses this idea to determine corrosion current densities. Figure 2 shows the high and low overpotential approximations of the Butler-Volmer equation.

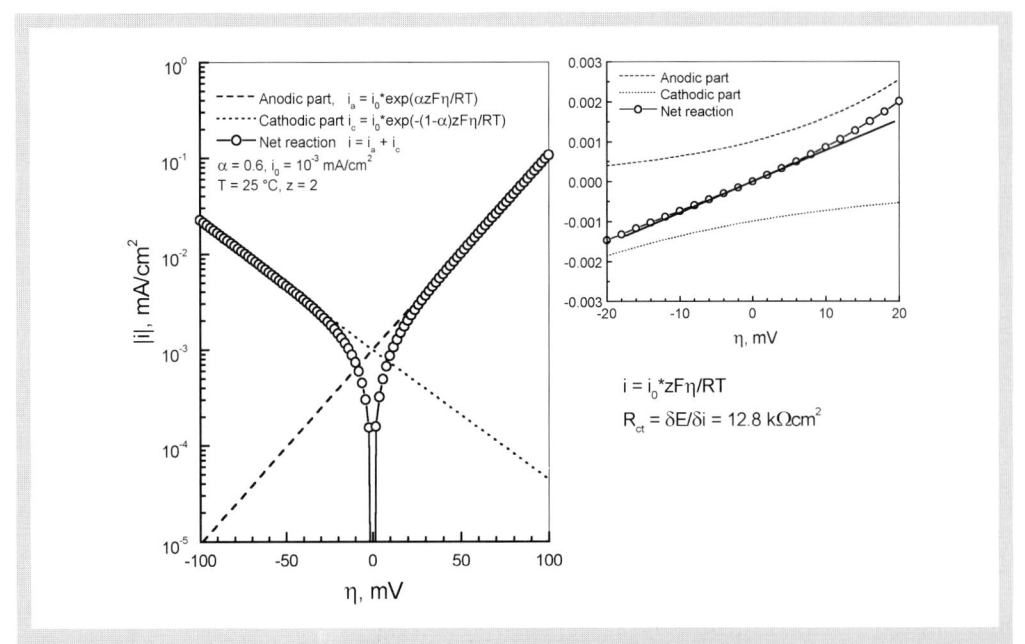

Figure 2. High and low overpotential approximations of Butler-Volmer equation, i.e., Tafel method and linear polarization.

Discussion will always exist whether potential changes cause current to flow or vice versa. For most practical and research applications, considering current density as a function of electrode potential is convenient. Without external current sources, the origin of current flowing through the electrode lies in the chemical potential of substances that makes them capable of undergoing electrochemical reactions. In spontaneous systems, the electrode potential therefore results from electrochemical reaction currents.

Different overpotentials have typical features. One can visualize the activation overpotential as a threshold to overcome before the reaction starts. An electrode reaction that has a low exchange current density also has a high activation overpotential. It will not start easily. This is beneficial in corrosion prevention. For example, the term "hydrogen overvoltage" describes the polarization required for hydrogen evolution to start. Some metals such as iron have low hydrogen overvoltage. Zinc has high overvoltage. The high hydrogen overvoltage of zinc means that hydrogen evolution on zinc does not begin easily and proceeds sluggishly. This explains the good corrosion resistance of very pure zinc in reducing solutions where the primary cathodic reaction is hydrogen evolution. If the zinc contains iron as an impurity, this will catalyze hydrogen evolution and lead to rapid corrosion.

Mass transfer overpotential results from a finite mass transfer rate from bulk electrolyte to electrode or vice versa. If the system is mass transfer controlled, a limiting current density exists. The limiting current density is the maximum reaction rate under mass transfer control. It increases as the concentration of the reacting species, their diffusion rate, temperature, or flow rate increase. In a system with limiting current density, the overpotential follows Eq. 10. The overpotential increases very rapidly when approaching the limiting current density.

$$\eta_{conc} = \frac{R \cdot T}{z \cdot F} \cdot \log\left(1 - \frac{i}{i_{lim}}\right) \qquad (10)$$

Reduction of oxygen is the most common cathodic reaction in natural environments. It is usually mass transfer controlled. Removal of oxygen from the bulk electrolyte or decreasing its diffusion rate to the surface with barrier coatings are common corrosion prevention practices. Both will decrease the limiting current density.

The resistance overpotential does not directly influence the electrochemical reaction, but it decreases charge transfer between anode and cathode. The resistance overpotential obeys Ohm's Law, Eq. 11.

$$\eta_\Omega = R_\Omega \cdot I \qquad (11)$$

As the solution resistance, measured current, or both increase, the effect of resistance overpotential becomes stronger. The resistance overpotential has more significance in systems where anodic and cathodic areas are clearly separate.

Measurement of the rate of an electrochemical reaction is electric current. To allow comparisons between different systems, the rate is usually measured as current density. This is the current flowing through the electrode divided by the electrode area. A

CHAPTER 3

plot relating current density to electrode potential is a polarization curve. Current density usually uses a logarithmic scale. The dependent variable can be current density or potential. Figure 3 shows polarization curves for different overpotentials. The current densities are in linear and logarithmic scale. The activation overpotential of the Butler-Volmer equation, Eq. 5, increases sharply at low current densities. It has no effect at high current densities. Concentration overpotential, Eq. 10, has no effect at low current densities. As current density approaches the limiting current density, it approaches infinity. Resistance overpotential, Eq. 11, increases linearly with increasing current density to obey Ohm's Law. The solution resistance causes an error called IR drop or ohmic drop in the polarization curves. The IR drop distorts polarization curves such that for anodic curves the true potential is always lower than measured potential. For cathodic curves, true potential is higher than measured potential.

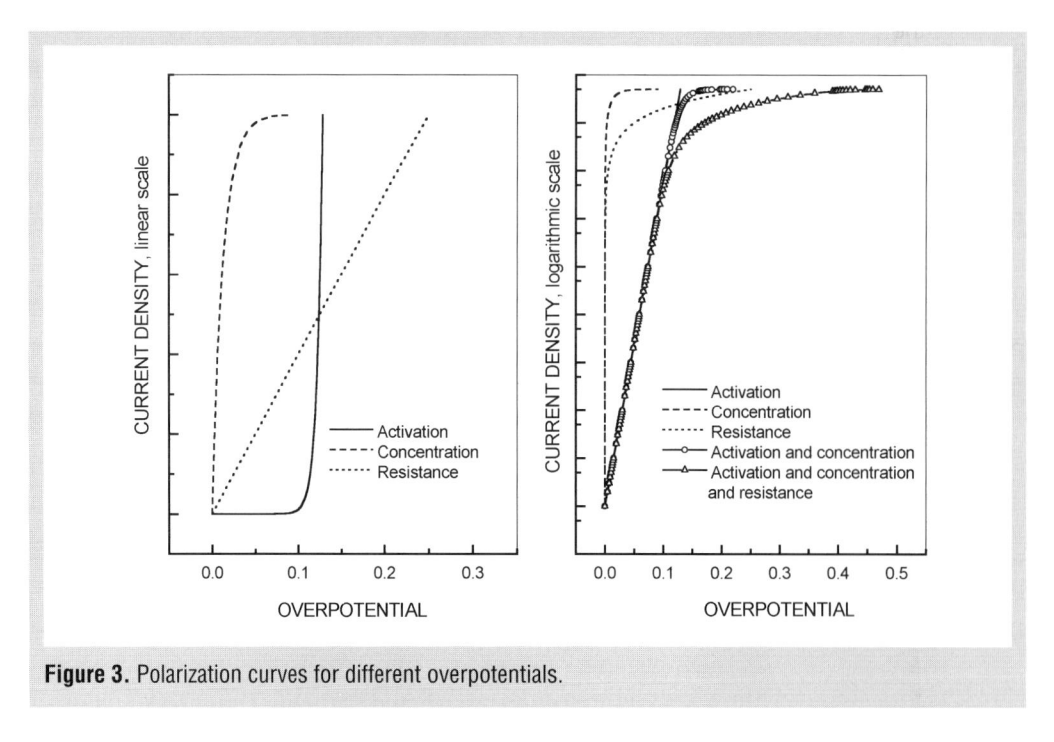

Figure 3. Polarization curves for different overpotentials.

Polarization is deviation from the equilibrium potential. The term sometimes describes a system whose activation overpotential is so high or limiting current density so low that no significant reaction rates occur, e.g., a "polarized electrode." The opposite term to this is "depolarization." Depolarization or depolarizing an electrode means any method to decrease activation or concentration overpotential to increase reaction rates. Reactants for cathodic reactions such as hydrogen and oxygen are "cathodic depolarizers." Since the polarization of the cathode results from accumulation of electrons, the reactants of a cathodic reaction that will consume the surplus electrons act as "depolarizers." Mixing the solution decreases concentration overpotential. It is therefore a depolarizing factor.

1.2 Electrode surface

The electrode surface is a complicated heterogenous system as Fig. 4 shows. Nearest the surface within a metal acting as an anode is a net negative charge. To maintain electroneutrality, next to the surface in the electrolyte side are adsorbed ions and water molecules. Toward the bulk electrolyte are cations surrounded by water molecules. This is the Helmholtz double layer. The inner Helmholtz layer contains water molecules and specifically adsorbed anions. The outer Helmholtz layer contains the second water molecule layer. From the Helmholtz double layer toward the bulk electrolyte are the diffusion layer and then the hydrodynamic layer. In the diffusion layer, the concentration of species changes from that of the bulk electrolyte to that of the electrode surface. The diffusion layer does not move, but its thickness will decrease with increasing bulk electrolyte flow rate to allow higher reaction rates. The diffusion layer thickness is inversely proportional to the square root of flow rate. The hydrodynamic layer or Prandtl layer has the same composition as the bulk electrolyte, but the flow of the electrolyte decreases from that of the bulk electrolyte to the stationary diffusion layer. The electrolyte next to the anode is the anolyte and that next to the cathode is the catholyte. The properties of anolyte and catholyte differ from those of the bulk electrolyte.

All electrochemical phenomena occur in these thin layers whose composition and properties differ from those of the bulk electrolyte. The inner Helmholtz layer is about 1 nm and the outer about 4 nm thick. The diffusion layer is about 10–100 µm, and the hydrodynamic layer is 0.1–1 mm thick. The double layer covers the entire metal surface and contains reactants and reaction products of all anodic and cathodic reactions. The composition and properties of the bulk electrolyte are usually known, but those of the electrode surface films are unknown. Specific adsorption of several different anions is a factor that is often unclear. Since this is a primary factor influencing surface film stabilities, the corrosion behavior of materials is often surprising. Simplified laboratory tests fail to describe actual process conditions.

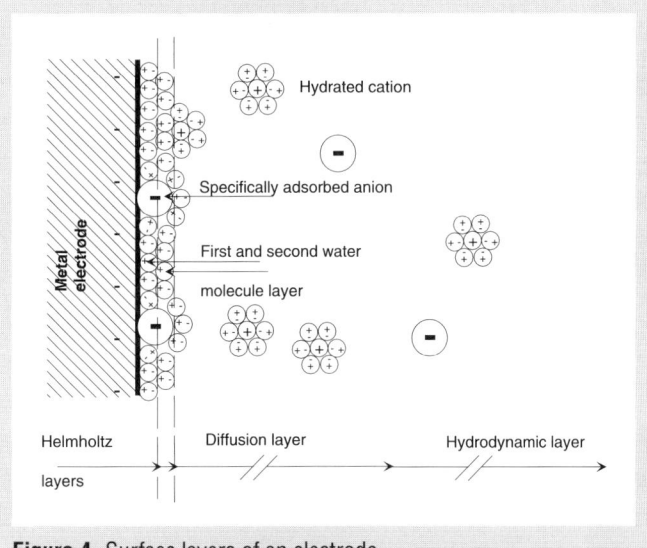

Figure 4. Surface layers of an electrode.

Corrosion phenomena begin with diffusion of cathodic reactants through the diffusion layer and adsorption on the metal surface. Thin diffusion layer, fast flow rate, high concentration, and high diffusion constants of the reactant will increase the cathodic reaction rate. After transport of the cathodic reactant to the surface, its reduction requires electrons. These electrons come from anodic dissolution reactions. Anodic reaction products such as metal ions will react with adsorbed water molecules and ions in the double layer or diffuse away as solvated cations. Figure 5 shows an example of the sequential steps in the corrosion of iron in a neutral aqueous solution. The ferrous ions produced by anodic dissolution will usually react chemically with the electrolyte. In natural waters, they may form ferrous or ferric hydroxides, oxides, or hydrated oxides. The rust layer on steel is a mixture of these reaction products. The solubility of oxygen in water at room temperature is about 8–10 mg/L O_2. Assuming that all the oxygen causes dissolution of iron and rust formation, then the oxygen in 5 L of water will cause thinning of 0.1 mm over an area of 1 cm². As Fig. 5 shows, the corrosion phenomena are complex with many steps. The final reaction rate and therefore the corrosion rate cannot be faster than the slowest step. This is the rate determining step.

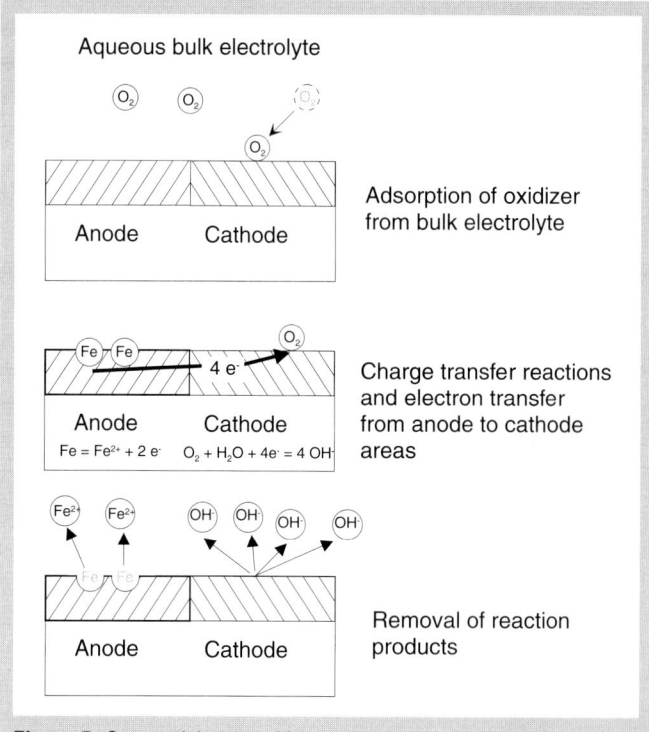

Figure 5. Sequential steps of iron corrosion in neutral aqueous solution.

2 Corrosion cell

A picture of a corrosion cell such as Fig. 6 often simplifies the corroding system. A corrosion cell contains the anode and cathode, an electrolyte, and an electrical connection between the electrodes. It is similar to a galvanic cell. An assumption is that only one anodic and only one cathodic reaction proceed in clearly separate areas. One electrode of the cell performs oxidation while the other performs reduction. The cathode where reduction occurs is the most positive electrode. It is therefore on the right. The electrons in the external circuit flow from left to right. The electrode on the left is the cell anode where oxidation occurs. Positive cations and negative ions carry charges in the electrolyte. The ions move to complete the circuit. Anions in the electrolyte move from right to left as electrons in the external circuit move from left to right. Cations move in the opposite direction because their charges have opposite sign.

The corrosion cell is misleading because it pictures a closed vessel. In most practical cases, the electrolyte is constantly replenished with a supply of reactants for cathodic reaction. The composition of the electrolyte at the anodic and cathodic regions does not remain unchanged during operation of the corrosion cell. Metal ions accumulate at the anodic regions, and alkalinity increases at the cathodic regions. When the reaction products of anodic and cathodic processes meet, precipitation of insoluble corrosion products can occur.

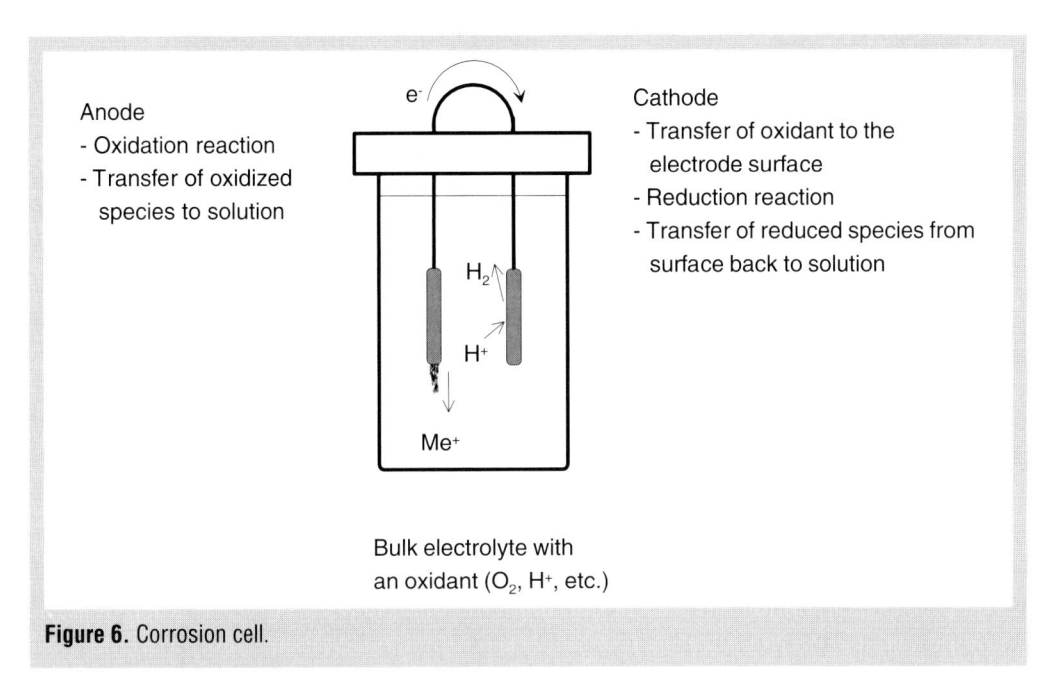

Figure 6. Corrosion cell.

The corrosion cell is a closed electrical circuit. Cathodic reactions consume electrons that release on the anode in anodic reactions and transfer by the connection to the cathode. Combining electrode potentials algebraically gives the cell potential. For a corrosion cell that operates spontaneously, the cell voltage will always be positive: $E = E_{cathode} - E_{anode}$. The current flowing in the cell obeys Ohm's law. A high cell voltage with low resistances in the system gives a high cell current. An anode or cathode can support more than one reaction, but the total anodic currents must equal total cathodic currents. This requirement leads to mixed potential theory. The following rules are important when studying the corrosion cell:

- Driving force of a corrosion cell is the cell voltage.
- Polarization phenomena will decrease this driving force and affect the corrosion rate.

CHAPTER 3

When a single metal corrodes, small anodic and cathodic areas form on the surface and distribute randomly. Figure 7 shows different versions of the resulting anodic and cathodic currents in a corrosion situation. A simple linear plot relating potential and current is the Evans diagram. Modifications that relate potential and current density as Fig. 7 shows are more informative. The equilibrium potentials are marked as E_{anode} and $E_{cathode}$. When the anode is polarized to higher potentials and the cathode to lower potentials from their respective equilibrium potentials, current starts to flow. The system settles to a state where total anodic current equals total cathodic current. At this point, the system is at corrosion potential, E_{corr}, and the current density is corrosion current density, i_{corr}. This situation is not an equilibrium situation because anodic and cathodic reactions proceed at some net rate. This leads to irreversible changes. Here it is consumption of the oxidant from the electrolyte and dissolution of metal from the anode.

In Fig. 7(a), both anodic and cathodic reactions are activation controlled. The cathodic reaction is the same for both anodes. The cell voltage between the

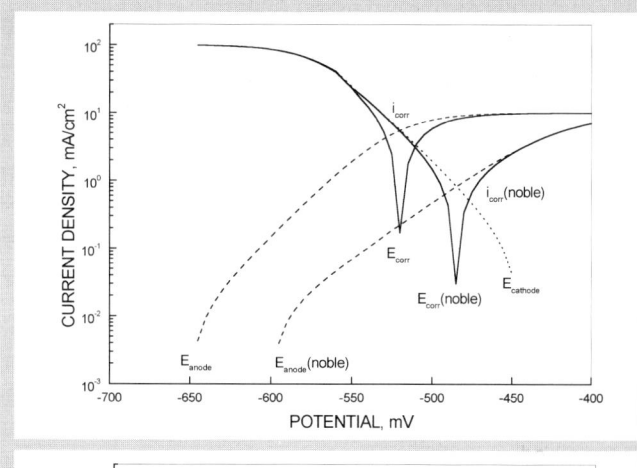

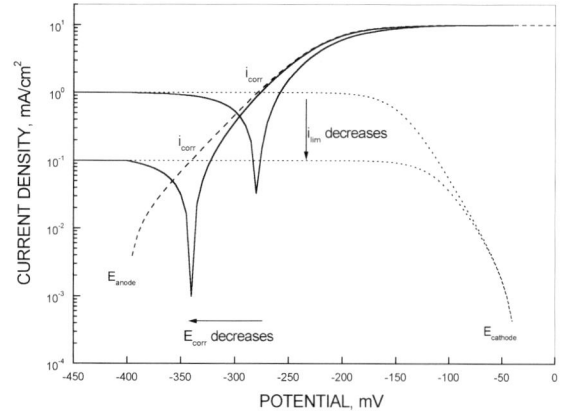

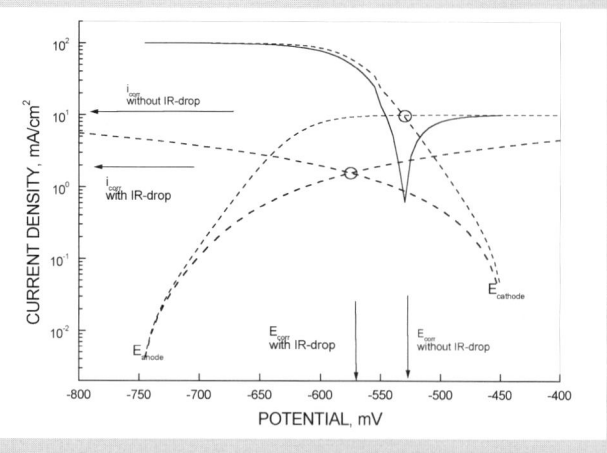

Figure 7. Formation of the corrosion potential and mixed potential theory. In (a), anodic and cathodic reactions are activation controlled, in (b) anodic reaction is activation controlled and the cathodic reaction is mass transfer controlled, and in (c) the presence of ohmic drop causes lower polarization of anodic and cathodic reactions.

less noble anode and cathode is higher than that between the more noble anode and cathode. The less noble anode will therefore polarize more, and the corrosion current density is higher. In acid solutions for example, iron that is less noble than nickel corrodes more rapidly. Although the corrosion potential of the more noble anode is higher, it has lower corrosion current density due to lower polarization. In Fig. 7(b), the anodic reaction is activation controlled metal dissolution. The rate of the cathodic reaction is mass transfer controlled. This shows a limiting current density. As the concentration of cathodic reactant decreases, the rate of cathodic reaction decreases, and the limiting current density decreases. This will then decrease the rate of the anodic reaction and the corrosion current density. As the corrosion current density decreases, the anode will not polarize as much and the corrosion potential will also decrease. In Fig. 7(c), the presence of ohmic drop causes lower polarization of the anodic and cathodic reactions. The additional resistance in the cell will consume some driving cell voltage as anodic and cathodic IR drops. Both reactions will still follow their respective polarization curves, but the steady state occurs at smaller deviations from the equilibrium potentials. This is one explanation for the lower corrosion rates found in low conductivity electrolytes.

Corrosion potential is a thermodynamic quantity indicating which reactions are possible. Corrosion currents are kinetic quantities associated with the dynamic nonequilibrium processes at electrodes. In corrosion, anodic reaction balances with one or more cathodic reactions. The favored cathodic reaction depends on the environment. Different cathodic reactions can occur simultaneously. Oxidizing species in the electrolyte such as dissolved oxygen or hydrogen ions will transfer from the bulk electrolyte to the cathode areas. The oxidizers have a strong tendency to reduce to a lower oxidation state. This occurs in a charge transfer reaction that consumes electrons. These electrons come from anodic dissolution reactions. Common cathodic reactions are the reduction of dissolved oxygen such as Eq. 12 in neutral or alkaline solutions, reduction of oxygen in acid solutions such as Eq. 13, and hydrogen evolution reaction in acid solutions such as Eq. 14.

$$O_2 + H_2O + 4e^- = 4OH^- \quad neutral, alkaline \tag{12}$$

$$O_2 + 4H^+ + 4e^- = 4H_2O \quad acid \tag{13}$$

$$2H^+ + 2e^- = H_2 \tag{14}$$

The corrosion of metals accompanied by hydrogen evolution shows little dependence on concentration polarization especially in strongly acidic or alkaline solutions. Concentration polarization may have a larger effect in neutral solutions. Hydrogen ions are charged ions having high diffusion rates and high migration rates in an electric field. The concentration of hydrogen ions is high in acid solutions. Hydrogen evolution will also produce gas bubbles that enhance mass transfer directly at the electrode surface.

The corrosion caused by hydrogen evolution depends on pH. The initial cathode potential becomes more positive with decreasing pH. This will then increase the maximum potential difference between anode and cathode. The cell potential is also highly dependent on the cathode material since different metals have different overpotentials for hydrogen evolution. Even impurities with low hydrogen evolution overpotential on the metal surface can significantly increase corrosion rate.

The corrosion with oxygen reduction as a cathodic reaction shows significant concentration overpotential. The reactants for oxygen reduction are neutral molecules of dissolved oxygen transported by diffusion and convection. The concentration of oxygen in aqueous systems is low because of its limited solubility. No additional stirring effect by gas evolution occurs. The oxygen usually comes from the surrounding atmosphere. To reach the cathode surface, it must pass through the air and electrolyte interface, major portions of the electrolyte by convection and diffusion, and finally diffuse through the Prandtl layer, diffusion layer, and possible cathode films to the surface. Frequently, the cathodic reaction rate is limited by the diffusion rate of oxygen trough the surface layers to the cathode surface.

Other possible cathodic reactions are metal ion reduction or metal deposition, reduction of sulfates by bacteria, etc. In oxidizing acids, the cathodic reaction is reduction of the acid molecule. Dissolved metal ions may deposit on a less noble metal in an exchange reaction. The electrons required in the deposition process are taken by anodic dissolution reaction of the less noble metal. The released metal ions go into the solution. This process is cementation or displacement. A common example is deposition of copper on steel when immersing steel in a copper containing solution.

The formation of a corrosion cell requires that a single metal surface or a larger structure in contact with the electrolyte have areas that can adopt different potentials. Potential differences may be due to internal factors such as compositional inhomogeneity, grain boundaries, structural defects, different grain orientation, locally different heat or other treatment, or surface factors such as discontinuous surface film, surface roughness, geometrical shape, different temperature, or mechanical stresses. Electrolyte inhomogeneity can also cause potential differences on the metal surface by local enrichment or dilution of dissolved substances or by differences in flow rates and temperature. Differential aeration is a common cause of heavy localized corrosion. It typically happens under partial immersion and on the waterline.

3 Passivation

The ability of metals to passivate is the most important reason that allows using metals in most circumstances. Passivation is a series of events that leads to formation of a protective surface film and causes the metal dissolution rate to decrease to a tolerable level. The metal first dissolves to produce metal ions. The concentration of the metal ions on the electrolyte surface film depends on their rate of formation and rate of mass transfer to the bulk electrolyte. Accumulation of metal ions and anions from the electrolyte into the surface film may start chemical reactions that produce stable compounds. If these compounds nucleate and grow on the metal surface, an adherent, dense, continuous, and protective surface film may form. The metal may then passivate. Examples of metals that show passivity are iron, chromium, nickel, titanium, and alloys containing these metals. Common corrosion product layers are usually not protective. They are merely deposits on the metal surface covering some areas but leaving others exposed. A Pourbaix diagram can estimate the possibility of passivation[1]. These diagrams show the stability domains of metals and their compounds using potential vs. pH scale. The diagrams take the name of their inventor, professor Marcel Pourbaix. Pourbaix diagrams are theoretical diagrams calculated from thermodynamic data. They show only theoretical stability domains and give no information on reaction rates. The diagrams can give a preliminary view on what can possibly happen in the system. Constructing a Pourbaix diagram requires calculation of the equilibrium states of selected reactions. Electrochemical reactions involving only charge transfer reactions will produce a horizontal line. Electrochemical reactions involving hydrogen ions or water molecules give a sloping line. Chemical reactions without charge transfer result in vertical lines. The Pourbaix diagrams often also show the stability area of water. This is between the sloping lines of oxygen evolution (often marked with "a") and hydrogen evolution (marked with "b"). When developing these diagrams, Pourbaix assumed the minimum ion concentration of a very dilute solution was 10^{-6} mol/L. When the metal ion concentration is over 10^{-6} mol/L, the metal is considered to corrode. For iron, the concentration of 10^{-6} mol/L corresponds to 55.9 µg/L or parts per billion of dissolved metal. Use of higher metal ion concentrations in a system construction will usually produce larger stability domains for metals and compounds.

The information in a Pourbaix diagram includes the areas of immunity, corrosion, and passivity. In the immunity area, the system is below the equilibrium potential of the anodic metal dissolution reaction. Reaction cannot occur. In the corrosion area, the potential is so high that anodic dissolution occurs, but the solution conditions do not allow formation of any metal compound. No surface films therefore form. In the passivity area, potential is still so high that the metal can dissolve. Under these conditions, reactions leading to formation of protective films can

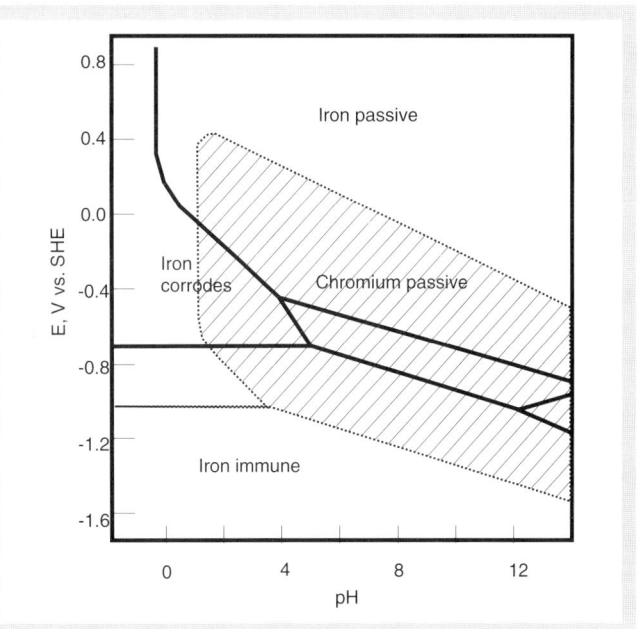

Figure 8. Pourbaix diagrams of iron and chromium where the hatched area is the passive region of chromium as Cr_2O_3.

occur. The formation of passive film is possible only if the initial corrosion rate is sufficiently high to supply metal ions and the solution conditions are such that film formation is thermodynamically possible and kinetically sufficiently rapid. Figure 8 shows Pourbaix diagrams of iron and chromium. The figure shows that the passivation zone of chromium is much larger than that of iron. This is the reason for chromium alloying of steel to make stainless steel.

Although estimation of passivation tendency often uses stability domains of oxides, the passive films are rarely simple oxide films. In practice, they are a mixture of amorphous and crystalline oxides, hydroxides, hydrated oxides, bound water, etc. The composition of a passive film is seldom uniform over its entire thickness. For example, the passive films of stainless steels are enriched with iron on the metal side and chromium on the electrolyte side.

Passivation is primarily a kinetic phenomenon. Figure 9 shows a theoretical anodic polarization curve of a passivating metal. The polarization curve has three potential ranges where the metal is in an active, passive, or transpassive state. E_{anode} indicates the equilibrium potential of the metal. As the electrode potential increases, the anodic current density increases. At some point described by passivation potential, E_{pass}, and critical current density, i_{crit}, the current density begins to decrease. This is due to formation of a stable reaction product layer. Passivation potential describes the thermodynamic possibility for passivation. Critical current density describes the kinetic conditions for passivation. Low passivation potential and low critical current density promote passivation. When the passive film is fully formed, current densities drop rapidly. The passive current density, i_{pass}, that is necessary to maintain passivity is usually 100–1 000 times lower than critical current density, i_{crit}, to start passivation. The passive potential range is described by zero or very small current density changes as potential increases. The dis-

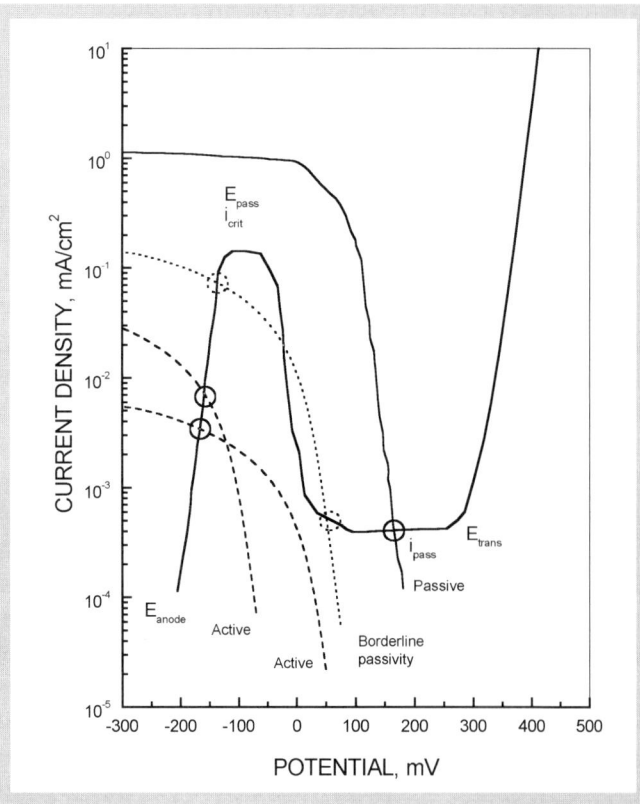

Figure 9. Polarization curve and mixed corrosion potential of a passive metal.

solution does not stop when the metal has passivated. A small dissolution rate is always necessary to maintain the passive film. When the potential is sufficiently high, current densities will again increase. At this transpassive potential, E_{trans}, the passive film begins to dissolve or fail due to other electrochemical reactions. Above the transpassive potential, the system is in a transpassive state, and a protective passive film no longer exists. In practice, the failure of the passive film is often localized. The potential for localized breakdown of passivity is the pitting potential.

Figure 9 also shows the effect of the cathodic reaction on passivation. If the equilibrium potential of the cathodic reaction is too low or the rate is too slow, it cannot polarize the metal over the passivation potential and critical current density. The metal will remain in an active state. If the equilibrium potential of the cathodic reaction is below passivation potential, the metal will never passivate. If the equilibrium potential of the cathodic reaction is higher than the passivation potential, the metal can passivate when the rate of the cathodic reaction exceeds critical current density. The metal can also show borderline passivity where the equilibrium potential of the cathodic reaction is sufficiently high but its rate is too slow. The metal can be passivated with addition of an oxidizing compound or use of an external current source to increase the metal dissolution rate to critical current density. In this case, the metal cannot passivate spontaneously, and any damages in the passive film will not repair themselves. The metal then reverts to the active state. For corrosion prevention, a passivating metal should have low passivation potential, low passivation current density, wide passive potential range, low passive current density, and high transpassive dissolution potential. The first two factors will ensure ease of passivation, and the others provide a wide application range.

Figure 10 provides a summary of the factors influencing corrosion. The most anodic and most cathodic reactions in the electrochemical series are likely to occur in the corrosion cell. Reaction products of these reactions may cause formation of surface films. Depending on the corrosive environment, materials, and reactions, many kinds of reaction products are possible. Corrosive environment and material will determine the corrosion form. Corrosion rates depend on the corrosion form and protective reaction product layers.

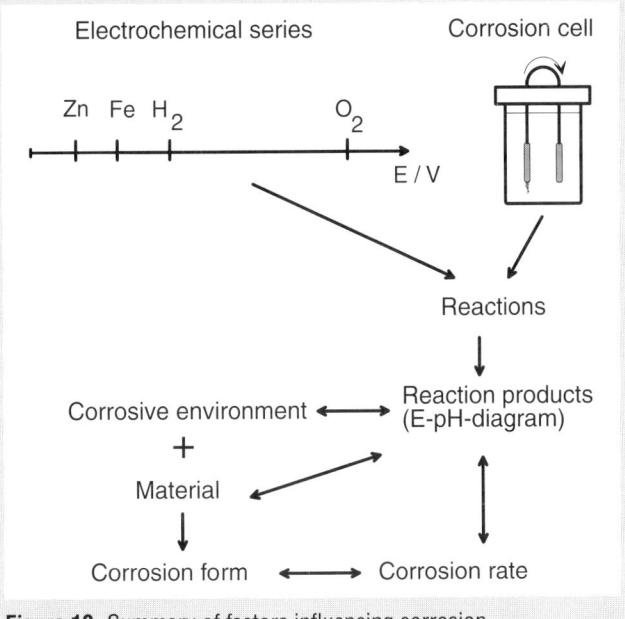

Figure 10. Summary of factors influencing corrosion.

The corrosion product layer is a result of its formation in contact with the metal, precipitation from solution, and dissolution into the solution. If the dissolution rate of the layer is smaller than the dissolution rate of the metal and compound precipitation rate, the layer is insoluble and potentially protective. Most protective layers are usually those that are insoluble and have slow cation diffusion rates[2].

4 Corrosion research methods

Corrosion tests can determine corrosion resistance of a metal, corrosivity of an environment, effectiveness of corrosion protection, or the contamination of environment or product by corrosion products. Corrosion test methods cover a wide range from simple weight loss coupons to sophisticated electrochemical techniques. In the strictest sense, the only "real" corrosion measurement methods are those that use direct physical measurement of weight loss, thickness, appearance, etc., in actual process conditions. Corrosion tests conducted under actual operational conditions are very reliable but sometimes tedious and time consuming. For example, atmospheric corrosion tests require months or even years. Corrosion tests conducted under simulated service conditions are more simple, but a risk always exists to forget some crucial factor. Accelerated corrosion tests use more severe conditions than actual service, but they provide information in a short time. Acceleration usually uses increased temperature or corrosive agent concentration, but the conditions should not change the corrosion mechanism. For example, heating a solution to reduce experiment time can remove dissolved gases that actually cause the corrosion. The results would therefore be completely false. Salt spray chamber tests and many electrochemical tests are typical accelerated tests. Electrochemical techniques will only give an instantaneous current density, resistance, or other value that is assumed to describe the corrosion rate or mechanism. Direct measurement of corrosion rate with electrochemical methods is usually impossible since anodic and cathodic areas cannot be separated. The measurements are therefore made by polarizing the sample according to a predetermined procedure, measuring resulting potentials and currents, and extrapolating these back to the original corrosion potential.

A corroding system is never deterministic. Calculating its future state by using the present state is therefore impossible. The number of microscopic variables is so large that measuring the value of every variable is not possible. On a macroscopic scale, the corroding system is essentially a "black box." The system is rarely in a state of equilibrium. It is in a steady state or a transient state changing from a steady state to another state. A corroding system has characteristic responses to external excitation signals. By measuring these responses, obtaining information on system properties is possible. With gradual changes in parameter values, the system usually changes smoothly. At certain critical points, the steady state of the system can change abruptly to a totally different state that may not be steady at all. These critical points usually relate to the starting points and rates of the corrosion reactions. A typical example is initiation of pitting corrosion when exceeding critical chloride concentration or temperature.

Studies of interaction between perfectly homogeneous solid and liquid phases that would give precise information are not possible since they do not exist. Estimating what could happen in a system using thermodynamic and kinetic information is possible. Thermodynamic information indicates the tendency for a reaction to happen. Kinetic information concerns the reaction rate. Considering one of these without regard for the other leads to false conclusions. The behavior of pure, homogeneous systems can extend to real systems assuming a statistical behavior. Corrosion phenomena also have a statistical nature. Although accurate predictions of a certain system may be impossible, the probability of corrosion initiation and propagation may provide useful information. Many design tables and curves contain information that show the general tendency and probabilities or areas of unpredictable behavior.

Corrosion rates and probabilities are merely a guide for materials selection. Maximum allowable corrosion rates should therefore not be defined too rigidly. Acceptance of a material by corrosion rate alone is not sensible. A better practice is to estimate the service life. In most constructions, selection of various materials with varying corrosion rates and material thicknesses is possible. These require balancing with each other. Corrosion test data from one set of conditions do not apply to another that seems or sounds similar. For example, corrosion rates in sodium sulfate, sodium sulfite, and sodium sulfide are seldom equal. Use of corrosion data of one alloy grade to evaluate corrosion behavior of another with almost equal composition is rarely useful. Changes in temperature, pH, or concentration of minor constituents in the environment or the alloy may totally change the corrosion behavior. A classic example is titanium in alkaline hydrogen peroxide solution. Initial laboratory tests with pure solutions indicate very high corrosion rates. In practice, natural scale-forming salts inhibit corrosion[3].

4.1 Physical measurements

Physical measurements will provide a direct answer how the corrosive environment affects the test sample. Typical tests are weight loss measurements, pit depth and pitting density, stress corrosion cracking tests, and controlled environmental cabinet tests for coatings. The measurements range from actual plant tests to accelerated corrosion tests. Evaluation of test results may involve various methods, but visual examination is frequently the best corrosion measurement technique. For ranking of materials, another factor such as weight loss has use. The result of any measurement should determine the probability of corrosion or the rate of corrosion. When poorly designed and conducted, all corrosion tests can give misleading results. Predicting the forms of corrosion that will occur or the primary influential factors is not always possible. Depending on the expected and most probable corrosion mechanisms of various materials, design of appropriate tests by using standard procedures is possible.

Immersion testing is very suitable for screening and eliminating unsuitable materials. Weight loss experiments are suitable for determining the rate of uniform corrosion. Special sample constructions can study various forms of localized corrosion. In weight loss experiments, small pieces of test materials with dimensions of centimeters are exposed to the corrosive environment with subsequent weight loss measurement. The samples should be sufficiently large to minimize the ratio of edges to the general surface area. The samples should be as large as possible consistent with the test vessel

CHAPTER 3

size, material availability, and capacity of the analytical balance. The samples must be sufficiently large that a measurable mass loss will occur in a meaningfully short time. Low corrosion rates require larger samples. In laboratory tests, the samples require weighing with an accuracy of at least ±0.5 mg, and the weight loss variation between duplicate species should not exceed ±10%. Temperature of the electrolyte must be within ±1°C. The minimum solution volume is 20–40 mL/cm^2 of sample area. The ±10% weight loss variation between duplicate specimens is often impossible to achieve in process environments.

Weight loss experiments will show only the average corrosion rate during the experiment. Assuming that corrosion is a linear function of time is usually incorrect. To overcome this problem, ASTM G31-72 "Laboratory immersion corrosion testing of metals" describes a planned interval corrosion test. The test allows monitoring of solution

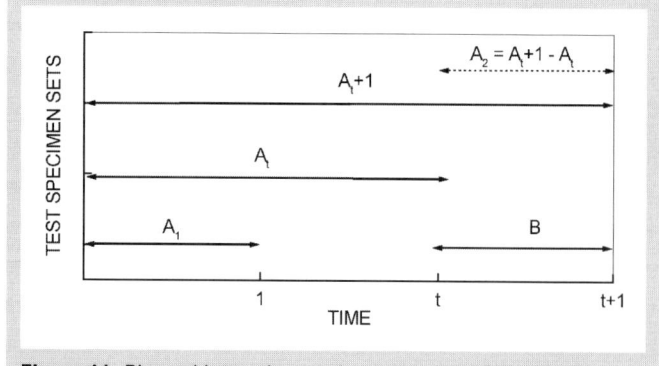

Figure 11. Planned interval corrosion test using ASTM G31-72.

corrosivity and sample corrosion rate. It uses a minimum of four sets of identical samples with use of duplicates for statistical validity. Calculation of test duration uses a "unit" time interval. When using four sample sets, three go into the test at the beginning. The first set is removed after the first time interval, the second set is removed one interval before the end of the experiment, and the third set is removed at the end of the test. A fourth set placed into the test simultaneously with removal of the second remains immersed until the end of the test. Letters describe the corrosion damage as weight loss or other relevant factor for each set of samples. A_1 indicates the first set, A_t is the set removed before one interval, A_t+1 is the set for the entire test. B is the last interval set. Figure 11 is a diagram showing the letters. A_2 is calculated by subtracting A_t from A_t+1.

The criteria in Table 1 provide an estimate of the changes in electrolyte corrosivity and metal corrosion rate. The occurrences can be combined freely to obtain changes in corrosivity and corrosion rate. For example, if the A_2 value is less than the B value while the B value is less than the A_1 value, a protective reaction product layer has decreased the corrosion rate, and the solution corrosivity has decreased. The results do not indicate causes for the corrosion rate changes. Usually, solution corrosivity, metal corrosion rate, or both will decrease during a test in a closed vessel. Solution corrosivity will seldom change in once-through plant tests or when using large solution volumes.

Table 1. Evaluation of planned interval corrosion test using ASTM G31-72.

	Occurrence	Criteria
Corrosivity	Unchanged	$A_1 = B$
	Decreased	$B < A_1$
	Increased	$A_1 < B$
Corrosion rate	Unchanged	$A_2 = B$
	Decreased	$A_2 < B$
	Increased	$B < A_2$

Weight loss and thinning are not the only parameters for study. Sensitivities to pitting corrosion, intergranular corrosion, or stress corrosion cracking have frequent study with immersion tests. Several methods are standardized. In most immersion methods, the test solutions contain an aggressive ion that is usually chloride and some oxidizing agent to increase solution redox potential and the corrosion potential of the sample. These accelerated methods therefore usually do not have use to predict corrosion resistance in actual environments. They only provide methods for ranking of materials. Immersion tests are usually simple and do not require expensive apparatus. ASTM 48–92 "Pitting and crevice corrosion resistance of stainless steels and related alloys by use of ferric chloride solution" describes tests in about 6% $FeCl_3$ solution. Variations in ferric chloride concentration and addition of NaCl or HCl are common. The method ranks the relative resistance of materials in chloride solutions. The samples are immersed for a reasonable length of time such as 72 h. Determination of corrosion as a function of time requires removal of samples at various times. The appearance of the sample is the first criteria to determine corrosion. Weight loss is a ranking criteria. Another immersion test method for pitting and crevice corrosion evaluation is the so-called "Green death." It uses an oxidizing solution containing boiling 11.5% sulfuric acid with 1.2% hydrochloric acid, 1% ferric chloride, and 1% cupric chloride to determine corrosion rate. The oxidizing acid 4% sodium chloride solution with 0.1% ferric sulfate and 0.01 M hydrochloric acid is a solution to determine critical pitting potential temperatures. The basic idea in these tests is to create an oxidizing solution with constant redox potential to polarize the sample above the pitting potential of the material. This causes pit initiation and propagation with sufficiently high chloride concentration and temperature. If the pitting potential of an alloy is higher than the redox potential of the solution, test times can exceed several days.

Weight loss is not the primary evaluation for the results of immersion tests. Pitted specimens can show negligible weight loss if the attack is very localized. In practice, more localized attack also means more severe pitting problems. Pit density on the planar surfaces and pit depths on cross sections evaluate pitting corrosion attack. For studying crevice corrosion, artificial crevices are created for example with PTFE blocks or rubber bands with the resulting attack evaluated according to corroded area, corrosion depth, and weight loss.

ASTM A262–91 "Detecting susceptibility to intergranular attack in austenitic stainless steels" describes six different immersion tests suitable to detect intergranular corrosion caused by chromium carbide precipitation. Some tests have specific names such as the boiling nitric acid test known as the Huey test and the boiling sulfuric acid and copper sulfate test known as the Strauss test, etc. ISO 3651–1:1976 "Austenitic stainless steels–Determination of resistance to intergranular corrosion–Part 1: Corrosion test in nitric acid medium by measurement of loss in mass (Huey test)" and ISO 3651–2:1976 "Austenitic stainless steels–Determination of resistance to intergranular corrosion–Part 2: Corrosion test in a sulfuric acid/copper sulfate medium in the presence of copper turnings (Monypenny Strauss test)" also describe these methods. The basic idea is the same as in the pitting and crevice corrosion tests described above. The test solution is made highly corrosive to the steel but this time only in the sensitized condition to avoid corroding the steel. This is possible by using highly oxidizing solutions without aggressive anions to dissolve chromium depleted areas and avoiding other forms of localized corrosion. Depending on the redox potential of the solution, some inter metallic phases or all those that cause intergranular corrosion will dissolve. The Huey test uses boiling 65% nitric acid that attacks chromium carbides and the α-phase. Copper sulfate solution with sulfuric acid attacks chromium carbides. Addition of metallic copper increases corrosion rate and reduces testing time. Nitric acid and copper sulfate tests primarily have use for austenitic and ferritic stainless steels. Ferric sulfate solutions with sulfuric acid with or without chloride containing compounds will attack carbides and inter metallic phases. These solutions also have use for highly alloyed stainless steels and nickel alloys. The results of intergranular corrosion tests use ranking by appearance and weight loss. The specimens in copper sulfate tests are bent after the test to reveal fissures. Microscopic examination in most tests indicates the extent of corrosion.

    Test methods and standard practices for sample preparation and testing of stress corrosion cracking resistance are available in various ASTM and ISO standards. Smooth test specimens include the U-bend, C-ring, bent beam, and direct tension specimens. Pre-cracked specimens have wide use, although obtaining reproducible results is more difficult. Traditionally, stress corrosion cracking tests involved measuring time to failure of smooth specimens exposed in the test environment under various levels of applied stress. The introduction of pre-cracked specimens led to acceptance of an incubation period for development of local crack tip chemistry. Smooth surface specimens should be used to simulate service conditions where propagation proceeds rapidly after a long incubation time, while pre-cracked specimens should be used to simulate service conditions in which the incubation time is brief and the time to failure is governed by the rate of crack propagation. Laboratory tests are available to produce accelerated stress corrosion cracking for a given alloy system or to simulate service conditions. Artificial solutions are more reproducible and easier to control than natural environments. The purpose of stress corrosion cracking testing is usually a need to predict occurrence of cracking in some real environment. Using the test results, material selection or rejection is possible. The metal and environmental combinations that result in stress corrosion cracking are relatively few, but certain solutions have become standards for certain alloy types.

The test environments are often more severe than the real environment to produce rapid results. For example, boiling MgCl$_2$ solution has had use with stainless steels since 1945, and boiling nitrate solution has had use for carbon steels. ASTM G36 "Evaluating stress corrosion cracking resistance of metals and alloys in a boiling magnesium chloride solution" defines test conditions as 155°C corresponding to 45% MgCl$_2$ solution at pressure of 1 atm. The method applies to wrought, cast, and welded stainless steels and related alloys. It can detect effects of composition, heat treatment, surface finish, microstructure, and stress on the susceptibility of these materials to chloride induced stress corrosion cracking. Any type of stress corrosion test specimen can be used. The samples are evaluated by time to crack initiation and crack growth rate. The boiling MgCl$_2$ solution is acidic and very corrosive. It may not represent actual operating conditions. ASTM G44 "Evaluating stress corrosion cracking resistance of metals and alloys by alternate immersion in 3.5% sodium chloride solution" covers a procedure for making alternate immersion stress corrosion tests applicable to aluminum and ferrous alloys. This is also an accelerated test that is suitable for alloy development and material quality control tests in certain applications. The method can induce severe pitting especially on high strength aluminum alloys that will interfere with stress corrosion initiation and may cause mechanical failure. Sample evaluation uses appearance with magnification if necessary. The criteria for resistance to stress corrosion cracking can be time to cracking or threshold stress below which cracking does not occur. Acceleration of the test can use a more aggressive environment and higher temperature or more severe mechanical stresses. The basic demand is to keep the failure mechanism the same as that occurring in practice. Representative failed specimens should therefore be examined metallographically to verify that the failure is due to stress corrosion cracking.

Some accelerated corrosion tests have use primarily for testing paints and other coatings. These tests use closed cabinets with control of exposure conditions to accelerate the corrosion rate without changing the corrosion mechanism. Tests often determine atmospheric corrosion resistance. Since materials can be resistant, the time under corrosive conditions and corrosion rates requires maximization. The simplest tests only keep temperature and humidity at constant values. Tests often use aggressive salts and gases to accelerate corrosion. Salt spray, sulfur dioxide, and CorrodKote tests in a cabinet are typical examples. The use of a cabinet test is often a compromise between allowable time and deviation from actual conditions. All cabinet test results require correlation with actual service performance such as using a reference material. Accelerated tests hardly ever correlate with actual coating performance. They are therefore more useful for comparison purposes only.

The most basic cabinet tests are those that control temperature and humidity. Several methods are standards, and commercial equipment is available. These tests have use to evaluate the effects of contaminants or corrosivity of absorbing materials such as thermal insulation or gasket materials. A variation of the humidity test is to include a corrosive agent as contaminant. ISO 4541:1978 "Metallic and other non-organic coatings–Corrodkote corrosion test" uses a paste containing cuprous nitrate, ferric chloride, and ammonium chloride placed on the sample before exposure in a cabinet. Condensing humidity tests also provide a rinsing action that makes them less severe. Corrosive gas tests combine humidity with controlled amounts of corrosive gases. The moist sulfur dioxide test described in ISO 6988:1985 "Metallic and other nonorganic coatings–Sulfur dioxide test with general condensation of moisture" simulates industrial atmospheric corrosion. It is suitable for detection of pores in protective coatings but not as a general guide to corrosion resistance. Corrosive gas tests can use a mixture of gases such as chlorine, hydrogen sulfide, and nitrogen dioxide. The most common standardized laboratory corrosion test is the neutral salt spray test where the samples have continuous exposure to a NaCl solution spray as described in ISO 9227:1990 "Corrosion tests in artificial atmospheres–Salt spray tests." Concentration of the salt solution varies in different standards from 0.5%–20%. This test is more severe when lowering the solution pH to 3 with acetic acid (ACSS test) or by increasing temperature to 50°C and adding cuprous chloride (CASS test). Although salt spray tests have frequent use, a direct relation between resistance in the test and resistance to corrosion in other media seldom exists. The evaluation criteria in these tests are appearance after the test, amount and size of corrosion damage after removal of visual corrosion products, and time to first corrosion damage[4].

4.2 Electrochemical techniques

Electrochemical measurements have frequent use to estimate corrosion resistance of metallic materials. The electrochemical nature of corrosion allows measurement of instantaneous corrosion rates. Monitoring corrosion rate over the time interval required for weight loss measurement is therefore possible. The mean corrosion rate obtained by integration of corrosion rate over experiment length should equal the rate from corresponding weight loss experiments. Although many methods are standards, frequent opportunities still remain for an inexperienced person to make improper interpretations and false conclusions. If one measures a current density in a laboratory cell, what does this mean in practice? With stainless steels for example, it usually does not mean a uniform corrosion rate. It is also not a measure of the penetration rate of localized corrosion. The operator must know experimental details such as IR drop and cell geometry. Computer controlled measuring systems are particularly risky unless the operator knows the logic of the measurement software, operation of the curve fitting procedures, etc. These points rarely receive consideration. Good theoretical background and considerable experience are necessary to decide whether a beautiful polarization curve or an accurate corrosion rate value is true.

  A major problem with electrochemical techniques is that they cannot separate currents originating from different phenomena. Dissolution of metal and oxidation of dissolved species can therefore give equal anodic behavior. With uniform corrosion, the life-determining factor of a structure is the average corrosion rate. Corrosion rate measurements are suitable. For localized corrosion, the life-determining factor is maximum penetration depth or minimum initiation time. With localized corrosion, test methods have primary use to determine the probability of initiation, propagation, or both for alloy composition or some environmental variable.

Electrochemistry of metallic corrosion and corrosion test methods

The simplest electrochemical technique to estimate corrosion is measurement of the corrosion potential of the sample. Potentials are usually determined with 10–20 mV accuracy for practical purposes. Sometimes 50 mV is sufficiently accurate. Measuring corrosion potentials in a corrosive environment will give the so-called galvanic series in that environment. Metals with low corrosion potential are active, and those with high corrosion potential are noble. The galvanic series is similar to the electrochemical series, but it gives an estimate of the corrosion probabilities in a real situation. Table 2 compares equilibrium potentials and corrosion potentials for some metals. As the table shows, the corrosion potentials of alloys are usually higher than the corresponding equilibrium potentials of pure metals. This is due to the polarization to the anodic direction caused by corrosion. The equilibrium state is no longer a pure metal and metal ion equilibrium. It reflects the equilibrium state of the metal and some of its compounds instead. These potentials are lower at higher pH values.

Table 2. Equilibrium potentials and corrosion potentials in V vs. hydrogen electrode of some engineering metals.

	E_0 Me/Me^{z+}		Tap water, Espoo, Finland		Ocean sea water[5,6]
Pt	1.2			Platinum	+0.4– +0.45
Ag	0.799			Silver	+0.08– +0.14
Cu	0.337	Copper	+0.22– +0.32	Cu pure Copper	-0.1 -0.14– -0.07
Pb	-0.126			Lead	-0.02– +0.05
Ni	-0.25			Ni pure Nickel	+0.07 +0.03– +0.15
		AISI 304	-0.3– +0.47	AISI 304	-0.15– -0.22
Fe	-0.44	Cast iron Carbon steel	-0.52– -0.15 -0.39– -0.2	Fe pure Carbon steel	-0.51 -0.47– -0.37
Cr	-0.74				
		Galvanized steel	-0.78– -0.68		
Zn	-0.763			Zinc	-0.8– -0.74
Ti	-1.63			Ti (passive)	+0.21– +0.3
Al	-1.66	Al, comm. pure	-0.85– -0.1	Al pure Al alloys	-1.2 -0.75– -0.5

Note that the nobility of many metals changes remarkably when immersed in different solutions. In the passive state, metals that passivate may have several hundred millivolts higher corrosion potentials than their respective equilibrium potentials. Using the galvanic series requires caution. The values are only suitable for use in the environment where measured. A potential risk is that no convention exists on how to present them. The most active or most noble materials may therefore be at the top of the diagram. Comparisons of materials being "higher" or "lower" in the series may refer to their actual corrosion potentials or mistakenly to the vertical arrangement order.

CHAPTER 3

Electrochemical measurement techniques use a controlled variation of electrode potential or current and recording of resulting current or potential changes. The measured current results from electrochemical charge transfer reactions at the electrode to electrolyte interface. For example, anodic currents can result from electrode dissolution or oxidation of dissolved compounds. Measuring the current does not explain its origin. Electrochemical measurements therefore require confirmation by measuring other system variables such as analyzing the reaction products. Another possibility is to combine different electrochemical measurements to determine if they give compatible results. Otherwise, one must rely on a "scientific guess."

4.2.1 Polarization curves

The basic test method in corrosion research is measurement of a potential/current density curve or polarization curve. A polarization curve provides the first view of the behavior of the system. Polarization curves are made by changing the electrode potential at a constant rate and recording potential and the corresponding current density. A common practice is to use semilogarithmic plotting of current density with units of mV or V for potential and mA/cm^2 or A/m^2 for current density. All currents are often treated as positive to simplify plotting. One must remember that negative current means the sample acts as a cathode. Positive current means it acts as an anode. Interpretation of the polarization curve is simple. Current density corresponds to a reaction rate according to Faraday's law. A sudden increase in current density usually corresponds to start of a new reaction, and a sudden decrease means passivation. When the curve becomes more horizontal, mass transfer limits the reaction rate (limiting current density) or ohmic resistance distorts the curve.

Figure 12 shows two examples of polarization curves measured in neutral and aerated natural water. Curve (a) is typical for a metal with active/passive transition such as stainless steel. The active peak often cannot be seen since the corrosion potential is in the passive range. Curve (b) is typical for actively corroding metal. Here the corrosion rate is mass transfer controlled, and it increases with increasing flow rate. Cathodic reactions are the same for both metals. In the area of corrosion potential, the cathodic reaction is oxygen reduction. As the electrode potential decreases, hydrogen evolution begins.

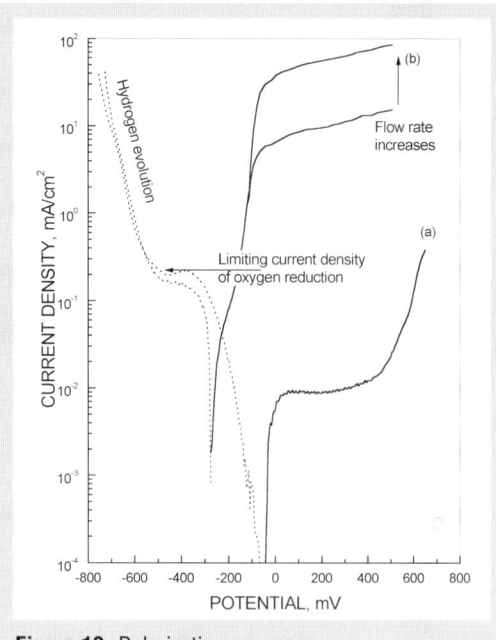

Figure 12. Polarization curves.

Several experimental variables such as potential sweep rate, step size, and the potentiostatic or potentiodynamic characteristics of the curve significantly influence the shape of the polarization curve. Potential sweep rate is the rate of potential change (for example in mV/min), and potential step size is potential change between two successive data points. In potentiostatic sweep, the potential is increased at discrete steps, and in potentiodynamic sweep the potential is changed continuously. When comparing different alloys or evaluating corrosivity of solutions, using the same experimental parameters during the test series is important. The polarization curve rarely shows a steady state. As sweep rate decreases, the polarization curve approaches the steady state. At low sweep rates, the system is theoretically described more accurately and the measured current density depends on the rate determining step. An increase in the sweep rate results in an increase of the measured current densities. At high sweep rates, the current density results from the fastest step. It is not the rate of the total reaction. Selection of the sweep rate is a compromise and depends on the system under study. It is usually 1–100 mV/min. When conducting a large series of similar tests, making a preliminary study on the effect of sweep rate and then making the measurements at the fastest sweep rate that does not give significantly different results than lower ones is prudent. When studying passivating metals, the use of an excessively fast scan rate can give current densities that are excessively high as Fig. 13 shows.

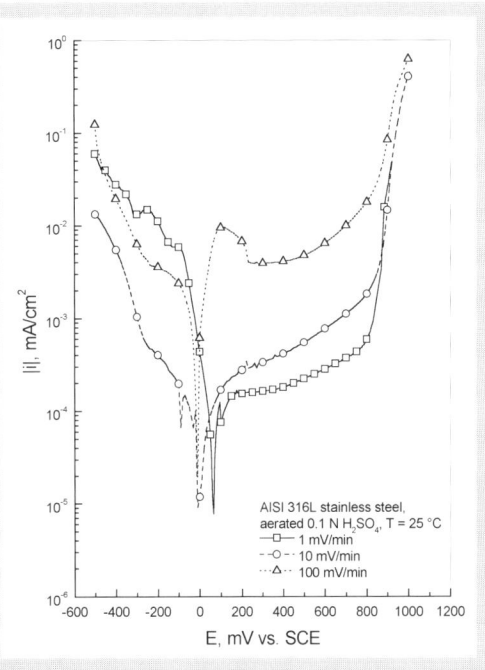

Figure 13. Effect of potential scan rate on polarization curves using AISI 316 stainless steel in aerated 0.1 N H_2SO_4 at 25°C.

The polarization curve can use a potentiostatic or potentiodynamic sweep. In potentiostatic sweep, the potential is increased at discrete steps. In potentiodynamic sweep, the potential changes continuously. This division easily leads to confusion. When using discrete steps, the current density increases sharply after every potential increase and decreases slowly to a constant value. When the step and time interval are sufficiently small, the current density does not reach a steady state before the next potential increase. A true potentiodynamic sweep will therefore lead to huge amounts of measurement points unless a potential interval for successively stored data points is selected. No defined maximum step size exists where potentiodynamic sweep changes to potentiostatic sweep. When using a potential interval of 1–20 mV, both sweep methods are equally accurate.

Polarization curves have use in several specific applications. The most common is determination of corrosion current density with the Tafel method. This method uses the assumptions that the polarization curve is linear and described with the Tafel equation, Eq. 8. This is valid at about 100–200 mV from the corrosion potential at current densities 100–1 000 times the corrosion current density. Constant *b* in Eq. 8 is the Tafel slope [mV/decade]. It has wide use in corrosion rate calculations. The Tafel method is a high overpotential range approximation of the Butler-Volmer equation describing a single activation controlled electrode reaction. It has constant use for mixed potential situations with several reactions. As the sample is polarized from its corrosion potential to the anodic direction, the rate of anodic reaction increases and the rate of the cathodic reaction decreases. Polarization to the cathodic direction results in an increase of cathodic reaction rate and decrease of anodic reaction rate. At sufficiently high overpotential, the sample acts solely as an anode or cathode. Potential vs. logarithm of current density shows a linear dependency. By extrapolating the linear ranges of anodic and cathodic polarization curves back to the corrosion potential, their intersections should meet and show the corrosion current density. Since the electrode reactions are never purely activation controlled, linear ranges may be difficult to locate. Extrapolated current densities are not too accurate. The linear part of the polarization curve should be at least one decade in current density to ensure reliable extrapolation. A lower number of electrons in the unit reaction requires a higher deviation from corrosion potential for reliable extrapolation. For a two-electron metal dissolution reaction, the extrapolation must use overpotential above 60 mV. For a three-electron reaction, the value must be above 40 mV. If the cathodic reaction is

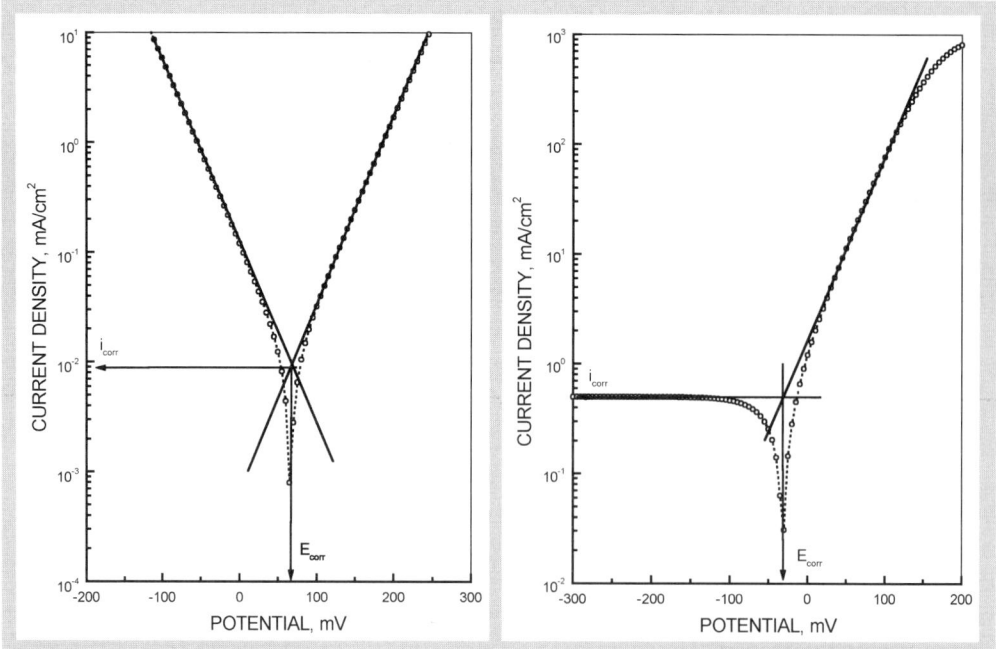

Figure 14. Examples from Tafel plots with (a) showing both reactions having activation controlled and (b) a mass transfer controlled cathodic reaction.

mass transfer controlled or the anode is in a passive state, current density on the horizontal part is an adequate estimate of corrosion rate.

The Tafel equation is strictly valid only for an activation controlled single electrode reaction. Using the Tafel method for a corroding system will therefore give experimental quantities not necessarily related to any particular theoretical reaction mechanisms. The plot may not even show a clearly linear part. Several reasons for this are possible. First, the polarization may be too low. In the vicinity of E_{corr}, the plot is never linear. An excessively rapid scan rate will prevent the specimen from reaching a steady state. Several charge transfer reactions, diffusion polarization, and ohmic polarization will also distort the inearity. Figure 14 shows two examples for Tafel plots. In (a), both reactions are activation controlled. The curves have clearly linear parts, and extrapolation to the corrosion potential is simple. In (b), the cathodic reaction is mass transfer controlled. The reaction rate is equal to the limiting current density. Here extrapolation of the anodic part may be useless since the corrosion current is obtained on the horizontal part of the cathodic reaction. Corresponding situations may occur when studying passivating metals or inhibited solutions. In both cases, the corrosion rate is simply the current density on the horizontal part.

Another important application for polarization curves is evaluation of pitting or crevice corrosion susceptibility with a cyclic polarization curve. This standard method is ASTM G61-86 "Conducting cyclic potentiodynamic polarization measurements for localized corrosion susceptibility of iron, nickel, or cobalt based alloys." The principle of the cyclic polarization test is to passivate the sample, create corrosion damages on the transpassive range, and then monitor the repassivation of these damages. Figure 15 shows a theoretical cyclic polarization curve. The susceptibility to localized corrosion initiation is the potential at which the anodic current increases rapidly. This potential is called transpassive potential, E_{trans}, critical potential E_{crit}, or pitting potential, E_{pit}. As this potential becomes more noble, corrosion initiation becomes less probable. After initiation, corrosion will propagate until electrode potential decreases to a sufficiently low value. This potential is the repassivation potential, E_{repass}, or protection potential, E_{prot}. At potentials above E_{crit}, pits initiate and propagate. At potentials between E_{prot} and E_{crit}, no new pits initiate, but existing pits propagate. At potentials below E_{prot}, corrosion in existing pits ceases. Note that crevices can initiate localized corrosion at potentials lower than the pitting potential, E_{pit}.

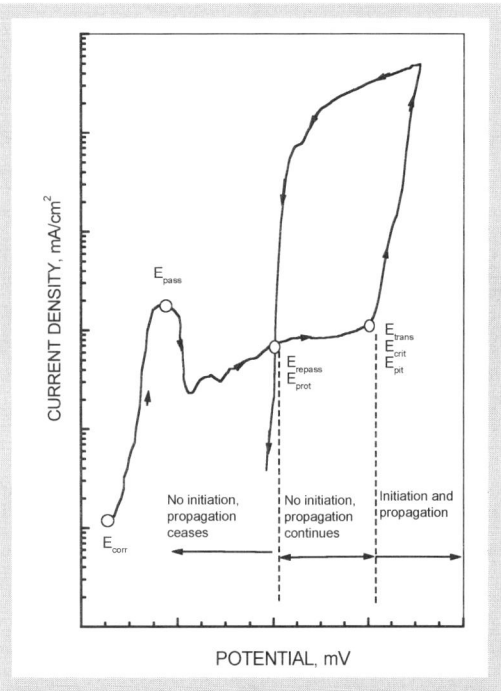

Figure 15. Cyclic polarization curve and its interpretation.

The following guidelines apply to the evaluation of pitting corrosion resistance from cyclic polarization curves. To ensure survival, the protection potential of the material should be higher than the redox potential of the solution. If the corrosion initiation potential is higher than the solution redox potential, good corrosion resistance will result. Material selection by using the protection potential is safer. The values of the pitting potential may show large scatter due to the sensitivity of pit initiation to the initial conditions. Since the protection potential describes the potential below which existing pits cease to propagate, this should have preferential use as a design parameter. Because unexpected process upsets can initiate pitting, long service life requires that pits on surfaces and in occluded areas do not propagate. A smaller difference between initiation and protection potentials means a more resistant material. This is because a small hysteresis loop indicates that the pits are easily repassivated. Potential difference below 50 mV is an indication of good corrosion resistance. The cyclic polarization curve is a reasonable method for preliminary comparison of alloys and environments. The main disadvantages are dependence of pitting and protection potentials on the scan rate and dependence of the hysteresis loop on the reversal current density and corresponding potential.

Another application of cyclic polarization curves is evaluation of sensitization of stainless steels. The sample is potentiostatically polarized on the passive range or scanned from its corrosion potential over the active range to the passive range at some predetermined potential. The scan is then made back to the corrosion potential. The first method is the single loop potentiokinetic reactivation method. The second method is the double loop potentiokinetic reactivation method. The single loop method description is in ASTM G 108 "Test method for electrochemical reactivation (EPR) for detecting sensitization of AISI type 304 and 304L stainless steels." The single loop test uses the assumption that dissolution occurs only at the grain boundaries. In the double loop test, the effect of intergranular corrosion can be separated from the total corrosion current. The results are evaluated by consumed charge during the downward scan in the single loop test and the ratio of maximum currents during reverse and anodic scans in the double loop test. A higher ratio of consumed charge to the grain boundary area or a higher ratio of currents gives a higher degree of sensitization.

4.2.2 Potentiostatic and galvanostatic tests

Potentiostatic and galvanostatic experiments provide a closer look at the phenomena exposed by a polarization curve. While a polarization curve is rapid and gives the first picture of the system, potentiostatic and galvanostatic experiments allow more detailed study of the system. In a potentiostatic experiment, the electrode potential is set to a constant value, and current density is recorded vs. time. In a galvanostatic test, the current density is set to a constant value, and potential variation is recorded. When potential or current is switched to the cell, the sample begins to move from a steady state to another state. As long as the measured current density or potential changes, the system is in a transient state. Potentiostatic tests are suitable for measuring steady state current densities, estimating passivity, evaluating coating quality, etc. Galvanostatic tests have use to determine corrosion probability, evaluate coating quality, etc. Examples of various tests follow.

The most simple application of potentiostatic tests is measuring the steady state corrosion current density. In this sense, a potentiostatic test is merely an extension of the polarization curve using a large potential interval. When the potential changes rapidly, the mass transfer controlled current density decreases linearly with the inverse square root of time. When the measured curve becomes horizontal, the steady state current density occurs. This method has use for estimating passivation tendency. If the material can passivate, the current density should decrease very rapidly to a constant value. When charge transfer typically controls reaction rate, the current density decreases linearly with the square root of time. If the current density decreases and then starts to increase, the reaction product may be porous or localized corrosion begins. Figure 16 shows an example of the use of potentiostatic tests where optimum anodic protection potential in sulfuric acid was necessary. A polarization curve showed large passive range with almost constant current density. The potentiostatic experiments showed that the minimum current density has a much narrower potential range. It increased rapidly when potential changed.

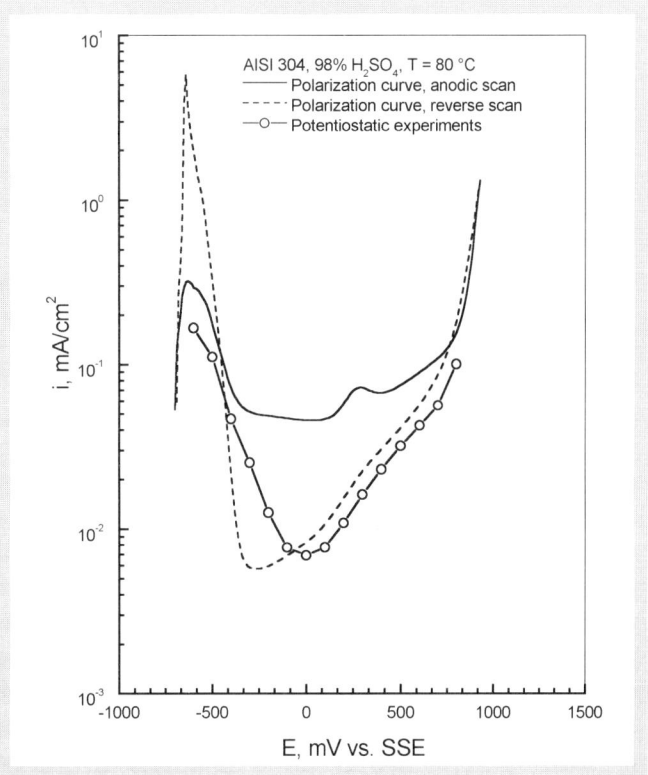

Figure 16. Anodic polarization curve and potentiostatic experiments for stainless steel in sulfuric acid.

For a potentiostatic experiment, linear dependence of current density on the inverse square root of time means diffusion controlled reaction. Linear dependence of current density on the square root of time indicates a charge transfer controlled reaction. In the latter case, current density decreases more rapidly than in the former. If the reaction is controlled by both, then charge transfer controls at the beginning with a change to diffusion control later. When studying localized corrosion or coatings, the sample does not necessarily follow either behavior. The beginning of new reactions often indicates increasing current density with respect to time. This can estimate pitting and protection potentials of passivating metals. These tests require some initial potential values that are available from a polarization curve. When measuring the pitting potential, the test begins with a passive, pit-free surface. This can result by polarizing the sample first cathodically for up to 30 min to remove all surface layers and then allowing it to stabilize to a steady state corrosion potential. The sample is then polarized to some potential on the passive range. If this set potential is below the pitting potential, no pits will initiate. The current density will decrease rapidly to values comparable to the passive current density. If the set potential is above the pitting potential, then the current density first decreases but increases after some initiation time. This initiation time decreases with increasing potential value. The increase in current density results from pit initiation and propagation. Figure 17 (a) shows this behavior. When measuring the protection potential, pits are first created on the surface by holding the sample above pitting potential for some time. The sample is then polarized to some potential value on the passive

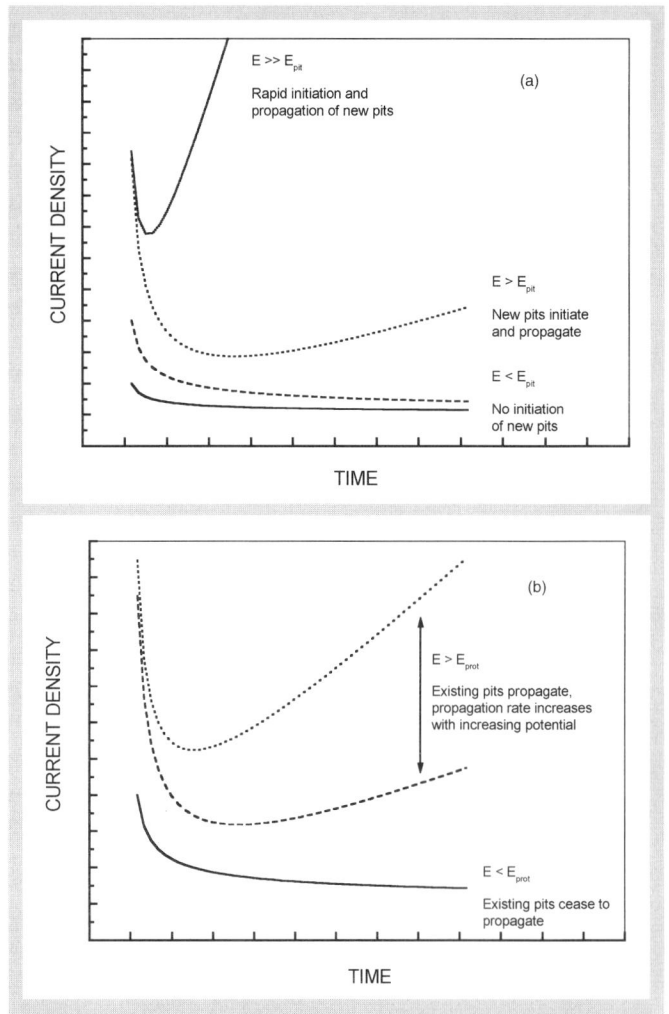

Figure 17. Potentiostatic tests for determining pitting and protection potentials of a passivating metal[7].

range. If the current density decreases rapidly to values comparable to the passive current density, the set potential is below the protection potential. If the sample is polarized to a potential above the protection potential, the current density first decreases but increases after some induction time. This is a result of continuing pit propagation. When measuring pitting potential, using a passive sample is necessary. When measuring protection potential, a sample with active pits is necessary. If the set potential value is below the pitting or protection potential, the current density decreases rapidly and remains low. If the set potential is above the corresponding potential, the current density first decreases and then begins to increase due to pit initiation and propagation of continuing propagation, respectively. Figure 17 (b) provides examples of these measurements.

Potentiostatic tests also determine critical pitting and crevice corrosion temperatures. In these tests, the sample is polarized to a potential that is the maximum corrosion potential expected in the system. The temperature is increased at predetermined intervals while monitoring current density for a reasonable time before the next temperature increase. Current densities that exceed some suitable criterion signify initiation of localized corrosion. These tests allow quantitative ranking of materials in resistance to pitting or crevice corrosion.

In a galvanostatic experiment, the current density, i.e., corrosion rate, is set to a predetermined value, and sample potential varies freely. This behavior can evaluate corrosion resistance. Figure 18 shows schematic examples of galvanostatic measurements. In Fig. 18 (a), one first selects a tolerable corrosion rate and then tests the material at the corresponding current density. A step increase in the potential indicates formation of a reaction product layer more noble than the base material. If the final potential is higher than the redox potential of the solution, the material is resistant. In this case, the redox potential of the solution is not sufficiently high to polarize the sample to potentials corresponding to the set current density and corrosion rate. If the sample potential first increases but starts to decrease and possibly oscillate, this may indicate a porous reaction product layer or possibly localized corrosion as Fig 18 (b) shows.

Galvanostatic experiments also have use to test organic coatings. Especially when using coatings with cathodic protection, they must be able to resist the resulting alkaline conditions and debonding effect of gas bubbles. The coatings are usually tested by polarizing the samples with cathodic current density sufficiently high to cause hydrogen evolution for long periods. Evaluation of the samples uses appearance.

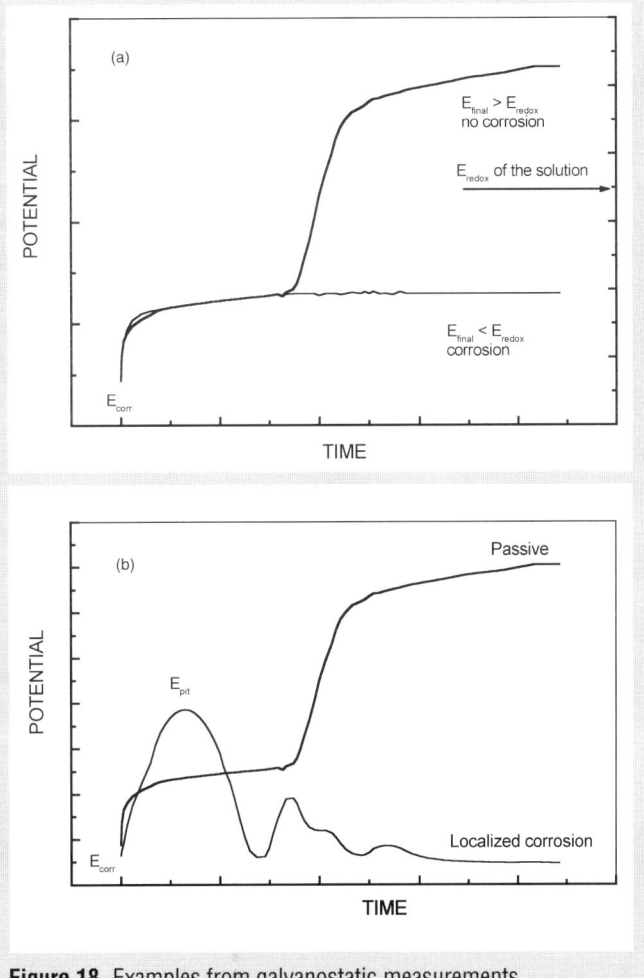

Figure 18. Examples from galvanostatic measurements.

CHAPTER 3

4.2.3 Polarization resistance measurement

Corrosion rates can sometimes be measured rapidly using methods that do not destroy the sample. One such method is measurement of polarization resistance as in ASTM G 59 "Conducting potentiodynamic polarization resistance measurements." Polarization resistance, $\Omega \cdot cm^2$, is the calculated slope of the polarization curve at corrosion potential, $R_p = (\partial E/\partial i)E_{corr}$. The method uses the low overpotential approximation of the Butler-Volmer equation. Another name for the method is the linear polarization method because the polarization curve is assumed linear in the area of corrosion potential. This assumption is valid if the corrosion potential is sufficiently far from the equilibrium potentials of the anodic and cathodic reactions and both reactions are activation controlled. The polarization resistance measured in a mixed potential case is inversely proportional to corrosion current density. It is analogous to the relationship between charge transfer resistance and exchange current density of a single electrode reaction. The corrosion current cannot be calculated from the polarization resistance by directly using the low overpotential approximation of the Butler-Volmer equation, Eq. 9. Corrosion current density in mA/cm^2 is instead calculated from Eq. 15 where b_a and b_c are anodic and cathodic Tafel slopes in mV. The factors involving Tafel slopes are often tabulated as B values.

$$i_{corr} = \frac{b_a \cdot b_c}{2.303 \cdot (b_a + b_c)} \cdot \frac{1}{R_p} = \frac{B}{R_p} \tag{15}$$

Assuming that the anodic reaction has theoretical slopes of 30–120 mV and cathodic reaction slopes of 120 mV for hydrogen evolution to infinity for mass transfer controlled oxygen reduction, the factor B would have values of 10–52 mV. A two-electron dissolution (b_a = 60 mV) with hydrogen evolution as the cathodic reaction (b_c = 120 mV) will produce B value of 17 mV. With mass transfer controlled oxygen reduction ($b_c = \infty$), B value would be 26 mV. Using a B value of 20–30 mV, one can usually estimate corrosion rate to at least a factor of 2. If the measured slopes do not obey Tafel behavior due to mass transfer and resistance effects, then the results are invalid. Poor correlation can also result with voluminous surface films, low conductivity, and presence of other redox reactions. There is an ongoing debate on calculating corrosion rates from linear polarization measurements. Some methods are valid from a theoretical viewpoint while others are more useful from the practical viewpoint.

Figure 19 shows examples of polarization resistance measurements. A higher polarization resistance gives a smaller slope for the i vs. E curve. The method cannot separate true charge transfer resistance from solution resistance. Especially in cases, where solution resistance is about the same magnitude as the charge transfer resistance, corrosion rates can be significantly underestimated. In Fig. 19, the additional solution resistance of R_Ω = 120 Ω almost doubles the value of the true charge transfer resistance, R_{ct} = 130 $\Omega \cdot cm^2$, since the measured polarization resistance is R_p = 250 $\Omega \cdot cm^2$. A smaller ratio of solution resistance to polarization resistance gives a smaller error due to solution resistance. The curves are seldom exactly linear, and the accuracy of the linear fit to determine the polarization resistance value can therefore strongly depend on the potential range used. At the lowest overpotentials, the curve is most probably linear but may show highest scatter especially when measuring low corrosion rates. If the corrosion rate is very low, the potential range must be large to obtain reliable current measurements. Then the curve may not be on the linear range any longer. One usually considers that the linear range occurs at overpotentials below 10 mV.

The most common errors in polarization resistance measurements are using excessively large overpotentials to invalidate the linear relationship, measurement during a transient state causing a change of corrosion potential, and currents from simultaneous side reactions. Concentration polarization and ohmic resistance will also cause errors, and the measured polarization resistance will overestimate the true charge transfer resistance associated with the corrosion reaction. This will cause an excessively low corrosion current density.

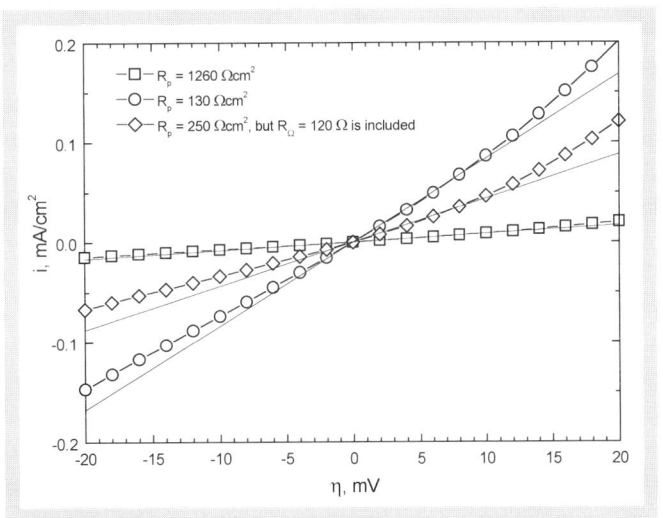

Figure 19. Determination of polarization resistance by the linear polarization method.

4.2.4 Electrochemical impedance spectroscopy

Electrochemical impedance spectroscopy (EIS) is a rather new method. All the previously described methods have been direct current methods. EIS is an alternating current method. The use of alternating current allows separation of charge transfer, mass transfer, and ohmic effects. In the method, a sinusoidal potential or current excitation signal is applied to the sample with recording of the resulting impedance magnitude and phase angle. Measuring the impedance usually uses several frequencies ranging for example from 10 mHz to 100 kHz depending on the system and information sought. The results of impedance measurements are presented as Bode plots showing impedance magnitude and phase angle as a function of frequency or as Nyquist plots showing the impedance reactive part on the vertical axis and the resistive part on the horizontal axis with frequency as a parameter. The main application of EIS in corrosion studies is measurement of polarization resistance in poorly conducting solutions.

The interpretation of impedance spectra uses the assumption that an electrified interface can be described with an equivalent circuit consisting of resistance and capacitances. The component values such as electrolyte resistance are assumed constant during the measurement. The most important values obtained from impedance mea-

CHAPTER 3

surements are solution resistance for IR drop correction and polarization resistance. The simplest equivalent circuit is a Randles circuit. Figure 20 shows the Randles circuit and its impedance spectra. It simplifies the electrified surface to parallel connected charge transfer resistance, R_{ct}, and double layer capacitance, C_{dl}. The solution resistance, R_Ω, is connected in series with this time constant. When the signal frequency is low, the capacitance will act as an insulator, and the system will behave as two resistances in series whose total resistance is $R_{ct}+R_\Omega$. When the frequency increases, the impedance of the capacitor becomes progressively smaller, and charge transfer resistance has less effect in the parallel circuit. At the highest frequencies, the system is a serial connection of ohmic resistance and double layer capacitance. Essentially, only the ohmic resistance is seen. The effect of the capacitance occurs at the middle frequencies. A higher capacitance value means that it appears at lower frequency. Comparing the impedance spectrum and equivalent circuit with the factors affecting the linear polarization method, one can see that the polarization resistance, R_p, is the sum of charge transfer resistance, R_{ct}, and ohmic resistance, R_Ω. Only R_{ct} is associated with the electrochemical reaction rate.

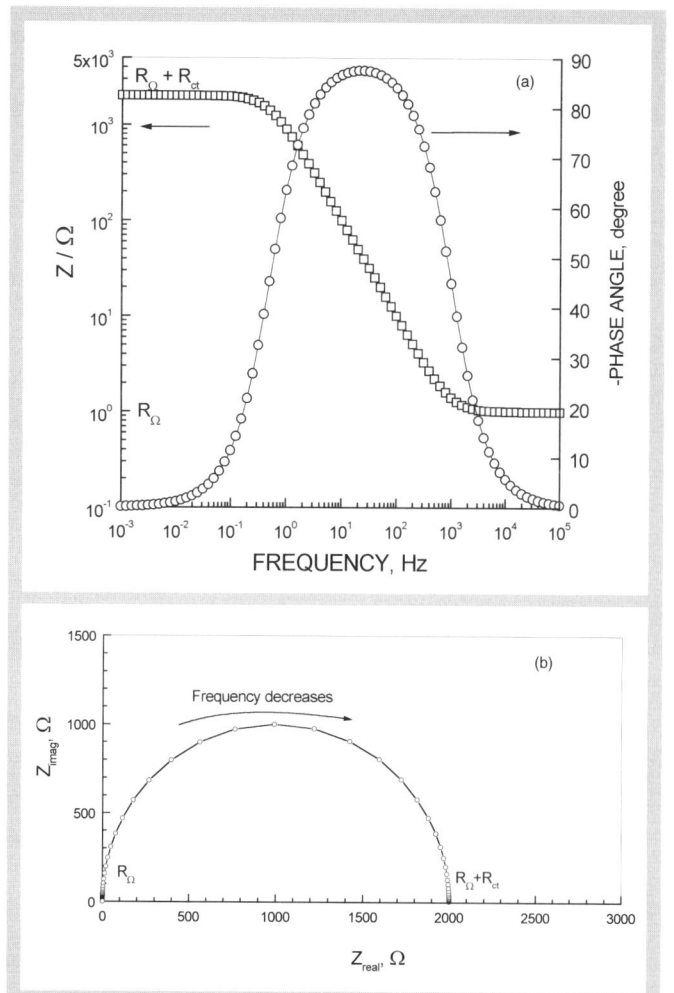

Figure 20. Randles circuit and its impedance spectrum. Figure 20 (a) is a Bode plot showing impedance magnitude and phase angle as a function of frequency, and Fig. 20 (b) is a Nyquist plot showing the real and imaginary parts of the impedance as a complex number.

The Randles circuit is often too simple to describe the actual system. If the system contains mass transfer effects, a so-called Warburg impedance is added in series with charge transfer resistance. Surface layers will cause additional time constants. The various time constants may connect in serial, parallel, or various combinations according to the preferred equivalent circuit. The possibility to fit different kinds of equivalent circuits and component values to one measured spectrum makes the interpretation difficult. Impedance spectra are measured with absolute resistance and capacitance values. Simple comparison of impedance spectra is not sensible unless the samples have exactly the same surface area. To compare separate measurements, calculating the equivalent circuit component values and relating them to the surface area is necessary. This results in units of $\Omega \cdot cm^2$ and F/cm^2. Only the ohmic resistance is useful as an absolute value.

Figure 21 shows some equivalent circuits used to describe a heterogeneous corroding surface. Starting from the simple parallel time constant of charge transfer resistance and double layer capacitance, the basic circuits have incorporated ohmic resistance and Warburg impedance. Systems with more than one phase in contact with the electrolyte require more time constants. These can be combined in parallel or series. No hard rules exist whether parallel or series combinations are preferable.

A problem with impedance measurements is that the interesting phenomena may not be in the measurable frequency window as Fig. 22 shows. If the values of the equivalent circuit are not suitable, the time constants in a complex circuit may overlap. The values of the time constants can be so large or so small that the capacitive ranges occur at such low or high frequency values that the equipment cannot measure them. The equivalent circuit in Fig. 22 contains two serially connected time constants and additional mass transfer effects in series with the charge transfer resistance that Fig. 21 shows. This kind of equivalent circuit has use for coated samples. The actual ohmic resistance cannot be measured since the maximum measurable frequency is still on the capacitive range. Depending on the value of the

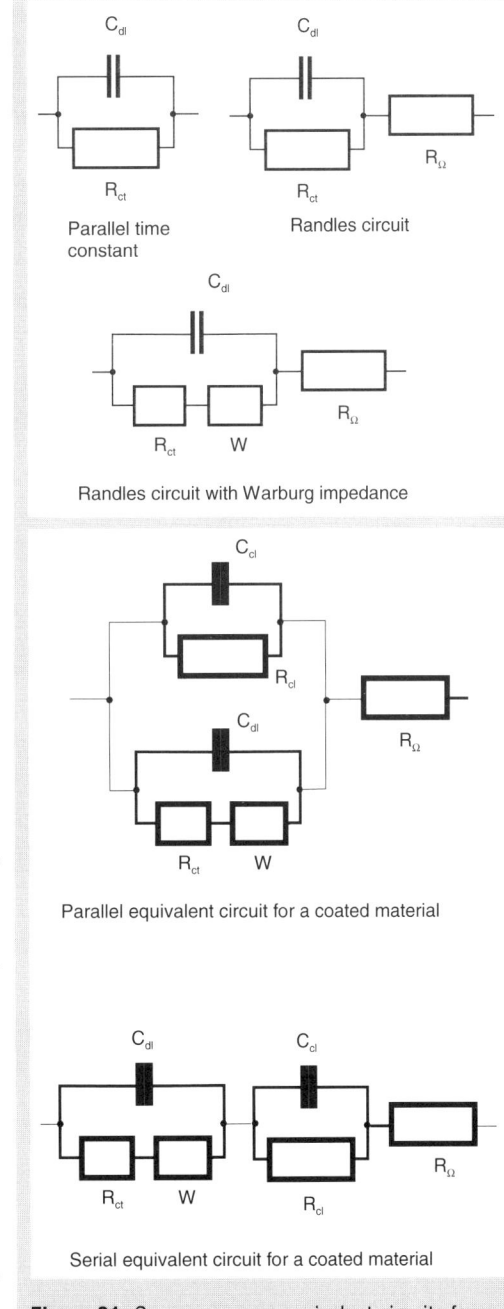

Figure 21. Some common equivalent circuits for single and multiphase systems.

ohmic resistance, the charge transfer resistance found at the high frequency horizontal part is usually overestimated. The low frequency horizontal part is a sum of coating resistance, charge transfer resistance, and ohmic resistance. Especially for such a complex system, the measurement of corrosion rate by the linear polarization method is questionable. With the linear polarization method, the measured polarization resistance is equal to the limiting impedance value as the frequency approaches zero. In the system shown in Fig. 22, the polarization resistance measured with the linear polarization method could be about 100 kΩ, but the value of the actual charge transfer resistance is about 100 Ω.

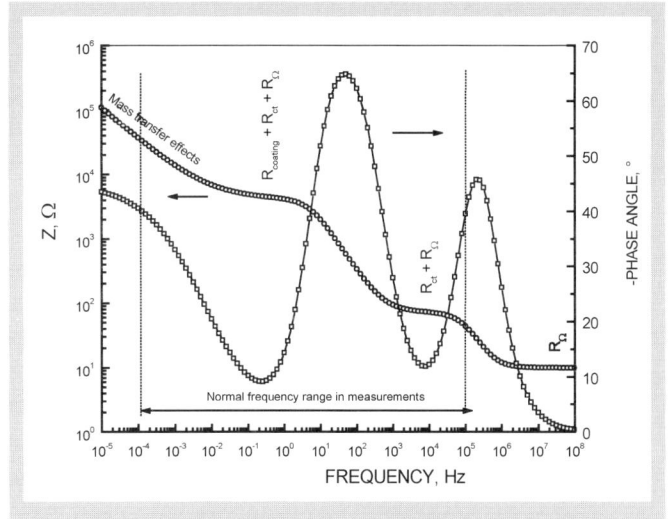

Figure 22. Frequency window of a complex equivalent circuit.

4.2.5 Calculating corrosion rates

Corrosion rate expressions use Faraday's law. Corrosion rate calculation can use weight loss, thinning, pit depth, etc. The measured current is converted for example to mass loss or thinning by using electrochemical equivalent and density of the metal. The electrochemical equivalent of an alloy can be estimated from its composition by the following:

$$Equivalent = \sum \frac{M_i}{z_i \cdot \%_i \cdot F} \qquad (16)$$

where M_i is molar mass of the ith element
        z_i number of electrons of ith element
        $\%_i$ atomic fraction of the ith element.

Usually, only elements with a mass fraction of more than 1% are considered. If the actual analysis is unknown, midrange approximations from the composition specification for each material can be used. Since the metals of an alloy can dissolve with varying numbers of electrons, the number of electrons per mole of alloy is not always an integer. The electrochemical equivalent shows the charge required to dissolve or deposit a known mass of the metal. Theoretically, pure iron dissolves as bivalent ions and has an electrochemical equivalent of 55.85 g/mol/(2 • 96500 As/mol) = 0.29 mg/As or 1.04 g/Ah. Values for copper are 0.33 mg/As or 1.18 g/Ah. Some values of electrochemical equivalent are shown in Table 3. The electrochemical equivalent values will

Electrochemistry of metallic corrosion and corrosion test methods

vary for metals and alloys with multiple oxidation states, i.e. number of electrons in the oxidation reaction. Calculated examples for stainless steel in Table 3 use two different oxidation states for metals.

Table 3. Electrochemical equivalents of some engineering materials.

	Molar mass, g/mol	Electrons	Equivalent, g/Ah
Iron	55.9	2	1.04
Chromium	52	3	0.65
Chromium	52	6	0.32
Nickel	58.7	2	1.10
Molybdenum	95.9	3	1.19
Copper	63.5	2	1.18
Steel	56	2	1.04
Zinc	65.3	2	1.22
Aluminum	27	3	0.34
Magnesium	24.3	2	0.45
Bronze CuSn6	65.4	2.07	1.18
Brass CuZn15	63.6	2	1.19
AISI 304 stainless steel, Fe^{2+}, Ni^{2+}, Cr^{3+}	58.9	2.17	1.01
AISI 304 stainless steel, Fe^{3+}, Ni^{2+}, Cr^{6+}	58.9	3.39	0.65
AISI 316 stainless steel, Fe^{2+}, Ni^{2+}, Cr^{3+}, Mo^{3+}	59.9	2.19	1.02
AISI 316 stainless steel, Fe^{3+}, Ni^{2+}, Cr^{6+}, Mo^{6+}	59.9	3.47	0.64

Calculation of corrosion rate requires knowing the charge that has passed through the electrode and the electrochemical equivalent. The charge must be associated with specific surface area. Its calculation therefore uses corrosion current density and time. When charge and electrochemical equivalent are known, corrosion rate can be calculated as mass loss per area and time. For example, dissolution of iron at current density of 1 mA/cm^2 results in the following mass loss:

Mass loss = 0.29 mg/As • 0.001 A/cm^2 • 86 400 sec/day = 25.0 mg/cm^2/day

Mass loss per area and time can be further converted to loss of thickness when the material density is known. The density of iron is 7.87 g/cm^3. When the mass loss of 25.0 mg/cm^2/day is divided by density, the loss of thickness becomes the following:

Loss of thickness = 25.0 • 10^{-3} g/cm^2/day / 7.87 g/cm^3 • 365 days/year = 1.16 cm/year

CHAPTER 3

For copper, the corrosion rates at 1 mA/cm^2 would be 28.4 mg/cm^2/day and 1.3 cm/year. The following is a good rule of thumb. For common engineering metals with molar weight about 60 g/mol, density about 8 g/cm^3, and dissolving with two electrons, current density of 1 mA/cm^2 equals 250–300 g/m^2/day or 11–13 mm/year.

Common expressions for corrosion rate are mm/year, m/year, g/m^2/day, and mg/dm^2/day. In the United States, mpy = milli-inches per year sometimes called mils. Corrosion rates expressed as weight loss must relate to area and time, and corrosion rates expressed as thickness loss must relate to time to make meaningful comparisons. From the design viewpoint, thinning or penetration rate has direct use to estimate equipment lifetime or necessary corrosion allowances. Table 4 gives approximate densities of metals and alloys, and Table 5 provides conversion factors for corrosion rates.

Table 4. Densities of metals and alloys[6].

Metal	Density, g/cm^3
Fe	7.87
Cr	7.23
Ni	8.9
Mo	10.22
Cu	8.94
Ti	4.5
Cast iron	7.1–7.3
Carbon steel	7.86
UNS S30400 (AISI 304)	7.94
UNS S31600 (AISI 316)	7.98
UNS J96200 (CF-8, cast SS)	7.75
UNS N08904 (904L Alloy)	8.0
UNS S31254 (254 SMO)	8.0
UNS N02200 (Nickel 200)	8.89
UNS N06022 (C-22 Alloy)	8.69
UNS R50400 (Ti grade 2)	4.54
UNS R05200 (Tantalum)	16.6
Brass	8.4
Bronze	8.7

Table 5. Corrosion rate conversion factors.

1 mA/cm^2 as corrosion rate equals factor K times the electrochemical equivalent	Factor K
89.2	mdd (mg/dm^2/day)
8.92	g/m^2/day
3.26/ρ	mm/year
129/ρ	mpy (mils per year)
0.129/ρ	in. per year

In Table 5, ρ is the density of metal as g/cm³. By using this information, the corrosion rates as weight loss or thinning can be calculated from measured current densities using Eq. 17. If the corrosion rate is expressed as mass loss, the material density is not necessary. When corrosion rate is expressed as thinning, the material density must be known.

$$Corrosion\ rate = i[mA/cm^2] \cdot electrochemical\ equivalent \cdot conversion\ factor \qquad (17)$$

4.3 Instrumentation

The hardware required in electrochemical measurements ranges from a simple reference electrode and multimeters to costly and complex computer controlled potentiostat and frequency response analyzer systems. The more sophisticated instrumentation and advanced measurement techniques are more apt to give misleading results. A potentiostat is the basic tool for electrochemical measurements. The potentiostat is basically an ideal current source that can keep the cell voltage constant regardless of the current drawn. A simple constant voltage source is not suitable. A potentiostat is basically a control amplifier with the test cell placed in the feedback loop. The objective is to control the potential difference between the test electrode and the reference electrode by the application of a current via the auxiliary electrode.

The potentiostat uses three electrodes: working, auxiliary, and reference. The current flows through the cell between working and auxiliary electrodes. The potential of the working electrode is measured with respect to the reference electrode. If the working electrode operates as an anode, the auxiliary electrode is a cathode and vice versa. The reference electrode measures the working electrode potential and compares it with the preset potential. If the reference electrode senses any potential difference, the potentiostat will pass more or less current through the cell to restore the working electrode potential. If the working electrode operates as an anode, excessively low measured potential will cause an increase in anodic current. The electrode is polarized to the anodic direction back to the set potential value. Excessively low measured potential will cause the potentiostat to decrease anodic current and restore the set potential value. The operational principle of the potentiostat causes some risks. If the reference electrode cannot sense any working electrode potential, the potentiostat will usually try to pass maximum available current through the cell. Since electrochemical protection systems also operate now with the potentiostatic mode, false readings or broken reference electrode cables can cause considerable damage.

The experimental parameters must be invariant to obtain meaningful and reproducible data. In electrochemical experiments, the reference electrode, cell construction, and specimen preparation are critical factors. The electrode potential is measured with respect to some reference electrode. A reference electrode is an electrode system with a known, stable potential. The hydrogen electrode consisting of platinum foil over which hydrogen is bubbling in sulfuric acid catalyzes the H^+/H_2 reaction and has by agreement an electrode potential of 0 V. The standard hydrogen electrode (SHE) provides the zero point for electrochemical potentials. In practice, other reference electrodes with metal/metal salt equilibrium are preferable. The effect of changing the reference electrode is to change the zero of a potential scale while the relative positions of all the potentials remain the same. Table 6 lists some reference electrodes and their properties. Electrode potentials are converted from one reference electrode scale to another by adding or subtracting, but confusion often occurs.

Table 6. Properties of reference electrodes[6, 8, 9].

Electrode	Filling electrolyte	E/mV vs. SHE at 25°C	Temperature range, °C	Temperature coefficient, mV/°C	Use
Ag/Ag$_2$S	4M Na$_2$S	-710		0.1	Kraft liquors
HgTl/TlCl	3.5M KCl	-507	0–150	0.1	Hot media
Cu/CuSO$_4$	Sat. CuSO$_4$	320		0.97	Soil, waters
Hg/HgO/OH$^-$	1M NaOH	+98 – +140			Alkaline media
Ag/AgCl	3M KCl	207	0–130	1	General, hot media
Hg/Hg2Cl2/Cl$^-$	Sat. KCl	242	0–70	0.65	General
Ag/AgCl	0.1M KCl	288	0–95	0.22	General
Hg/HgSO$_4$	Sat. K$_2$SO$_4$	+616 – +640	0–70	0.09	Sulfate containing media, alkaline media

All reference electrodes are not suitable for every environment. Electrodes should usually perform well when they have a common anion with the electrolyte, e.g., calomel and silver chloride electrodes for chloride containing solutions and mercury sulfate for sulfate containing media. Use of a simple system, e.g., by immersing an Ag/AgCl wire directly in the chloride containing solution, may be risky since the electrode potential varies with the test solution concentration. The

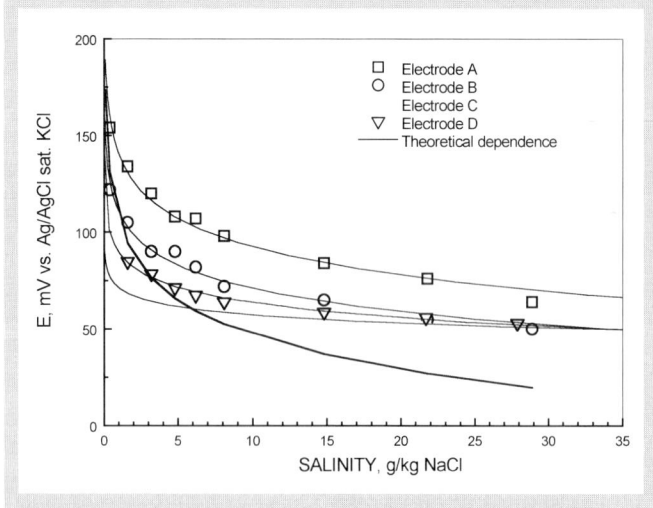

Figure 23. Variation of Ag/AgCl reference electrode potential on chloride concentration.

electrode potential increases logarithmically with decreasing anion concentration. If the concentration is sufficiently high, even this arrangement can be sufficiently stable. Figure 23 shows the variation of four commercial Ag/AgCl reference electrodes with the chloride concentration. As Fig. 23 shows, the theoretical variation is larger than that with the actual electrodes. With these commercial electrodes, significant variation was expected when chloride concentration was less than 1 000–2 000 ppm depending on the electrode.

Special reference electrodes have been applied in process environments. For example, silver/silver sulfide, molybdenum sulfide, and stainless steel reference electrodes have had use in laboratory and field studies of corrosion in kraft pulping liquors. The use of a corroding piece of metal as a pseudo reference electrode is sometimes justifiable. The metal should then corrode actively at a constant rate, and changes of the corrosive environment should not influence its corrosion potential. The use of a passive metal is more doubtful. Its corrosion rate is low, and its potential may wander hundreds of millivolts on the passive range.

Electrode potential measurements can use a multimeter with sufficiently high input impedance. According to Ohm's law, 1 V potential measured with a multimeter with 10 kΩ input impedance draws 0.1 mA current. Such a high current will polarize the reference electrode and the working electrode leading to false results. Multimeters with input impedance over 10 MΩ should be used. A resolution of 0.1 mV is usually sufficient for corrosion research. Current measurements must use a multimeter with a low input impedance.

Electrochemical measurements almost exclusively use a three-electrode arrangement. The electrodes are a working electrode, reference electrode, and counter electrode. The working electrode is the unit under study. Its potential is measured with respect to the reference electrode. Current flows through the cell between working and counter electrodes. The counter electrode is usually larger than the working electrode. To ensure proper current flow, the electrodes should have symmetrical arrangement. Planar electrodes facing each other or a cylindrical working electrode surrounded by a ring-shaped counter electrode are suitable arrangements. A large distance between electrodes will also provide more even current distribution.

The tip of the reference electrode should be as close as possible to the surface of the working electrode, but it should not cause any shielding effects. This occurs best with a Haber-Luggin capillary (also known as Luggin capillary). The capillary is a glass tube with a sharp tip held firmly at a fixed distance from the working electrode as Fig. 24 shows. A Haber-Luggin capillary reduces potential gradient between working electrode and reference elec-

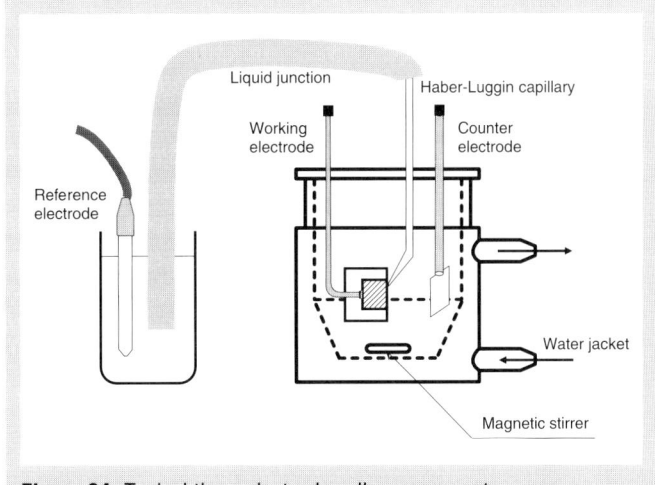

Figure 24. Typical three electrode cell arrangements.

CHAPTER 3

trode to reduce the IR drop. The capillary is often lengthened with plastic tubing to allow positioning the reference electrode in a separate beaker. The optimum distance of the tip of the capillary from the sample is twice the inner diameter of the tip.

Figure 24 shows a typical three-electrode cell arrangement. Working electrode and counter electrode are immersed into the test cell. The reference electrode, e.g., a saturated calomel electrode (SCE), is in a separate beaker. The reference electrode with a Haber-Luggin capillary measures the working electrode potential. The connection between capillary and reference electrode beaker is a salt bridge although a liquid junction would be a more precise term since the beaker holds the same test solution as the cell. Mounting the SCE reference electrode in saturated potassium chloride gel protects it from the test solution. Such a potassium chloride salt bridge can be made by dissolving 13 g KCl and 3 g agar-agar into 100 mL boiling distilled water. The solution is allowed to cool. While it is still pourable, it is poured into a glass tube with a porous ceramic plug.

Special cell constructions for various purposes are available. The rotating disk and rotating cylinder electrodes rotate the sample at constant speed. By measuring with different rotating speeds, elimination of mass transfer effects is possible. Rotating cylinder electrodes are valuable for example in quantifying effects of flow rate on corrosion in pipes and tanks.

A very effective cell construction to eliminate unwanted crevice corrosion when studying pitting corrosion is available[10]. This is the Avesta cell of Fig. 25. The crevice corrosion between sample and sample holder is eliminated by constantly flushing the contact area with distilled water. In all cell constructions, good electrical connections to electrodes, continuity of capillary and tubing, and shielding from external electrical and electromagnetic noise sources require consideration.

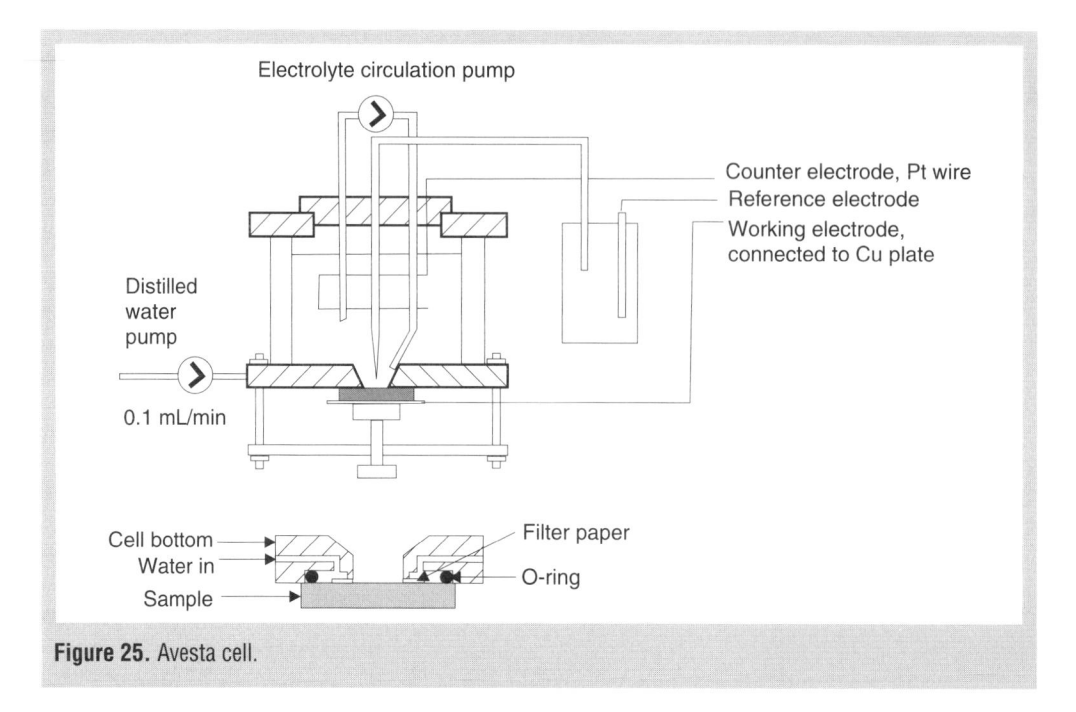

Figure 25. Avesta cell.

A counter electrode uses any material that is conductive, has a relatively large surface area, and does not contaminate the electrolyte. A platinum sheet is an ideal counter electrode material, but it is expensive. A carbon rod is also suitable.

Cell volume depends on the electrode size and expected reaction rates. The principle is that electrolyte composition should not change during the experiment. The test solution should not become contaminated from reaction products, and its oxygen content, pH, etc., must remain constant. Various ASTM standards give 20–40 mL per cm^2 of sample area as the minimum solution volume. Stirring the electrolyte is necessary. This usually requires a magnetic stirrer. Gas purging is a common requirement usually to remove oxygen with N_2 or Ar or ensure constant oxygen concentration with aeration. The gas often passes through a sintered glass bubbler that disperses the gas into fine bubbles. Gas purging can also provide solution stirring. The purge gas may become saturated by bubbling it through a vessel with the same test solution before passing it into the cell. If the solution will be purged, the cell must be a closed vessel to avoid atmosphere back-flow into the cell. Toxic purge gases or gases passed through toxic solutions need special scrubbing procedures.

Every electrochemical measurement is affected by IR drop or ohmic drop. The IR drop is a linear systematic error caused by electrolyte resistivity. Since the electrode potential is not measured from the electrode itself, the current flowing through the cell causes an additional potential drop between the electrode and potential measuring point located in the solution some distance from the electrochemical double layer. The presence of IR drop causes actual polarization that is always smaller than that indicated by potential measurement. The polarization curves become more round, and current densities corresponding to a potential value are often underestimated. The effect of IR drop can be estimated from the measured current and solution resistance using Ohm's law. IR drop usually causes appreciable errors in systems with high current densities or low conductivity and low current densities. A special case is measurement of polarization resistance when polarization resistance and solution resistance are the same order of magnitude or the solution resistance is higher. Solution resistance then causes overestimation of R_p and underestimation of corrosion rate. Placing a reference electrode or a Haber-Luggin capillary near the electrode surface can minimize the IR drop. Computer-controlled potentiostat systems can compensate for the IR drop. This uses interrupt methods since positive feedback compensation can cause oscillation. The measured results can also be corrected later if the solution resistance is measured before the experiment, e.g., with impedance spectroscopy. For every measured point, the product of measured current and ohmic resistance is then subtracted from the measured potential value:

$$E_{true} = E_{measured} - I \cdot R_\Omega \tag{18}$$

A spreadsheet program easily performs this calculation. The measured current or current density values remain unaltered. The main disadvantage of the method is that the solution resistance does not always remain constant during the entire experiment.

CHAPTER 3

Figure 26 shows an example from the IR drop correction of a polarization curve. The potential difference at any current density is the product of total measured current times solution resistance. Using too low ohmic resistance value for correction results in undercompensated IR drop. Using an excessively large ohmic resistance value causes the curve to bend backwards as Fig. 27 shows. If the correct ohmic resistance value is unknown, using trial and error to eliminate most of the IR drop is possible. This is only suitable for simple systems with one or two clear current peaks.

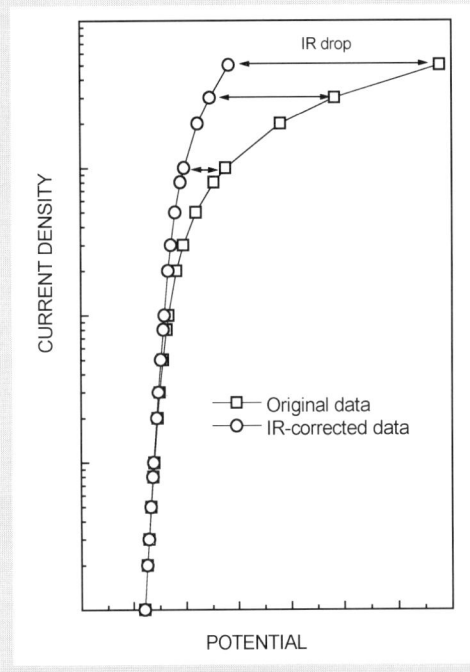

Figure 26. Mathematical correction of polarization curves to eliminate the effect of ohmic drop.

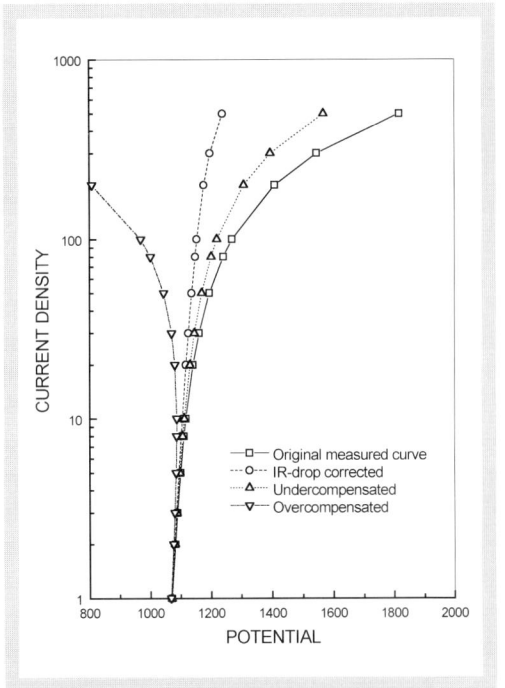

Figure 27. Overcompensation of polarization curves when eliminating the effect of ohmic drop.

Accuracy and reproducibility of electrochemical measurements are often overestimated. When measuring a polarization curve under identical conditions, potential variation of ±20 mV and current density variation by a factor of two are sufficient. Because the current density variation is rather large, small variations in electrochemical equivalent or alloy density values are not significant when calculating mass loss or thinning rates. Extremely accurate critical potentials or corrosion rates measurements are not necessary since the accelerated tests never describe the real situation. A spreadsheet program can very effectively analyze the measured data points from an electrochemical measurement. The analysis does not require extensive computing power. With a personal computer, one can do almost everything. The usual ways to analyze measured points from an electrochemical experiment are curve plotting and linear fit to some part of the curve. Tafel slopes can be obtained by a linear fit to the E,log(i) data points on a suitable range. Integration to obtain consumed charges from a potentiostatic measurement simply uses the trapezoidal rule. The IR drop correction method described above is not a tedious task. Impedance can be analyzed by a deconvolution method after selecting a suitable equivalent circuit. In this case, the effects of components and time constants are subtracted successively from the measured spectrum. The procedure is tedious and requires experience.

4.4 Sample preparation

The working electrode can be made in several ways. Typically, electrode size is 1–10 cm^2. If the studied material is a bulk material, it can be mounted in a nonconductive resin such as metallographic and dental resins. Electrical contact must be made on the back side of the specimen using solder, spot weld, screw, conductive tape, or adhesive before mount-

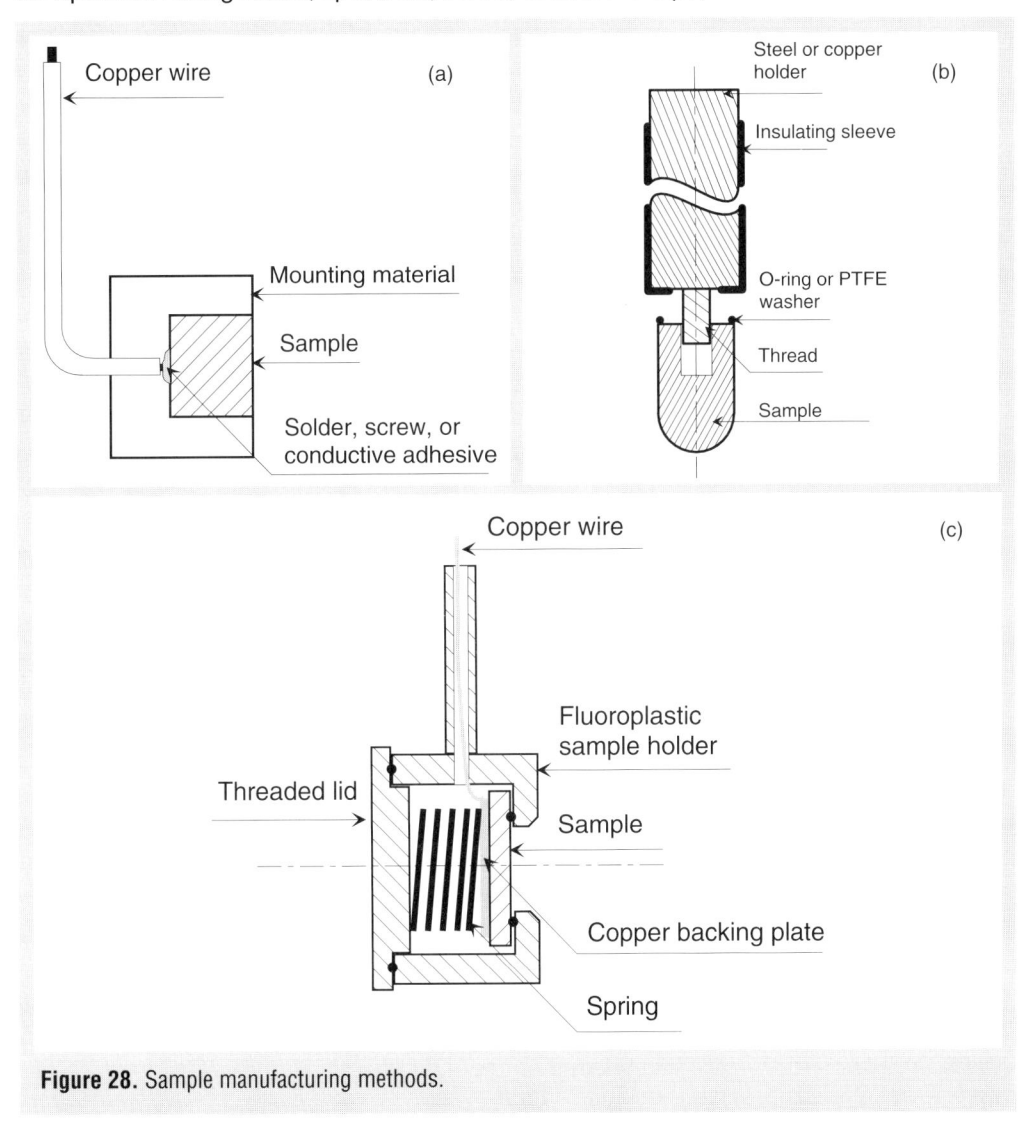

Figure 28. Sample manufacturing methods.

ing. Before measurement, the sample is ground as for metallographic study that can alter its surface state. When studying localized corrosion of passivating metals, sample manufacture is critical. Making a sample with electrical contact and controlled exposed surface without creating a crevice between the sample and mounting material is difficult. Samples with surface films or coatings are more difficult to make because grinding would destroy the surface of interest. Protective lacquers for electroplating have use to cover electrical contact and sample edges.

Figure 28 shows some methods for sample manufacture. The simplest manufacturing method is mounting in a nonconductive resin as Fig. 28 (a) shows. When selecting the mounting material, the most important factors are good adhesion to the sample to prevent crevices and extremely low conductivity. Some resins have use in scanning electron microscopy (SEM) studies. They are therefore conductive and not suitable for electrochemistry. The Stern-Makrides method of Fig. 28 (b) uses cylindrical samples with a threaded connection. A crevice in the joint is prevented by a machined fluoroplastic washer. When assembling the sample, the idea is that any sharp edges will cut into the plastic and prevent crevices[11]. The washers therefore have use only once. If crevice corrosion is not a problem, plastic washers or suitable rubber o-rings can be used several times. The sample can be assembled into a special holder such as Fig. 28 (c) shows. The sample holder is made from a fluoroplastic. The sample is pressed against the opening by using a threaded backing plate or a spring. These sample holders are suitable for sheet or plate samples. Using mounting materials is not necessary. This makes weight loss and microscopic studies easier. The samples and sample holders require precise machining to ensure reproducibility.

The composition and metallurgical state of the sample require good definition. The specimen should represent the material actually used in practical applications. Commercially produced materials are seldom homogeneous. The surface of a material can differ from the bulk due to oxidation, decarburizing, etc. In such cases, one must decide whether the altered surface or bulk material properties are of interest. Fontana compares the identification of specimens to the foundation of a house[12]. Chemical composition, fabrication history, metallurgical history, and positive identification are all necessary. When the equipment uses wrought materials, tests must use wrought materials. For cast structures, cast samples are necessary. Weld materials and welding procedures are often tested. They require even more careful documentation and identification than single material samples. The welds made for testing purposes should be identical to the process equipment in question. Ideally, the surface preparation should be the same for the samples and actual process equipment. The surface conditions of commercial metals and fabrications vary. For example, stainless steels can have a specific surface treatment. When running a test series, the surface preparation must be the same for every sample. Usually, tests for practical purposes require less prepared surface than tests for corrosion mechanism studies. Polishing must use wet polishing machines. The same papers must not be used for dissimilar metals like copper alloys and stainless steels. Polishing with 120 abrasive paper is sufficient to remove carburized layers to reach original metal[12]. While wet polishing will remove affected surface layers, it will not influence internal areas such as the sensitization of high carbon grades of stainless steels.

Sample manufacturing for crevice corrosion and stress corrosion cracking tests requires special attention. The problem in crevice corrosion tests is obtaining reproducible results. The occurrence of crevice corrosion depends strongly on sample geometry especially on the crevice gap. A number of crevice assemblies and stressed specimens are available in the standards. Figures 29 and 30 show some examples of these. The geometry and nature of the crevice and the technique used in the exposure test influence the extent of crevice corrosion. Typically, tighter crevices promote greater localized attack. Serrated fluorocarbon plastic or ceramic washers have common use for obtaining reproducible crevice dimensions. These washers are bolted to the specimen using a corrosion resistant bolt with constant applied torque. The rate of crevice attack during the incubation period is very low. Since corrosivity inside the crevice increases with exposure time, the crevice attack rate usually increases with time in the test. Multiple exposure periods may therefore be necessary to determine crevice corrosion rates accurately[13]. The size and geometry of the damage produced by crevice corrosion is random. Data from corrosion tests using artificial crevices should only be used to obtain qualitative comparisons of the corrosion resistance.

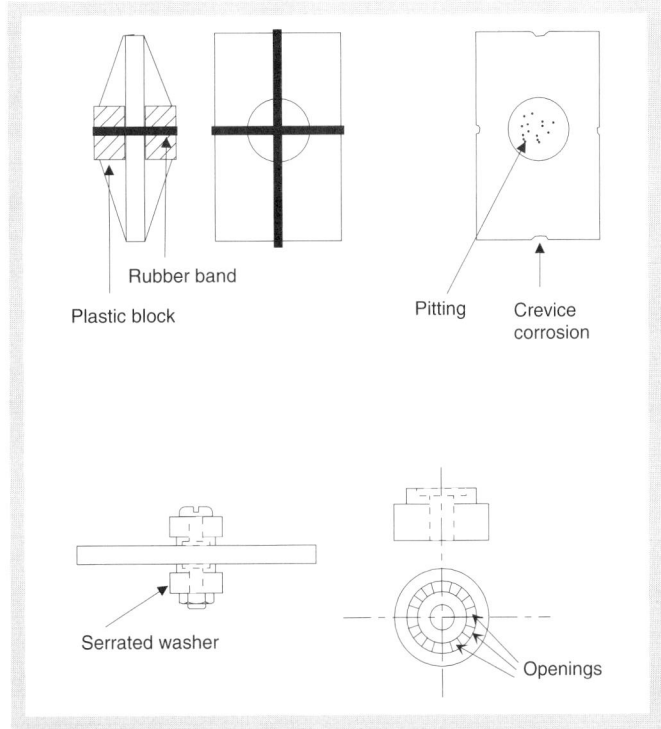

Figure 29. Samples for crevice corrosion tests.

Applying stress to stress corrosion samples can use constant strain, constant or sustained load, or dynamic strain. Various methods including U-bend, C-ring, and point loaded bends can provide constant strain. Constant strain samples are simple and inexpensive, but the reproducibility of stresses can be poor. Stress levels are difficult to quantify. A calibrated spring or a weight attached to the sample can apply a constant load. In constant-load testing, the stresses are well defined. These tests are usually performed to find a threshold stress below which stress corrosion cracking does not occur in a certain environment. Dynamic strain can be applied to the test specimen similar to the measurement of tensile strength. The method is slow strain rate testing (SSRT). The material is strained slowly and plastically deformed until it ruptures. SSRT tests are rapid compared with constant strain or constant load tests, but they may be unreliable

CHAPTER 3

since the materials are strained to failure. A material that could be resistant to SCC in a certain environment can fail in the SSRT test giving a false negative result. Relative ranking of similar alloys such as austenitic or duplex stainless steels or the influence of environmental parameters on the resistance of an alloy are suitable applications for slow strain rate tests[14].

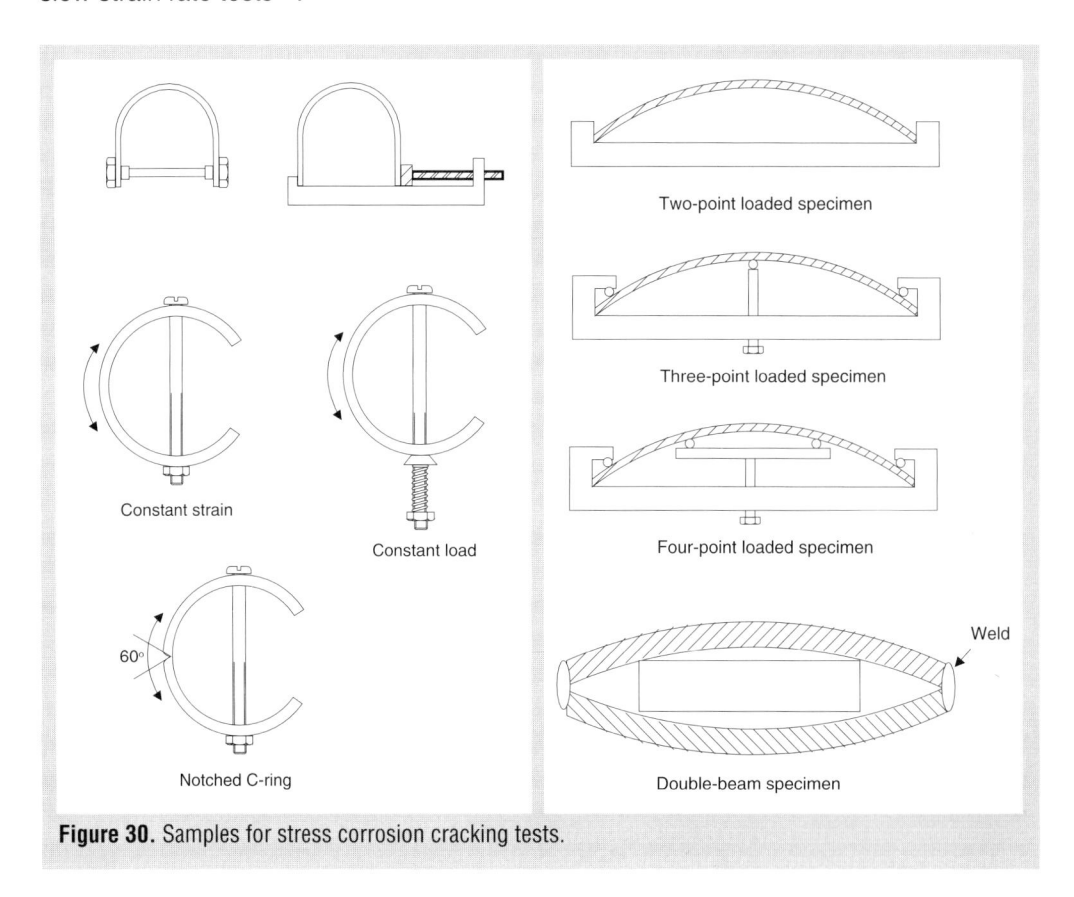

Figure 30. Samples for stress corrosion cracking tests.

5 Field testing

Field testing often has different objectives than laboratory testing. Field testing usually includes corrosion monitoring, detection of process changes, and diagnosis of sudden corrosion problems. The methods are usually conventional methods involving more rugged equipment than laboratory testing. Most sophisticated laboratory test methods are too sensitive for plant use. Difficult process conditions, debris, high flow rates, vibration, and electric noise problems may limit their use. Depending on the test method, the corrosion tests conducted in a plant will give metal loss during the test period or the instantaneous corrosion rate. Test coupons and electric resistance probes will indicate the mass loss. The response time of test coupons is usually weeks to months depending how often they are removed and analyzed. Electric resistance probes change their response when the metal loss increases the electric resistance of the specimen. These sensors change their response when corrosion reduces the minimum cross section of the metallic material in the exposed portion of the probe. They will respond to changes in corrosivity in a few hours to a few days.

The electrochemical measurement methods will continuously measure the corrosion potential or corrosion rate as faradaic current with the limitations described above. Polarization resistance, corrosion potential, corrosion current, and galvanic current measurements are common methods. Their response times are usually minutes. Since the corrosion is rarely uniform even in a localized area of a plant, using more than one measurement method is preferable. This improves confidence and limits the probability of false results. The most efficient measuring locations are those where corrosion problems are most severe.

Corrosion coupons are an inexpensive solution for determining average corrosion rates over periods of a few months. They are not suitable for detecting changes in the corrosivity during the exposure period. Electric resistance probes are real-time corrosion coupons since the electric resistance can be measured without removing the samples from the plant environment. Since the corrosion decreases the cross-sectional area of an electric resistance probe, its resistance will increase to give an estimate of the cumulative mass loss. Corrosion rate changes can be detected as changes in the slope of metal loss vs. time. The electric resistance probes are more suitable for measuring uniform corrosion since the various forms of localized corrosion can cause significant damages before a clear change in the resistance value occurs. Conductive deposits on the probe surface may cause a false indication on reduction of the corrosion rate.

Electrochemical methods can determine corrosion rates in a few minutes. They will measure current or another variable that is assumed to describe the rate of metal dissolution reaction. They have use only in sufficiently conductive solutions. Linear polarization probes with or without ac component for solution resistance estimations are the most common. The measured variable must be converted to corrosion rate by using essentially constant factors including experimentally measured Tafel slopes, electrochemical equivalent, and density of the material.

6 Summary

Corrosion theory uses two basic elements: mixed potential theory and formation of surface films. In a corrosion situation all the electrons released in one reaction must be consumed in another reaction. This results in a closed circuit and leads to mixed potential theory. Mixed potential theory is visualized in a corrosion cell. The components of the corrosion cell include anode, cathode, a electrical connection between them, and a corrosive solution. Removal of one of these components will stop the function of the corrosion cell. The corrosion resistance of any material depends on its nobility and its capability to form stable surface films. Noble materials are less prone to react. Stable surface films will form a barrier between the metal and its environment to prevent further reactions. This is passivity.

CHAPTER 3

The electrochemical reactions in corrosion are complex. For practical applications, note that the corrosive environment affects electrochemical reaction rates and formation of surface films. High electrochemical reaction rates equal high corrosion rates unless protective surface films can form. Corrosion rate also depends on the form of corrosion. Uniform corrosion is often less damaging than localized corrosion.

Several methods can measure the probability and rate of corrosion. Some methods use changes in the physical properties of the sample. Electrochemical methods measure the rates of electrochemical reactions on the sample surface and translate these to corrosion probabilities and corrosion rates. Many test methods are accelerated tests and are suitable for comparison purposes only. The use of electrochemical techniques especially requires experience and common sense when interpreting the results.

References

1. Pourbaix, M., Atlas of Electrochemical Equilibria in Aqueous Solutions, Pergamon Press, Oxford, 1966, Chap. 4.

2. Crolet, J. -L., in Modelling Aqueous Corrosion (K. R. Trethewey and P. R. Roberge, Ed.), Kluwer Academic Publishers, Dordrecht, 1993, pp. 1–28.

3. Andreasson, P. and Troselius, L., The corrosion properties of stainless steel and titanium in bleach plants, KI Rapport 1995:8, Swedish Corrosion Institute, Stockholm, 1995, Chap. 4.

4. Haynes, G., in Corrosion Tests and Standards: Application and Interpretation (R. Baboian, Ed.), ASTM, Philadelphia, 1995, Chap. 8.

5. Groover, R. E., Smith, J. A., and Lennox, T.J., Corrosion 28(3):101(1972).

6. Treseder, R. S., NACE Corrosion Engineer's Reference Book, NACE International, Houston, 1991, pp. 60–100.

7. Szlarska-Smialowska, Z., Pitting Corrosion of Metals, NACE, Houston, 1986, Chap. 3.

8. Ives, D. J. G., in Reference electrodes – Theory and Practice (D. J. G. Ives and G. J. Janz, Ed.), Academic Press, New York, 1961, Chap. 7.

9. Eriksrud, E. and Heitz, E., in Guidelines on electrochemical corrosion measurements (A. D. Mercer, Ed.), The Institute of Metals, London, 1990, Chap. 4.

10. Qvarfort, R., Corrosion Science 28(2):135(1988).

11. Kelly, R. G., in Corrosion Tests and Standards: Application and Interpretation (R. Baboian, Ed.), ASTM, Philadelphia, 1995, Chap. 18.

12. Fontana, M. G., Corrosion Engineering 3rd ed., McGraw-Hill, New York, 1987, Chap. 4.

13. Kearns, J. R., in Corrosion Tests and Standards: Application and Interpretation (R. Baboian, Ed.), ASTM, Philadelphia, 1995, Chap. 19.

14. Lisagor, W. B., in Corrosion Tests and Standards: Application and Interpretation (R. Baboian, Ed.), ASTM, Philadelphia, 1995, Chap. 25.

CHAPTER 4

Corrosion forms

1	Uniform corrosion	124
2	Galvanic corrosion	128
3	Flow-induced corrosion	132
4	Pitting and crevice corrosion	137
5	Intergranular corrosion	149
6	Stress corrosion cracking and corrosion fatigue	153
7	Selective leaching	159
8	Hydrogen damage	161
9	High temperature corrosion	163
10	Summary	170
	References	172

CHAPTER 4

Jari Aromaa

Corrosion forms

Classification of corrosion forms usually uses the appearance of damage they cause on a metal. Visual examination can identify most corrosion forms although the corrosion cause may remain unclear. The difference between corrosion form and corrosion cause is important. The corrosion form will usually be a basic type discussed later in this chapter, but the corrosion cause is the fundamental reason for corrosion damage. For example, the corrosion form might be pitting corrosion, but cause might be chloride ions introduced into the system. Recognizing corrosion forms in industrial equipment provides valuable clues about the cause of corrosion. Corrosion effect is merely a change in any part of the corrosion system caused by corrosion. Corrosion damage is corrosion effect that causes impairment of the function of the construction material, the environment, or the technical system. Corrosion failure is damage characterized by total loss of function of the technical system. No universally accepted list of corrosion forms exists since many of them interrelate and the terminology has changed over the years. Obsolete historical terms relating to specific cases may describe basic corrosion forms. Fontana and Greene introduced in the 1960s the "eight forms of corrosion." This classification is still valid and probably the most frequent[1].

The most fundamental classification is general or localized corrosion. In general corrosion, the entire metal surface is subject to a corrosive environment causing a uniform weight loss or thinning. If no single area of the metal surface corrodes preferably, general corrosion is called uniform corrosion. Uniform corrosion causes the greatest material loss on a tonnage basis, but localized corrosion causes more rapid failures. Localized corrosion is generally a problem of passive metals. The passive surface film protects most metal from uniform corrosion, but localized corrosion occurs in places where the passive film is weaker or the corrosive environment is more severe. For example, an inclusion on a stainless steel surface can result in a weak surface film. Dirt on a metal surface or a crevice in a flange joint can result in a more severely corrosive environment. Localized corrosion can have macroscopic and microscopic forms. Macroscopic pits and crevices are visible on the surface, but microscopic localized corrosion forms are barely visible to the naked eye. They proceed primarily inside the metal. Figure 1 shows one classification of corrosion forms. Uniform corrosion and some localized corrosion forms are characterized by uniform access of the oxidant to the surface, but most localized forms develop specific solution chemistries inside the corroding area. Many special cases of corrosion do not fall under any main category. In the following text, discussion of the forms of corrosion will begin with the most general cases and proceed to more specific areas.

CHAPTER 4

Figure 1. Corrosion forms.

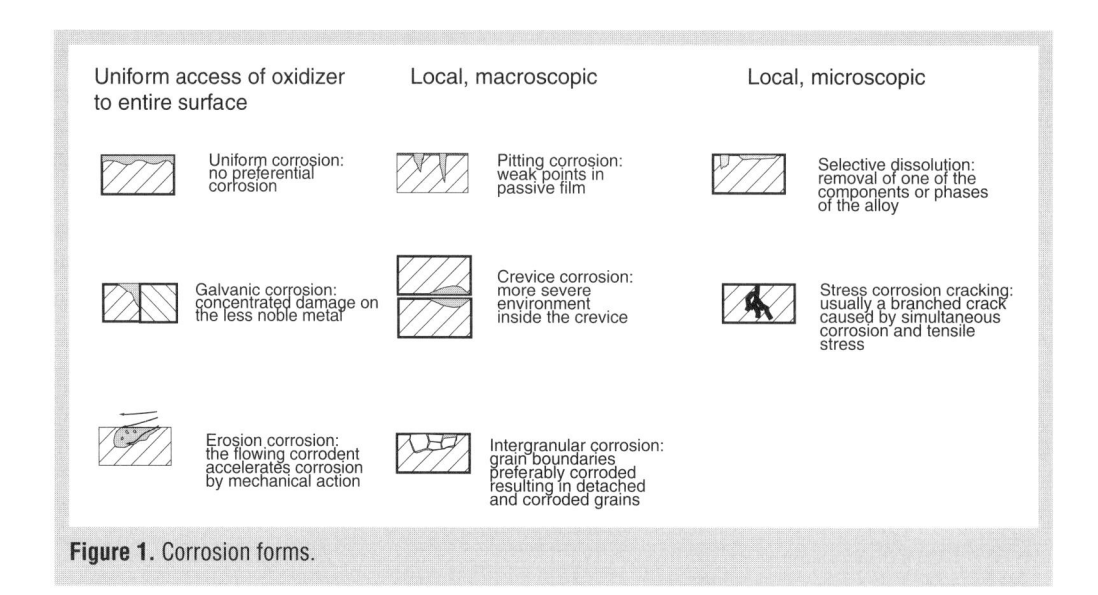

Statistical surveys of corrosion damage in the process industries are available[2-3]. About 50% of the failures in the chemical industry usually relate to corrosion. For the pulp and paper industry, the authors have estimated that the percentage of corrosion failures is about 30%–40%. Approximately 50% of these relate to some type of environmentally induced cracking.

1 Uniform corrosion

The term uniform corrosion means a simple electrochemical dissolution of a metal in corrosive solution. Uniform corrosion occurs over a large area of the metal surface. The surface of the metal is divided into microscopic anodic and cathodic areas resulting in miniature corrosion cells. The operation of these cells changes constantly. New cells start, old cells stop reacting, and anodic and cathodic areas switch places. This is primarily due to preferential corrosion of microstructural active areas and accumulation of corrosion products in some areas. The oxidant has good access to the entire surface area so no preferential anodic reaction occurs. The initially less resistant areas first dissolve. As the corrosion process progresses, the environment may change due to material dissolution or precipitation from the solution. The corrosion products may have a large volume. Corrosion products form on the cathodic areas and may be more noble than the original metal. This can cause the change of anodic and cathodic areas. Then previously cathodic areas that were not attacked begin to corrode as Fig. 2 shows. The corrosion therefore seems uniform to the naked eye. The most typical example of uniform corrosion is atmospheric corrosion. Active dissolution in acids is also often uniform since no surface films or reaction product deposits form. A common case of uniform corrosion is dew point corrosion caused by condensing acidic vapor. Transpassive dissolution caused by incorrect anodic protection or external leakage currents may also show rather uniform morphology.

Corrosion forms

Uniform corrosion is usually not a difficult problem for the designer. Knowing the corrosion rate allows calculation of the necessary corrosion allowance. Corrosion allowance is the planned design life multiplied by the expected annual corrosion rate. Adding the corrosion allowance to the minimum wall thickness means that sufficient material thickness remains even at the end of the planned design life. Corrosion allowance has use only if the metal undergoes truly uniform corrosion. Specifying corrosion allowance may not be feasible for small wall thicknesses, if corrosion results in pits deeper than the uniform corrosion damage. In practice, design of all structures, tanks, and equipment uses a corrosion allowance, but actual corrosion rates are rarely available for correctly specifying the corrosion allowance. This can cause unexpected corrosion failures. Laboratory tests will usually give higher corrosion rates than those actually found. An acceptable corrosion rate for an inexpensive material such as carbon steel may be about 0.2–0.3 mm/year. As a rule of thumb, a corrosion rate of 0.01 mm/year indicates that no harmful corrosion will happen, 0.1 mm/year requires maintenance or corrosion allowance, 1 mm/year should eliminate a material unless good economical reasons exist for its use, and 10 mm/year is totally unacceptable. With proper design and corrosion allowance, little maintenance service is necessary.

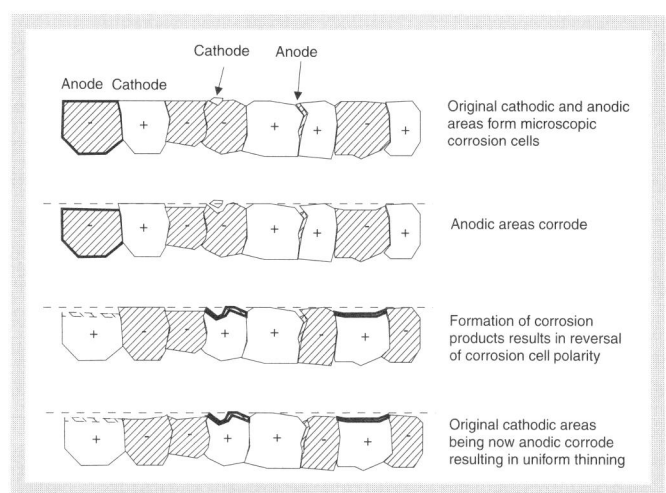

Figure 2. Schematic expression of uniform corrosion with random distribution of microscopic anodic and cathodic areas.

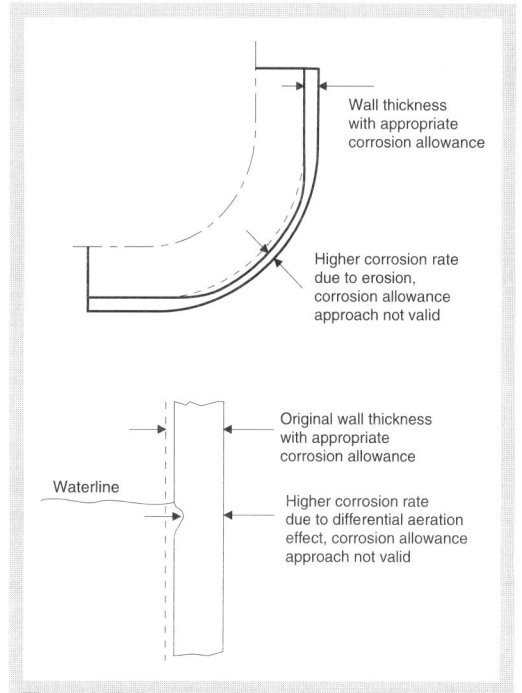

Figure 3. Flow induced corrosion or differential aeration can result in higher corrosion rates and cause the corrosion allowance approach to be invalid.

For more resistant and therefore more costly materials, the uniform corrosion rates must be lower for economical application. Specifying a more alloyed material is not a trouble-

CHAPTER 4

free solution, since more alloyed material may show higher localized corrosion rates than the uniform corrosion rates of less alloyed materials.

Uniform corrosion must be truly uniform for the corrosion allowance approach to be valid. Flow accelerated uniform corrosion in a pipeline elbow or differential aeration at waterline can result in higher corrosion rates as Fig. 3 shows.

The general corrosion of iron and structural steel may be considered uniform, although it is not always uniform in the classical sense. This is because the entire surface does not corrode at an equal rate. Predictable corrosion rates exist for atmospheric corrosion and single phase corrosive environments such as cooling waters, storage and handling of some industrial chemicals, etc. The uneven corrosion that follows the mechanism of uniform corrosion but results in clearly separated, deeply corroded areas is pitting corrosion. Do not confuse this with the autocatalytic corrosion process associated with passivating metals. This type of corrosion can occur in acid solutions and steam condensates contaminated with oxygen for example. Condensing combustion gases can form acidic solutions that cause uniform corrosion. This is dew point corrosion. Stainless steels are usually not susceptible to uniform corrosion since the passive film protects them. In strong solutions of halogen acids and in some organic acids, stainless steels may not passivate. In very oxidizing solutions, transpassive dissolution can occur. With very hot alkaline media, stainless steels can corrode uniformly. Nickel alloys may corrode uniformly in unsuitable applications. Ni-Mo alloys developed for reducing media may therefore show rapid corrosion in the presence of oxidizing species. As another example, titanium may show uniform corrosion in hot formic acid[4].

In pulp and paper processing, most outdoor constructions will show uniform corrosion without proper coating. The same applies to indoor machinery, steel foundations, and framing. The most common example of uniform corrosion in pulp and paper manufacture is corrosion of mild steel in kraft digesters. The corrosive environment is a complex mixture of hydroxide and sulfur compounds. The mixture of sodium hydroxide and sodium sulfide is reducing, but dissolved oxygen, thiosulfate, and polysulfides will act as oxidizers. Depending on the oxidizers and alkalinity of the solution, carbon steel can passivate. The high alkalinity of the solution helps to maintain the passive state. The complex balance of caustic and sulfur chemistry is the main reason for the variances in reported corrosion rates. Uniform corrosion can occur if the alkalinity is too high. Passive films will not form, or they will dissolve. If the concentration of oxidizing sulfur compounds is too low to start passivation, uniform corrosion may result. Brownstock washing equipment can use carbon steel if alkalinity remains sufficiently high to enable passivation. Carbon steel used in green liquor piping may show uniform corrosion if the liquor is well aerated and has a high temperature and chloride content. In this case, the cathodic reaction is strong, and chlorides prevent formation of protective films. Corrosion resistance of carbon steel in white liquor also depends on the complex caustic and sulfur chemistry. Risk of uniform corrosion has resulted in stainless steel replacement.

The example in Fig. 4 uses a brownstock washer. The construction material of the manhole side and support chain were carbon steel. The corroded parts show typical deep grooves caused by an acidic environment. The cause of corrosion in this case was not excessively acidic conditions. Analysis of the steel composition showed that the silicon content was high. The construction material was common killed steel not a specified low-silicon grade.

The corrosion of carbon steel in digesters is intrinsically uniform. The carbonate scale precipitation, liquor circulation, and loading and emptying procedures will cause certain areas of the digester to corrode more rapidly than others. Acid cleaning to remove carbonate scales from a digester can cause corrosion as large pits on areas where cleaning acid remains or washing is insufficient. Noncorrosive or inhibited cleaning acids should be used[5].

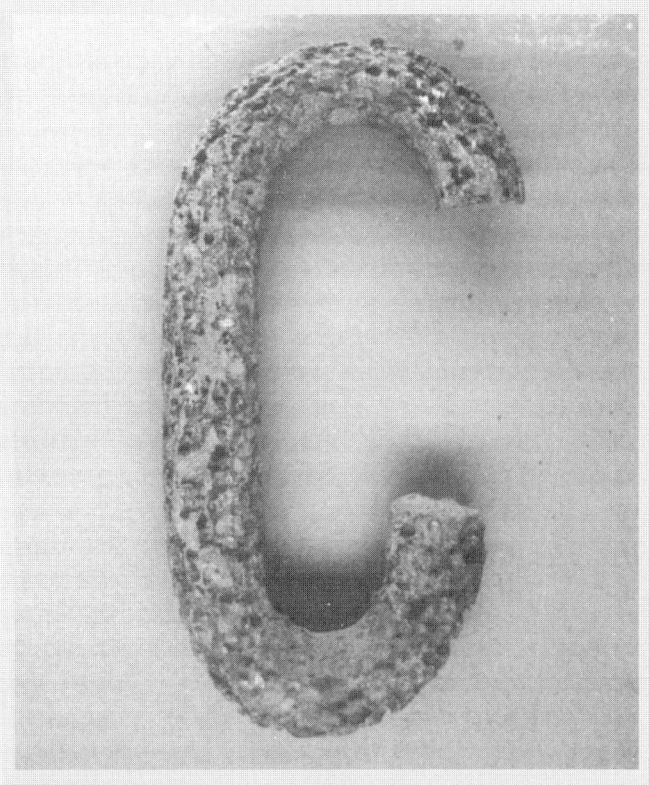

Figure 4. Uniform corrosion in a brownstock washer.

Black liquor evaporators made from less alloyed austenitic stainless steels such as Types AISI 304 and AISI 316 may show uniform corrosion due to the hot alkaline media. The corroded surfaces can be smooth and even shiny after removal of deposits. The corrosion rate of a stainless steel may increase rapidly when approaching the boiling point of the alkaline solution. More concentrated alkaline solutions boil at higher temperatures. Then high temperature with high alkalinity can cause uniform corrosion.

The prevention of uniform corrosion includes all common and straightforward methods. The initial, obvious methods are to select a more corrosion resistant material and use a coating such as a sacrificial hot dip galvanizing material or paint. Changing the environment is sometimes possible by removing the corroding material such as oxygen or by using inhibitors. Cathodic and anodic protection are often suitable corrosion prevention methods. For example, using polysulfides as passivating inhibitors and anodic protection has prevented uniform corrosion of digesters.

2 Galvanic corrosion

The mechanism of galvanic corrosion is similar to uniform corrosion, but the corrosion proceeds at preferred active areas. The term galvanic corrosion has often been restricted to the action of bimetallic corrosion cells where corrosion occurs primarily at a dissimilar metal joint. In this book, "galvanic corrosion" means bimetallic corrosion or contact corrosion where corrosion rate of the less noble material increases due to galvanic coupling. The corroding active metal can also couple to a conductive nonmetal such as a carbon impregnated gasket material. As in uniform corrosion, the supply of the oxidant is uniform to the structure surface. If the metals were uncoupled, they would corrode uniformly at their typical rates. Galvanic corrosion assumes that the cathodic current density remains the same as in the uncoupled case. Due to the coupling, the less noble metal delivers all or most anodic current. The more noble material acts solely as a cathode. To maintain the equality of anodic and cathodic currents, the anodic dissolution rate must increase. Since the structure consists of different metals, the most active of these will become anodic and corrode at an increased rate. Simultaneously, the corrosion rate of the remainder of the structure decreases or becomes zero.

The electron flow in the galvanic couple causes the more active and more noble members to polarize and their potentials approach each other. Depending on the materials and environment, either can polarize more easily. If the potential of the more noble material changes easily, the shift of the more active material is small and galvanic corrosion is not as great as it might otherwise be. If the potential of the more noble material does not change easily, the potential of the more active material changes significantly and galvanic corrosion becomes appreciable. The principles of galvanic corrosion are very important since they extend to various forms of localized corrosion such as pitting and crevice corrosion, selective leaching, and intergranular attack.

A special case of galvanic corrosion is the exchange reaction between two dissimilar metals known as cementation or displacement. When process liquor passes from one vessel to another, carryover of corrosion products from a noble material vessel and their deposition on a downstream vessel constructed from less noble material may produce cementation problems. In this case, dissolved noble metal ions reduce and precipitate on the more active metal. This precipitation reaction consumes electrons that come from the dissolution of the more active metal. The resulting noble metal deposits provide cathodic sites for further galvanic corrosion. Typically, deposition of copper on ferrous material surfaces may lead to severe galvanic corrosion. The small copper deposits cause miniature galvanic cells and thus increase corrosion rate of ferrous metal around them.

The principle of different nobilities does not apply only to solid materials. Various types of concentration cells can produce areas with different electrochemical potential on the same surface. The most common types of concentration cells are oxygen and metal ion concentration cells. For example, an oxygen concentration cell may develop at a stagnant vapor and liquid interface to accelerate corrosion attack. This type of corrosion will produce waterline corrosion. The concentration cell will develop since oxygen is readily available at the vapor and liquid interface. The oxygen has increasingly restricted access to levels further from the surface. This leads to a preferential dissolution of the tank wall immediately below the waterline. Filiform corrosion is also a unique type of galvanic corrosion that occurs under painted or plated surfaces with poor adhesion and under gaskets. Thin lacquer films and rapidly drying paints are most susceptible. This type of corrosion occurs when moisture permeates the coating. The morphology of filiform corrosion is a radial "worm-like" corrosion path starting from a central initiation point. The head of the corrosion filament is under low oxygen content, and the tail of the filament has access to air. This causes a potential difference and results in corrosion propagation at the filament head.

The severity of galvanic corrosion depends on the corrosion potentials and uniform corrosion rates of the metals, relative surface areas of active and noble metals, and the conductivity of the electrolyte. A larger difference between corrosion potentials of the active and noble metal usually causes a higher risk of galvanic corrosion. Using the galvanic series, i.e., tabulated corrosion potential values in a known environment, often estimates this risk. The galvanic series assumes freely corroding materials that are not in contact with other materials. No hard rules on the value of potential difference and risk of galvanic corrosion are available. For example, an actively corroding metal could have a Tafel slope of 50–60 mV. A potential shift of 50–60 mV would then cause the corrosion rate to change by a factor of ten. If the more noble metal polarizes easily, a potential difference of several hundred millivolts will not necessarily lead to significantly higher corrosion rate. Metals and alloys that form passive films will show varying potentials. Their corrosion rates are difficult to quantify. The initial potential difference for galvanic corrosion will normally decrease with time because of polarization at the anodes and cathodes. This decrease in the potential difference then reduces the galvanic corrosion current density and lowers anode corrosion rate. Galvanic series do not give information on the initial polarization behavior or long term behavior in contact.

Figure 5 shows the theoretical effect of varying surface areas and conductivities. If the area of active metal is large compared with the noble metal, the increase of total anodic current is small and that of the anodic current density even smaller. In this case, the galvanic coupling does not cause a significant increase in the corrosion rate of the active metal. If the area of the active metal is small compared with that of the noble metal, the anodic current increases and anodic current density increases even more. This results in much higher corrosion rate since current density and not absolute amount of current determines the corrosion rate as Eq. 1 shows:

$$I_a = I_c \rightarrow i_a \cdot A_a = i_c \cdot A_c$$
$$i_a = i_c \cdot \frac{A_c}{A_a} \quad (1)$$

The conductivity of the electrolyte affects the severity of corrosion attack. A low conductivity solution will cause corrosion only near the joint. This usually results in deep, narrow cavities. High conductivity solution results in a shallower attack over a large area. The mass loss in a high conductivity electrolyte may therefore be higher, but

CHAPTER 4

the corrosion attack may be more dangerous in a low conductivity solution due to deeper penetration. The attack caused by galvanic corrosion decreases exponentially when moving away from the joint of dissimilar metals. The decrease of the attack depends on conductivity. To some extent, estimating the effect of conductivity is possible with the Wagner number. This is the polarization resistance of the active metal divided by solution resistivity, $W = R_p/\delta$. The Wagner number has units of length, $R_p[\Omega \cdot cm^2]/\delta[\Omega \cdot cm] = cm$. The Wagner number is the characteristic distance over which current will flow. A large value indicates that potential does not change much along the surface. High polarization resistance means low current density so the active metal does not "draw" much current. A low resistivity means that potential difference is not wasted in solution resistance, and the current can "see" longer over the surface. A high Wagner number (10^3–10^4 or more) therefore suggests wide, shallow attack, and a low number implies narrow, deep attack.

Galvanic corrosion of the more anodic material in a couple may be general or localized depending on the geometrical configuration, nature of materials, and nature of corrosion product films. Galvanic corrosion may occur between various material combinations. A combination of dissimilar metals is common in engineering design. Metallic coatings also provide a type of dissimilar metal couple. Noble metal coatings require consideration as cautiously as coupling of massive metals. Various nonmetallic materials are conductors and may act as cathodes in galvanic couples. Carbon and carbon containing materials are typical examples. Conductive surface films can also create a galvanic couple. For example, mill scale on carbon steel or heat tint on stainless steel are non-

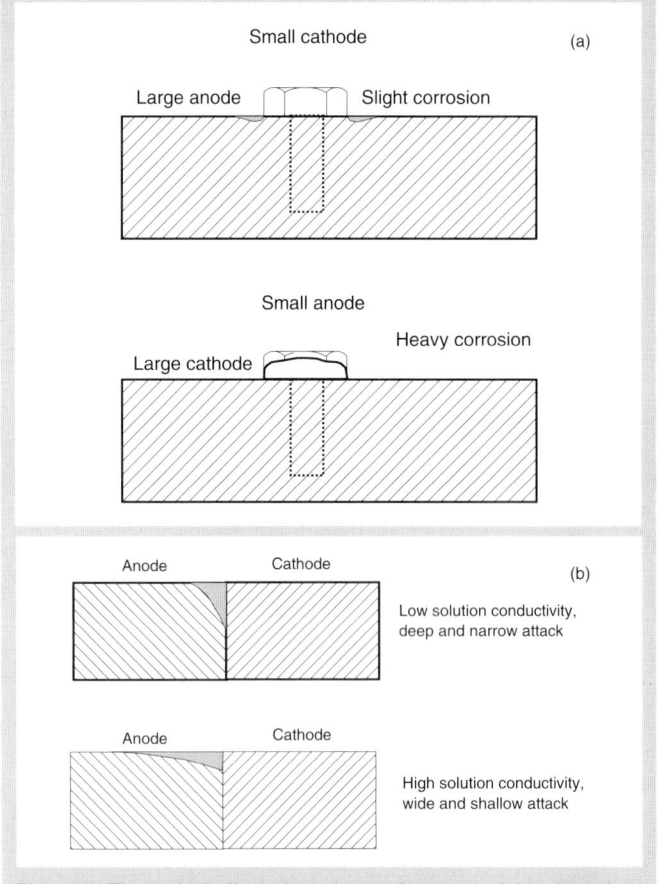

Figure 5. Theoretical effect of varying surface areas (a) and conductivities (b) on galvanic corrosion.

protective oxide layers more noble than the respective metal. These scales may become effective cathodes in acidic solutions to accelerate the total hydrogen evolution rate that requires compensation by higher metal dissolution rate. Iron sulfides on steel behave similarly.

Iron and structural steels are active materials and are usually anodic in galvanic couples. Minor compositional differences between ferrous materials may result in galvanic corrosion due to differences in anodic polarization behavior. Although the materials would corrode at about equal rates under cathodic control, coupling of the two results in a galvanic cell. The behaviors of stainless steels depend on their state. A stainless steel in a passive state is often the more noble metal. Protective film decomposition caused by aggressive anions may change stainless steel to the more active species. Stainless steels in a passive state will polarize easily. Their galvanic effect on more active materials can therefore be small. Nickel alloys are usually very noble and cause corrosion of less noble metals, but Ni-Cr alloys behave like stainless steels. Some nickel alloys adopt corrosion potentials that lie between the active and passive potentials of stainless steels. They can therefore cause localized corrosion of stainless steels. Titanium is a very noble metal. It is almost always the cathode in a galvanic pair. Titanium alloys are usually not susceptible to galvanic corrosion. When acting as cathodes, the strong hydrogen evolution on titanium alloys may cause hydrogen embrittlement.

As in every field of engineering, galvanic corrosion is possible in the pulp and paper industry whenever dissimilar materials connect. Common cases are noble metal parts connected in massive carbon steel. This usually gives a less dangerous surface area ratio, but corrosion of carbon steel next to a stainless steel fastener could result in dislodging. Repair and retrofit with higher alloyed steel grades can result in galvanic corrosion of original, less alloyed parts. Figure 6 shows a classic example of galvanic corrosion. The shaft of the clarifier originally used stainless steel. Lengthening used a welded piece of carbon steel tube neglecting the risk of galvanic corrosion. Figure 6 shows that corrosion is highest near the joint of dissimilar metals.

Figure 6. Galvanic corrosion at a joint of carbon steel and stainless steel.

CHAPTER 4

The first step in galvanic corrosion prevention is proper design. Galvanic couples must be avoided especially if the corrosion potential difference is large. The galvanic series most often published shows materials in seawater. This galvanic series is specific to seawater at nearly ambient conditions. Temperature and the presence of other chemical species can greatly influence the order of materials. Differences in environmental conditions can even reverse the polarities of galvanic couples. If different metals must be used, metals with nearly equal corrosion behavior should be selected. The corrosion behavior of the dissimilar metals can be tested beforehand with electrochemical and immersion tests. If galvanic couples cannot be avoided, the parts made from the less noble material should be thicker or easily replaceable. Large cathode areas must be avoided. Painting can decrease cathodic areas. Painting only the anodic areas is risky, since damages in the paint film will expose a very small anodic area where corrosion rates can be very high. The distance between dissimilar metals can be increased, but this is usually not practical. Threaded joints between materials far apart in the electrochemical series can be risky. A preferential method is to use welding or brazing with an alloy more noble than at least one of the metals to be joined. Porous materials may absorb moisture that can possibly create new galvanic pairs. Active metals have use to give sacrificial cathodic protection, but this is not a trouble-free solution. Even galvanizing can cause problems because zinc may become more noble than iron at temperatures more than 60°C. Inhibitors have use to decrease corrosivity of the environment. This will also reduce galvanic corrosion. Cathodic protection is a useful method against galvanic corrosion.

A common error in isolation of dissimilar metals happens in bolted joints. Insulating washers under bolt heads and nuts are assumed to provide insulation, but electrical contact remains since the shank of the bolt touches both metals. Putting an insulating sleeve over the bolt shank solves the problem. Figure 7 shows proper insulation. If complete insulation is not possible, a resistive paint or plastic coating on the cathodic areas will help somewhat.

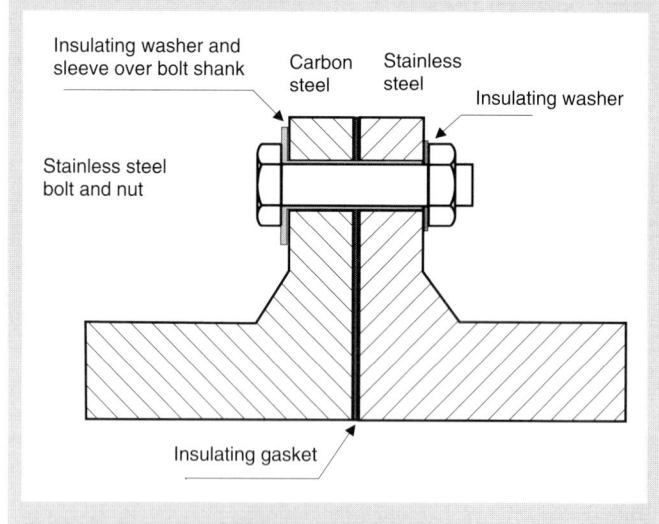

Figure 7. Proper insulation of a bolted joint.

3 Flow-induced corrosion

The flow of corroding material or relative movement of solid surfaces can increase corrosion rates significantly. Erosion corrosion and cavitation are due to flowing solution. Small relative movement of contacting surfaces with a flow film between them can cause fretting. In erosion, the metal erodes mechanically. In erosion corrosion, erosion removes the protective oxide film from the metal surface resulting in a repeating reoxidation and film removal cycle. This process can lead to local thinning or penetration of the wall. Erosion corrosion or flow-induced corrosion is due to essentially uniform corrosion that the high velocity of the electrolyte flow enhances. If the electrolyte were stagnant or flowing slowly, corrosion rates would be modest or low. The rapid movement of the electrolyte physically removes possible protective films exposing the metal again to accelerate corrosion. Suspended particles in the solution will enhance the eroding effect. Solid particles in a flowing solution or droplets in a gaseous flow will increase the extent of mechanical attack on the surfaces. Another term for this form of corrosion is impingement attack. The fluid actually bombards the surface of the equipment.

The relative importance of corrosion and mechanical wear can be difficult to evaluate. Figure 8 shows schematically how flow rate can influence corrosion rate. The figure uses the assumption that the cathodic reaction is mass transfer controlled and corrosion rate is therefore controlled by cathodic reaction. In this case, the corrosion rate depends on the supply of the cathodic reactant to the surface. This rate is proportional to the square root of flow velocity. Up to a certain critical velocity, only electrochemical reactions cause corrosion. At this velocity, the mechanical stress caused by the flow damages the surface film. Small damages in the surface film cause strong galvanic effects. The extent of damage increases rapidly with increasing flow rate causing an increase in the corrosion rate. Corrosion rate increases with the flow rate until reaching a constant value in the film-free situation. At this velocity range, repassivation is not possible.

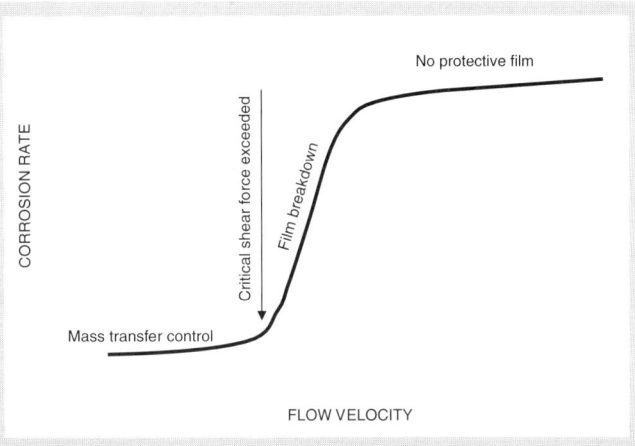

Figure 8. Corrosion and erosion dependence on flow rate assuming that the corrosion rate is controlled by cathodic reaction.

Attack usually follows the surface flow direction and turbulence. Typical places for flow-induced corrosion are tube constrictions, tube joints, elbows, valves, and pumps where surface flow velocity changes rapidly. Two-phase flow can also cause severe corrosion when gaseous phase and condensate droplets are present together. The damage types of flow-induced corrosion are numerous. They may be grooves, waves, drop-shaped or horseshoe-shaped pits, and depressions. The damage may cut under the metal surface in an upstream or downstream direction. Impingement corrosion usually occurs near the entrance to the tubes. This takes the form of pitting or grooving with the remainder of the tube showing no sign of corrosion. Similar attack may occur in the vicinity of an obstruction inside a pipe.

CHAPTER 4

Fluid flow is laminar or turbulent. Laminar flow is smooth, and the flow rate is slowest at the metal surface with increasing value when moving toward the bulk electrolyte. As the electrolyte flows, it receives constant replenishment. The supply of cathodic reactants and aggressive ions is constant, and dissolved metal ions are removed from the surface. These factors can increase dissolution rates and suppress formation of protective films. An increased supply of the cathodic reactant may increase the corrosion rate above the critical current density of a passivating metal allowing it to passivate. Laminar flow usually does not cause excessive mechanical stresses to surface films unless the solution contains solid particles. Increase in laminar flow rate can also remove poorly adhering and weak corrosion product films such as steel in flowing white liquor. As the laminar flow rate increases, the flow becomes turbulent after exceeding a critical rate. Flow around the surface becomes unpredictable, and impact of fluid and particles will cause significant stress to the material surface. Turbulent flow may be due to sudden diameter or direction change, surface discontinuity such as a badly fitting joint, deposit, or corrosion damage. A typical example of the effect of turbulent flow is damage to heat exchanger tubing. When the flow is forced from the head to tubes, sudden diameter change causes turbulence. The inlets of the tubes are under turbulent flow and corrode until the fluid develops laminar flow again farther in the tube. Figure 9 shows schematic examples of poor design causing turbulence. In all these cases, turbulence occurs when solution moving at high flow rate meets a discontinuity. A low flow rate could prevent turbulence formation.

Cavitation is a special form of erosion corrosion. It results from extremely high flow velocity or changes of surface temperature in a liquid and gas system. A sudden increase in the velocity causes pressure reductions that nucleate vapor bubbles that then implode on the surface when pressure increases again. Implosion produces a

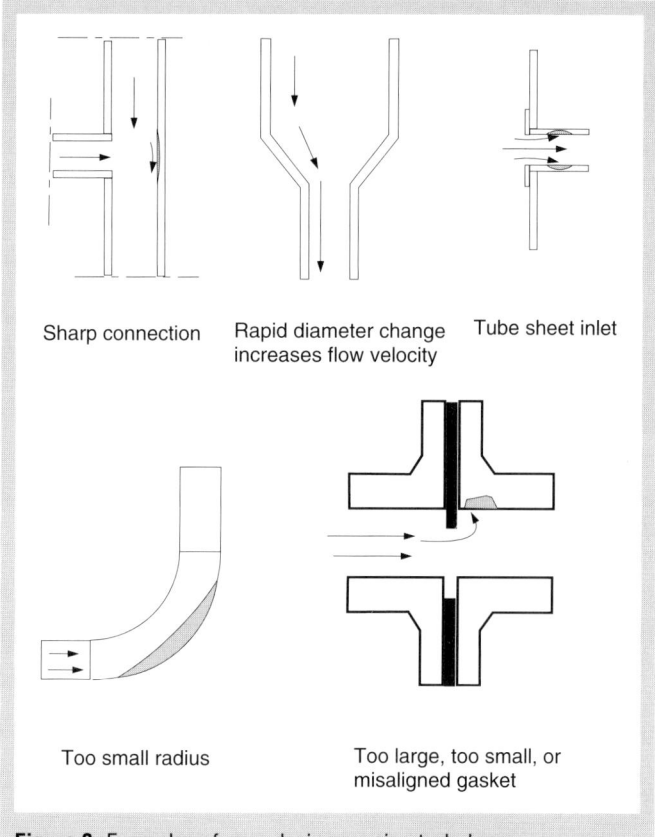

Figure 9. Examples of poor design causing turbulence.

pressure burst that breaks the surface film. The implosion can even remove particles. To cause damage, the bubbles must adhere on the surface before collapsing. Cavitation is a major problem in the chemical industry because it often affects pumps. It occurs in pump impellers, turbine and propeller blades, and pipes where large pressure changes occur. The morphology of cavitation damage is rough pits. Cavitation often appears as rough cavities on the low pressure side, and erosion appears as smoother cavities on the pressure side. Corrosion can increase cavitation, but corrosion is not necessary as with normal erosion corrosion.

Fretting is a wear process accelerated by corrosion. It is the opposite to normal erosion corrosion. The wear results from contact between two solid surfaces. Oxide films resulting from corrosion processes deform under pressure. Even slight motion causes abrasive wear. A corrosive environment may accelerate fretting by increasing the corrosion product formation rate. Fretting corrosion primarily occurs on surfaces whose design does not allow them to move relative to each other and start to oscillate during operation. The relative movement between the mating surfaces can be as small as 1 nm.

Alloys that are susceptible to erosion corrosion generally have low strengths and often have low corrosion resistance even if fluid flow does not affect them. Carbon steel and copper alloys are typical examples. Erosion corrosion is also possible with passivating metals if erosion or cavitation continuously removes the passive film. Most metals will be susceptible to flow-induced corrosion under particular flow conditions. Resistance to erosion corrosion is a combination of the natural corrosion resistance of the metal and hardness and ductility of the metal and surface film. If the metal cannot give sufficient backing to the surface film, the film will crack more easily. The material must first resist corrosion caused by the environment. To protect from corrosion, a protective film must form. Protection from erosion also requires withstanding mechanical wear. The flow-induced damages to the protective film depend on film hardness, adherence, and fracture toughness. If the film can form rapidly with low surface concentration of metal ions and is sufficiently strong mechanically, it can protect the metal. The resistance to cavitation depends on alloy hardness, microstructure, and fracture toughness. Factors influencing adhesion of vapor bubbles to the surface may also be important. If the bubbles are not easily adsorbed to the surface, they cannot cause stresses when collapsing. Flow-induced corrosion frequently attacks structural steels. Stainless steels and nickel alloys are seldom attacked unless the design is faulty or service conditions change. Titanium alloys are also usually resistant. If the passive film breaks, the reactivity of these metals can cause rapid attack[6].

Erosion corrosion can happen anywhere solution flow is too rapid or its velocity or direction changes abruptly. For example, the top dome in pulp digesters can suffer from erosion corrosion due to turbulent liquor flow. On carbon steel, this may cause grooves typical to erosion corrosion and wide and shallow pits. Inlets and outlets subject to rapidly streaming liquor often become corroded. Erosion corrosion frequently attacks blow tanks and blow plates. In green liquor containing high amounts of solid particles, erosion corrosion can occur in pumps and other moving equipment. Increase of production

CHAPTER 4

rates above the designed rate will sometimes increase flow rates. This can lead to erosion and corrosion in a digester shell for example. Erosion corrosion can cause serious failure of power plant boiler feed water piping due to wall thinning. A serious aspect of erosion corrosion is that it can lead to conditions of catastrophic failure, i.e., break before leak, when a large section has weakened uniformly. This is especially dangerous when it combines with other forms of material degradation such as creep and forming of graphite. Erosion corrosion has been a recognized problem in steam generation systems for more than 50 years. The first problems were in steam condensate return lines where low pH caused grooving and premature failure. The problem has increased in magnitude because velocities have steadily increased in an effort to improve plant economics. Factors that can influence erosion corrosion include steel composition, water and steam composition, operating temperature, design and component geometry, and flow velocity.

Classical cavitation occurs when the absolute pressure of a moving liquid decreases to or below the vapor pressure of the liquid. Bubbles form because of this pressure drop. In centrifugal pumps, the lower pressures in the impeller eye are due to variations in velocity of the fluid and friction losses as the fluid enters the impeller. The bubbles then sweep outward along the impeller. On the other side of the impeller vane, the pressure may exceed the vapor pressure and cause the bubbles to collapse. The impacts caused by the collapsing bubbles are sufficiently strong to cause minute areas of fatigue on the impeller surfaces. Most process applications require a pump to handle a mixture of components that will have a different vapor pressure or boiling point. Cavitation damage to a centrifugal pump may range from minor pitting to total failure. Most damage usually occurs within the impeller nonpressure side of the vanes. This is the area where the bubbles will normally begin to collapse. The result will be the rough surface of Fig. 10. The original construction of the component used AISI 316 type stainless steel. The impeller usually failed within 2–14 days. After identifying the corrosion cause as cavitation, a very hard titanium nitride coating solved the problem. The coating prevented cavitation damages almost totally. The service life of the impeller was extended to months.

Figure 10. Centrifugal pump impeller damaged by cavitation and improved part coated with titanium nitride, TiN.

Design or material selection can prevent erosion corrosion. The equipment and piping require design such that surface flow velocities are low and turbulence is avoided. Elbows with a large radius, pipes with a large diameter, and smooth dimension changes are preferable. In some cases, the cost of larger piping, valves, and pumps can be higher than the cost of replacing smaller diameter pipe. Modification of electrolyte corrosivity may also prevent erosion corrosion. Sometimes excessive design with corrosion allowance or easy replacement are the most economical measures. Upgrading the material to a more corrosion resistant one will often stop erosion corrosion. The intrinsic corrosion resistance of old and new alloys need not differ much, but the more alloyed grade often develops a mechanically stronger surface film. Hard facing alloys and welded overlays can sometimes be suitable. Cathodic protection can prevent erosion corrosion if the flow rates are so low that electrochemical corrosion is a more important factor than mechanical erosion.

Cavitation prevention principles are identical. The system design should minimize pressure drops. Alloys with harder surface films and very hard metallurgical coatings such as nitrides may be successful. Removal of dissolved gases in closed systems can be effective by reducing bubble nucleation. At high flow rates and in cavitation, the damage is primarily mechanical. Electrochemical protection methods will therefore not help.

Interrupting surface contact and relative motion prevents fretting. Using lubricants and surface coatings to limit contact wear between metals can also prevent fretting. Barrier coatings that reduce transport of a corrosive environment and corrodents such as oxygen to mating surfaces are sometimes useful. Using a more corrosion resistant material will usually not prevent fretting since the nature of abrasive particles and wear resistance of the alloys control this form of corrosion damage. Materials with increased surface hardness may prevent fretting. Reducing the bearing loads on mating surfaces will decrease the wear rate.

4 Pitting and crevice corrosion

Pitting and crevice corrosion are typical corrosion forms of passive metals such as stainless steels. The most common examples of materials susceptible to pitting and crevice corrosion are stainless steels in neutral or acid chloride solutions. Discussing pitting corrosion and crevice corrosion together is convenient because they have many similarities in their initiation and propagation phases. For stainless steels, pitting is the preliminary stage of all localized corrosion. An important factor for the initiation and propagation of pitting and crevice corrosion is formation of a concentration cell, i.e., anodic and cathodic areas surrounded by electrolytes with different compositions. While pitting corrosion occurs on small localized areas on the surface, crevice corrosion occurs in crevices, under deposits, etc. Any shielded area on a metal surface containing a small volume of solution can be a crevice. The critical conditions for pitting and crevice corrosion are high halogen salt concentration, high temperature, and oxidizing substances.

In pitting corrosion, the protective film breaks down in weak areas resulting in a corrosion cell where the exposed metal in the pit is anodic to the undamaged part. The solution inside the pit is usually an acidic concentrated metal chloride solution. In crevice corrosion, the restricted flow conditions lead to a more corrosive solution in the crevice interior compared with the bulk solution outside. The crevice solution will finally become so aggressive that the protective film breaks down. The inside of the crevice becomes anodic to the rest of the structure. Pitting and crevice corrosion are autocatalytic processes. The corrosion process itself produces conditions that will further stimulate the corrosion activity inside the pit or crevice. An important factor in crevice corrosion is the arrest temperature. Crevice corrosion can start if the process temperature increases too much. This corrosion can stop if the temperature decreases sufficiently. Once crevice corrosion begins, the problem is that the operating temperatures are higher than the arrest temperature required to stop it. The corrosion rates inside pits and crevices are several orders of magnitude higher than those estimated by using passive current density. Much structure is often free from corrosion, but pits can penetrate the wall or joints start to leak due to crevice corrosion.

A suggested analogy is that a pit is a miniature crevice. Several theories exist for pitting and crevice corrosion mechanisms. In most cases, the passive film is locally weak or the corrosive environment locally aggressive so the protective film breaks down. Typical features for the initiation of pitting and crevice corrosion are the following:

- Concentration of aggressive species above a critical level
- Critical nucleation or breakdown potential above which corrosion can nucleate and propagate
- Induction period during which the initiation phase produces conditions for sufficiently aggressive passive film breakdown and further corrosion propagation (The induction period for pit initiation is the time to form the first pits on the passive metal.)
- Breakdown of the metal at localized sites.

Potential has a major effect on pitting and crevice corrosion. Figure 11 shows the schematic effect of critical potentials. The critical potential, E_{crit}, gives the susceptibility to localized corrosion initiation. As this potential becomes more noble, corrosion initiation becomes less probable. After initiation, corrosion will propagate until electrode potential decreases below protection potential, E_{prot}. At potentials between E_{prot} and E_{crit}, no new pits will initiate, but existing pits will propagate. Figure 11 shows examples of cyclic polarization curves used for pitting and crevice corrosion testing. In a solution without chloride, the current density increase at high potentials is due to transpassive behavior. This means oxygen evolution by decomposition of water associated with some metal dissolution. In solutions containing chloride, the current begins to increase at lower potentials than the transpassive potential. Formation of corrosion pits accompanies this current density increase. As the chloride concentration increases, the pitting potential moves to lower values, passivation potential moves to higher values, and passive range current density and passivation current density increase. The shift of corrosion potential with increasing chloride concentration is seldom as large as the shift of repassivation potential.

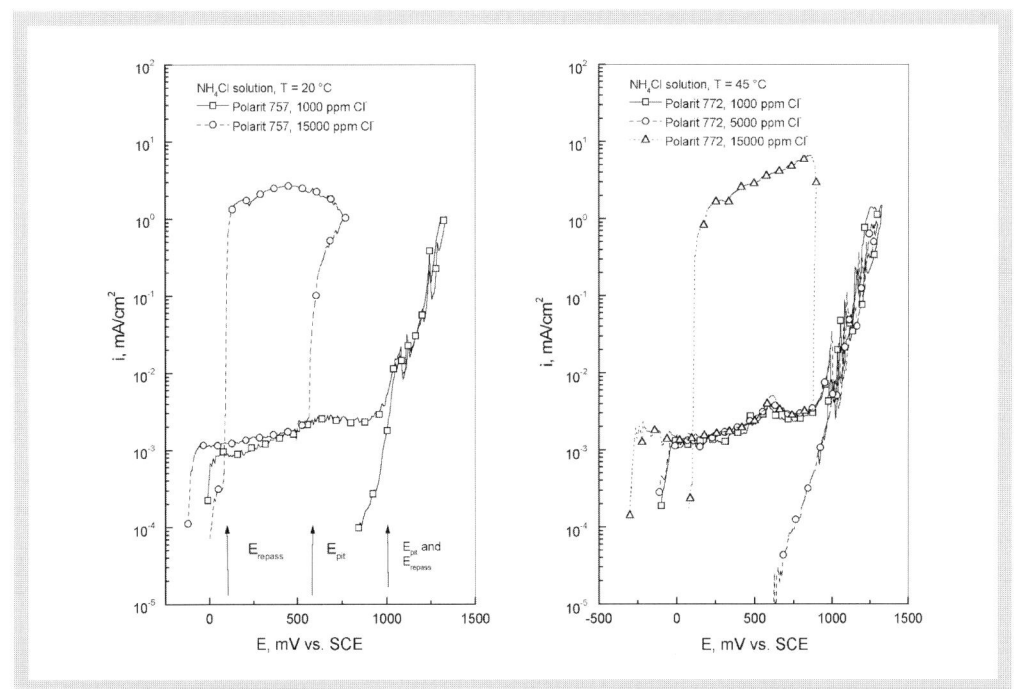

Figure 11. Cyclic polarization curves for 16.0% Cr, 10.5% Ni, and 2.5% Mo minimum stainless steel approximately equivalent to AISI 316 and 16.5% Cr, 12.5% Ni, and 4.0% Mo minimum stainless steel in NH_4Cl solution.

Figure 12 shows the combined effect of aggressive anion concentration, temperature, and potential. For a particular environment, corrosion resistance of a metal may be characterized by very narrow ranges of temperature, anion concentration, and potential above which pitting initiates and below which it will not initiate. The aggressiveness of the anion depends on the metal involved. The chloride ion is the most common aggressive anion causing pitting of iron, nickel, titanium, and their alloys. Sul-

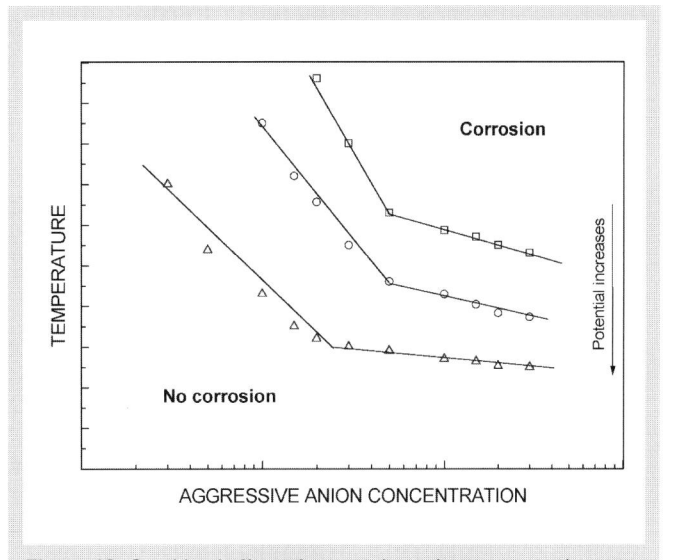

Figure 12. Combined effect of aggressive anion concentration, temperature, and potential on pitting corrosion.

fate may cause pitting of low carbon steels. The role of halides relates to microscopic local destruction of passive film and counteracts its healing. Halide ions strongly influence the anodic reactions enabling metal passivation. With low concentrations, only passive currents increase. Larger concentrations result in localized corrosion, and very high concentrations cause such dense pitting that the corrosion attack becomes uniform. As anion concentration increases, maximum service temperature, pitting, and protection potentials decrease. The potentials often decrease as the logarithm of aggressive ion concentration. Maximum service temperature may also follow this behavior, but usually below some anion concentration the solution becomes clearly less corrosive. At a constant potential higher than the pitting potential, the induction time decreases as chloride concentration increases. At a constant chloride concentration, the induction time decreases as potential increases.

The severity of pitting can decrease due to inhibiting anions in the solution. Several compounds can act as inhibitors in solutions with chloride such as sulfates, nitrates, and hydroxyl ions for steels, stainless steels, and nickel alloys and molybdates for stainless steels. Usually, the compounds that are effective against general corrosion will also decrease pitting corrosion. Cations do not usually affect pitting. The inhibiting anions require a minimum concentration to be effective. The relationship between anion concentrations follows this equation:

$$\log[aggressive\ anion] = a + b \cdot \log[inhibitive\ anion] \qquad (2)$$

In the presence of inhibiting anions, the concentration of the aggressive anion must be higher to produce pitting. The inhibition mechanism is either competitive adsorption or repair of passive film. The adsorption of inhibiting anions decreases surface concentration of aggressive anions or repairs the passive film by increasing surface film pH, formation of insoluble compounds, or enhanced passivation of the metal surface[7]. Depending on complexity of the solution composition, estimating solution aggressiveness can be difficult.

Initiation of pitting probably happens by adsorption of aggressive anions followed by displacement of oxygen in the passive film, by migration and penetration of aggressive anions to the metal and film interface, or by aggressive anions affecting the competing breakdown and repair of passive film. Higher anion concentration in the bulk solution and higher anodic potential enhance anion adsorption. Initiation of crevice corrosion follows the models of pitting. A difference is that the environment becomes more aggressive with time. A crevice can operate as a corrosion site only when it is sufficiently large to allow liquid to enter but not sufficiently wide to allow liquid replenishment. Pit propagation probably depends on dissolved metal hydrolysis reactions with the pit growth rate controlled by hydrogen evolution rate, on formation of a salt film layer inside the pit with salt film dissolution rate controlling pit growth, or on mass transfer rates in and out of the pit[8]. Propagation of crevice corrosion depends on the rate of cathodic reaction outside the crevice. If the pH is sufficiently low, it also depends on hydrogen evolution inside the crevice. As with pitting, the propagation rate depends on reaction kinetics, mass transfer, and geometry.

The traditional crevice corrosion mechanisms have used the assumption that acidification and accumulation of aggressive anions cause the breakdown of passive film. This model is the deoxygenation-acidification mechanism. Initially, the solution in the crevice has the same composition as the bulk solution outside. The corrosion reactions are metal dissolution and oxygen reduction. They consume equal amounts of electricity. The separation between anodic and cathodic sites is negligible. Since transport of oxygen to the crevice is limited, oxygen is consumed in the crevice faster that it can be replenished. The cathodic reaction rate inside the crevice will decrease, and the interior of the crevice will become progressively anodic. Finally, the crevice will become anodic while cathodic reaction occurs outside the crevice. Anodic reactions inside the crevice produce metal ions that hydrolyze to produce metal hydroxides and hydrogen ions. To maintain electric neutrality, the positive hydrogen ions must balance the negative anions such as harmful chloride ions that migrate to the crevice. This hydrolysis and anion migration causes acidification and ion concentration increase in the crevice solution[9,10]. Figure 13 shows the deoxygenation-acidification mechanism.

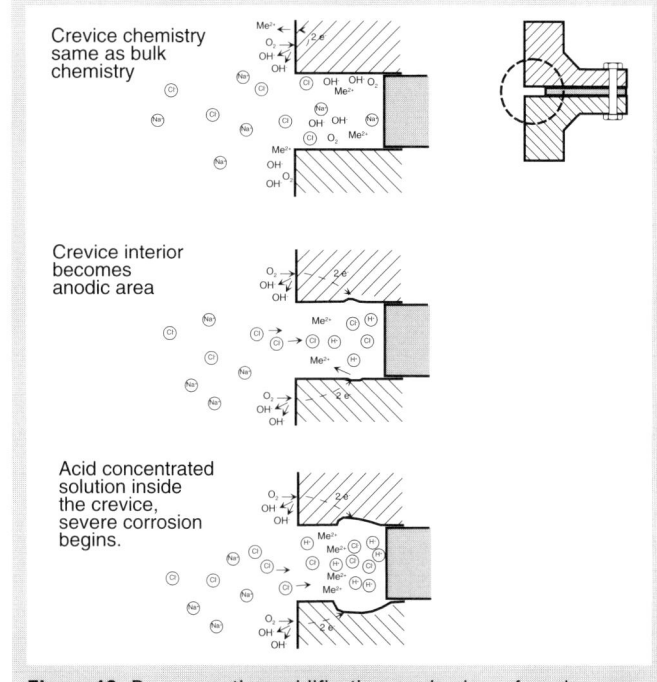

Figure 13. Deoxygenation-acidification mechanism of crevice corrosion in neutral aerated solution.

The extent of acidification depends on the dissolving metal ion diffusion, hydrolysis rates, and the bulk solution pH. Acidification of the crevice solution commonly occurs for neutral or alkaline solutions but seldom for strongly acid solutions. Table 1 gives an explanation for this. Different metals have different minimum pH values that the hydrolysis reaction can theoretically produce. Ferric iron and chromium ions have a significantly higher hydrolysis constant than most other metals found in the constructional alloys. They therefore cause strong acidification. The increase of metal ion concentration and subsequent acidification will continue until the solubility product of the hydroxide, chloride, or other salt occurs. The pH values in Table 1[11] can be calculated from the following:

$$a\ Me^{z+} + b\ H_2O = c\ metal\ hydroxide + d\ H^+ \tag{3}$$

CHAPTER 4

Calculation of the equilibrium constant of Eq. 3 assumes that the only reaction product is solid metal hydroxide:

$$K = \frac{[metal\ hydroxide]^c \cdot [H^+]^d}{[Me^{z+}]^a \cdot [H_2O]^b} = e^{\frac{-\Delta G}{R \cdot T}} \qquad (4)$$

Rearranging Eq. 4 gives the following theoretical equilibrium pH:

$$pH = \frac{-\log(K) - [Me^{z+}]^a}{d} \qquad (5)$$

Note that the equilibrium pH values will vary significantly depending on the reaction products and thermodynamical data used. If the reaction product is not hydroxide but some hydroxide ion instead, its concentration will also affect the equilibrium pH. The reaction of hydroxide ions to a hydroxide requires hydroxyl ions that will further decrease pH of the crevice solution[9]. Exact calculation of solution pH inside the crevice is hardly possible since ascertaining the actual reaction products is difficult, diffusion rates of ions are unknown, and reactions of H^+ ions may occur inside the crevice. The estimated concentration of metal ions inside the crevice ranges from $1M$ solution[11] to a saturated metal chloride solution[12].

Table 1. Hydrolysis of metal cations at 25°C using data from various references.

Reaction	Equilibrium constant	Equilibrium pH	Reference
$Fe^{2+} + H_2O = FeOH^+ + H^+$	log k = -8.3		9
$Fe^{2+} + H_2O = FeOH^+ + H^+$	log k = -9.5		13
$Fe^{2+} + 2 H_2O = Fe(OH)_2 + 2 H^+$		6.64–1/2 • log[Fe^{2+}]	11
$Fe^{3+} + H_2O = FeOH^{2+} + H^+$	log k = -2.7		13
$Fe^{3+} + 3 H_2O = Fe(OH)_3 + 3 H^+$		1.61–1/2 • log[Fe^{3+}]	11
$Ni^{2+} + H_2O = NiOH^+ + H^+$	log k = -9.5		9
$Ni^{2+} + H_2O = NiOH^+ + H^+$	log k = -9.86		13
$Ni^{2+} + H_2O = NiOH^+ + H^+$	log k = -9.5		14
$Ni^{2+} + 2 H_2O = Ni(OH)_2 + 2 H^+$		6.09–1/2 • log[Ni^{2+}]	11
$Cr^{3+} + H_2O = CrOH^{2+} + H^+$	log k = -3.8		9
$Cr^{3+} + H_2O = CrOH^{2+} + H^+$	log k = -4.0		13
$Cr^{3+} + H_2O = CrOH^{2+} + H^+$	log k = -3.8		14
$Cr^{2+} + 2 H_2O = Cr(OH)_2 + 2 H^+$		5.50–1/2 • log[Cr^{2+}]	11
$Cr^{3+} + 3 H_2O = Cr(OH)_3 + 3 H^+$		1.60–1/3 • log[Cr^{2+}]	11
$Mn^{2+} + H_2O = MnOH^+ + H^+$	log k = -10.59		13
$Mo^{6+} + H_2O = MoO_4^{2-} + H^+$	log k = -25.7		14

Since a high equilibrium constant for the hydrolysis reaction value suggests strong acidification, this can compare the effect of different metals. The values in Table 1 indicate that ferric iron and chromium ions will cause most acidification. Ferrous iron and nickel have low equilibrium constants. Their equilibrium pH values are therefore higher and will not cause as strong acidification. Hexavalent molybdenum will hardly cause any acidification. The most corrosion resistant alloys often produce crevice solutions that are the most acidic and contain the highest chloride concentrations. This can be due to higher chromium concentration that results in stronger hydrolysis and acidification.

Crevice corrosion also occurs in strongly acidic solutions in the absence of chlorides and in buffered solutions where pH remains constant. The deoxygenation-acidification mechanism cannot explain these cases. To explain this behavior, the traditional mechanism has been modified by adding potential gradient along the crevice as a parameter. This modified mechanism is the IR mechanism or IR drop depassivation. According to this mechanism, the potential along the crevice wall decreases when moving away from the mouth of the crevice due to IR drop. The magnitude of the potential shift depends on the current flowing from the crevice and resistance that in turn depends on electrolyte resistivity and crevice dimensions. For example, the potential gradient for stainless steel may be such that at the mouth of the crevice steel is on the pitting corrosion range, passive range is some distance from the mouth, and the bottom of the crevice is on the active potential range. This results in corrosion damage deep in the crevice. Figure 14 shows the principle of IR drop depassivation[10, 15].

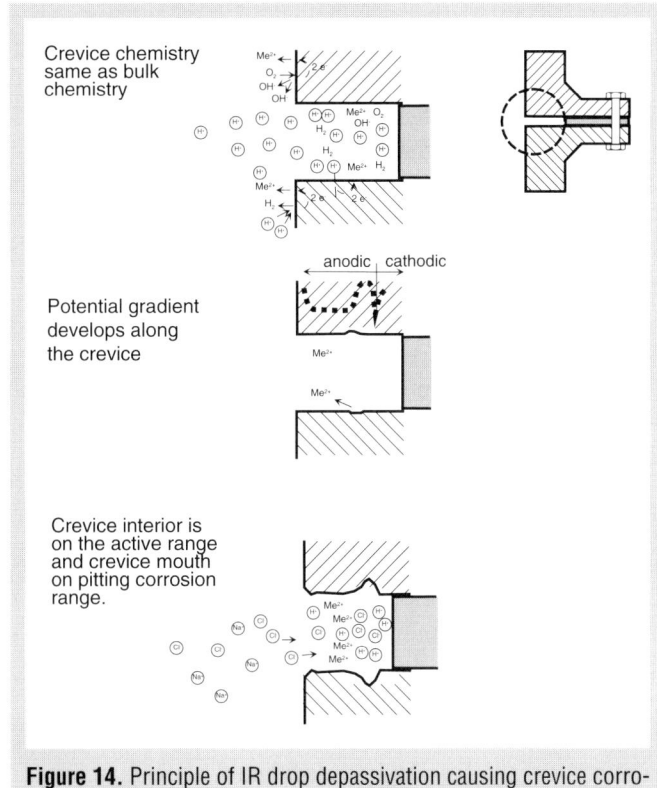

Figure 14. Principle of IR drop depassivation causing crevice corrosion.

Influences on pitting and crevice corrosion include material, electrolyte, and structural factors. The main effect of alloy composition is the resulting passive film. Alloys with low passivation and passive current densities are more resistant. A narrow active potential range and wide passive potential range are also preferable. The composition also influences corrosion propagation since dissolved metal ions hydrolyze to give local pH decrease. For example, higher alloyed stainless steel grades are more resistant to pit and crevice initiation, but they can corrode faster after corrosion initiation. This is partly due to the chromium alloying giving stronger passive films. When dissolved, this will lead to more rapid pH decrease. Solution pH, redox potential, aggressive ion concentration, temperature, etc., will determine electrolyte corrosivity. Nucleation and protection potentials decrease when aggressive ion concentration and temperature increase resulting in more probable corrosion. High specific resistance of the solution can promote crevice corrosion by the IR drop mechanism. Structural geometry has less effect on pitting corrosion than crevice corrosion. Narrow and deep crevices are usually more dangerous than wide and shallow ones. A critical crevice gap can sometimes occur, and corrosion occurs if the gap is tighter than this. If the crevice depth is constant, a crevice with a tighter gap requires a more alloyed steel to ensure corrosion resistance. In very narrow crevices, corrosion is less probable since diffusing significant amounts of corrosive agents and corrosion products in and out of the crevice is difficult.

The morphology of pitting varies. Pits may be deep and narrow or large and shallow. Undercutting is common. Very severe cases of pitting corrosion with numerous small pits may have a surface appearance like that of uniform corrosion. In most cases, the pits are small with the surface diameter about the same or less than the depth. Pits usually grow in the direction of gravity down from horizontal surfaces. This is because the pit solution is more dense than the bulk solution. It comes out from pits on vertical surfaces or those that open downward more easily. Chloride-induced pitting of stainless steel produces a large number of small pits with sharp edges, but sulfur and thiosulfate induced pitting cause a small number of large pits. Figure 15 using ASTM standard practice G46-92 "Examination and evaluation of pitting corrosion" shows some pitting morphologies. Pit rating charts of ASTM G46-92 may estimate the severity of pitting.

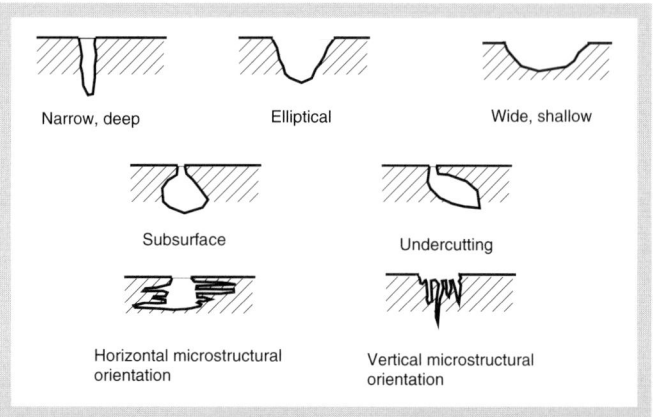

Figure 15. Pitting morphologies from ASTM G46-92 "Examination and evaluation of pitting corrosion."

Corrosion forms

Pitting factor can also quantify pitting severity. This is the ratio of deepest penetration to average penetration measured by weight loss as Fig. 16 shows. A pitting factor of one means uniform corrosion and a large number indicates higher probability for failure. One must use common sense when calculating pitting factors since very small pitting or average corrosion rates will give meaningless results.

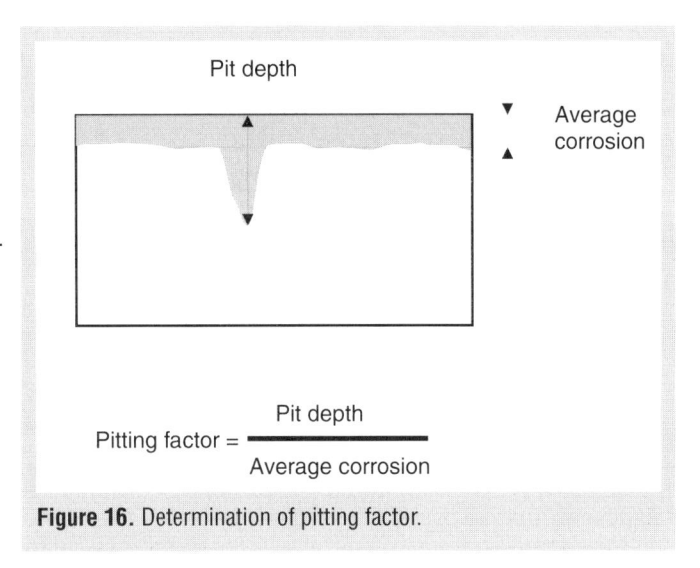

Figure 16. Determination of pitting factor.

$$\text{Pitting factor} = \frac{\text{Pit depth}}{\text{Average corrosion}}$$

The morphology of crevice corrosion depends on the crevice corrosion mechanism. Typically, crevice corrosion by a deoxygenation and acidification mechanism attacks both walls of the crevice. The attack is largest some distance from the crevice mouth. Crevice corrosion caused by the IR drop depassivation mechanism will attack the interior of the crevice. The strongest attack occurs at the distance where the potential is on the active range as Fig. 14 shows. Crevice corrosion due to different metal ion concentrations will attack the mouth of the crevice. The potential of a metal is higher in solutions with a higher concentration of corresponding ions. The surface outside the crevice contacts the dilute solution, adopts a lower potential, becomes anodic compared with the crevice interior, and corrodes. A special case of crevice corrosion is under-deposit corrosion. The mechanism of under-deposit corrosion uses the concentration cell as in the deoxygenation and acidification mechanism. A differential aeration cell is created between clean and shielded areas that are under deposits of sand, pulp, debris, etc.

The surface of every engineering metal has physical or chemical inhomogeneities that are more prone to corrode than the general surface. Pitting is likely to initiate on these weak spots. One can consider the passive film as a thin, ideal ceramic coating. This film cannot form on defects in the base material. For example, manganese sulfide inclusions are commonly favored sites for pitting initiation on carbon and stainless steels. Carbon steels are seldom subject to pitting in the classical sense. If these steels have an air-formed, very thin oxide layer as a covering, corrosion may initially begin as pitting corrosion but will soon change to uniform corrosion. In certain cases where carbon steel is protected by a true passive film, pitting corrosion can occur if the environment becomes more corrosive such as during acid cleaning of carbonate scales in digesters. Crevice corrosion of carbon steels can frequently occur in cases where steel can passivate, but deposits, joints, etc., result in more aggressive solution conditions. Crevice corrosion of carbon steels is possible under atmospheric exposure with very aggressive chemicals.

CHAPTER 4

Stainless steels are the largest alloy group subject to pitting and crevice corrosion. The resistance of stainless steels depends on the composition. With increasing chromium and molybdenum contents, the steels become more resistant to pitting and crevice corrosion. The pitting resistance equivalent (PRE) calculated from nominal composition gives an estimate of the resistance to localized corrosion. These values are best for giving an approximate ranking between different steels. The most commonly used expressions consider the effect of chromium, molybdenum, and nitrogen as Eq. 6 shows:

$$PRE = \%Cr + a \cdot \%Mo + b \cdot \%N \qquad (6)$$

The first expression of pitting resistance equivalent determined that 1% molybdenum could replace 3.3% chromium in the case of austenitic stainless steels. A relative resistance to pitting and crevice corrosion could be calculated from the alloy composition[16]. Considering the effect of nitrogen, the index is often expressed as PRE_N. The PRE value usually has a linear relationship with the critical pitting temperature (CPT) or critical crevice temperature (CCT) above which corrosion starts in some tests. The PRE values are merely rough estimates of the corrosion resistance, and small differences are irrelevant in practice. Using the PRE value, repassivation potential, or critical temperature, one can estimate the corrosion resistance.

Ferritic grades are often more resistant than austenitic grades with equal amounts of chromium and molybdenum. Due to the poor ability to form and weld for ferritic steels, they seldom have use in the process industries. If the corrosion resistance of an austenitic grade is not sufficient, replacement with a duplex grade is possible. Nickel alloys are intrinsically resistant in reducing media. Their resistance to oxidizing media is the result of chromium addition. Compared with stainless steels, nickel alloys with equal amounts of chromium and molybdenum are more resistant but also more expensive. Nickel alloys also contain sulfide inclusions that can act as pit initiation sites. Titanium is very suitable in oxidizing media but has variable resistance to strong reducing acids. Commercially pure titanium and its alloys perform very well in chlorine and chloride solutions. Pitting will often only initiate at potentials up to 8–10 V. This never occurs in spontaneous corrosion. Intermetallic phases are common sites for pit nucleation in some titanium alloys. In hot oxidizing chloride solutions, crevice corrosion is possible with very tight crevices.

Pitting and crevice corrosion are common problems in every stage of the papermaking process. Stainless steels are attacked in sulfite digesters, bleach plant equipment, paper machine headboxes, and piping. In kraft digesters, stainless steels do not usually suffer from pitting corrosion. An incorrect acid cleaning process or incorrectly operating anodic protection system can cause pitting of stainless steels. Crevice corrosion can occur in pipe joints, under deposits, and in storage vessels where material can settle on the bottom. Closure of the systems will increase the solids content of process

Corrosion forms

solutions and increase the risk of deposits and crevice corrosion. If these deposits were dense and continuous, they could actually reduce the corrosion rate by acting as protective scales. Typical pitting and crevice corrosion examples follow. These examples also show that changes in process environment can cause unexpected corrosion damage to materials believed to be resistant.

The closure of mills has resulted in lower pH, increased temperature, and increased chloride content. Materials supposed to be resistant are corroding since the environments have become more severe. For example, AISI 304 stainless steel was long considered resistant in white water systems, but heavy pitting can occur today. Figure 17 shows a typical example of corrosion caused by environmental changes. The figure shows SIS 2333 (18% Cr and 9% Ni equivalent to AISI 304) stainless steel pitting in a white water storage tank. The pits were 0.6–1.3 mm and hemispherical or small, rough, and covered with deposit. The morphology of pits in Fig. 17 (a) is typical for sulfur induced pitting and in (b) for chloride induced pitting. In Fig. 17 (b) the linear array of pits suggests that corrosion may be due to incorrect welding that has lowered the corrosion resistance of steel.

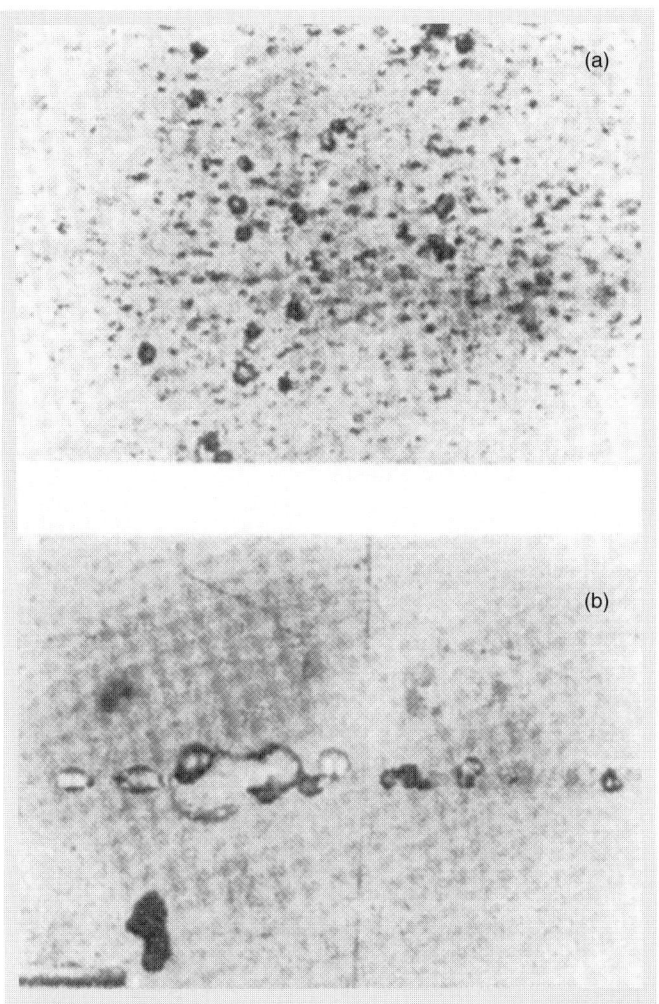

Figure 17. SIS 2333 (18% Cr and 9% Ni) stainless steel pitting in a white water storage tank with (a) results typical for sulfur induced pitting and (b) for chloride induced pitting morphology.

Figure 18 shows pitting inside a stainless steel tube. The material was AISI 304 type stainless steel and the corrosive environment was cooling water in a closed system. The cooling system used hypochlorite addition for prevention of biofouling. The small, rough pits were surrounded by concentric rings of rust-colored stain that can indicate microbiologically influenced corrosion (MIC). The same rings could result from differential aeration cells causing crevice corrosion.

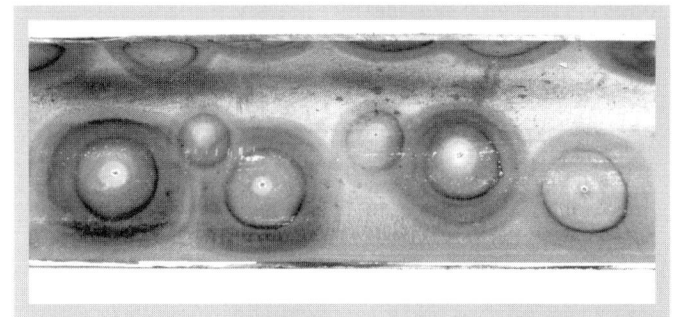

Figure 18. Pitting corrosion of AISI 304 stainless steel.

Formation of deposits is a common cause of crevice corrosion. The example in Fig. 19 shows stainless steel tubing of an evaporator in a semichemical pulping mill. The thin black deposits were the result of excessively high temperature and consistency in chemical recovery. This caused solids to adhere on the surfaces. Analysis of the organic material showed more than 60% carbon. The deposits that contained high concentrations of carbon could also have acted as efficient cathodic sites and promoted localized corrosion by a mechanism similar to galvanic corrosion.

The methods for prevention of pitting and crevice corrosion are similar. Prevention requires attention to system design, materials selection, and solution properties. The system design

Figure 19. Crevice corrosion of stainless steel under deposits.

should minimize any possibilities for development of stagnant and concentrated solutions. The solution flow should be adequate and uniform at all points. Suspended solids should be removed from the solution flow. Surface cleaning will reduce accumulation of deposits. Design of vessels and piping should ensure complete drainage. The number of crevices should be minimum, not too tight, and sealed if possible. Nonabsorbing gasket materials are preferable since wick and capillary actions will certainly cause a stagnant solution. A solid material or welded construction is preferable to bolting or riveting.

Material selection is an obvious way to combat pitting and crevice corrosion. In the pulp and paper industry, the common materials are stainless steels with increasing amounts of Cr and Mo, nickel alloys with increasing amounts of Cr and Mo, titanium alloys, and nonmetallic materials. The resistance to pitting and crevice corrosion increases in the order listed. Repassivation potentials, critical temperatures, and PRE numbers are useful aids for material selection. Increased wall thickness will increase time for penetration. Lowering aggressive anion concentration, increasing pH, or decreasing temperature and solution redox potential can reduce solution corrosivity. They will decrease the probability of pitting or crevice corrosion to start. These measures can occur separately or together.

5 Intergranular corrosion

Intergranular corrosion is preferential attack at or adjacent to the grain boundaries of the metal. Typically, intergranular corrosion results from depletion of a protective alloying element form or enrichment of a reactive impurity element at the grain boundaries. As a result, the grain boundaries or adjacent regions are less resistant. If the grain boundaries were only slightly less resistant than the grains, uniform corrosion would result. The preferential attack along grain boundaries can detach metal grains and cause rapid decrease in effective material thickness. The mechanism of intergranular corrosion can be simplified as galvanic corrosion on a microscopic scale. Intergranular corrosion happens most frequently on metals that are not sensitive when properly heat treated but have become sensitized by exposure to some improper temperature range. The thermal exposures required can be short as in welding or very long as in high temperature service. Examples of intergranular corrosion causes are chromium carbide, $Cr_{23}C_6$, precipitation in stainless steel, zinc enrichment in brass, and iron accumulation in aluminum alloys. Intergranular corrosion results in loose grains that gradually fall from the metal surface or in lifting of elongated subsurface grains in plates or forged products. This is exfoliation. Exfoliation is generally oriented in the direction of rolling, extrusion, or primary deformation. The most common example of intergranular corrosion is sensitization of austenitic stainless steels. Aluminum alloys can show intergranular corrosion of the exfoliation type.

The morphology of intergranular corrosion depends on the relative corrosion rates at grain surfaces and grain boundaries. If the corrosion rate at grain boundaries is high compared with that at the surfaces, corrosion can penetrate deeply without significant weight changes. If the grains do not undergo attack and remain in place, the metal may appear the same although intergranular corrosion has penetrated through the entire cross section. When the grain surfaces are also attacked, the grooves at the corroded grain boundaries widen. This results in grain dislodgement and noticeable weight loss. Cast alloys with large grain size can corrode more deeply with smaller weight loss than their wrought counterparts. Figures 20 and 21 show some schematic appearances of intergranular attack. The structures in Fig. 20 occur on stainless steels after an electrolytic oxalic acid etching test. The etched specimens are examined with magnification for identification as follows[17]:

- Step structure: steps between the grains, no evidence of ditches (The step structure includes very uniform dissolution of grain surfaces, and no intergranular attack is expected.)

- Ditch structure: ditches at grain boundaries with one or more grains completely encircled by ditches (The ditch structure has deep grooves caused by preferential dissolution of grain boundaries.)

- Dual structure: some ditching at grain boundaries, but no grain completely encircled by ditches (The dual structure occurs when chromium carbides are present at grain boundaries but not as a continuous phase.).

The "step" structure indicates properly heat treated material, but the "ditch" structure indicates susceptibility to intergranular corrosion. Figure 21 shows the cross-cut appearance of intergranular corrosion. Some of the top grains are loose and several grains inside the metal structure are missing.

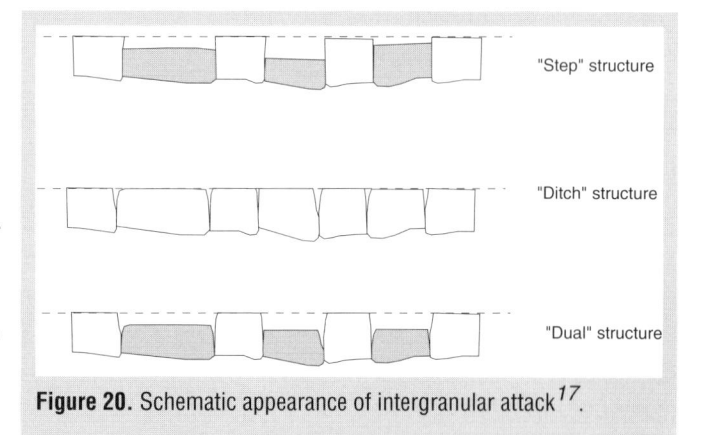

Figure 20. Schematic appearance of intergranular attack[17].

The formation of precipitates at the grain boundaries may also influence other forms of corrosion. In stainless steels, carbide precipitation can decrease resistance to pitting and crevice corrosion. Carbides can act as initial damage in stress corrosion cracking.

Sensitization usually means loss of corrosion resistance caused by chromium carbide precipitation at grain boundaries. This is frequently a problem with stainless steels, but nickel alloys may be even more prone to sensitization. Chromium carbides form at about 540°C–700°C. This temperature range can occur during heat treatment or welding. Due to its low diffusion rate, chromium migrates from areas next to grain boundaries, reacts with carbon, and forms continuous chromium-depleted zones. Chromium carbide precipitation does not occur if carbon content is 0.03% or less. Molybdenum containing steels may still behave as sensitized in highly oxidizing solutions although no visible carbides or other precipitates are visible. Stainless steels alloyed with titanium or niobium can prevent chromium carbide precipitation. This is stabilization of steels. When steel is heated, the carbon reacts with Ti or Nb at high temperatures rather than with Cr at low temperatures. Stabilization leads to formation of discontinuous carbide precipitates, and grain boundaries are not vulnerable to preferential corrosion.

Weld decay in austenitic stainless steels or certain nickel alloys is another example of intergranular corrosion. Weld decay is a specific term for intergranular corrosion that occurs as a result of sensitization in the heat-affected zone during the welding operation. During welding, a zone on each side of the weld reaches 500°C–800°C and may become sensitized. Knifeline attack is also a form of intergranular corrosion in welds. This may occur in titanium or niobium stabilized stainless steels. During welding of thin sheets, the cooling may be so rapid that TiC and NbC do not have sufficient time to form at the high temperatures. Chromium carbides do then precipitate at lower temperatures. Impurities concentrated at grain boundaries can also cause intergranular corrosion. Phosphorus and silicon are harmful impurities, and their effect is visible in highly oxidizing solutions. Precipitation of Cr_7C_3, chromium carbide, is possible in nickel-chromium alloys, and molybdenum carbide may precipitate in high molybdenum alloys of type C-276 (15% Mo). Molybdenum carbides decrease corrosion resistance of molybdenum-depleted areas in reducing media. In oxidizing media, the carbides are attacked[18].

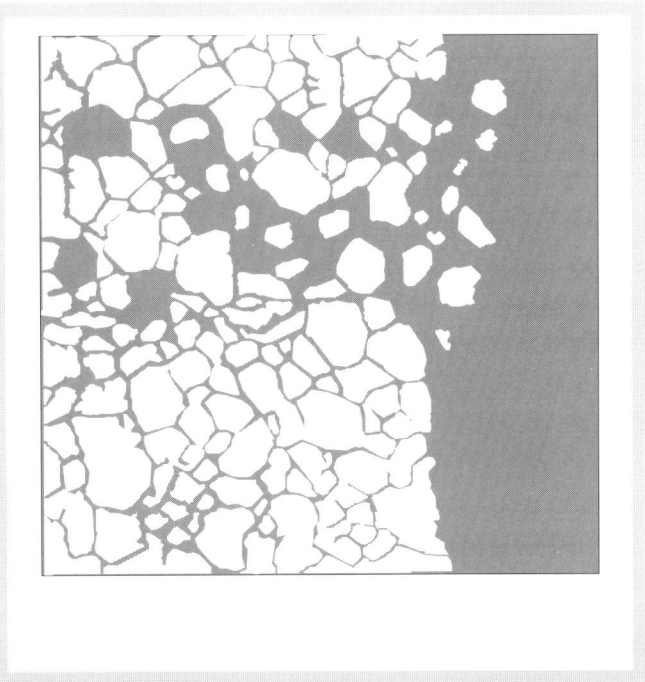

Figure 21. Schematic cross-cut appearance of intergranular corrosion.

In structural steels, intergranular corrosion rarely occurs. Ferritic and austenitic stainless steels may suffer from intergranular corrosion. In some ferritic alloys, chromium carbide precipitation is possible. During welding, carbide or nitride precipitation is also possible in stabilized alloys. Austenitic stainless steels with a basic composition of 18% Cr and 8% Ni are subject to intergranular corrosion unless the carbon content is sufficiently low or they are stabilized. Even low carbon grades may show intergranular corrosion if sigma or chi phases have formed because of prolonged heating. High nickel stainless steels may be more susceptible because stabilizing compounds require lower temperatures than common stainless steels. Modern Ni-Mo alloys are usually not susceptible to intergranular corrosion. Ni-Cr and Ni-Cr-Mo alloys may become sensitized due to chromium, molybdenum, or both carbide precipitation. Titanium is usually not susceptible to intergranular attack[18].

Intergranular attack is usually due to improper welding. Field welding of stainless steel can be especially difficult because structures cannot be heat treated. The low carbon versions of stainless steels eliminate the possibility of sensitization. Process piping and equipment in a bleach plant and paper mill are subject to intergranular corrosion since the redox potentials are often high. Weld overlays in digesters can also show intergranular corrosion. Heavy intergranular corrosion frequently attacks grain surfaces and grain boundaries. Figure 22 shows intergranular attack of stainless steel in an alkaline environment.

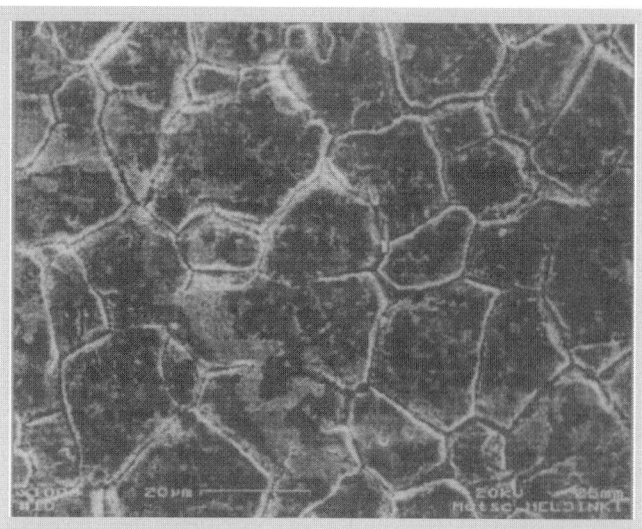

Figure 22. Intergranular attack of stainless steel in an alkaline environment.

Prevention of intergranular attack begins by assuring proper heat treatments and welding procedures during the entire manufacturing process. This includes annealing and quenching at mills and shops. These treatments will dissolve chromium carbides and nitrides, molybdenum carbides, and inter metallic phases. Austenitic stainless steels with high carbon content are heated to 1040°C–1080°C followed by water quenching. Ferritic stainless steels require treatment at 650°C–800°C and quenching. Quenching is important because slow cooling would sensitize the entire metal structure. The temperature to dissolve carbides completely decreases with decreasing carbon content. Low carbon grades can therefore be treated at lower temperatures than high carbon grades. Low carbon austenitic grades can be heat treated at about 950°C if necessary. Heat treatment times are 2–4 h. In duplex stainless steels, the desirable 50% ferrite and 50% austenite microstructure is obtained after a heat treatment, hot working, or a combination of these. Normal heat treatment is in the range of 1030°C–1150°C that causes some original delta ferrite to transform to the more stable austenite. Prolonged heating at temperatures above 315°C and slow cooling rates can cause the precipitation of many phases that will reduce toughness. Inappropriate heat treatment of duplex stainless steels can also cause formation of second phase precipitates that lower the corrosion resistance. High temperatures such as during welding can cause chromium and molybdenum to segregate between the austenite and ferrite phases. As more Cr and Mo move into the ferrite phase, the corrosion resistance of the austenite phase lowers. The addition of nitrogen reduces the risk of inter metallic secondary phases. Nitrogen lowers the content of Cr and Mo in the ferrite phase. This means less susceptibility for inter metallic precipitation. Localized corrosion can initiate in chromium-depleted zones surrounding nitride precipitates.

The material may change to a more resistant one. Stainless steels should have 0.03% carbon maximum if the material thickness is more than a few millimeters thick. Decrease of nitrogen content also prevents precipitation of interstitial compounds. Titanium and niobium in the stabilized grades will react with carbon to form carbides and lower the free carbon content to such low values that chromium carbides do not precipitate. Electrochemical protection is usually not effective against intergranular corrosion. Proper welding practices are most important in intergranular corrosion prevention. External carbon sources should not be present. Parts and electrodes require cleaning from grease, etc., before welding. Lower heat input can reduce the degree of sensitization.

6 Stress corrosion cracking and corrosion fatigue

Stress corrosion cracking or environmental sensitive cracking is a dangerous form of corrosion. The environment usually causes only minimal corrosion, but the simultaneous action of tensile stress and corrosion will sometimes result in brittle fracture of an alloy that would normally be ductile. Three simultaneous conditions are necessary to cause stress corrosion cracking: susceptible alloy, corrosive environment, and tensile stress. These factors are not always present together. As operational conditions vary over time, dangerous combinations can occur. Through evaporation, dilute and therefore nonaggressive environments can concentrate critical components and become aggressive. Tensile stresses can result from bolting, welding, etc., during maintenance operations. Carryover of corrosive solutions to subsequent process steps or countercurrent washing can also change a solution composition to sufficiently corrosive to cause SCC. Cracking can be transgranular, intergranular, or mixed mode. It usually occurs perpendicular to the tensile stress. If the stress is static tensile stress, the corrosion form is stress corrosion cracking. Cyclic stress and corrosion will result in corrosion fatigue.

Most alloys are susceptible to stress corrosion cracking, but dangerous combinations of alloy and corroding material are relatively few. The environment and electrode potential where cracking occurs are such that the metals show borderline passivity. The crack tip remains active due to stress concentration, but the crack sides will passivate. As the crack sides repassivate, the crack preserves its sharp geometry. Residual stresses may arise from cold work, heat treatment, or welding, and external stresses may be applied during service. Many years may pass before cracks become visible. Then they propagate rapidly. The cracks often propagate until the remaining cross section does not sustain the design load, and the final rupture is purely mechanical. The earliest example of stress corrosion cracking is probably cracking of carbon steel riveted boilers in alkaline media. The term for the phenomenon is caustic embrittlement. Caustic embrittlement is due to the escape of steam at riveted joints. This might happen today in rolled joints in boiler bank tubes. Escape of steam concentrates chemicals dissolved in boiler water including free alkali. When the concentration of alkali is sufficiently high, the steel becomes susceptible to caustic stress corrosion cracking. In early cases, the failures start at rivet holes that are subject to high residual stresses.

Stress corrosion cracking has two phases. In the initiation phase, the cause of the stress forms. The crack grows during the propagation phase leading to failure. The stress corrosion cracking mechanism is a combination of chemical, physical, and mechanical processes that destroy metal bonds in the crack tip leading to crack growth. The crack tip is a special chemical environment that develops during initiation and propagation. At the crack tip, fresh metal is constantly being exposed. Before it has time to passivate, it remains active. In contrast, the metal on the sides of the crack has had time to become passive. This galvanic cell between the large cathodic area on the crack sides and the small anodic area at the crack tip increases the dissolution rate at the tip.

In the preliminary stage of stress corrosion, some form of localized corrosion such as pitting or intergranular corrosion occurs. If the corrosive environment is dangerous from the SCC viewpoint, then the preliminary defect is slow to repassivate and the passive film eventually breaks again. This allows preliminary corrosion damages acting as stress causes to form. The crack propagation involves simultaneous action of corrosion and stress. The point of the crack becomes anodic and the surrounding area becomes cathodic. Corrosion concentrates on the point of the crack. The attack is a narrow often branched crack with a sharp tip in the metal. Oxidation and corrosion products usually adhere to the crack surfaces. The crack surfaces appear cleavage-like. The cracks can follow intergranular (between grains) or transgranular (through grains) paths as Fig 23 shows. Transgranular failures are less common than intergranular ones, but even a single specimen may show both types of attack. Intergranular stress corrosion

Intergranular stress corrosion cracking	Transgranular stress corrosion cracking

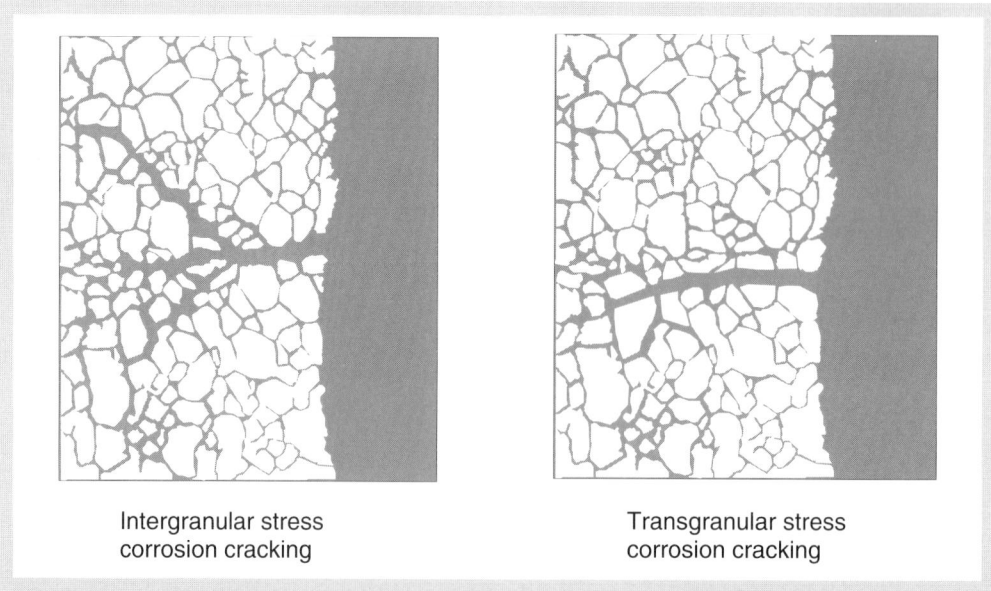

Figure 23. Schematic cross-cut appearance of intergranular and transgranular stress corrosion cracking. Compare the appearance with intergranular corrosion in Fig. 21.

cracking may be associated with normal intergranular corrosion without stress. The rate of crack propagation depends on the rate of film reformation and film breakdown cycle.

The time required for stress corrosion failure depends on stress level, microstructure and composition of the metal, composition of the environment, and temperature. Solution corrosivity, temperature, and stress level are critical factors. Exceeding a threshold value of only one of these factors may result in cracking. Residual stresses caused by welding, machining, bending, or heat treatment can approach or exceed the yield point and therefore be particularly dangerous. Residual stress is the stress remaining in a material after removal of all applied loads. High levels of residual stress develop in welding due to the thermal gradients and differential shrinkage that occur during cooling. These stresses range from tensile stresses near the yield strength of the material to highly compressive stresses. If the structure being welded is unable to withstand the forces due to shrinkage during cooling, the structure will show distortion. The residual stresses will be relieved. If the structure is sufficiently stiff to withstand these forces, minimum distortion will occur, but high levels of residual stress will develop particularly along welds. Thick structures develop a higher degree of residual stress, and thin structures develop a higher degree of distortion. The existence of tensile residual stresses in welded structures is usually undesirable. It can lead to cracking that can ultimately result in structural failure. Some types of cracking that can occur due to the presence of residual stresses are solidification and hot cracking, hydrogen cracking, stress corrosion cracking, fatigue cracking, and corrosion fatigue cracking.

Alloy composition directly influences general corrosion performance, but minor variations may affect stress corrosion cracking. If the metal is prone to localized corrosion, time to failure can be short. Alloys hardened by heat treatment are more susceptible to SCC than the same alloys in the annealed condition. Notches and other surface discontinuities will act as stress causes. In the absence of stress, alloys do not usually

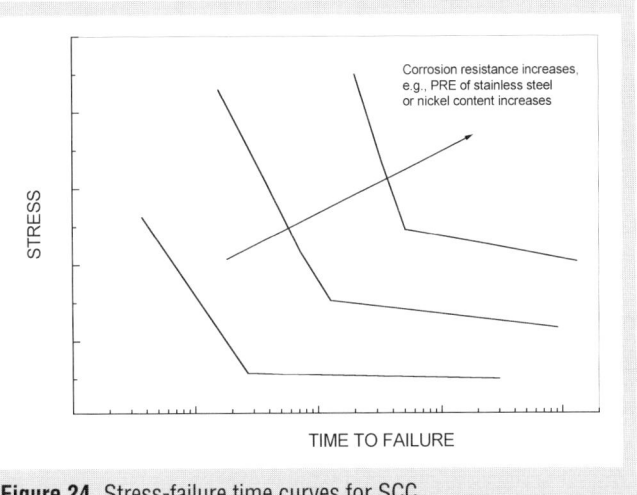

Figure 24. Stress-failure time curves for SCC.

suffer from SCC in the same environment where they would otherwise crack. A stress increase decreases the time to cracking. A threshold stress sometimes exists below which cracking does not occur as Fig. 24 shows. The times for cracking and threshold stress depend on alloy corrosion resistance and solution corrosivity. An interesting point for stainless steels is that the high nickel alloys are often more resistant to SCC. The commonly used PRE equations do not include nickel. From a practical standpoint,

CHAPTER 4

material selection using a threshold stress can be misleading because it varies with alloy, strain, degree of cold work, environment, and time.

Corrosion fatigue is a brittle failure caused by cyclic stress and corrosion, but it requires no specific environment. Typically, a rotating shaft with slight misalignment will vibrate causing cyclic stresses, but it will not be subject to corrosion fatigue unless moisture penetrates to the surface. Lubricants constitute an important class of environments in which metals can be exposed to potential corrosion fatigue damage. The lubricants may contain several additives such as antioxidants, anti-wear agents, dispersing agents to prevent particulate deposits, corrosion inhibitors, viscosity modifiers, etc. Contaminants result from degradation of the lubricant base material such as oxidation of mineral oil and from the surroundings.

All metals will crack if they experience high cyclic stresses for many cycles. Wöhler curves or S-N curves describe fatigue resistance. If the system is not corroding, the number of cycles for failure decreases as the stress increases. Below a threshold stress known as endurance limit or fatigue limit, no fatigue develops. Ferrous alloys have clear fatigue limits, but many nonferrous alloys do not. Corrosion will decrease the threshold stress above which failure will occur. Often, no clear fatigue limit exists as Fig. 25 shows. Fracture due to corrosion fatigue generally occurs at stresses below the fatigue limit found in noncorrosive conditions although the corrosion damage can be extremely small. Damage from corrosion fatigue is therefore greater than the sum of the damage from cyclic stresses and corrosion. If the sum of residual stresses and cyclic stresses is very low, the time to corrosion fatigue failure may be very long.

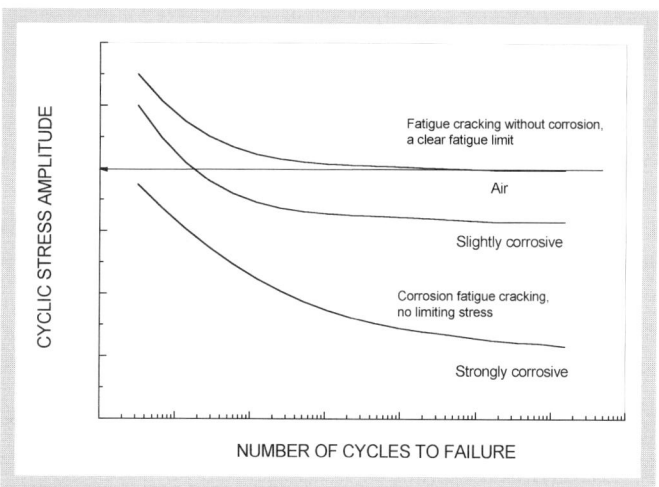

Figure 25. Stress-failure time curves for fatigue cracking and corrosion fatigue.

Corrosion fatigue occurs in two stages. During the first stage, the combined action of corrosion and cyclic stresses damages the metal by localized corrosion and crack formation to such a degree that cyclic stressing fracture would occur even if the corrosive environment was absent. In the second stage, failure proceeds by crack propagation. Stress concentration effects and the physical properties of the metal primarily control this. Corrosion fatigue is therefore more probable in cases where the material is prone to other forms of localized corrosion such as pitting. The cracks will then initiate in damage acting as stress causes. The crack growth follows transgranular paths, and cracks are not branching. The width of the wedge-shaped damage profile depends on the stress frequency. Fine cracks result from high frequency loading, and broad cracks result from low frequency loading. The tip of the crack is usually blunt, and crack surfaces show marks or striations. In rotating parts, the marks are irregular elliptical rings starting from an origin at the outer surface. They show the successive positions of the moving crack front. Striations are microscopic fracture features that show the position of the crack front after each stress cycle. They also show the direction of local crack propagation. A structure that has failed because of corrosion fatigue will usually have visible corrosion products on the fracture surface, but a fatigue crack leaves a smooth bright surface. Corrosion fatigue is most common in rotating parts, but it may be due to pressure changes, temperature changes, or thermal expansion.

A metal is susceptible to stress corrosion cracking or corrosion fatigue if its corrosion potential is in the transition range from active to passive or from passive to transpassive as Fig. 26 shows. In Zone 1, the form of corrosion is uniform corrosion. It is pitting in Zone 2. Besides the dangerous potential range, the tensile stress must be near yield strength or stress must be cyclic to cause failure. In some cases, potentials near the equilibrium potential are also dangerous. The corrosion potential is sometimes below the passivation potential and slightly lower than the transition range from active to passive behavior. Then a small increase of oxidant concentration may move the corrosion potential to the dangerous zone.

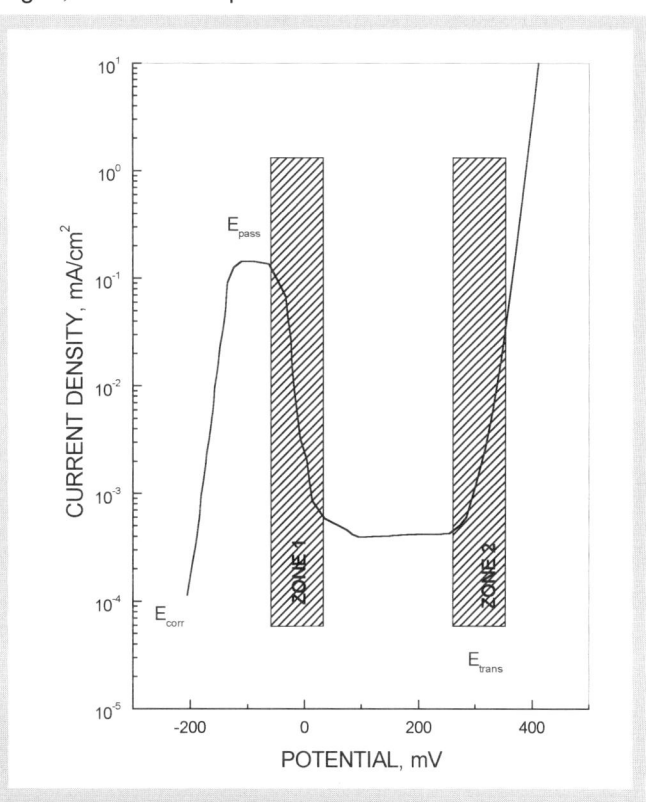

Figure 26. Potential ranges causing SCC or corrosion fatigue with Zone 1 active to passive transition range and Zone 2 to transpassive dissolution range.

Most environments do not cause stress corrosion cracking. Even in the dangerous combinations of metal and environment, failure will not always develop. In the pulp and paper industry, common cases are alkaline SCC of carbon steel and austenitic stainless steel and chloride-induced SCC of austenitic stainless steels. Many stress corrosion cracking cases have specific terms often relating to the corrosive environment. For example, sulfide stress cracking refers to the failure under tensile stress and corrosion in the presence of water and hydrogen sulfide. Carbon steels are susceptible to stress corrosion cracking by caustic solutions, anhydrous ammonia, and nitrates. Transgranular and intergranular cracking are possible. High strength steels with more than 900 MPa yield strength are also susceptible to SCC in the environments mentioned above. Austenitic stainless steels can show transgranular SCC in chloride solutions and hot caustic solutions. The more nickel alloyed grades are less susceptible to chloride-induced SCC, but nickel addition has no effect with caustic cracking. Hydrogen sulfide present in chloride solutions can cause transgranular cracking at ambient temperatures and very low chloride concentrations. Ferritic grades are usually resistant to chloride-induced stress corrosion cracking but will corrode in hot caustic solutions. As the nickel content of the ferritic grades increases, the resistance to stress corrosion cracking decreases. The duplex grades have the ductility of austenitic grades combined with the resistance to SCC of the ferritic grades. They find constant use in severe conditions where austenitic grades are susceptible to pitting, crevice corrosion, and stress corrosion cracking. The presence of a high ferrite content in duplex stainless steels improves their resistance to stress corrosion cracking by comparison with pure austenitic grades. Chromium-nickel alloys are usually suitable replacements for stainless steels in chloride media[19].

Stress corrosion cracking has been a problem in several instances in the pulp and paper industry. The continuous kraft digesters introduced in the 1950s operated about twenty years without significant corrosion problems. Cracking of welds in carbon steel digesters began to appear in the late 1970s. Cracking was associated with caustic stress corrosion cracking after the Pine Hill, Alabama, rupture of a digester that resulted in separation of the top of the vessel in 1980. Cracking occurred in the nonstress-relieved and stress-relieved welds[20, 21]. Austenitic stainless steels have shown chloride-induced stress corrosion cracking in heat exchangers, etc. Cleaning with hydrochloric acid and

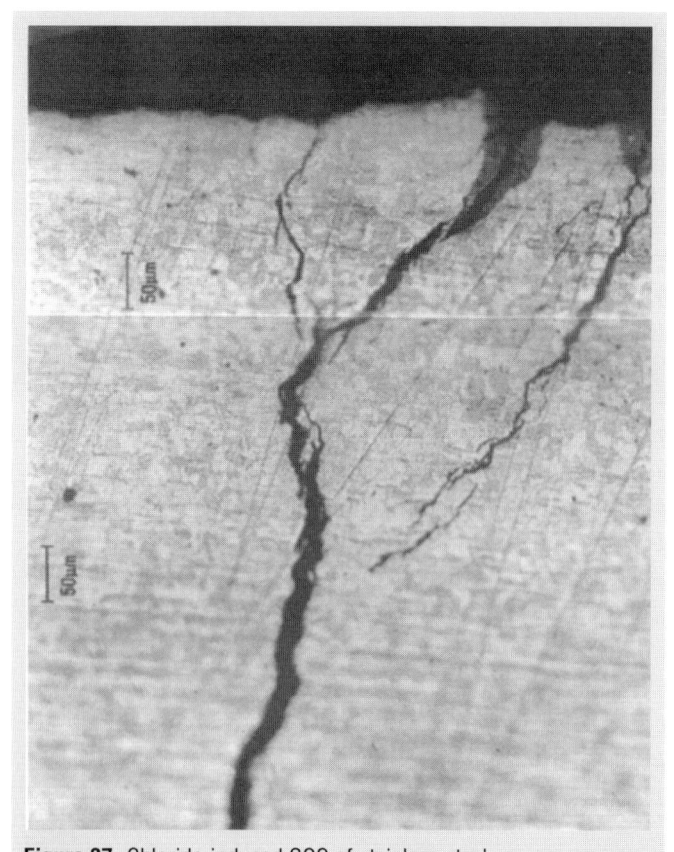

Figure 27. Chloride-induced SCC of stainless steel.

insufficient spooling are common causes of chloride-induced SCC. Corrosion fatigue can occur in rotating parts and heat exchanger tubing subject to vibration.

Figure 27 is an example from chloride-induced SCC of stainless steel in a spent cooking liquor evaporator. The initial corrosion damage that caused cracking is clearly visible as a darker area around the mouth of the crack.

Stress corrosion cracking damage cannot generally be repaired. Proper design, material selection, coating, reducing solution corrosivity, and electrochemical protection can prevent stress corrosion cracking. The design should minimize tensile stresses to below threshold stress. The total stress level resulting from residual stresses, thermal stresses, and operation require consideration. Sharp corners and notches that can cause stress must be avoided. Equally important is avoiding geometries that would result in stagnant solution pockets where concentration could occur. Putting surfaces under compressive stresses by shot-peening, sandblasting, etc., can reduce existing tensile stresses. Stress-relieving heat treatment will also reduce residual stresses present after welding. Material selection is a common remedy for a stress corrosion cracking problem. Failed alloys require replacement by more alloyed grades. For example, an austenitic stainless steel can be replaced with a more alloyed austenitic grade, a duplex stainless steel, or nickel alloy. When considering alloy replacement, identifying the critical solution factor causing stress corrosion cracking is important. Lowering the concentration of the critical component, changing solution redox potential by removing oxidizing species, etc., can control stress corrosion cracking. Since the corrosion required to start corrosion fatigue can be very small, protection of parts subject to alternating stresses should occur wherever practical even if the environments are only slightly corrosive. Cathodic protection can be very effective against stress corrosion cracking and corrosion fatigue. Anodic protection can decrease SCC susceptibility by moving the corrosion potential from a possible risk range to a clearly passive range. Figure 28 shows a summary of potential approaches to prevent stress corrosion cracking. To prevent stress corrosion cracking, the metallurgical state, environment, or tensile stress levels must change.

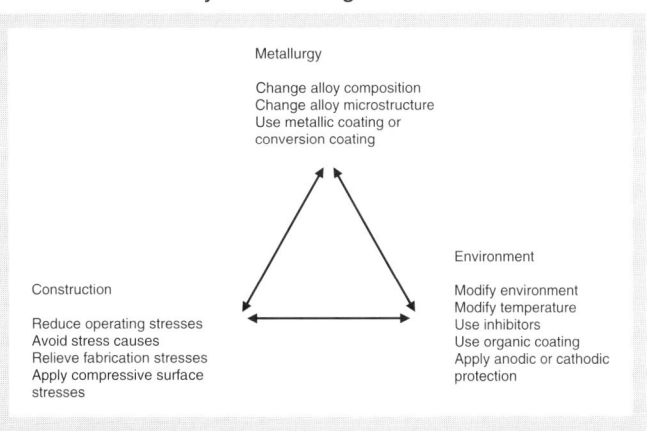

Figure 28. Potential approaches to prevent stress corrosion cracking.

7 Selective leaching

Selective leaching is a form of corrosion where the alloy components react in proportions that differ from their proportions in the alloy. Dealloying and parting are alternative terms for selective leaching. Most common examples are removing zinc from brass and graphite from cast iron. Selective leaching may occur with different mechanisms, and sometimes different mechanisms may occur simultaneously. The less noble alloying component dissolves selectively leaving the more noble one in place. Selective leaching is galvanic corrosion on a microscopic scale. Alloys that consist of components with highly different nobilities such as copper and zinc in brass are especially susceptible. Graphite (carbon) is also electrochemically active. It forms a galvanic couple with iron in cast iron. The dissolution of one element leaves a porous material that has no more mechanical strength. Uniform corrosion followed by redeposition of an alloy constituent also causes porous material. The attack may be layer or plug formed. In the first case, the entire surface may be corroding, and the shape of the structure does not change. For the latter case, corrosion is localized and may very rapidly lead to perforation. Layer type selective leaching occurs more often in solutions where uniform corrosion is common, and plug type occurs in solutions where localized corrosion is common. Selective dissolution can also occur in alloys with several different phases such as duplex stainless steels. Selective leaching is often visible as a color change. Brasses will turn from yellow to reddish, and cast iron will change from silvery gray to matte dark gray. Since the corroded parts are porous and soft, they may also be detected by scraping with a sharp tool. Deposits and colored solutions can make inspection difficult. Most nondestructive test methods cannot detect selective dissolution since the corroded area has approximately the same density and same molar volume as the uncorroded material.

Gray cast irons are subject to selective dissolution of iron leaving a residual grain boundary graphite as a long continuous phase. Free graphite in cast iron is electrochemically more noble than ferrite. It can accelerate corrosion especially in acid or salt solutions by increasing the rate of the cathodic reduction reaction. Rapid corrosion of ferrite next to the graphite flakes results in a porous structure. The ferrous corrosion product may mechanically interact with the free graphite flakes hiding outward appearance of corrosion damage but reducing mechanical strength. The apparently better corrosion resistance of cast iron compared with low carbon steels in some environments probably relates to the somewhat protective nature of the porous graphite-iron corrosion product. Other cast irons do not have the same microstructures and are therefore more resistant. For example, the lack of free graphite in white cast iron results in immunity to graphitic corrosion. Alloyed cast irons are also immune to forming of graphite.

Copper alloys show various forms of selective leaching problems. Brasses with more than 15% zinc may corrode by zinc removal unless they are inhibited by alloying of small amounts of arsenic, antimony, or phosphorus. Tin bronzes have been subject to selective dissolution of tin, and aluminum bronzes have been subject to selective dissolution of aluminum. Cupronickels suffer selective dissolution of nickel in reducing media. Stainless steels, nickel-chromium alloys, and titanium are usually not subject to selective dissolution. A notable exemption is selective dissolution of the ferrite phase from stainless steels in very oxidizing media or chloride solutions containing hydrogen sulfide. The ferrite can be residual ferrite found in welds or castings or ferrite in duplex grades[22].

Figure 29 shows preferential dissolution of the ferrite phase from duplex stainless steel UNS S31803. The test solution was sulfuric acid at pH = 3 bubbled with an ozone and oxygen mixture. The redox potential of the solution was about 1200 mV vs. SCE. During the immersion test, the ferrite grains dissolved much faster than the austenite grains.

Welding of stainless steels requires that a small amount of ferrite remains in the weld metal. Preferential attack of a ferrite phase has occurred in bleach plant environments. Cast stainless steels can also be subject to selective leaching of the ferrite phase. A small amount of ferrite is deliberately maintained in CF8M and CF3M castings to provide a sound product. Usually, this is harmless, but ferrite can be preferentially attacked under acidic conditions.

Figure 29. Preferential dissolution of the ferrite phase from duplex stainless steel UNS S31803 in sulfuric acid at pH = 3 bubbled with ozone and oxygen mixture.

Prevention of selective leaching includes material selection, decrease of environment corrosivity, and cathodic protection. Selective leaching is a special and rare form of corrosion. Very minor changes in material or environment can therefore solve the problem. Preventing removal of zinc uses inhibited brasses containing a small amount of arsenic (As), antimony (Sb), or phosphorus (P). Formation of graphite is a problem with gray cast irons only because of the galvanic couple between iron and graphite. Nodular or malleable cast irons do not have a continuous carbon network, and white cast iron has essentially no free graphite. Material selection is probably the most cost-effective method for prevention of selective leaching since the prevention of zinc removal may require complete oxygen removal.

8 Hydrogen damage

The terms hydrogen damage or hydrogen attack include all damage associated with hydrogen. Hydrogen damage is a general term for the embrittlement, cracking, blistering, and hydride formation that can occur when hydrogen is present. Forms of hydrogen damage are hydrogen blistering, hydrogen embrittlement (HE), hydrogen-induced cracking (HIC), hydrogen stress cracking (HSC), or stress-oriented hydrogen-induced cracking (SOHIC). The terminology and abbreviations used for various types of hydrogen damage can be very puzzling.

All forms of hydrogen damage are due to internal action of hydrogen. Before hydrogen can enter metal, it must adsorb on the metal surface as atomic hydrogen. Atomic hydrogen can result from hydrogen evolution as a cathodic corrosion reaction, cathodic protection, decomposition of hydrogen containing gases, pickling, and electroplating. When atomic hydrogen adsorbs on the metal surface, it may recombine to form hydrogen gas or enter the metal structure. Atomic hydrogen can diffuse through the metal lattice because its size is very small, but hydrogen molecules cannot diffuse. Atomic hydrogen in a lattice moves to preferential sites such as dislocations or structural defects. These are inclusion sites, micro cracks, and voids. At these sites, atomic hydrogen can recombine to form hydrogen molecules that remain trapped in the internal defect. Since hydrogen recombines inside the metal forming molecular gas, the resulting internal pressure creates voids or cracks.

The appearance of hydrogen damage depends on the mechanism involved. Hydrogen blistering causes surface bulges when atomic hydrogen recombines in subsurface voids producing gaseous hydrogen with a greater volume. As the molecular hydrogen forms in the defect area, the pressure increase causes separation and growth of the defect. Blistering usually occurs in low strength alloys such as carbon steels. A clad metal or lined construction can suffer from blistering if the alloy on the hydrogen side allows hydrogen diffusion but the other side is impermeable to hydrogen. Hydrogen embrittlement causes loss of ductility and tensile strength in a metal. Internal hydrogen embrittlement can occur when hydrogen enters molten metal becoming supersaturated immediately after solidification. Environmental hydrogen embrittlement results from solid metals absorbing hydrogen. Structural steel is the most frequent alloy subjected to hydrogen embrittlement. With a tensile stress exceeding a specific threshold, the atomic hydrogen induces crack growth leading to fracture. In practical applications, the hydrogen embrittlement can cause brittle cracking at loading well below the yield strength of the steel. Hydrogen induced cracking and hydrogen stress cracking are forms of hydrogen blistering. In hydrogen-induced cracking, stepwise internal cracks destroy the integrity of the metal. Hydrogen-induced cracking can occur even when little or no applied or residual tensile stress exists. It looks like blisters or blister cracks oriented parallel to the surface. Hydrogen stress cracking results from hydrogen in combination with tensile stress. It occurs most frequently with high strength alloys[23]. Stress-oriented hydrogen-induced cracking causes blister cracks that link in the perpendicular direction to the surface by transgranular, cleavage cracks.

Metals will dissolve more hydrogen at elevated temperatures and pressures. The penetration rate of hydrogen also increases with temperature and pressure. Hydrogen absorbed in steel may result in a loss in ductility, and failure may occur at stresses significantly below the yield strength. The susceptibility to this form of failure increases with an increase in yield strength. Low strength steels may show embrittlement particularly in the presence of cathodic poisons that will catalyze hydrogen evolution or retard hydrogen ion recombination to gaseous hydrogen. Some compounds considered as cathodic poisons are hydrogen sulfide, cyanides, phosphorus, and arsenic compounds that act as recombination poisons. The surface oxide film on titanium prevents penetration of hydrogen. When the solubility limit of hydrogen in titanium is exceeded, hydrides begin to precipitate. Absorption of several hundred ppm of hydrogen results in embrittlement and the possibility of cracking. A serious situation exists when induced currents, i.e., cathodic overprotection or stray currents, generate nascent hydrogen directly on the metal surface.

The common examples of hydrogen damage in the pulp and paper industry relate to welds and high temperature boiler water. Prevention of hydrogen induced damage first reduces the possibilities of formation of atomic hydrogen. The origin of hydrogen is usually in the cleaning and pickling processes, welding with moist coated electrodes, overprotection during cathodic protection, or corrosion with hydrogen evolution as the cathodic reaction. Hydrogen evolution can diminish by reducing the corrosion rate. In galvanic couples, the more noble metal will act as a cathode, and hydrogen evolution will concentrate on that metal. Cleaning and pickling processes use acid solutions. The mechanical effect of hydrogen gas bubbles enhances the removal of oxide scales. These processes can introduce significant amounts of hydrogen in the metal lattice. The use of inhibited acids will lower the risk of hydrogen entry to the metal. Moisture in coated welding electrodes is a common hydrogen source. Hydrogen entry during welding can cause cracking in the weld metal or the heat-affected zone.

In high temperature water, hydrogen damage is a secondary effect to caustic corrosion. When concentrated caustic dissolves the protective magnetite layer, water can react directly with iron evolving hydrogen. Hydrogen evolution occurs also when the caustic itself reacts directly with iron. The hydrogen liberated in atomic form can diffuse into the steel. Inside the steel, atomic hydrogen can recombine to molecular hydrogen or react with carbon to form methane, CH_4. Molecular hydrogen and methane are unable to diffuse through the steel. They concentrate on certain locations such as grain boundaries. Eventually, the gas pressure will cause separation of metal grains. Accumulation of micro cracks will decrease tube strength until internal pressure exceeds the tensile strength of the remaining intact metal. The resulting damage is usually a thick walled, longitudinal burst resembling that resulting from creep rupture.

Prevention of internal cracking or blistering uses steel with low levels of impurities, i.e., sulfur and phosphorus and use of coatings and inhibitors. Using lower strength and lower hardness alloys can prevent hydrogen embrittlement. The environment can be modified to reduce hydrogen charging. Heat treatment can remove absorbed hydrogen. Using cathodic protection requires extreme caution when risk of hydrogen induced damage is possible. Overprotection, i.e., polarizing the protected structure to excessively low potentials, will start hydrogen evolution reaction and make the problem more severe.

9 High temperature corrosion

High temperature corrosion is oxidation of metals in a gaseous atmosphere. Using aqueous environments above the boiling point of water at normal pressure also creates a high temperature environment. Many corrosion phenomena differ from those found in normal aqueous environments. "High temperature" is not an unambiguously defined

CHAPTER 4

temperature because different alloys have specific mechanical and corrosion behavior at different temperatures. A high temperature for a material is any temperature where it begins to creep or starts to react with low humidity atmosphere, molten salt, or liquid metal. The rate of attack often increases with increasing temperature.

High temperature corrosion is direct oxidation of metals by dry gases, i.e., without any moisture film, but it still has some similarities to aqueous corrosion. Other terms for this type of corrosion are scaling or tarnishing. Most metals react with oxygen to form an oxide film. Oxidation can occur at the gas and scale interface or at the metal and scale interface. The reaction rate can be limited by the rate of diffusion of anions, cations, or electrons through the surface scale. The surface scale behaves as a solid electrolyte. As the oxide film grows in thickness, it restricts mass transfer between the metal and surrounding atmosphere to decrease corrosion rate. The rate determining step becomes metal or oxygen diffusion through the film.

In aqueous solution, concentration of dissolved compounds in the solution describes the environment. In gaseous high temperature systems, the partial pressure of gases describes the environment. The partial pressures, i.e., the relative concentrations, of gases will determine which reaction products will form. Contrary to aqueous corrosion at ambient temperatures, high temperature corrosion is more dependent on thermodynamic factors than kinetic factors, and theoretical calculations can estimate the various corrosion products. Compound stabilities can be estimated as the minimum partial pressure of the gaseous component required for compound formation.

Oxidation in clean, nonpolluted gases like air and oxygen results in a thin oxide film that may grow to a thicker scale. In polluted gases, oxide films are often porous. The reaction product may not be an oxide. It can be a sulfide, carbide, or other compound. The protective scales are usually oxides, but they may also be sulfides, carbides, or mixtures of these. Usually, only oxides are protective. The rate of attack increases with increasing temperature. If the scale is continuous and nonporous, ionic transport through the film will be the rate determining step. Properties of the oxide scale will determine the high temperature corrosion resistance of the metal. The desired properties for a protective oxide scale are high thermodynamic stability, high melting temperature, good adherence to the surface, and low conductivity and diffusion coefficients of reactive species so the scale grows slowly. The important environmental factors affecting scale formation are temperature, thermal cycle with respect to time, concentrations of reactive species and their chemisorption and diffusion characteristics, and static and cyclic stresses.

Figure 30 shows a simplified version of high temperature oxidation and its analogy with aqueous corrosion. On a clean metal surface, the first step is adsorption of the oxidant that is usually oxygen. This causes reactions with the metal and nucleation and growth of the first reaction products. These initial products often grow laterally to form a continuous surface scale. Simultaneously, some oxidant dissolves into the metal. Since the continuous surface film forms a barrier between metal and gas, further reaction depends on partial pressure of the oxidant and transport rates of reactants and electrons through the film. Mass transfer rates will increase with an increasing number of structural defects such as interstitial sites, micro and macro cracks, cavities, porosity, etc.

Corrosion forms

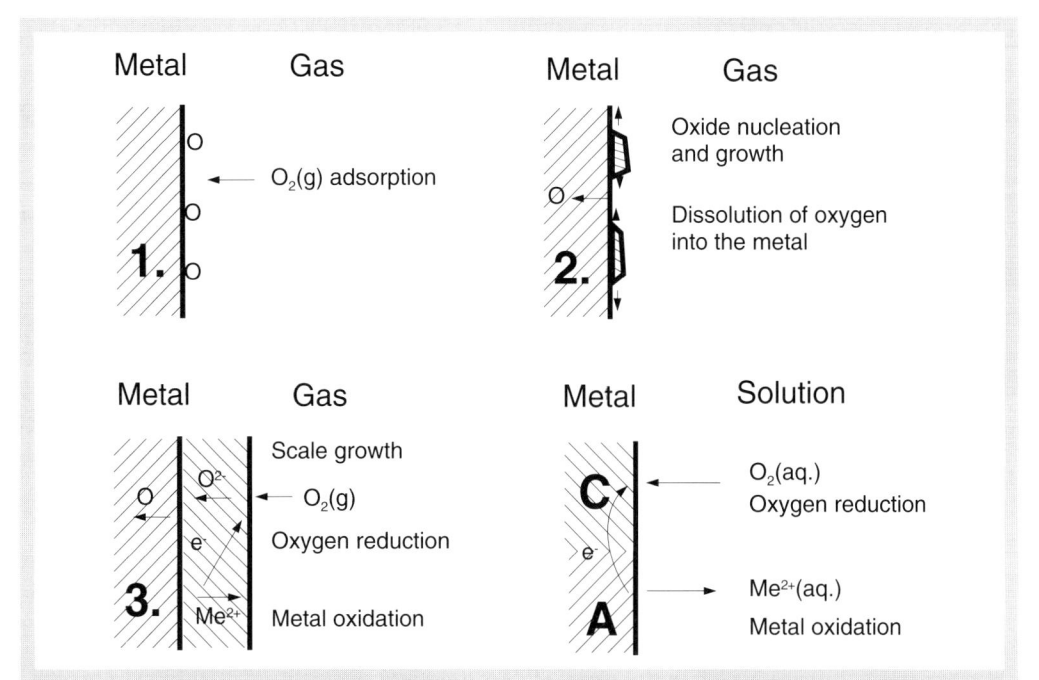

Figure 30. Comparison of the electrochemistry of high temperature oxidation with the electrochemistry of general corrosion in aqueous media[24].

Scale formation often follows some rate law with the general form shown in Eq. 7 with $0 < p \leq 1$.

$$Cumulative\ weight\ change = k \cdot t^p$$

$$Corrosion\ rate = \frac{\partial(weight\ change)}{\partial(time)} = k \cdot p \cdot t^{p-1}$$

(7)

As Eq. 7 shows, rate laws with factor p values less than one will show decreasing corrosion rates with time indicating formation of a protective scale. Usually, the initial oxidation is very rapid. If this initial scale is nonporous, the transport mechanism through the film will determine the oxidation rate law. The rate laws are expressed in mass gain or film thickness with respect to time or some function of time. The common rate laws are linear, parabolic, cubic, and logarithmic laws. If the scale growth follows a linear law (p = 1), the mass gain is a linear function of time, the waste rate is constant, and the film does not protect the metal. If the square (p = 1/2) or the cube (p = 1/3) of mass gain is linear when plotted vs. time, waste rate decreases with time and the film is protective. The parabolic rate law indicates that the rate determining step is diffusion of ions through the oxide layer. When the oxide grows thicker, diffusion length increases and diffusion rate decreases. Cobalt, nickel, copper, and tungsten often follow the parabolic rate law. The high temperature oxidation of high temperature resistant alloys containing aluminum or chromium as oxide forming elements also follows the parabolic rate law. Logarithmic and inverse logarithmic rate laws describe behavior where the logarithm of mass gain or its reciprocal are linearly dependent on time. These two rate laws are usually observed only at low temperatures and thin oxide films in the initial stages of oxidation. The logarithmic rate law occurs when the diffusion paths for ions and vacancies decrease with time. Figure 31 shows theoretical waste vs. time curves for different rate laws. The corrosion rates in atmospheric corrosion often follow the same laws. Short successive periods of the parabolic rate law give an impression of a linear rate law called a paralinear rate law. In most practical cases, any given rate law lasts only a limited time.

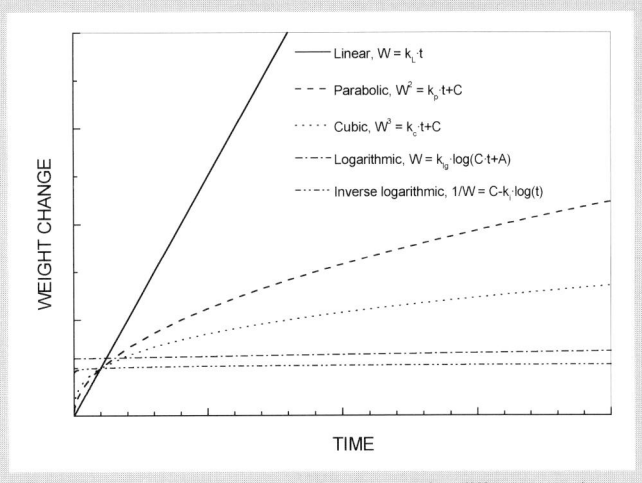

Figure 31. Theoretical waste vs. time curves for different rate laws.

Curves shown: Linear, $W = k_l \cdot t$; Parabolic, $W^2 = k_p \cdot t + C$; Cubic, $W^3 = k_c \cdot t + C$; Logarithmic, $W = k_{lg} \cdot \log(C \cdot t + A)$; Inverse logarithmic, $1/W = C - k_i \cdot \log(t)$.

Two forms of high temperature corrosion do not follow any of the previously mentioned rate laws. Breakaway corrosion occurs when numerous cracks form and extend rapidly to the metal surface. This leaves the bare substrate continuously exposed. The system may first follow the parabolic law and the linear law after breakaway of the scale. Mass change measurements are meaningless since scales form and break rapidly. Catastrophic oxidation or hot corrosion refers to the situation where a liquid phase forms on the oxide scale. This liquid phase penetrates through the scale and at the interface between substrate and scale. Then it spreads by capillary action. This leads to detachment of the scale. Prediction of corrosion rates is impossible since different areas of the substrate are subject to continuously changing attack. Hot corrosion is due to impurities such as sulfur, vanadium, and sodium that may form low melting compounds. These compounds can form molten slags on surfaces and flux the protective oxide layers. This makes the metal again vulnerable to corrosion. Hot corrosion is an accelerating oxidation process. For example, sulfide scales can convert to more stable oxides, but the released sulfide ions penetrate to the metal matrix to form more sulfides that then oxidize. Sequential oxidation and sulfidation cause more rapid corrosion than either alone.

High temperature water above 100°C has very different properties from that at ambient or elevated temperatures. When subjected to elevated temperature and pressure, the compositions of most aqueous products change. Solubilities of salts are exceeded, and the effects of dissolved gases become stronger. The temperature increase does not create any new forms of corrosion but will affect corrosion resistance and scale formation in ways that are not always predictable from results obtained from low temperature experiments. Solubilities of salts can increase up to the melting point of the salt, or the solubility decreases up to the critical temperature of water. Deposits on surfaces in contact with high temperature water can form by solid phase crystallization from a supersaturated solution or by adhesion of solid particles on the surface. In the former case, the deposits are usually strong and adherent. In the latter case, they are comparatively porous, unstable, and poorly adhered.

The rates of electrode reactions will increase with increasing temperature. Dissolved gases and salts may cause more severe corrosion than at ambient temperatures. Carbon dioxide and oxygen dissolved in water will greatly increase corrosion. At the temperatures found in a boiler system, even small amounts of oxygen can cause severe damage as pits. Carbon dioxide will decrease water pH at high temperatures and cause destruction of protective films such as magnetite (Fe_3O_4) on steel. Carbon dioxide causes rather uniform thinning. Scale and corrosion product deposits can indirectly cause more dangerous environments. When deposited on one side, they can locally decrease thermal conductivity leading to increased surface temperatures on the other side of the wall. Common forms of corrosion in high temperature water are uniform corrosion caused by dissolved gases, pitting and crevice corrosion, stress corrosion cracking, and hydrogen damage.

The term boiler corrosion has use for various corrosion phenomena that occur in treated high temperature water and steam. In a boiler, energy in the form of heat transfers from a fuel source to the water across a heat transfer surface. In water tube boilers, water changes to steam inside the tubes. Hot combustion gases or recovery boiler smelt are on the outside. The tubes connect to common water distribution, collection drums, and steam piping. Carbon and low alloy steels are the primary materials of construction for boiling equipment in steam generation. Under steam generating operating conditions, dissolved oxygen in the boiler feed water represents the single most aggressive species. The dissolved oxygen concentration should be below 0.01 ppm. Supercritical boilers with very high pressure sometimes use different feed water compositions with much higher oxygen concentrations. Oxygen pitting corrosion is a randomly occurring, highly localized form of attack that is often very destructive to the boiler system. Oxygen corrosion is not very common in operating boilers but is often found in idle boilers. Oxygen corrosion will usually produce deep, distinct, and almost hemispherical pits. Tubercles may cover them. In an operating boiler, oxygen corrosion usually links to a poorly operating deaerator or improper oxygen scavenging.

Secondary factors relating to boiler corrosion are pH and protective oxide scales. The corrosion process on the steam side between iron and chemically pure water virtually free from oxygen results in the formation of a protective magnetite film according to the following:

$$3\ Fe + 4\ H_2O = Fe_3O_4 + 4\ H_2 \qquad (8)$$

At temperatures above 550°C, the reaction of iron with steam becomes very rapid. At those temperatures, a nonprotective wüstite (FeO) film forms instead of protective magnetite. Chromium additions to steel have decreased attack by steam at these high temperatures.

The natural corrosivity of water varies with the impurities in the raw water supply. One common indicator of corrosivity is the pH. A lower pH equates to more corrosive water. Other factors that can influence the corrosivity of water are the quantity of dissolved minerals and gases, temperature, and velocity. From the corrosion standpoint, a boiler is a thin film of magnetite supported by steel. Preserving this film is the primary consideration for operating and idle boilers. A good water chemistry program should minimize the failures caused by corrosion. It should also minimize the amount of scale forming deposits that enter the equipment. The accumulation of scale on the water-side surfaces causes many problems from decrease in operating efficiency to corrosion. An effective boiler water treatment program can decrease the amount of energy wasted by blowdown. In modern boiler practice, the main methods to prevent corrosion are pretreatment of feed water to prevent acid corrosion through controlled buffered alkali additions and mechanical and chemical treatments to remove oxygen. A boiler water treatment program should include the following:

- Control of dissolved oxygen, carbon dioxide, and other noncondensible gases
- Control or removal of calcium and magnesium hardness
- Control of sludge conditioners to ensure that suspended solids remain in solution until blowdown
- Control of appropriate levels of alkalinity to maintain passivity of steel. (The level of alkalinity required in a boiler system depends upon several factors, and it usually relates to the operating pressure of the boiler).

A specific problem may occur in steam condensate lines due to air infiltration. The effect of oxygen is particularly damaging if CO_2 has not been completely removed from the boiler feed water and is carried into the condensate to produce acidic corrosion conditions. Figure 32 is an example of corrosion caused by carbon dioxide in high temperature water. The resulting acidic conditions have caused deep, sharp-edged pits on carbon steel piping.

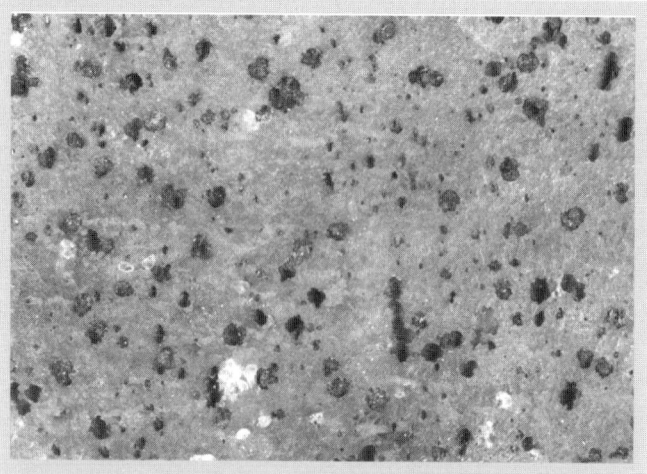

Figure 32. Corrosion of carbon steel caused by carbon dioxide in high temperature water.

Cast irons find some application in the construction of boilers and condensate return systems in steam generating equipment. Pressure vessel codes restrict the use of cast irons in equipment under pressure. They generally perform well in high temperature applications due to the adherence of the oxide films. Carbide stabilizing elements such as chromium and vanadium provide useful life at temperatures up to 550°C. Large amounts of chromium (20%–30%) give resistance to scaling up to 950°C. The resistance to scaling and growth of low alloy cast irons in high temperature steam results from addition of small amounts of nickel and chromium[25].

Carbon and low alloy steels are common materials for plant construction. Typical equipment includes steam generators, boilers, reactor vessels, incinerators, and flue gas systems. The most common limitation of these materials is the loss of strength due to high temperature. In practical use, optimum material selection requires a balance between satisfactory oxidation resistance with respect to service life and ease of component fabrication. Carbon steels are not useful at temperatures above 400°C due to loss of strength and oxidation resistance[26]. In oxidizing conditions, additions of chromium promote formation of protective films. At sufficient levels, nickel promotes austenitic microstructure and minimizes embrittlement. Low alloy steels containing chromium, silicon, and aluminum find many uses although usually limited to maximum temperatures of 550°C–580°C. High temperature environments that contain oxygen, water, carbon dioxide, and oxides of nitrogen can accelerate oxidation and require more than 17% Cr and more than 8% Ni. Breaking of iron carbides and formation of graphite causes loss of strength and ductility above about 470°C. Molybdenum and chromium alloyed grades will be more resistant at high temperatures. Molybdenum additions up to 2% reduce the oxidation rate at temperatures greater than 510°C. For particularly severe conditions, alloys with additional niobium, tungsten, and silicon have use. Silicon allows the formation of an SiO_2 layer that increases resistance to formation of graphite. Alloys containing 5%–10% aluminum have excellent oxidation resistance, but their poor mechanical properties preclude widespread use. Lower concentrations of aluminum (1%–2%) with small additions of silicon give better oxidation resistance without decreasing mechanical properties[27].

Mild steel has use in steam up to about 450°C, but low alloy steels with Cr and Mo are resistant to much higher temperatures. Temperature fluctuations increase corrosion rates of mild steels, but addition of chromium above about 6% reduces spalling. The condensate lines are usually low alloy steel and operate at temperatures around 100°C–120°C. Although condensate is relatively pure water, traces of dissolved O_2 and CO_2 can cause heavy corrosion.

The austenitic stainless steels can resist oxidation by clean air and steam up to about 850°C–1100°C, and ferritic steels can go to 700°C–1100°C. If the structure is subject to stresses, creep resistance and not oxidation resistance will determine the maximum operating temperature. In this case, the maximum operating temperatures are lower. They are about 700°C for austenitic stainless steels and 350°C for ferritic grades. The low maximum temperature of ferritic grades results from various embrittlement phenomena. Hot corrosion is a common form of corrosion found in gas turbines, fireside coal and oil fired boilers, furnaces and steam super heaters, and waste incinerators. Contaminants in the combustion environments such as V, Na, and Cl can greatly add to the severity of corrosion. In many applications at temperatures of 450°C, austenitic stainless steels are necessary. At higher temperatures, the use of higher Cr alloys is necessary all the way to straight Cr-Ni alloys. Nickel does not oxidize easily, and the protective oxide film allows operation at temperatures up to about 1100°C–1200°C. Nickel is not suitable for an alternating oxidizing and reducing atmosphere. Chromium containing nickel alloys have better oxidation resistance, and their resistance increases with increasing chromium content.

Sulfur is a common impurity in many combustion environments that leads to an increased corrosion rate at high temperatures. Such environments can contain hydrogen, water vapor, oxides of carbon, and hydrogen sulfide under reducing conditions. In most cases, Cr-Mo steels or Cr-Ni-Mo steels have use to about 500°C. For sulfur containing oxidizing environments, Ni alloys with at least 25% Cr and possibly Si are necessary. Sulfur and sulfur compounds are usually harmful to nickel alloys and will decrease the maximum operational temperature[27].

Material selection for boiler tubing considers oxidation or corrosion resistance, strength, and cost. Unalloyed carbon steel is the most suitable material if the environment is not too corrosive. If carbon steels are not suitable, low alloyed steels are the next material group for consideration. Unalloyed carbon steels are suitable for boiler, economizer, and superheater tubing if maximum wall temperatures are less than about 450°C. Molybdenum alloyed steels have use in boiler, economizer, and superheater tubing at temperatures up to 525°C. Chromium and molybdenum alloyed steels are suitable to about 550°C. Higher temperatures require use of creep resistant, finely grained steels, vanadium alloyed grades, or austenitic Cr-Ni steels.

10 Summary

Classification of corrosion forms usually uses the appearance of damage they cause on a metal. Visual examination can identify most corrosion forms although the corrosion cause may remain unclear. Recognizing corrosion forms in industrial equipment provides valuable clues about the cause of corrosion.

The most fundamental classification is general or localized corrosion. In general corrosion, the entire metal surface is subject to a corrosive environment causing a uniform weight loss or thinning. Galvanic coupling of dissimilar metals and fluid flow can increase corrosion rate. General corrosion or uniform corrosion causes the greatest material loss on a tonnage basis, but localized corrosion causes more rapid failures. Localized corrosion is generally a problem of passive metals. The passive surface film protects most metals from uniform corrosion, but localized corrosion occurs in places where the passive film is weaker or the corrosive environment is more severe. Pitting and crevice corrosion and stress corrosion cracking are common forms of localized corrosion.

Uniform corrosion is usually not a difficult problem for the designer. Knowing the corrosion rate allows calculation of the necessary corrosion allowance. For most forms of localized corrosion, corrosion rate estimations are not useful. Here the probability of corrosion has more importance. This probability depends on the corrosion resistance of the material and corrosivity of the solution. Stresses, vibration, etc., will also affect the corrosion probability. High corrosion probability means that corrosion can initiate easily. Many forms of localized corrosion have an autocatalytic nature. This means that once initiated corrosion will result in local changes that promote further corrosion. Often high initiation rates equate to high propagation rates. Corrosion will then be rapid.

For general corrosion, lowering corrosion rate is possible by lowering the rates of electrochemical reactions. For example, this occurs by removing corrosive agents from the solution. For localized corrosion, corrosion prevention is maintaining the protective passive film on metal surface. This is possible by using more alloyed metals or by decreasing environment corrosivity. Lowering temperature, oxidative power (redox-potential), or aggressive anion (Cl^-) concentration lowers the corrosivity.

CHAPTER 4

References

1. Fontana, M. G. and Greene, N. D., Corrosion Engineering, McGraw-Hill, New York, 1967, Chap. 3.

2. Collins, J. A. and Monack, M.L., Materials Protection and Performance 12(6):11(1973).

3. Spähn, H., Wagner, G. H., and Steinhoff, U., 1973 Conference on Stress Corrosion Cracking and Hydrogen Embrittlement of Iron Base Alloys, Vol. NACE-5, NACE International, Houston, p. 80.

4. Verink, E. D., in Forms of Corrosion – Recognition and Prevention (C. P. Dillon, Ed.), NACE International, Houston, 1982, Chap. 1.

5. Crowe, D. C., Corrosion on acid cleaning solutions for Kraft digesters, 1992 Proceedings of the 7th International Symposium on Corrosion in the Pulp and Paper Industry, TAPPI PRESS, Atlanta, p. 33.

6. Glaeser, W., in Forms of Corrosion – Recognition and Prevention (C.P. Dillon, Ed.), NACE International, Houston, 1982, Chap. 5.

7. Szlarska-Smialowska, Z., Pitting Corrosion of Metals, NACE, Houston, 1986, Chap. 12.

8. Oldfield, J. W., International Material Reviews 32(3):1(1987).

9. Oldfield, J. W. and Sutton, W. H., British Corrosion Journal 13(1):279(1978).

10. Betts, A. J. and Boulton, L. H., British Corrosion Journal 28(4):279(1993).

11. Peterson, M. H., Lennox, T. J., and Groover, R. E., Materials Protection 9(1):23(1970).

12. Hakkarainen, T. J, Varjonen, O., and Mahiout, A. M., 1992 Proceedings of the 12th Scandinavian Corrosion Congress, The Corrosion Society of Finland, Espoo, vol. 1, p. 71.

13, Turnbull, A., Corrosion Science 23(8):833(1983).

14. Watson, M. and Postlethwaite, J., Corrosion 46(7):522(1990).

15. De Force, B. and Pickering, H., Journal of Metals (9):22(1995)

16. Lorenz, K. and Médawar, G., Thyssenforschung 1(3):97(1969).

17. Majidi, A.P. and Streicher, M. A., Corrosion 40(8):393(1984).

18. Streicher, M. A., in Forms of Corrosion – Recognition and Prevention (C. P. Dillon, Ed.), NACE International, Houston, 1982, Chap. 6.

19. Payer, J. H., in *Forms of Corrosion – Recognition and Prevention* (C. P. Dillon, Ed.), NACE International, Houston, 1982, Chap. 4.

20. Bennett, D. C., 1993 *Proceedings of 4th International Symposium on Corrosion in the Pulp and Paper Industry*, Swedish Corrosion Institute, Stockholm, p. 2.

21. Wensley, D. A., 1989 *Proceedings of 6th International Symposium on Corrosion in the Pulp and Paper Industry*, The Finnish Pulp and Paper Research Institute, Helsinki, p. 7.

22. Heidersbach, R. H., in *Forms of Corrosion – Recognition and Prevention* (C. P. Dillon, Ed.), NACE International, Houston, 1982, Chap. 7.

23. Warren, D., *Materials Performance* 26(1):38(1987).

24. Kofstad, P., 1981 *Proceedings of High Temperature Corrosion Conference*, NACE International, Houston, vol. NACE-6, p. 123.

25. Collins, H. H. and Gilbert, G. N. J., in *Corrosion*, Vol. 2 (L. L. Shreir, R. A. Jarman, and G. T. Burstein, Ed.), 3rd edn., Butterworth-Heinemann, Oxford, 1994, Chap. 7.3.

26. Pinder, L. W., in *Corrosion*, Vol. 2 (L. L. Shreir, R. A., Jarman, and G. T. Burstein Ed.), 3rd edn., Butterworth-Heinemann, Oxford, 1994, Chap. 7.2.

27. Wright, I. G., in *Forms of Corrosion – Recognition and Prevention* (C. P. Dillon, Ed.), NACE International, Houston, 1982, Chap. 8.

CHAPTER 5

Corrosion in different environments

1	**Aqueous corrosion**	**176**
1.1	Acids and alkalis	177
1.2	Oxidizing and reducing agents	181
1.3	Dissolved salts	184
1.4	Dissolved gases	185
1.5	Sulfur and chlorine chemistry	187
1.6	Temperature and flow rate	191
1.7	Corrosion in pulp preparation	192
1.8	Kraft pulping	193
1.9	Sulfite pulping	198
1.10	Mechanical pulping	201
2	**Corrosion in the bleach plant**	**202**
3	**Corrosion in the paper mill**	**210**
4	**Corrosion in chemical recovery**	**214**
5	**Corrosion under thermal insulation**	**218**
6	**Atmospheric corrosion**	**221**
7	**Microbiological corrosion**	**224**
7.1	Biofilm formation	227
7.2	MIC mechanisms	228
7.3	MIC problems	230
8	**High temperature corrosion**	**231**
8.1	Mixed atmospheres	235
8.2	Recovery boiler corrosion	236
8.3	Corrosion in high temperature water	239
9	**Summary**	**243**
	References	244

CHAPTER 5

Jari Aromaa

Corrosion in different environments

Modern pulp and paper mills are highly complex industrial facilities. The entire papermaking process includes wood chip preparation, pulp cooking, washing and bleaching, pulp refining, papermaking, and chemical recovery. Most raw material arrives as logs. The logs are soaked in water, and the bark is removed. Debarked logs are then cut into pieces in a chipper. The cooked pulp is washed to remove solid impurities, dissolved lignin, and other naturally occurring materials. To brighten the pulp, some but not all paper grades use bleaching. The fibers are mechanically treated by beating or refining before making the feedstock for the paper machine. Pigments and fillers are added to the stock, and the pulp is diluted with water to make a lean slurry. Stock entering the paper machine is more than 98% water. This makes it easy to spread and form the web in the forming section of the machine. The entire papermaking process contains many unit processes. Since no two mills are alike, giving detailed solutions to specific corrosion problems is impossible. Over the years, experience has shown that only a few generic materials have common use for most applications with good results. Only when corrosion and maintenance costs become too high is replacement with a more corrosion-resistant material a consideration. Metal manufacturers are constantly developing more resistant grades. To obtain the most benefit from these grades, finding the detailed reason for corrosion is necessary. One can then compare the properties of old and new grades in this regard.

Most corrosion phenomena have identical principles. Simultaneous anodic and cathodic electrochemical reactions result in net reactions at the corrosion potential, and possible formation of surface films gives protection. Reaction probabilities and rates and formation of surface films depend on environmental conditions. Papermaking processes are basically alike, but every mill has its own characteristics. The operational parameters of a mill will often vary with production rate changes, new chemical introduction, or change in raw material composition. Almost every factor used to increase the production rate such as increase of temperature, flow rate, or chemical concentration also increases the corrosivity of the environment. Many corrosion problems occur because original materials do not resist new conditions. Wood pulping processes will face new modifications and closure of the processes. Closure means maximum recycling of process water. This will undoubtedly lead to higher concentrations of dissolved species. Improved or alternate construction materials or methods to handle the changing chemistry of the process are constantly necessary. The following discussion will provide an understanding of the effects of a corrosive environment with examples from processes.

CHAPTER 5

The key variables for characterizing corrosive environments are solvent, pH, redox-potential, temperature, dissolved salts and gases, and flow conditions:

- Solvent is the main component of the solution. It can be aqueous, organic, or a combination of these.
- pH describes the acidity or alkalinity of a dilute aqueous solution. pH will affect corrosion mechanism and corrosion rate.
- Redox-potential is a measure of the oxidizing or reducing potential of the solution. Redox-potential will affect corrosion mechanism and corrosion rate. It often determines the suitability of various alloys. For example, passivating alloys are useful in oxidizing environments, but copper and nickel alloys are useful in reducing environments.
- Temperature will affect reaction rates. Corrosion is usually initiated or accelerated by temperature increase.
- Dissolved salts in the solution will often affect formation and stability of protective surface films. Salts can also affect electrochemical reaction rates. Dissolved gases are often reactants for cathodic reactions and can affect formation and stability of protective surface films and scales.
- Fluid flow conditions can affect reaction rates and formation and stability of surface films. Increased flow rate can accelerate general corrosion rate and cause erosion corrosion, but stagnant conditions can cause pitting, differential aeration corrosion, and crevice corrosion. A minimum supply of oxidizing compound is necessary for passivation and maintaining passivity.

1 Aqueous corrosion

Most corrosion phenomena occur in aqueous solutions. Aqueous environments range from thin moisture films to bulk solutions and include natural environments and chemicals. The corrosivity of an aqueous solution depends on the amount of dissolved gases, salts, and organic compounds and the presence of microorganisms. The concentrations of dissolved species, pH value, hardness, conductivity, etc., describe aqueous solutions. Aqueous solutions are acidic or alkaline and oxidizing or reducing. An acidic solution contains more H^+ ions than OH^- ions, and an alkaline solution is the opposite. If the activity of dissolved oxygen is higher than that of dissolved hydrogen in pure water, the solution is oxidizing. When the activity of hydrogen is higher, the solution is reducing. This gives four different environmental conditions as Fig. 1 shows.

The situation is never that simple because aqueous solutions also contain dissolved salts. When a salt dissolves in water, it dissociates to positive cations and negative anions and forms an electrolyte solution. The term electrolyte originally described a salt that dissociated into ions in solution, but an electrolyte solution is also simply called an electrolyte. If the salt dissociates easily, it is a strong electrolyte. Sodium chloride and sodium sulfate are typical strong electrolytes. Weak electrolytes such as acetic acid dissociate only partially. Electrolyte solutions conduct electricity because the charged ions can move through them. Electrolyte solutions are ionic conductors as distinguished from the electronic conductors in which movement of electrons carries the charge.

Several methods can give the composition of an electrolyte solution. Composition can relate to moles of dissolved substance. Molar solutions contain a given number of moles of ion or substance in one liter of solution. For example, dissolving one mole of sodium chloride, NaCl (atomic weight 58.44 g/mol), in water and adjusting the volume to one liter will give a 1M solution. Since chemical reactions theoretically happen in whole number proportions, molar solutions are preferable in research work. More field-oriented ways to give the solution composition are grams per liter or weight percent. Note for example that 10% by weight sodium chloride solution contains 100 g NaCl and 900 g water not 100 g NaCl in one liter of water. That would be a 100 g/L NaCl solution. A known mass in a known volume such as grams per liter is preferable to weight percent solution because composition calculation is simpler and less ambiguous.

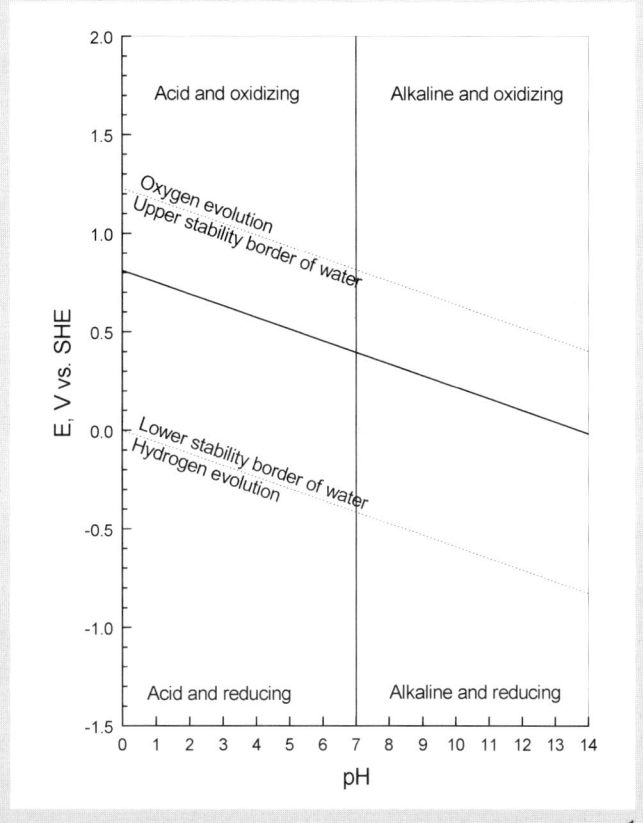

Figure 1. Properties of aqueous solutions (H_2O Pourbaix diagram)[1].

1.1 Acids and alkalis

The water molecule (H_2O) consists of a combination of a hydrogen ion (H^+) and a hydroxyl ion (OH^-). A neutral aqueous solution has equal amounts of hydrogen ions and hydroxyl ions. Hydrogen ions as free protons, H^+, do not exist since they immediately solvate to H_3O^+. The completely hydrated proton, $H^+(aq)$, is also a hydrogen ion. Protons are not localized at individual oxygen atoms but change their places continually. The pH value describes the acidity or alkalinity of a solution. Origin of pH lies in the fact that water dissociates to a very small extent. One mole of pure water or dilute aqueous solution contains 10^{-7} moles of hydrogen ions and 10^{-7} moles of hydroxide ions. This is neutral. The product of hydrogen and hydroxide ion concentrations is always 10^{-14}.

CHAPTER 5

Increase in hydrogen ions changes the water on the acid side, and increase of hydroxide ions makes water more alkaline. Mathematically, the pH is the negative logarithm of the hydrogen ion concentration: pH = $-\log_{10}[H^+]$. A change of one unit in the pH scale means a tenfold change in H^+ ion concentration.

Normality of acid or alkaline solutions refers to the molarity of hydrogen or hydroxide ions. Dissolving one mole of monoprotic (one hydrogen atom) acid such as hydrochloric acid, HCl (36.5 g/mol) to a total volume of one liter therefore gives a 1N solution. Dissolving one mole diprotic (two hydrogen atoms) acid such as sulfuric acid, H_2SO_4 (98 g/mol), gives a 2N solution. In concentrated solutions, the activity of ions changes. This influences pH measurements. Compositions of concentrated acid or alkaline solutions are often given in weight or volume percent solutions. This is due to the low amount of acid or alkali required to produce an acid solution with pH = 0 or an alkaline solution with pH = 14. If the compounds dissociate completely, a solution with 0.5 moles or 49 g/L H_2SO_4 will theoretically have pH = 0. A solution with 1 mole or 40 g/L NaOH has pH = 14. The pH is also temperature dependant and decreases with increasing temperature. This is due to the changing activity of hydrogen ions.

An acid is a substance that increases the hydrogen ion concentration of an aqueous solution. It decreases the pH to less than 7.0. Acids can be inorganic or mineral acids and organic acids. Common mineral acids include sulfuric (H_2SO_4), nitric (HNO_3), phosphoric (H_3PO_4), hydrochloric (HCl), and hydrofluoric (HF). The groups of inorganic acids are hydrogen acids and oxyacids. In hydrogen acids, the acid hydrogen atom is directly bound to the central atom such as Cl in hydrochloric acid. Most acids are oxyacids that have a hydroxyl group, i.e., the hydrogen atom is bound to the central atom via an oxygen atom. Under normal temperature and pressure, halide acids such as hydrogen chloride and hydrogen fluoride are gaseous. They are used as water solutions of the gaseous compound. For example, hydrogen chloride dissolved in water forms hydrochloric acid. The corresponding oxyacids form when oxides of chlorine dissolve in water such as HClO, $HClO_2$, and $HClO_3$. The oxyacids usually form when their anhydrides dissolve in water. Dissolving sulfur trioxide, SO_3, gives sulfuric acid, H_2SO_4.

Of the common mineral acids, nitric acid is strongly oxidizing. Pure sulfuric acid is oxidizing when concentration is more than 90% and temperature is high. Otherwise, it is reducing. Phosphoric acid is oxidizing or reducing depending on concentration and impurities. Hydrochloric acid is strongly reducing. In reducing or nonoxidizing acids, the cathodic reaction is hydrogen evolution. Reducing acids are sensitive to aeration or oxidizing impurities. The presence of oxidizing compounds will often increase the corrosion rate in reducing acids. In oxidizing acids, the primary cathodic reaction is usually reduction of the acid anion such as NO_3^- in nitric acid. Aeration and oxidizing compounds do not usually influence corrosion in oxidizing acids, but the presence of reducing impurities like halogens can increase their corrosivity. If a metal has ions with different oxidation states, corrosion in reducing acids usually results in lower oxidation state ions and corrosion in oxidizing acids gives higher state ions.

The most important organic acids are formic acid (HCOOH) and acetic acid (CH_3COOH). Organic acids are weak, reducing acids, and they become weaker as the carbon chain becomes longer. Formic acid is most strongly dissociated followed by acetic acid, etc. Organic acids typically are not handled as pure chemicals but as mixtures with inorganic acids, organic solvents or salts, or mixtures of several organic acids. Totally anhydrous organic acids can be very corrosive, but a few percent water will decrease corrosivity to many alloys. For most organic acids, the impurities will determine the oxidizing or reducing capacity of the acid solution.

Alkalis are compounds that release hydroxyl groups (OH⁻ ions) when dissolved in water increasing the pH over 7.0. Alkalis can neutralize acids to form a salt and water. The activity of the hydroxide ion depends on the associated cation. Caustic hydroxides are strong alkalis. The most common alkalis are sodium hydroxide (caustic soda, NaOH), potassium hydroxide (caustic potash, KOH), calcium hydroxide [$Ca(OH)_2$], and ammonium hydroxide (NH_4OH). Compounds and salts of weak acids such as sodium carbonate (Na_2CO_3) and sodium bicarbonate ($NaHCO_3$) are weak alkalis although they cannot release OH⁻ ions. When these salts dissolve in water, they form undissociated weak acids such as carbonic acid (H_2CO_3) causing water to hydrolyze and OH⁻ concentration to increase. From a corrosion point of view, "alkali" means many kinds of alkaline compounds such as caustic hydroxides, anhydrous ammonia, and the previously mentioned salts. Even organic alkaline compounds such as amines are sometimes classified as alkalis.

The effect of pH and redox potential on the corrosion probability of a metal can be estimated with a Pourbaix diagram, but it gives no information on reaction rates. Figure 2 shows the principal types of pH dependence. The solution pH affects the solubility of corrosion products and possibility of protective film formation. The corrosion rate of noble metals such as gold is intrinsically low and generally independent of environmental pH. The oxides of some active metals are soluble in acids and alkalis. Examples are zinc, aluminum, lead, and tin. They are amphoteric metals. In acid solutions, the dissolution leads to metal ions and in alkalies metal complexes such as zincate, ZnO_2^{2-}, or aluminate, AlO_2^-, ions. The corrosion rate of amphoteric metals has a minimum at some neutral pH range with corrosion rate increasing in the acidic and alkaline directions. Metals such as

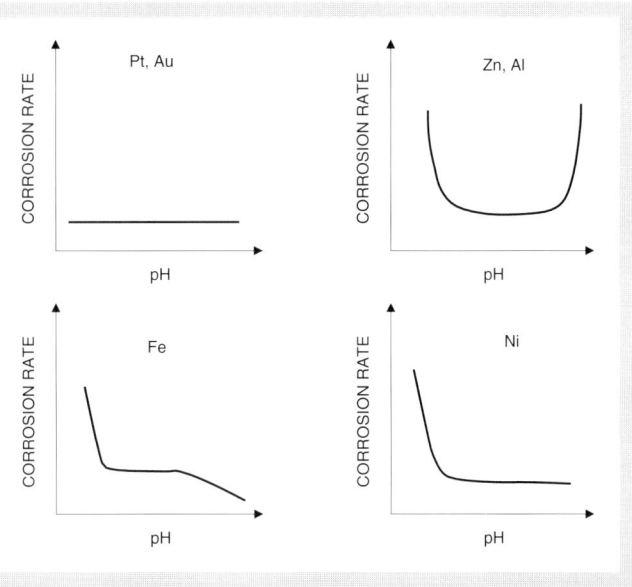

Figure 2. Principal types of relationships between pH and corrosion rate.

iron, nickel, and chromium form oxides that are easily soluble in acids but not in alkalies. In nonoxidizing acid solutions, the dissolution rate is high due to hydrogen evolution. It adopts a nearly constant rate in neutral solutions and begins to decrease in alkaline solutions. At very high alkali concentrations, some metals such as iron may dissolve again to form complexes.

The pH dependence of iron corrosion belongs to the third group. With decreasing pH, the hydrogen evolution reaction becomes a more probable and more rapid cathodic reaction. For example, corrosion of steel is rapid below pH = 4.5–5 due to hydrogen evolution as the cathodic reaction. No surface film will form. At pH > 5, oxygen reduction is the main cathodic reaction, and the corrosion rate depends on the mass transfer rate of oxygen.

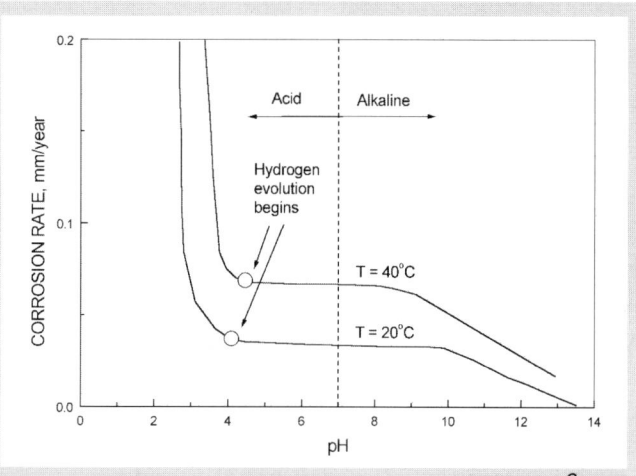

Figure 3. Corrosion rates of steel as a function of solution pH[2].

The pH value also has no direct influence on steel dissolution rates, but pH affects the reaction product layers. In the nearly neutral pH range of 5–10, a porous nonprotective film forms. At pH > 10, surface film thickness increases and porosity decreases with increasing pH causing corrosion rate to decrease. Figure 3 shows the changes of iron corrosion as a function of pH in water with 5 cm³/L oxygen. The pH of the solution is adjusted with NaOH or HCl[2]. The temperature does not have a large effect on the hydrogen evolution range, but the effect becomes clearer with the formation of deposits. At higher temperature, the formation of deposits begins at lower pH values.

The oxidation of metals and formation of reaction product layers usually occur in several steps although reaction equations show the total equations. The reaction products are therefore seldom homogeneous films of a single compound. For example, the reaction in atmospheric corrosion of steel is often written as follows:

$$2\ Fe + O_2 + 2\ H_2O = 2\ Fe^{2+} + 4\ OH^- \tag{1}$$

The reaction products Fe^{2+} and OH^- will react further. Ferrous ions will react with hydroxide ions and oxygen to produce magnetite (Fe_3O_4) or goethite (FeOOH):

$$3\ Fe^{2+} + 2\ OH^- + \frac{1}{2} O_2 = Fe_3O_4 + H_2O$$

$$Fe^{2+} + OH^- + \frac{1}{2} O_2 = FeOOH \tag{2}$$

Corrosion in different environments

The iron in magnetite has two oxidation states. Some iron is in the ferrous (FeO) and the rest in the ferric (Fe_2O_3) state. These can react further to produce goethite, and the result is rust layer. That is a thin oxide film of magnetite covered by a thicker outer layer of goethite.

Another example is oxidation of iron in an alkaline environment. Wensley and Charlton suggested that oxidation in kraft white liquor has two stages[3]. The first is oxidation of iron to ferrous hydroxide, Eq. 3, and the second is oxidation of hydroxide to magnetite, Eq. 4:

$$Fe + 2\ OH^- = Fe(OH)_2 + 2\ e^- \qquad (3)$$

$$3\ Fe(OH)_2 + 2\ OH^- = Fe_3O_4 + 4\ H_2O + 2\ e^- \qquad (4)$$

The total equation is the following:

$$3\ Fe + 8\ OH^- = Fe_3O_4 + 4\ H_2O + 8\ e^- \qquad (5)$$

These reactions came from studies with pure lithium hydroxide[4] and were assumed to occur in alkaline cooking liquors. Electrochemical measurements can show current peaks for both reactions.

1.2 Oxidizing and reducing agents

The concentration of oxidizing and reducing agents in a solution determine the oxidizing or reducing conditions. Common oxidants are oxygen, ozone, chlorine, and sodium hypochlorite. Gaseous hydrogen, sulfur dioxide, hydrogen sulfide, and sodium bisulfite are reducing compounds. Oxidants will oxidize other compounds meaning they accept electrons and reduce. Reducers will reduce other compounds, donate electrons, and become oxidized. Pure water is oxidizing if the activity of dissolved oxygen is higher than that of dissolved hydrogen as Fig. 1 shows. Water is usually a weak oxidizer due to dissolved oxygen. In corrosion research, reducing environments are often considered to be those where corrosion is due to hydrogen evolution as a cathodic reaction. The ability of a solution to oxidize or reduce another material is determined by measuring the solution redox potential. The actual redox potential value depends on the concentration and activity of the oxidant present. In most applications, this is more useful information than actual oxidant or reducer concentration. When several oxidants are present, the redox potential measurement cannot determine the effect of a single oxidant. The effect of oxidants on the corrosion potential of metal can vary over a range of several hundred millivolts. In electrochemical corrosion, the intrinsic nobility of the metal determines whether it will dissolve in oxidizing or reducing solutions. Noble metals like nickel have high equilibrium potentials and will dissolve only under oxidizing conditions. Active metals such as iron or zinc have low equilibrium potentials and will dissolve under oxidizing and reducing conditions.

CHAPTER 5

All the cathodic reactants in electrochemical corrosion are oxidizing agents. A higher equilibrium potential of the corresponding redox reaction means a stronger oxidation potential of the element. Corrosion with hydrogen evolution as the cathodic reaction is often called corrosion under reducing conditions, but the hydrogen ion is also an oxidant. The reducing conditions mean that the equilibrium potential of the cathodic reaction is so low that passivation cannot occur. Most bleaching chemicals have high equilibrium potentials and are therefore strong oxidants giving highly oxidizing solutions. The solubility of oxygen in pure water at room temperature is about 8 mg/L and decreases with increasing temperature and concentration of dissolved salts. Oxygen has use in the pulp and paper industry for a variety of applications including pulp bleaching, black liquor oxidation, and enrichment of the lime kiln atmosphere. At room temperature, ozone has higher solubility than oxygen. Increasing temperature decreases dissolved ozone concentration. Chlorine gas dissolved in water gives a solution of chlorine compounds. The composition of the solution depends on pH. Hydrogen peroxide and its acid solutions known as peracids are new bleaching compounds. Peracetic, performic, and peroximonosulfuric acid or Caro's acid are all strong oxidizers. The theoretical oxidizing effect of a compound can be estimated from the equilibrium potentials in Table 1 of the corresponding redox reaction. As the right column of Table 1 shows, the redox potential of most bleaching chemicals decreases with increasing pH and increases with increasing chemical concentration.

Table 1. Equilibrium potentials of cathodic reactions supported by bleaching chemicals and other relevant chemicals in the pulp and paper industry.

Chemical	Reaction	Equilibrium potential (mV vs. SHE) at 25°C
Oxygen pH<7	$O_2 + 4 H^+ + 4 e^- = 2 H_2O$	$E_0 = 1228 - 59 \cdot pH + 14.7 \cdot \log(pO_2)$
Oxygen pH>7	$O_2 + 4 H^+ + 4 e^- = 4 OH^-$	$E_0 = 401 - 59 \cdot pH + 14.7 \cdot \log(pO_2)$
Ozone	$O_3 + 6 H^+ + 6 e^- = 3 H_2O$	$E_0 = 1501 - 59 \cdot pH + 9.8 \cdot \log(pO_3)$
	$O_3 + 2 H^+ + 2 e^- = O_2 + H_2O$	$E_0 = 2076 - 59 \cdot pH + 29.5 \cdot \log(pO_3/pO_2)$
Hydrogen peroxide	$H_2O_2 + 2 H^+ + 2 e^- = 2 H_2O$	$E_0 = 1776 - 59 \cdot pH + 29.5 \cdot \log(H_2O_2)$
	$HO_2^- + 3 H^+ + 2 e^- = 2 H_2O$	$E_0 = 2119 - 88.6 \cdot pH + 29.5 \cdot \log(HO_2^-)$
Chlorine acidic conditions	$Cl_2 + 2 e^- = 2 Cl^-$	$E_0 = 1395 - 29.5 \cdot \log(Cl_2)/(Cl^-)^2$
Hypochlorous acid near neutral conditions	$HClO + H^+ + 2 e^- = Cl^- + H_2O$	$E_0 = 1494 - 29.5 \cdot pH + 29.5 \cdot \log(HClO)/(Cl^-)$
Hypochlorite alkaline conditions	$ClO^- + H_2O + 2 e^- = Cl^- + 2 OH^-$	$E_0 = 890 - 59 \cdot pH + 29.5 \cdot \log(ClO^-)/(Cl^-)$
Chlorine dioxide	$ClO_2 + 4 H^+ + 5 e^- = Cl^- + 2 H_2O$	$E_0 = 1511 - 47.3 \cdot pH + 11.8 \cdot \log(pClO_2)/(Cl^-)$
Sulfur	$S + 2 H^+ + 2e^- = H_2S$	$E_0 = 141 - 59 \cdot pH - 29.5 \cdot \log(pH_2S)$
Thiosulfate	$S_2O_3^{2-} + 6 H^+ + 4 e^- = 2 S + 3 H_2O$	$E_0 = 499 - 88.7 \cdot pH + 14.7 \cdot \log(S_2O_3^{2-})$
Hydrogen	$2 H^+ + 2 e^- = H_2$	$E_0 = 0 - 59 \cdot pH - 29.5 \cdot \log(pH_2)$
Dithionite	$S_2O_4^{2-} + 2 H_2O = 2 SO_3^{2-} + 4 H^+ + 2 e^-$	$E_0 = 416 - 118.2 \cdot pH - 29.5 \cdot \log(SO_3)^2/(S_2O_4^{2-})$

Usually, an increase of oxidant concentration increases the redox potential of the solution only to a certain level. At higher oxidant concentrations, the redox potential remains essentially constant. Figure 4 shows this for a mixture of ozone and oxygen in pH = 3 sulfuric acid solution at room temperature. The environment corresponds to that of ozone delignification.

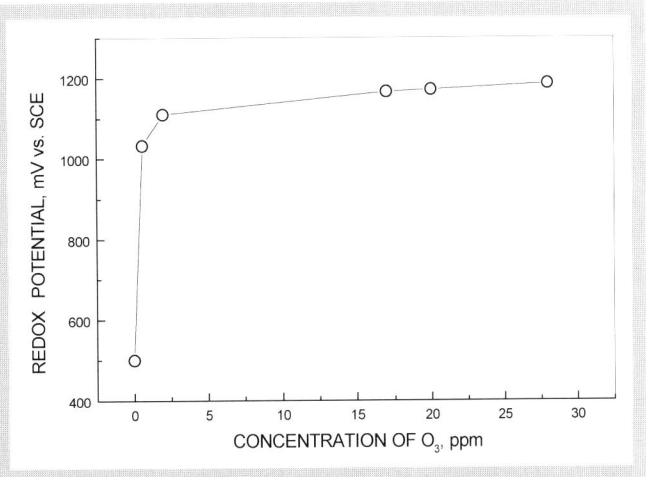

Figure 4. Redox potential for a mixture of ozone and oxygen in pH = 3 H_2SO_4 solution at room temperature.

The oxidant concentration is often monitored by measuring its residual level. In principle, the residual oxidant concentration is a measure of surplus chemical that did not react in the process. A high residual level indicates poor process control and may lead to unwanted solution conditions since the redox potential usually increases with residual chemical concentration. A higher redox potential often gives higher corrosion potentials and corrosion rates of metals. For example, the use of chlorine dioxide to replace elemental chlorine resulted in higher corrosion rates of stainless steels. Chlorine dioxide is a more powerful oxidizer than chlorine gas. Since the corrosion mechanism remains essentially the same, using more alloyed stainless steels prevents corrosion. Figure 5 shows the effect of residual chlorine on redox potential and corrosion rate of stainless steel. Higher residual chlorine concentration means higher redox potential[5].

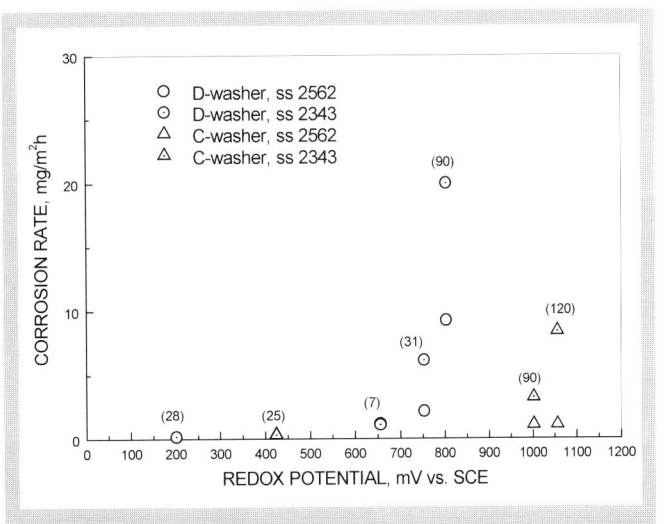

Figure 5. Effect of residual chlorine on potential and corrosion rate of stainless steel[5]. The numbers in brackets indicate measured residual chlorine concentrations in mg/L.

CHAPTER 5

1.3 Dissolved salts

Dissolved salts increase the conductivity of a solution. Equivalent ionic conductance describes the mobility of individual ions. Higher equivalent ionic conductance equates to higher conductivity of a salt solution. As salt concentration increases, the increase in conductivity is less than linear, but the extent of nonlinearity is different for each salt. For strong electrolytes, the relative increase of conductivity with increasing concentration is almost linear. For weak electrolytes, the corresponding change is initially very rapid. At higher conductivities, increasing concentration has very little effect on the equivalent ionic conductance. Table 2 shows typical conductivities of various treated and untreated waters.

Table 2. Conductivities of treated and untreated waters.

Type of water	Conductivity, S/cm
3.5% Sodium chloride solution	53 000
Ocean water	30 000–50 000
Untreated natural water	>200
Soft tap water	100–200
Commercial deionized water	10–100
Fully deionized water	≈2
Deionized and ultrafiltered water	≈0.2

The conductivity of the solution will not influence the reaction equilibria but can influence corrosion morphology. In soft natural waters with low amounts of dissolved salts, the corrosion morphology is small corroded sites near each other. The corrosion appears uniform. As the conductivity increases, the number of corroding sites decreases and corrosion concentrates on few sites far from each other. This is valid only if no reaction product films form. Many dissolved ions have a direct influence on the corrosion resistance of materials. Most common soluble salts increase corrosion rates up to some critical concentration. In sodium chloride solution for example, the critical concentration for steel is about 3% by weight sodium chloride. This is equivalent to ocean seawater[6].

The dissolution of salts can change the pH of a solution. Salts of strong bases and weak acids increase pH, and salts of strong acids and weak bases decrease pH. Several dissolved salts are common in all natural and industrial solutions. The dissolved salts have three categories. The first group contains salts that do not form insoluble compounds on anodic or cathodic sites. These are typically alkali chlorides, sulfates, and nitrates. The most common salt in this group is sodium chloride. The salts in the second group can form a sparingly soluble compound on cathodic sites. Examples are calcium bicarbonate and many zinc, nickel, chromium, and manganese salts. They can work as cathodic inhibitors. The third group contains salts that form a sparingly soluble compound on anodic sites such as sodium phosphate, silicate, and chromate. They may work as anodic inhibitors. Some salts are strong oxidants. When dissolved in water, they give very oxidizing solutions due to metal ions with a high valence state. Typical examples are cupric and ferric chlorides, $CuCl_2$ and $FeCl_3$. The metal ions, Cu^{2+} and Fe^{3+}, have high equilibrium potentials. They can therefore cause dissolution of solid metals by the cementation mechanism.

Localized corrosion of passive metals strongly depends on the anion concentration of the solution. The effect of aggressive anions like chloride and other halogens is well documented. They will decrease the corrosion resistance of stainless steels and other passivating alloys. Corrosion will initiate at lower temperatures and potentials when the aggressive anion concentration is higher. Some anions reduce pitting. In chloride solutions for example, anions like SO_4^{2-}, OH^-, ClO_3^-, CO_3^{2-}, and CrO_4^{2-} can increase the pitting potential. Their inhibiting tendencies depend on their concentrations and concentration of chloride ions in the solution. Solutions with higher molar concentrations of inhibiting anions in comparison with chloride concentration can be less aggressive to stainless steels[7].

In natural waters, the most important dissolved ions are calcium and magnesium cations. These give water its hardness. Hard waters with high calcium and magnesium concentrations are less corrosive because they can form a protective carbonate film on the metal surface. Expression of the water hardness can use different scales such as ppm $CaCO_3$. A medium hard water contains 50–100 ppm $CaCO_3$, soft water contains less than 50 ppm, and hard water has more than 100 ppm $CaCO_3$. Water hardness is usually not a problem in pulp and papermaking unless the water is so hard that it can cause carbonate scale precipitation. Most mills are periodically acid cleaned to remove carbonate scale that accumulates on digester walls, piping, and screens. This can cause severe corrosion. Cleaning is usually done with hydrochloric acid, but formic and sulfamic acids also have use. Even nitric acid has use in fully stainless steel clad digesters[8]. The acid concentration must be sufficiently high to dissolve deposits. Higher temperature increases descaling reaction rates. The corrosion rates also increase with acid concentration and temperature. Velocity will accelerate corrosion by increasing mass transfer of the reactants and removal of inhibitors from the surface. The corrosivity of the cleaning acid decreases when lowering the acid temperature, improving acid circulation, and using higher inhibitor concentrations.

1.4 Dissolved gases

Dissolved gases are often reactants for cathodic reactions. The solubility of gases usually decreases with increasing temperature and dissolved salt level. The most important dissolved gas is oxygen, since the reduction of oxygen is a common cathodic reaction and oxygen is necessary for oxide film formation. The effect of dissolved oxygen on corrosion rates can be corrosion enhancing or corrosion decelerating. If no stable surface film forms, the corrosion rate is directly proportional to the oxygen mass transfer rate to the metal surface. Loose and porous reaction product films cannot protect from corrosion, and corrosion rate increases with increasing oxygen content until mass transfer through the film becomes the rate determining step. The corrosion rate increases until a critical oxygen concentration occurs after which it begins to decrease. This critical con-

CHAPTER 5

centration depends on the solution flow rate. It is lower in rapidly flowing solutions. If the oxygen concentration is higher than the critical concentration, the surplus oxygen not consumed in cathodic reactions will have use in formation of a protective film. If the metal can passivate, this phenomenon will occur above some oxygen concentration, and corrosion rate will decrease sharply. The protective layers should not be too thick since they can easily crack. This causes local variations in oxygen transfer to the surface. If the oxygen transport rate to different areas on the surface is not equal, concentration cells will form. In a concentration cell, areas with high oxygen concentration will be cathodic compared to areas with low oxygen concentration. The areas with low oxygen concentration, e.g., under deposit areas, will corrode.

Carbon dioxide from air can dissolve in water. It is present in many chemical processes and has various effects on the solution and solid compound equilibria. Carbon dioxide is in equilibrium with air at about 1–10 mg/L in natural surface waters. Dissolved carbon dioxide forms carbonic acid that dissociates to H^+ and bicarbonate, HCO_3^-, ions to reduce the solution pH as Eq. 6 shows.

$$CO_2 + H_2O = H_2CO_3 = H^+ + HCO_3^- \tag{6}$$

Carbon dioxide dissolved in water is a weak acid that can act as a corrosive agent at elevated temperatures. Carbon dioxide can especially attack iron in oxygen-free systems that have insufficient alkali for neutralization. Corrosion is primarily uniform.

The formation of calcareous deposit depends on carbon dioxide content. If calcium is present in water, the precipitation reaction of Eq. 7 occurs.

$$CO_2 + H_2O = CaCO_3 = Ca(HCO_3)_2 \tag{7}$$

The Langelier index provides an estimate of the possibility for precipitation of calcareous scales in fresh water. Calculation of the index uses measurement of the original pH and the pH when the water is in equilibrium with solid calcium carbonate, pH_s. The definition of saturation index is $pH - pH_s$. A positive index value indicates scale precipitation. Additions of calcium hydroxide or sodium carbonate can adjust it. The carbonate deposit might create an effective barrier for oxygen diffusion to metal. Corrosion then stops. If carbon dioxide concentration is too high, the index value is negative, and water will dissolve existing carbonate scale. Negative index values might be preferable in heat exchange applications where scale precipitation would decrease heat transfer rates. No protective scale will form if carbon dioxide content is excessively low and the saturation index is highly positive. The rapidly forming deposit will be porous and nonprotective. If the carbon dioxide concentration is only slightly too low to keep the bicarbonate in solution, calcium carbonate will deposit on the cathodic areas. The preferred saturation index value is also slightly positive at about 0.6–1.0. The saturation index is a thermodynamic value and does not indicate scale formation rates[9].

The Ryznar Stability Index is an empirical method for predicting scaling tendencies of water. This index has frequent use combined with the Langelier index to improve accuracy in predicting the scaling or corrosion tendencies of an aqueous solution. The Ryznar Stability Index is calculated from the Langelier saturation pH. It is the pH when the water is in equilibrium with solid calcium carbonate, pH_s. The Ryznar Stability Index is $2 \bullet pH_s - pH$, where pH is the measured pH. The values of Ryznar Stability Index relate to scaling and corrosion tendencies as Table 3 shows.

Table 3. Scaling and corrosion tendencies estimated by using the Ryznar Stability Index.

Ryznar Stability Index	Water tendency
4.0–5.0	Heavy scale
5.0–6.0	Light scale
6.0–7.0	Little scale or corrosion
7.0–7.5	Significant corrosion
7.5–9.0	Heavy corrosion
9.0 and higher	Intolerable corrosion

1.5 Sulfur and chlorine chemistry

Sulfur and chlorine compounds and their chemistry are especially important for corrosion. Sulfur compounds occur extensively in pulp and paper manufacture. In aqueous solutions, sulfur has several oxidation states from -2 to +6. Simple sulfur compounds are elemental sulfur; sulfur dioxide, SO_2; hydrogen sulfide, H_2S; and sulfide, S^{2-}; hydrosulfide, HS^-; and sulfate, SO_4^{2-}, ions. More complex compounds found in process liquors are thiosulfate, $S_2O_3^{2-}$; tetrathionate, $S_4O_6^{2-}$; sulfite, SO_3^{2-}; and dithionate, $S_2O_6^{2-}$, ions. The sulfate ion has the highest oxidation state. Since it is very stable, it is not oxidizing and will not affect the corrosion rate in aqueous solutions. The sulfite ion is the next most stable. It is a weak oxidizer. Sulfite, SO_3^{2-}; bisulfite, HSO_3^-; and sulfurous acid, H_2SO_3, are hydrolysis products of sulfite salts. Their relative amounts depend on pH. In acid solutions, H_2SO_3 is dominant. In alkaline solutions, the sulfite ion is dominant. H_2SO_3 is a strong, reducing acid. Hydrogen sulfide and its ions have the lowest oxidation states, and they are reducing compounds. In the pH range 11–14 for kraft cooking, sulfur compounds exist as an equilibrium mixture of S^{2-} and HS^-. Sulfide solutions are usually alkaline and reducing, and they cause a slight, uniform corrosion. Sodium thiosulfate accumulates in the kraft system due to oxidation of the sulfide ion during green liquor causticizing. Oxidation converts thiosulfate to polysulfide. Polysulfides are compounds of several sulfide ions with general composition of S^{2-}_x where $x = 2-5$. Polysulfides are weak oxidizers. In the presence of alkali, polysulfides can decompose to thiosulfate and hydrosulfide. The hydrosulfide ion, HS^-, is an active chemical in pulping. Dithionates are reducing, bleaching chemicals. Figure 6 shows the relationship between various sulfur compounds.

Sulfur dioxide and hydrogen sulfide will dissolve easily in water. Dissolved sulfur dioxide will form sulfur-containing acids. The simple hydrolysis of sulfur dioxide results in sulfurous acid, H_2SO_3. The corrosivity of water with high concentrations of SO_2 is comparable to sulfuric or sulfurous acids. Dissolved hydrogen sulfide will form a weakly acid solution. The general corrosion in H_2S containing waters is usually slight, but various forms of localized corrosion can proceed rapidly.

Chlorine compounds are common in industrial solutions. Dry gaseous chlorine will not attack most metals except titanium that will react violently at water concentrations less than 0.1%. Moist chlorine gas is a very corrosive environment. The most stable ion of chlorine is chloride ion. Chlorides are salts of hydrochloric acid. They decrease solution pH when dissolved. Chlorides are present in natural water and are therefore often impurities in process solutions. Chloride ion does not necessarily participate in corrosion reactions, but it is very harmful for the formation and stability of passive films. Chlorine forms many compounds with oxygen. The most stable chlorine and oxygen compound is perchlorate ion, ClO_4^-, that is usually not corrosive. Chlorate ion, ClO_3^-, is also stable in aqueous solutions. It is not corrosive except in very acid solutions. Four extremely reactive compounds are strong oxidizers: gaseous chlorine, Cl_2; hypochlorite ion, ClO^-; chlorine dioxide, ClO_2; and chlorite ion, ClO_2^-. These compounds have use in bleaching. The corrosivity of a chlorate solution is comparable with a chloride solution. Chlorite and especially the hypochlorite solutions are more aggressive.

Gaseous chlorine in water hydrolyzes very rapidly to form hypochlorous acid, HOCl, as Eq. 8 shows.

$$Cl_2 + H_2O = HOCl + HCl \tag{8}$$

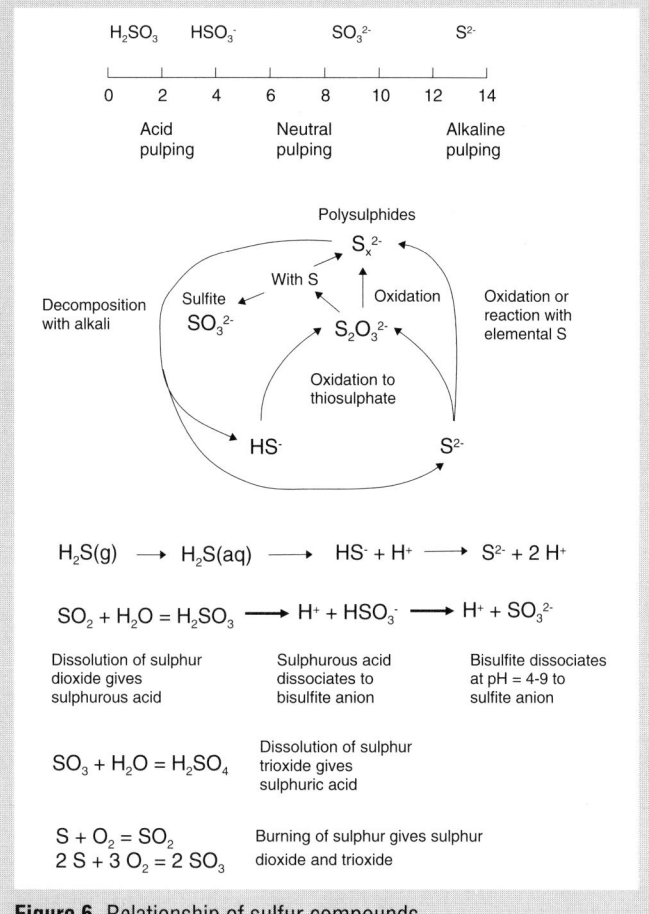

Figure 6. Relationship of sulfur compounds.

Hypochlorous acid is a weak acid that will dissociate further to hydrogen and hypochlorite, OCl⁻, ions. The dissociation depends on pH. The resulting solution is a mixture of hydrochloric and hypochlorous acids that is oxidizing and very corrosive. The relative concentrations of OCl⁻, HOCl, and Cl_2 are functions of pH and total chlorine concentration. Hypochlorites are salts of hypochlorous acid. Addition of hypochlorites such as NaOCl or $Ca(OCl)_2$ to water results in formation of hypochlorous acid with the corresponding alkali.

Chlorine dioxide is a more powerful oxidizer than chlorine. It is soluble in water without dissociation and stable in aqueous solutions for a long time. High pH, light, impurities, and heat will accelerate its dissociation. Chlorine dioxide can explode at a concentration above 30 g/L. Storage or transportation as a concentrate is therefore not possible. This requires preparation at the point of use[10]. Preparation of chlorine dioxide uses acidification or oxidation of chlorite ion (ClO_2^-) or reduction of chlorate ion (ClO_3^-) in the presence of an acid. Production from chlorite is uneconomical since the reaction is reversible. Industrial preparation uses chlorate as raw material. Commercially used reducing agents for production from chlorate are sulfur dioxide (SO_2), methanol (CH_3OH), chloride ion (Cl⁻), and hydrogen peroxide (H_2O_2). The choice of reducing agent for chlorine dioxide generation from chlorate affects reaction conditions, by-products, and economics. The Mathieson, Solvay, and R2 processes operate at atmospheric pressure. They use sulfur dioxide, methanol, and sodium chloride, respectively, as reducing agents. They produce a significant amount of spent sulfuric acid. Attempts to eliminate the waste acid led to the vacuum evaporator processes. The first vacuum process was R3. It used the same chemistry as the atmospheric R2. Further developments resulting in new processes have decreased sodium, sulfur, and chlorine by-products[11].

Electrolysis of aqueous solutions of sodium chloride with the anode and cathode in close proximity gives chlorine gas. Depending on the solution parameters (pH and temperature especially), various final products will result. At pH of 6.5–7.0, Eq. 9 shows the overall reaction.

$$Cl^-(aq) + 3\ H_2O \rightarrow ClO_3^-(aq) + 3\ H_2(g) \tag{9}$$

Under appropriate conditions of aqueous electrolysis, the chlorate ion, ClO_3^-, can oxidize further to the perchlorate ion, ClO_4^-. The total reaction for this oxidation is Eq. 10.

$$ClO_3^-(aq) + H_2O \rightarrow ClO_4^-(aq) + 2\ H^+(aq) + 2\ e^- \tag{10}$$

CHAPTER 5

Evaporation of the solutions following electrolysis produces chlorate and perchlorate salts. Both are powerful oxidizing agents producing hazardous combinations with metals or organic compounds. Reaction of the oxidizing chlorine compounds with organic substances results in oxidation of the organic material and chloride ions. Figure 7 shows the relationship between various chlorine compounds.

The term "active chlorine" has frequent use with bleaching compounds. Active chlorine means the amount of chlorine in percent by weight evolved on addition of hydrochloric acid. The relative amounts of chlorine, hypochlorous acid, and hypochlorite in a solution depend on pH as Fig. 8 shows. At a pH greater than about 3, virtually no dissolved molecular chlorine exists in the solution. With addition of hydrochloric acid, pH decreases, and all the compounds convert to chlorine gas that will then evolve from the solution.

The combined effect of chlorine and sulfur compounds is very complex. Thiosulfate pitting corrosion demonstrates this very well. Thiosulfate pitting unlike chloride induced pitting occurs below rather than

	0	2	4	6	8	10	12	14
	Cl_2		HOCl		ClO^-		$ClO_2^- + ClO_3^-$	
					ClO_2			
	Acidic chlorination		Cellulose degradation		Hypochlorite bleaching			

$\begin{cases} ClO_2 \\ Cl_2(aq) \\ HOCl \end{cases}$ + organic compound = oxidized compound + Cl^-

$2\ NaClO_3 + SO_2 + H_2SO_4 = 2\ ClO_2 + 2\ NaHSO_4$ — Chlorine dioxide is produced from chlorate salts

$NaClO_3$
$NaClO_4$ ← Evaporation produces chlorate and hypochlorite salts

$ClO_3^- + H_2O = ClO_4^- + 2\ H^+ + 2\ e^-$ — Electrochemical oxidation of hypochlorite to perchlorate

$2\ HOCl + OCl^- = ClO_3^- + 2\ H^+ + 2\ Cl^-$ — Chlorate formation from hypochlorous acid and hypochlorite

$Cl_2(g) + H_2O = HOCl + HCl$ — Dissolution and hydrolysis of gaseous chlorine

$HOCl = H^+ + ClO^-$ — Dissociation of hypochlorous acid gives hypochlorite

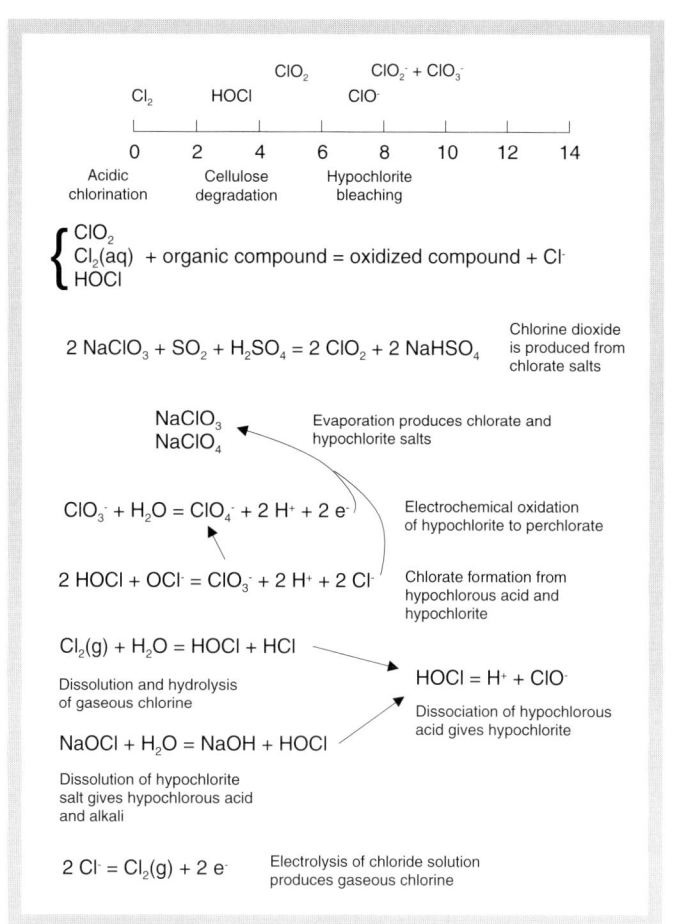

$NaOCl + H_2O = NaOH + HOCl$ — Dissolution of hypochlorite salt gives hypochlorous acid and alkali

$2\ Cl^- = Cl_2(g) + 2\ e^-$ — Electrolysis of chloride solution produces gaseous chlorine

Figure 7. Relationship between various chlorine compounds.

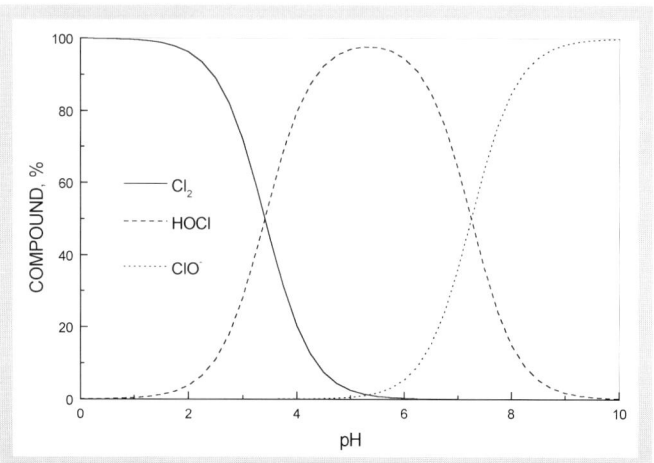

Figure 8. Relative amounts of chlorine, hypochlorous acid, and hypochlorite as a function of pH.

above a certain potential. In this case, this is the thiosulfate reduction potential. When hydrogen ions reduce thiosulfate in acid solution, a sulfur monolayer adsorbs on the metal surface. Adsorbed sulfur activates anodic dissolution and hinders repassivation. High concentration of hydrogen ions is also necessary to produce acidification of the pit. Thiosulfate pitting especially occurs with stainless steel grades that do not contain molybdenum. The pits are very stable and will not repassivate. Experimental results have shown that the worst cases of thiosulfate pitting occur within the molar concentration ratio shown in Eq. 11[12].

$$\frac{[SO_4^{2-}]+[Cl^-]}{[S_2O_3^{2-}]} = 10-30 \qquad (11)$$

Above this range, insufficient thiosulfate reaches the pit nucleus. Below this range, excessive thiosulfate reduction prevents acidification of the pit. Thiosulfate pitting is also potential dependent. At excessively high potentials, thiosulfate does not reduce sufficiently fast to form the adsorbed sulfur monolayer. At excessively low potentials, the metal dissolution is too slow to retain the acid concentrated pit environment.

1.6 Temperature and flow rate

Temperature influences rates of electrochemical reactions and properties of the electrolyte. A rule of thumb for an activation controlled reaction is that a change of 10°C will double the reaction rate. For a mass transfer controlled reaction, a change of 30°C will double the reaction rate. Two common observations on the effect of temperature increase are a very rapid or exponential rise in corrosion rate or a negligible effect followed by a rapid acceleration of corrosion as Fig. 9 shows.

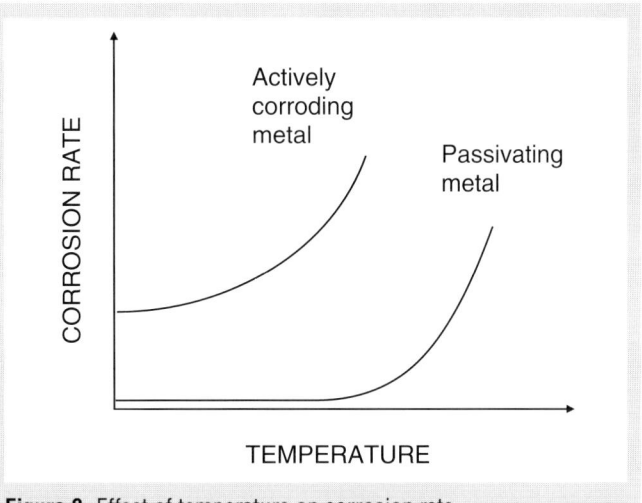

Figure 9. Effect of temperature on corrosion rate.

The first case is common for an actively corroding metal. The second case occurs for passive metals with corrosion potential close to transpassive potential. The oxidation power of the corroding material increases with increasing temperature. At some point, the metal begins to corrode actively on the transpassive range[13].

CHAPTER 5

The pH of the solution will decrease with increasing temperature. Solubilities of dissolved gases decrease with increasing temperature. The effect of temperature can be very complex. In closed systems, the corrosion rate of iron increases linearly with temperature. In open systems, a maximum corrosion rate occurs when increasing iron dissolution rate is overcompensated by decreasing cathodic reactant (O_2) solubility. In high temperature applications, the hydrogen evolution overpotential may decrease with increasing temperature accelerating corrosion. At temperatures well above atmospheric boiling point, reaction product layers often become more dense and decrease corrosion rates.

The flow rate will directly influence the mass transport rates of reactants and reaction products. An increase in the electrolyte flow rate will usually increase the corrosion rate. If the system is under activation control, the flow velocity does not affect the reaction rate. Under mass transport control, the reaction rate depends linearly on the square root of flow rate. Turbulent flow can cause even faster corrosion rates since the protective film will suffer damage. The increase in flow rate then has no effect or a slight effect on the corrosion rate until the film actually breaks down.

If corrosion is due to oxidizer present in small amounts such as oxygen reduction in natural water, an increasing mass transport of oxygen may overcome the kinetic barrier for passivation with decreasing corrosion rate. The flow rate also influences corrosion morphology. In stagnant or slowly moving solutions, corrosion is local, and the morphology is pitting or grooves. In more rapidly moving solutions, corrosion appears more uniform. This is due to removal of corrosion products that would otherwise attach on the surface and create concentration cells.

1.7 Corrosion in pulp preparation

Pulp is a wet slurry of fibers and water. Pulping separates cellulose fibers adhered with lignin. The raw materials for pulp manufacture are usually hardwoods and softwoods. Softwood pulp comes from evergreen trees such as pines and firs. Hardwood pulp results from deciduous trees that drop their leaves. Softwood pulp has long fibers that give paper strength. Hardwood fibers are short and provide smoothness, bulk, and body. Different raw materials require different pulping processes or pulping environments. Making pulp can use a mechanical or chemical process or a combination. Mechanical pulping grinds logs or wood chips to break them into short fibers. The lignin that causes paper to yellow and disintegrate over time remains in the pulp. Thermomechanical pulping uses steam under pressure to soften wood chips before mechanical grinding. Chemical pulping cooks wood with chemicals to separate the cellulose fibers. This process also dissolves most lignin. Chemicals used in most wood pulping also helps to dissolve impurities. Pulping processes should produce a pulp with low residual lignin content to decrease the amount of bleaching. Unfortunately, dissolving lignin below a certain level during pulping will adversely influence pulp properties. Pulping modifications to achieve low residual lignin level and preserve pulp properties can change the process conditions to become more corrosive.

In cooking, the digester environment is an alkaline, neutral, or acid solution. The most important pulping process is the kraft process that uses a hot alkaline mixture of sodium sulfide and sodium hydroxide. Another name for the kraft process is the sulfate process. Kraft pulping began very early in the twentieth century. Continuous digesters appeared in the 1950s. The solutions used in the kraft process are usually oxygen-free alkali solutions with varying concentration of sulfide species. Before development of the kraft process, the main pulping process was the sulfite process. The sulfite process used acid and calcium bisulfite until about 1950. Due to raw material limitations of acid cooking, other chemicals such as sodium, magnesium, and ammonium hydroxide were introduced. The sulfite cooking liquor consists of free sulfur dioxide and bisulfite dissolved in water. The pH in the sulfite process cooking liquor can range from 1 for acid cooking to 10 for alkaline cooking. The kraft process produces stronger pulp and allows more efficient recovery of cooking chemicals than the sulfite process, but lignin removal is less efficient in the kraft process. This requires more bleaching. Table 4 compares the environments of major pulping methods[14].

Table 4. Chemical environment of major pulping methods[14].

	Kraft	Acid sulfite	Bisulfite	Neutral semichemical
Chemicals	NaOH Na$_2$S	H$_2$SO$_3$ M(HSO$_3$) M = Ca,Mg,Na,NH$_4$	M(HSO$_3$) M = Ca,Mg,Na,NH$_4$	Na$_2$SO$_3$ Na$_2$CO$_3$
Cooking time	2–4 h	4–20 h	2–4 h	1/4–1 h
Liquor pH	over 13	1–2	3–5	7–9
Temperature	170°C–180°C	120°C–135°C	140°C–160°C	160°C–180°C

1.8 Kraft pulping

The kraft process is the predominant pulping method. About 80% of the world pulp production uses the kraft process. It has rapid pulping rates, adapts to many types of wood feedstock, produces strong pulp, and has low chemical costs. The process also has many deficiencies. The yield of pulp is low, pulp requires extensive bleaching, sulfur chemicals lead to malodorous emissions, and high capital investment is necessary for the chemical recovery system and environmental processing technology. Traditional kraft pulping consists of chemical impregnation at moderate temperature followed by delignification at higher temperature. Wood chips are heated in a digester with hot cooking liquor. The lignin and other components dissolve releasing the cellulose fibers as pulp. The digester is a vertical, cylindrical pressure vessel usually constructed of carbon steel rings with some parts made from stainless steel. Passivation of carbon steel in hot alkaline solution requires some oxidizing compound such as oxygen. Increases in temperature, alkalinity, or sulfide content will disturb passivation. The design criterion of the carbon steel structure is strength corresponding to maximum operating pressure. This requires some corrosion allowance. Today, the digester often has a lining of a more corrosion resistant alloy at least in critical places. A stainless steel lining such as a weld overlay or cladding or even solid duplex stainless steel can be feasible.

CHAPTER 5

Figure 10 shows the unit processes and mass balances of the kraft pulping process. After cooking, the black liquor is washed from pulp and treated to regenerate the valuable cooking chemicals. The new cooking liquor contains sodium hydroxide (NaOH) and sodium sulfide (Na_2S). The spent cooking liquor contains the sodium compounds as inorganic compounds and bound to organic acids. In addition, the spent cooking liquor contains dissolved lignin, carbohydrates, etc. The spent cooking liquor is separated from the pulp in a series of washing operations producing lean black liquor. This is concentrated to facilitate its burning in the recovery boiler. The recovery boiler converts spent cooking chemicals back to usable ones. The smelt, i.e., the molten mixture of sodium carbonate and sodium sulfide, of the recovery boiler is dissolved in water and treated with calcium carbonate to regenerate cooking liquor with sodium hydroxide and sodium sulfide. In the kraft pulping process, many more steps are used in chemical recovery than in pulping.

Digestion can occur in small batch digesters or larger continuous digesters. Batch digestion is the oldest method. In this technique, all cooking steps occur in the entire digester. The digester is loaded with chips and heated according to a predetermined procedure. In a batch digester, the corrosive environment changes in cycles. With continuous digestion, the digester has vertical zones with various purposes. In a continuous digester, the environment depends on the elevation but remains essentially the same as Fig. 11 shows. The wood chips first undergo impregnation at the top followed by cooking in the middle and washing at the bottom of the unit. Cooking, extrac-

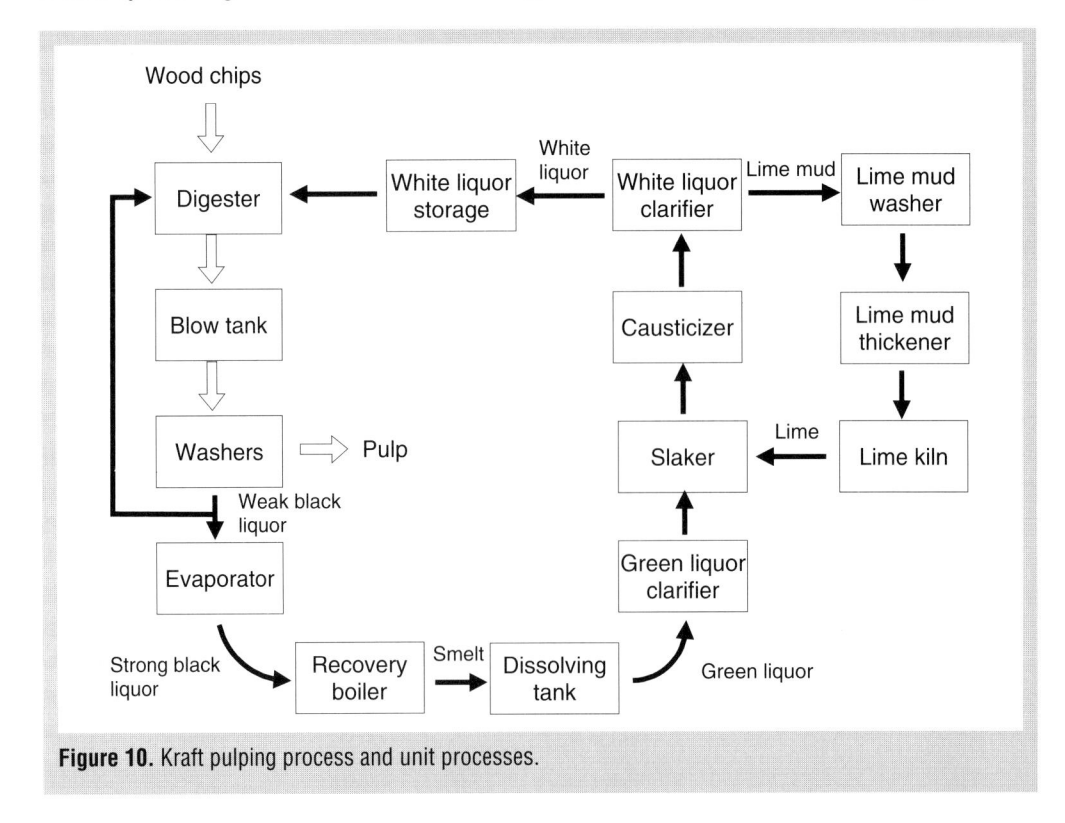

Figure 10. Kraft pulping process and unit processes.

tion, and washing liquors circulate through the charge using internal screens and piping. With a continuous digester, the chips are usually impregnated first with white liquor or sometimes with a mixture of white and black liquor before loading into the digester. The impregnation vessel has about the same environment as the digester.

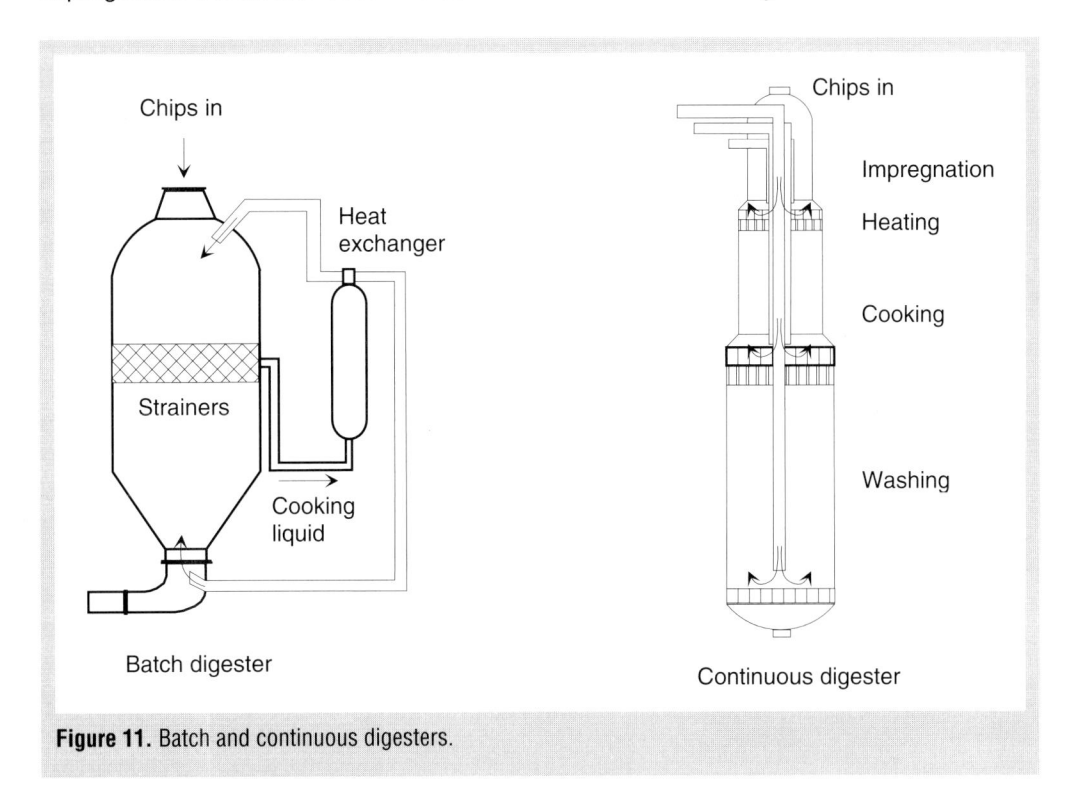

Figure 11. Batch and continuous digesters.

The corrosion rate in batch digesters is generally higher than in continuous digesters. This is because the cooking liquor in batch digesters is more corrosive in the beginning and protective films can undergo damage during blowdown. Closure of mills has resulted in use of bleaching stage solutions for washing and making cooking liquor. Solutions from oxygen delignification are more oxidizing, and solutions from chlorine or chlorine dioxide delignification have more chlorides than solutions normally used in pulping. Both factors can result in a more corrosive environment inside the digester.

The cooking liquor in the kraft process is a mixture of sodium hydroxide (NaOH) and sodium sulfide (Na_2S) known as white liquor. Besides the active cooking components, the cooking solution contains sulfates, carbonates, thiosulfates, and chlorides. The cooking liquor is usually a mixture of white liquor and spent black liquor from previous cooking. The composition of the cooking liquor varies, and the industry has developed various conventions to describe the liquors. A common term is the sulfidity (S) of Eq. 12 that is the percentage ratio of Na_2S to active alkali NaOH + Na_2S. Active alkali is another widely used term[15].

CHAPTER 5

$$Sulfidity\ [\%] = \frac{Na_2S}{[NaOH]+[Na_2S]} \cdot 100 \tag{12}$$

All the chemicals are expressed as their equivalent weight of Na_2O. Two moles of NaOH are equivalent to one mole of Na_2O, and one mole Na_2S is equivalent to one mole Na_2O. Typically, the white liquor has approximate molar proportions of 2.5:1 for hydroxide and sulfide[15,16]. The normal sulfidity of white liquor is 25%–30%.

The pH of the white liquor is 12–14. It contains considerable organic compounds but no dissolved oxygen. The environment is therefore reducing. The cooking temperature is about 150°C but varies between 80°C–180°C. A typical mixture of 100–120 g/L sodium hydroxide and 20–50 g/L sodium sulfide in the absence of oxidizing species is not very corrosive to carbon steel. The corrosive agents in cooking liquor are excessive levels of sodium hydroxide and oxidizers such as oxygen, elemental sulfur, polysulfides, and sodium thiosulfate. The white liquor of a modern kraft mill may contain as little as 0.015 mol/L chlorides (less than 0.3 g/L calculated as NaCl)[17], but the chloride content has no effect up to 25 g/L NaCl[18].

Stronger cooking liquors with higher amounts of NaOH and Na_2S are usually more corrosive. For example, the corrosivity is highest in the beginning of batch cooking. If sodium hydroxide concentration increases, pH increases and the passivity of iron is lost. In continuous digesters, spontaneous passivation usually occurs in areas of lean liquors. High sulfide concentration restricts passivation that would happen in the later stages of batch cooking since it increases the critical current density for passivation.

Thiosulfate has an oxidizing nature, but it will not assist passivation of carbon steel. Instead, it will increase corrosion rate and form a nonprotective iron sulfide scale or soluble iron and sulfur complexes. The primary sulfur compounds, sulfide and thiosulfate, will react to form polysulfides such as Na_2S_2. Polysulfides will function as oxidizers increasing corrosion until a critical concentration for passivation occurs. This concentration is about 3–6 g/L for carbon steel[19]. The carbon steel may also suffer from alkaline stress corrosion cracking. This happens at a relatively narrow potential range at the active to passive transition. Stress corrosion cracking requires high tensile stresses most often found at the welds. High sulfide content may increase crack growth rates.

Since the 1950s, empirical equations have estimated the corrosivity of kraft cooking liquors to carbon steel. Stockman and Ruus developed the first equation. They showed that corrosion of carbon steel depends on the concentration of sodium sulfide, Na_2S; thiosulfate, $Na_2S_2O_3$; and NaOH according to Eq. 13[20]. The corrosivity, C, in this equation means mass loss in g/cm^2 per one cook in a batch digester. Concentrations are in g/L.

$$C = -3.6 + 0.03 \cdot [Na_2S] + 0.11 \cdot [NaOH] + 0.04 \cdot [Na_2S] \cdot [Na_2S_2O_3] \tag{13}$$

Their results also indicated that secondary factors such as reaction product layers on surfaces, properties of the metal, and heating method will significantly influence corrosion. The idea of Eq. 13 is that sodium hydroxide acts as an activator that raises solution pH to such high levels that the protecting iron oxide film dissolves. Sodium sulfide, Na_2S, and thiosulfate, $Na_2S_2O_3$, will react to form polysulfide, Na_2S_2, and sulfite, Na_2SO_3, as Eq. 14 shows. Polysulfide is an oxidizer, and an increase in its concentration will increase the dissolution rate of iron.

$$Na_2S + Na_2S_2O_3 = Na_2S_2 + Na_2SO_3 \tag{14}$$

Roald[21] created two equations to describe the corrosivity of cooking liquors. At low concentrations, he assumed that polysulfide diffusion rate controls the corrosion rate. At higher concentrations, only a portion of these ions reaching the steel surface will react as Eq. 15 shows. His equations use sodium sulfite concentration, and corrosivity relates inversely to the sodium sulfite concentration. At low sulfite concentrations, the corrosion rate relates linearly to sulfite concentration. At high concentrations, it relates to its square root. These equations suggest that low sulfite content is detrimental since it will shift the equilibrium of Eq. 14 to the product side and therefore increase polysulfide concentration.

$$Corrosion\ rate = constant \cdot \frac{[Na_2S] \cdot [Na_2S_2O_3] \cdot ([NaOH] + 0.4 \cdot [Na_2S_2])}{[Na_2SO_3]} \tag{15}$$

$$Corrosion\ rate = constant \cdot \frac{[Na_2S] \cdot [Na_2S_2O_3] \cdot ([NaOH] + 0.4 \cdot [Na_2S_2])}{\sqrt{[Na_2SO_3]}}$$

In kraft white liquor, Wensley and Charlton[3] concluded that sodium sulfide and sodium thiosulfate will accelerate corrosion. Sodium sulfide is the major corrosive agent. Sodium polysulfides control the corrosion potential analogous to oxygen. Sodium sulfite and sodium sulfate have no effect.

Some modified continuous cooking processes with high pH, high temperature, and low lignin content can be more corrosive than conventional ones. High pH and temperature will naturally cause a more corrosive environment, but the lignin has passivating properties. Higher extraction of lignin in the top of the digester will cause low lignin content in the washing stage on the bottom of a continuous digester[22,23].

Batch and continuous digesters often have stainless steel linings. Compound plates also have use. Low carbon grades AISI 304L and AISI 316L can tolerate high chloride concentrations in high alkalinity environments. Liquor heater tubes made from these common grades may suffer from stress corrosion cracking: ferritic or duplex grades are more resistant. Washing, screening, and dewatering equipment can use the common grades AISI 304L and 316L, but higher chloride contents may require use of more alloyed grades. This depends on pH. Recovery of spent cooking chemicals begins with evaporation where stainless steel today is the most common material. If the alkali content and temperature are not too high, common stainless steel grades perform well. Use of duplex stainless steels as solid metal or cladding is becoming more popular in

CHAPTER 5

batch and continuous digesters, blow tanks, etc. The typical grade is S31803 (common name 2205) with 22% Cr and 3% Mo[24–26].

Direct injection of steam or indirect heating of the cooking liquor heats batch digesters. Indirect heating always heats continuous digesters. The cooking liquor circulation system contains heat exchangers heating the black liquor to about 170°C–180°C, pumps, and piping. Batch digesters usually have one heat exchanger, and continuous digesters have several. Most heat exchangers are tube units with cooking liquor in the tubes and steam in the shell. Stainless steel use is common. Originally, the heaters used ferritic grade AISI 410, but they suffered from general corrosion and fatigue. Austenitic grades AISI 304 and 316 were then introduced, but they can suffer from chloride-induced and alkaline stress corrosion cracking. This happens in most cases on the steam side when cooking liquor leaks onto the steam side. Common austenitic grades AISI 304 and 316 can show general corrosion if pH and temperature are too high. High nickel alloys and duplex stainless steels are good choices if austenitic grades fail[24,25].

The digester empties under pressure to a blow tank, and the softened chips disintegrate into fibers. Blow tanks and especially blow plates experience erosion corrosion: hard stainless steels perform best. The fibers then separate from the spent cooking liquor in several brownstock washing stages. The brownstock washing system has a screening system; "knotter" to remove uncooked chips, knots, and debris; and a series of rotary drum filters with intermediate repulpers for dilution and agitation or a continuous washing machine. The system usually has countercurrent flow. During washing, the wash water is forced between fibers where it dilutes the remaining cooking liquor. It is then pressed out in the filter. The corrosivity of the brownstock depends on chloride level, temperature, pH, and level of aeration, i.e., redox potential. Brownstock washing equipment is usually carbon steel or AISI 304 and 316 stainless steel[24,26]. Alkaline, aerated solutions with low chloride levels are less corrosive. Use of filtrate from chlorine or chlorine dioxide bleaching stages for washing and dilution will affect the pH and increase the chloride level. Vapors removed from the digester during blowing undergo heat recovery with the condensate pumped to brownstock washing. The remaining gases are combusted in kilns or boilers for pollution control.

The spent cooking liquor containing dissolved wood constituents is black liquor. From the brownstock washing, the black liquor goes to chemical recovery. A series of evaporators usually made of austenitic stainless steels first concentrates the black liquor. Since the alkali concentration and temperature are lower than in the cooking chemical heating system, corrosion problems are less frequent. Concentrated black liquor goes to a recovery boiler.

1.9 Sulfite pulping

Sulfite digestion has many similarities to the kraft process although the cooking liquor chemistry differs. Both processes can use batch or continuous digestion and recover spent cooking chemicals. Process steps following the digestion such as brownstock washing, bleaching, etc., can be identical. The product of sulfite digestion is red stock. The acid, calcium bisulfite based solution with excess dissolved sulfur dioxide only has

Corrosion in different environments

use for some species of wood. The pH of the solution must be low due to the limited solubility of calcium[27]. By using more expensive chemicals based on magnesium, sodium, and ammonia, higher pH values are possible. This makes the pulping method useful for most wood species. Calcium began to disappear in the 1940s as other so-called soluble bases were introduced. Chemical recovery soon became part of the sulfite process due to environmental reasons and the higher cost of new bases.

The sulfite process is the most corrosive method of pulping. An acid sulfite solution resulting from dissolution and hydrolysis of sulfur dioxide gives sulfurous acid with pH below 2. Sulfite cooking liquor contains free sulfur dioxide dissolved in water and sulfur dioxide in the form of bisulfite. The ratios of sulfurous acid, bisulfite, and sulfite depend on the solution pH. Addition of an alkali increases solution pH and makes bisulfite ions more stable. At a pH of 2–7, the solution is buffered. In alkaline solutions, the sulfite ion is stable. Following are the types of sulfite pulping:

- Acid sulfite: pH is approximately 1–2. The solution contains H_2SO_3 and bisulfite $(X)HSO_3$.

- Acid bisulfite: pH is approximately 1.5–4. True bisulfite pH is approximately 4–5. The solution contains bisulfite $(X)HSO_3$.

- Neutral sulfite: pH is approximately 6–9. The solution contains sulfite $(X)SO_3$ and carbonate $(X)CO_3$.

- Alkaline sulfite: pH is higher than 10. The solution contains sulfite $(X)SO_3$ and hydroxide $(X)OH$.

(X) denotes the cation of the base used in the process, e.g., Ca^{2+}, Mg^{2+}, Na^+, and NH_4^+.

Figure 12 gives an overview of the sulfite pulping process. Burning elemental sulfur and adsorbing the SO_2 into an alkaline solution prepares the cooking liquor. The raw cooking acid is fortified in accumulators with SO_2 from the digesters. Fortification is necessary because of the limited solubility of SO_2 in water at atmospheric pressure. The cooking usually occurs in a batch digester.

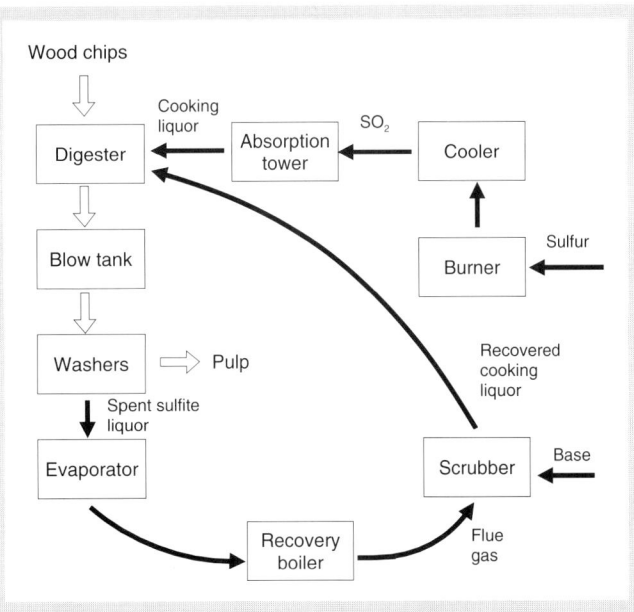

Figure 12. Overview of the sulfite pulping process.

CHAPTER 5

The chips and hot acid are heated according to a predetermined procedure by forced circulation of cooking liquor. During cooking, sulfurous acid reacts with lignin to produce insoluble lignosulfonic acid that converts to more soluble salts in the presence of base. At the end of the cooking, heating ceases and pressure begins to drop. When the pressure is sufficiently low, the digester is emptied into a blow tank and pulp is washed from residual cooking liquor[27].

The sulfite pulping liquor is made by burning elemental sulfur at about 1100°C to form sulfur dioxide, cooling it to about 200°C, and dissolving it in a solution of calcium, sodium, magnesium, or ammonium hydroxide. The pH of the resulting cooking liquor depends on the base and amount of dissolved SO_2. Depending on the process, the pH can vary from 1–10[27]. The redox potential of the cooking liquor depends on the sulfur dioxide concentration. Pulping liquors may contain thiosulfates that cause pitting of stainless steels. Thiosulfate forms spontaneously. Austenitic stainless steels are generally not suitable for this purpose. Ferritic or duplex grades are necessary for burning. Cooling and washing towers are often acid resistant bricks or lined with such bricks. Stainless steel use is possible, but the steel should have high molybdenum content. An improved type AISI 317L (for example UNS S31725 with 18% Cr, 15% Ni, and 4.5% Mo) or even more alloyed 904L is necessary[26].

Cooking temperatures are usually below 140°C for calcium-based cooking, but 160°C or higher is possible with other bases. The rate of pulping increases with decreasing pH, increasing temperature, and higher sulfur dioxide concentration. Generally, higher pulping rate corresponds to higher corrosivity. Pulping rates also depend on the base used and increase in the order Ca → Na → Mg → NH_4 [27]. In sulfite cooking, the digester may have a stainless steel lining, or it can be made with stainless steel clad plate. Stainless steel type AISI 316L is no longer sufficiently resistant, and type AISI 317L or 18% Cr, 15% Ni, and 4.5% Mo grades are preferable. Localized corrosion and stress corrosion of type AISI 316L have occurred with chloride levels of about 1 g/L[16]. Condensing gases may cause dew point corrosion due to drops of concentrated sulfuric acid. Type 904L is then necessary. In process piping and associated equipment, the trend is to specify type AISI 316L or AISI 317L. For the most difficult parts, type 904L is useful. Washing, screening, and dewatering equipment can use types AISI 316L or 317L, but type AISI 304L is not always resistant due to small amounts of bisulfite present in the residual liquor. Under-deposit corrosion may require Type AISI 317L or even higher alloyed materials[25]. If the molybdenum alloyed stainless steel types fail, use of nickel alloys or titanium is necessary.

The sulfite pulping industry was an initial large scale user of stainless steel. The standard material to handle fresh and spent liquors soon became type AISI 316, i.e., molybdenum alloyed austenitic stainless steel. This type is not sufficiently resistant today. In the sulfite process, the digester and auxiliary equipment use highly alloyed stainless steel. Digesters in the sulfite process use stainless steel or linings of stainless steel. High molybdenum content is essential with the minimum content often considered 4.5%. Duplex stainless steel S31803 has also had use in sulfite digesters[25,26]. The most important corrosion problems of stainless steels in sulfite mills relate to sulfur dioxide concentration. Sulfur dioxide can maintain the passivity against acidic cooking liquor or it can form sulfuric acid by decomposition or SO_3 oxidation. The redox potential of the cooking liquor depends on the SO_2 concentration. Lower solution pH requires higher SO_2 concentration to passivate the steel. If stainless steel is in an active state, an increase of SO_2 concentration will increase corrosion rate as with other oxidizers. In batch cooking, the conditions are less severe than in continuous cooking since the passive film can repair itself as temperature decreases and fresh SO_2 enters[26].

1.10 Mechanical pulping

Mechanical pulp is made by grinding logs or refining chips. Mechanical pulp contains all the natural tree substances. Some manufacturing processes are simpler than those used to produce chemical pulp partly because a recovery system for chemicals is not necessary. Using higher pressures, temperatures, or chemical treatment often enhances the pulping methods. The two primary types of mechanical pulp are groundwood pulp and refiner pulp. Both types have variations. Groundwood was once the predominant means of mechanical pulping. In the simplest version, debarked logs press against a rotating grindstone. Water sprayed on the stone prevents heating and burning of fibers and removes them as a slurry. In the absence of chemical addition, the solution is usually slightly acidic at pH = 4–6 due to fatty and resin acids[16]. Corrosion is usually not a problem in these units.

Disc refining is a newer version of mechanical pulping introduced in the 1960s. It can produce higher strength pulp and use chips and residues as raw material. The refiner plates are cast iron, steel, or stainless steel often coated with some wear-resistant metallic or ceramic coating. A screw feeder introduces the raw material to the eye of the refiner. Water supplied through the eye controls pulp thickness[26]. Thermomechanical pulping is a modification of disk refining. The raw material undergoes steaming for a short period before and during refining. The use of steam softens the chips producing a higher percentage of long fibers. Heating and refining usually occur under pressure. The maximum temperature is about 140°C[28]. To separate impurities and long stiff fibers, a hydrocyclone screens and cleans the pulp. The rejected fibers undergo further refining. Corrosion is a more severe problem in disk refining than in groundwood due to higher temperatures and pulping chemicals present in

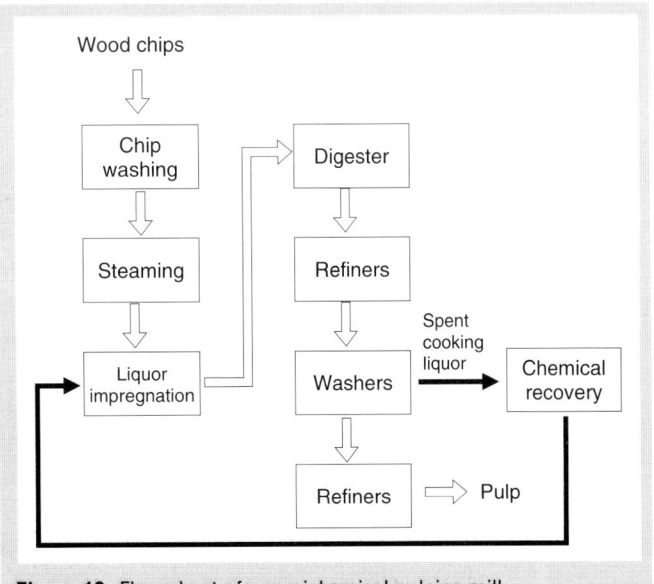

Figure 13. Flow sheet of a semichemical pulping mill.

some versions. Wear can occur due to abrasion and high velocities of steam and water[16,26].

Use of chemicals and mechanical treatment to separate the cellulose fibers of wood is chemimechanical or semichemical pulping. The wood chips are first partially digested by chemical means, and the remainder is done mechanically usually using disc refiners. Figure 13 shows the flow sheet of a semichemical mill. The chemical stage in the chemimechanical pulping is often similar to the kraft or sulfite processes. Corrosion problems and material selection are therefore identical.

The pulping equipment in semichemical pulping is subject to various corrosive environments. The solution in mechanical pulping is slightly acid due to organic acids dissolved from the wood. Addition of sodium carbonate will increase solution pH reducing plate corrosion and preventing accumulation of metal ions in the pulp[16]. The chlorine content depends on the raw material and degree of closure. Temperatures are 40°C–60°C in cold grinding and 60°C–90°C in hot grinding[16]. The semichemical pulping processes contain two process stages: partial dissolving of lignin with chemicals and mechanical refining. The selection of construction materials depends on the chemicals used in the dissolution stage. The neutral semichemical pulping processes contain 120–200 g/L sodium sulfite and sodium carbonate, bicarbonate, or hydroxide to maintain pH level at 7.2–9.0. Cooking temperatures are about 170°C–190°C[25]. The liquors are more corrosive than kraft liquors requiring use of stainless steels of Type AISI 316L up to and including the washing stage.

2 Corrosion in the bleach plant

Pulp consists of cellulose fibers and about 5% lignin. Lignin is a natural resin that gives pulp a brownish color. Bleaching the pulp with a series of treatments increases the brightness of pulp. In general, bleaching sulfite pulp is easier than kraft pulp. Hardwood pulps are easier to bleach than softwood pulps due to the nature of the residual lignin. The nature of the pulp influences the compositions and corrosiveness of bleach plant solutions. Dissolution and removal of lignin begin in the digestion process and finish in the bleaching process to give white fibers. At the beginning of bleaching, lignin dissolves. At the end, colored compounds are removed or made colorless. Brightness formerly resulted by dissolving the lignin molecule using chlorine compounds. Dissolution and removal of lignin makes the remaining cellulose fibers appear white. Oxygen has use today to break down lignin molecules and to bleach the dark spots created by non-cellulose components of wood. Between bleaching stages, alkaline treatment extracts the reaction products from the pulp.

Three broad classes of traditional bleaching systems use the following chemistry:

- Oxidizing chlorine-based systems
- Chlorine-free oxidizing systems
- Chlorine-free reductive systems.

Corrosion in different environments

The chemicals used in bleaching can include chlorine, chlorine dioxide, sodium or calcium hypochlorite, oxygen, ozone, hydrogen peroxide and peracids, caustic soda (NaOH), or quicklime (CaO). New variations of bleaching processes and chemicals are continuously being introduced. For example, chlorine dioxide has substituted for molecular chlorine. This reduces the amount and degree of chlorinated product release. For any pulp bleaching stage, the important process variables are pH, temperature, and retention time. Considering the variety of chemicals, one can easily understand that the most severe corrosion environments occur in bleaching plants. For example, stainless steels have wide use as construction materials in bleach filters and their support structures, suction pipes, suction heads, and pumps. They can suffer from pitting and crevice corrosion, erosion corrosion, stress corrosion cracking, corrosion fatigue, and sometimes also cavitation. The basic problem in bleach plant corrosion prevention is to define clearly the effect of factors controlling environment corrosivity such as the following:

- Chloride level
- Residual oxidant level (Cl_2, O_2, O_3, etc.)
- pH
- Temperature
- Flow conditions
- Deposits
- Temperature and concentration gradients.

The general tendencies are usually known. For example, Fig. 12 of Chapter 4 shows the effect of redox-potential, chloride concentration, and temperature schematically. The problem is to define clearly the critical values of chloride concentration, redox-potential (residual oxidant level), temperature, etc., for different materials. Another subject is incompatible material and environment combinations such as titanium in alkaline peroxide. These should be avoided unless verification has occurred under conditions that corrosion will not occur. Chapter 6 in this volume discusses material selection.

A typical bleaching system has three different unit processes: delignification with an oxidant, alkaline extraction between bleaching stages for removal of lignin and non-cellulose carbohydrate, and oxidation for final brightness development. These unit processes can combine in different ways to obtain the desired products. This complicates material selection since the corrosive environment in two successive steps is very different. Selection of the best possible material for a single stage could therefore result in heavy corrosion when using the same material in the next stage. Several problems have occurred when pulp enters from one stage to another creating a rather unpredictable corrosive environment. Bleaching proceeds in separate stages. Each stage requires a mixer to blend fibers, chemicals, and steam; a reaction tower; and a washer separating treated fibers from waste as Fig. 14 shows. Some bleaching sequences require use of pulps with different consistency necessitating dilution and thickening of pulp between

CHAPTER 5

stages. The construction materials have usually been selected so that they can handle the assumed corrosive environments. Changes and gradients in temperature, concentration of species, splashes, deposits, etc., can cause more aggressive conditions.

Two approaches are possible in the chemical bleaching of pulp. One approach is to remove residual lignin almost totally. Some chemicals do not attack the lignin but destroy some compounds contributing to pulp color. This is brightening to distinguish it from true bleaching. Since brightening does not remove lignin, it will rapidly discolor these paper grades. Table 5 shows some abbreviations used for various bleaching stages. Chlorination and extraction usually occur in sequence to delignify the pulp since very little brightening happens there. The oxygen stage has use primarily for delignification. A hypochlorite stage adds flexibility to the system compared with a chlorine stage. A chelation stage to remove metal ions is often necessary with a peroxide stage[29].

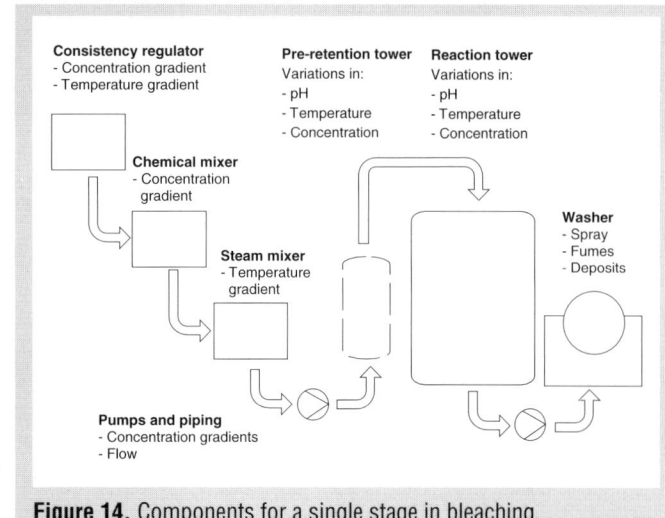

Figure 14. Components for a single stage in bleaching.

Table 5. Abbreviations used for various bleaching stages.

Process stage	Abbreviation	Chemicals
Acidification	A	H_2SO_4, SO_2
Chlorine delignification	C	Cl_2
Chlorine dioxide delignification	D	ClO_2
Alkaline extraction	E	NaOH
Alkaline extraction with O_2	E_O	NaOH, O_2
Alkaline extraction with O_2 and H_2O_2	E_{op}	NaOH, O_2, and H_2O_2
Hypochlorite	H	NaOCl, NaOH
Oxygen delignification	O	O_2, NaOH
Peroxide	P	H_2O_2, NaOH
Dithionite	Y	$Na_2S_2O_4$
Chelation, complexation of metal ions	Q	DTPA, EDTA
Ozone	Z	O_3, O_2

World production of kraft pulp bleached entirely without elemental chlorine (ECF pulp) or chlorine compounds (TCF pulp) is gradually increasing to meet demand. Residual delignification typically occurs in an oxygen reactor. Ozone and hydrogen peroxide have use as brightening chemicals. Totally chlorine free (TCF) pulping reduces the chemical oxygen demand of the effluent and produces no chlorinated organic compounds. The wastewater from a TCF bleaching process resembles the effluent from a typical paper mill. Bleaching with oxygen chemicals, i.e., oxygen, ozone, and peroxide, is also less corrosive toward the equipment than bleaching with chlorine chemicals since chloride ions are absent[30].

Delignification uses chlorine, chlorine dioxide, or oxygen. These chemicals are much more selective in the delignification process than any cooking liquor in the digestion. The cooking liquors cannot remove lignin below a certain level irrespective of chemical concentration or retention time. Oxygen cannot remove all lignin since it attacks only certain chemical bonds in the lignin molecule. Oxygen delignification has use as a preliminary lignin removal step before bleaching. For higher lignin removal, acid treatment with chlorine or chlorine dioxide is necessary. To achieve brightness, cleanliness, and strength, bleaching must use several stages.

Chlorine bleaching (C stage) generally provides the best performance with the least damage to fibers. During chlorination of the pulp, the lignin undergoes oxidization or chlorination by a substitution reaction between hydrogen in lignin and chlorine. The substitution reaction primarily influences the lignin, but oxidation also influences carbohydrates and decreases pulp strength. Both reactions change lignin to a more alkali-soluble form. The reaction products are hydrogen and chloride ions[29]. Acid conditions favor a substitution reaction. Chlorination usually uses low consistency to ensure dissolution of chlorine and thorough mixing.

Delignification with chlorine is a strongly acidic and oxidizing environment. The pH is usually 2–3, but it may be less than one. Chloride content is about 1–10 g/L. Residual chlorine content can be as high as 0.1 g/L. Temperatures were previously about 20°C–30°C to avoid decrease of pulp quality, but addition of chlorine dioxide has allowed a temperature increase up to 60°C. Chlorine bound in organic compounds causes problems in waste water treatment, and chlorides prevent direct use of filtrates in other stages. The bleaching towers normally use steel and concrete lined with acid resistant bricks or plastic. The corrosive environment in a C stage is due to the following:

- High acidity
- High redox-potential due to chlorine
- High chloride concentration
- High temperature when using ClO_2.

The corrosivity of the C stage can most efficiently be controlled by controlling pH and residual chlorine concentration. Factors that favor complete reaction of chlorine and low residual level are adequate retention time in the chlorination tower, thorough mixing of pulp and chlorine, higher temperature within the conventional range, and controlled

CHAPTER 5

addition of chlorine to minimize high levels of residual chlorine. The lined chlorination tower itself can stand temporarily higher corrosivity, but bare metals in the following piping, washer, etc., will corrode. If high residual chlorine levels and low pH values are frequent, the remaining option is to select more corrosion resistant alloys.

The chlorine dioxide (D stage) environment has temperatures of 60°C–80°C, pH from 2 to about neutral, high redox potential, and presence of chloride ions formed during reduction of chlorine dioxide. Chlorine dioxide has 2.5 times the oxidizing power of chlorine gas on a molar basis[29]. Therefore, chlorine dioxide bleaching usually has lower chloride ion concentration than chlorine bleaching. Chlorine dioxide is a very selective bleaching chemical that attacks lignin but not cellulose or hemicellulose. Retention times are usually 3–4 h. Most delignification occurs relatively fast. Then the process slows. An increase in temperature will increase the reaction rate allowing lower retention times or giving higher brightness at constant retention time. Chlorine dioxide acts most efficiently in neutral or slightly acid solutions, but delignification rates are higher in strongly acidic solutions. Excessively acidic solutions will decrease pulp strength. The corrosive environment in the D stage is due to the following:

- High redox-potential
- High temperature
- Chlorides that are usually less than in chlorine bleaching
- High acidity in some cases.

The gas phase in both C and D stage washers is more corrosive than the solution phase due to lower pH and higher redox potential caused by residual chlorine. Chlorine dioxide is usually made from sodium chlorate, $NaClO_3$. Chlorate ions are reduced to chlorine dioxide by various reducing agents in a strong acid solution. Residual chlorine dioxide must be decomposed to avoid corrosion problems in subsequent stages. For example, the use of an antichlor agent between retention tower and washer can help prevent D stage washer corrosion. In addition, adjusting pH to a higher value can help. Then the pH should be close to neutral (pH = 6–8). Materials that have shown satisfactory corrosion resistance in chlorine dioxide include stainless steels, nickel alloys, titanium, and glass reinforced plastic. The introduction of wash water recycling has made D stage washer environments more aggressive to the extent that 2%–4% molybdenum containing grades such as AISI 316L and 317L do not survive. The preferred material for a chlorine dioxide tower is now stainless steel with high molybdenum content such as 4%. Suitable materials are austenitic UNS S31254 or duplex S32570. Titanium also finds use. The tower can have a lining of acid resistant brick, and glass reinforced plastics have use in various places.

Many pulp mills use caustic soda, a reducing agent, or both to remove chlorine and chlorine dioxide residues from bleached pulp. Sodium hydroxide, white liquor, and sulfur dioxide gas have common use for these applications. Sodium hydroxide and white liquor increase the alkalinity of the solution and shift the equilibrium of chlorine compounds to make the hypochlorite ion more stable. These solutions require application at a sufficiently high rate to be effective. This can result in significant equipment and energy costs. The high alkalinity can cause precipitation of calcium carbonate scales. Sulfur dioxide gas works by reducing the residual chlorine and chlorine dioxide to chloride ions. Sulfur dioxide has frequent use because it is cost-effective, but it adds potential safety and corrosion hazards.

Bleaching with oxygen, hypochlorite, or peroxide usually uses alkaline conditions. Stainless steels are therefore less susceptible to pitting and crevice corrosion than in chlorination stages. Oxygen delignification (O stage) cleans fibers before actual bleaching[29]. Oxygen delignification can lower the lignin content of the pulp by 50% from that after the digester. This is an effective method when trying to eliminate chlorine in bleaching chemicals or bleach plant effluents. Another advantage is that less corrosive chemicals are necessary and therefore concentration of their reaction products (such as Cl⁻) are lower. Most oxygen bleaching uses 20%–30% high consistency pulp. The chemicals in oxygen delignification are gaseous oxygen or a mixture of oxygen and peroxide, sodium hydroxide or oxidized white liquor, and magnesium sulfate to minimize cellulose degradation[29]. The temperature is 90°C–130°C, pressure is 0.2–0.4 N/mm² (2–4 bar), and retention time is about 1 h. Dissolved organic matter and sodium added require removal before the next bleaching stage. Oxygen reactors often use high nickel stainless steel such as 33% Ni to prevent stress corrosion cracking[26]. The corrosive environment in oxygen bleaching is primarily due to the following:

- High temperature
- High redox-potential
- Remaining chlorides.

Oxidation with hypochlorite (H stage) uses alkaline conditions with calcium or sodium hydroxide. During hypochlorite bleaching, residual lignin decomposes, and pulp brightness increases. The hypochlorite treatment leaves a yellow shade in the pulp that is stronger at higher pH. The hypochlorite stage is sensitive to pH. At low lignin concentrations, hypochlorite also attacks cellulose to decrease pulp strength. The pH in the H stage is initially 11–11.5 and drops as low as 8. The optimum pH range is 8–10. If pH is too low, hypochlorite will react to hypochlorous acid to attack cellulose. If pH is too high at the start of bleaching, the initial reaction rate can be too slow[29]. Temperature is normally about 35°C–45°C. The environment in the hypochlorite stage is not as aggressive as in the chlorine stage primarily due to the alkalinity. The retention time in a hypochlorite stage can be 2–6 h requiring large towers that are concrete or steel construction lined with tiles. At the end of the hypochlorite stage, addition of sulfur dioxide, thiosulfate, or other compound decomposes the residual hypochlorite. Residual hypochlorite would cause corrosion in a paper machine. High alkalinity allows higher tolerance for chloride concentration. The corrosive environment in H stage is described by the following:

- High pH
- Moderate temperature
- Sulfur compounds after residual hypochlorite removal.

CHAPTER 5

Lignin preserving bleaching maintains the high yield of mechanical pulps. Bleaching chemicals may be oxidizing (primarily hydrogen peroxide), reducing (primarily the dithionites), or a combination. Lignin preserving bleaching uses conversion or stabilization of light absorbing functional groups called chromophores. For example, dithionites reduce quinones to hydroquinones. The pH range in hydrogen peroxide bleaching is high. Since hydrogen peroxide is a strong oxidizer, the conditions can be very corrosive. The concentration of hydrogen peroxide in bleach solutions is about 0.3%–0.5% higher for chemical pulp than mechanical pulp. At low temperatures of 35°C–55°C, hydrogen peroxide is an effective lignin preserving agent. At 70°C–80°C, it increases brightness in the later stages and improves brightness stability[29]. Hydrogen peroxide has use in several locations. Typically, it is added with oxygen to the first or second extraction stage. Hydrogen peroxide also has use as a stand-alone stage toward the end of the bleaching sequence. H_2O_2 can also be added to the bleached pulp storage towers providing more stable brightness and also increased brightness. Hydrogen peroxide bleaching solutions can be acidic or alkaline. Alkaline hydrogen peroxide solutions can be made by feeding alkali and hydrogen peroxide directly to the process stream. Acidic hydrogen peroxide solutions must be generated first and then fed to the system. The corrosive environment of P stage depends on the following:

- pH
- Temperature
- Redox potential
- Dissolved solids (especially as inhibitors for titanium).

Hydrogen peroxide has three reacting species during the bleaching of kraft pulp. Hydroxonium ion (HO^+) is stable under acidic conditions, perhydroxyl ion (HO_2^-) is stable under alkaline conditions, and hydroxyl radical ($HO·$) occurs especially in the presence of transition metals such as iron and manganese. These species react with lignin and cellulose to some extent. Desired reactions are delignification and brightening, but hydroxyl radicals can result in nonselective reaction with pulp. pH is the primary reaction parameter for optimizing peroxide use. Peroxide effect improves at a pH value of 10.5 and continues up to approximately 12. High temperature also has a positive effect on extraction stage efficiency. An optimum metal ion composition will improve brightness and viscosity while minimizing the hydrogen peroxide consumption. A satisfactory level of alkaline earth metals such as magnesium and calcium with minimum concentration of transition metals such as manganese, iron, and copper is preferable. This is possible by including a chelation stage (Q stage). The conditions in the chelating stage are typically 40°C–80°C, pH of 4–7, and concentration of active chelating agent such as EDTA less than 0.5 g/L.

Storage of concentrated hydrogen peroxide can use tanks constructed of aluminum or type 304L or 316L stainless steel. Rubber lining or rubber seals are not suitable for peroxide use. Fluorinated hydrocarbons are suitable gasket materials. Metallic impurities can cause hydrogen peroxide decomposition. Hydrogen peroxide bleaching solutions do not corrode conventional stainless steels or lined equipment. Use of titanium in hydrogen peroxide environments is a complicated issue. Conversion of titanium equipment used in chlorine and chlorine dioxide stages to include peroxide injection has had disastrous results. Laboratory and field tests have shown extremely high and extremely low corrosion rates. The ions of calcium, barium, and strontium inhibit titanium corrosion, but the required ion concentration may be so high that it causes scaling problems. Lowering the pH to about 9 can also decrease titanium corrosion rate[31].

Another bleaching method for mechanical pulp uses dithionite. Hydrosulfite bleaching is another term for dithionite bleaching. The active compound is usually sodium dithionite, $Na_2S_2O_4$. If the brightness only requires a slight increase, dithionite can bleach the pulp. To achieve higher brightness values, hydrogen peroxide and sodium hydroxide have use. The dithionite bleaching process has some similarities with peroxide bleaching. For example, both chemicals require the use of a complexing agent. The main difference is that peroxides are oxidizing and dithionites are reducing compounds. The reaction product of dithionite is sulfite ion that can oxidize to thiosulfate. Dithionite solutions are not very stable, and they cannot be stored for long periods. The corrosive conditions are due to the following:

- Low pH
- Thiosulfate.

Ozone delignification is an alternate for eliminating or partially replacing the use of chlorine containing chemicals. Ozone as a strong oxidizer is especially replacing elemental chlorine. Oxygen in an ozone generator produces a gaseous mixture containing 7%–13% O_3. The mixture of ozone and oxygen is compressed and fed to a reactor or mixer. The ozone concentration depends on the temperature and partial pressures of ozone and oxygen. It is usually less than 100 mg/L. Ozone bleaching of chemical pulps uses an acid pH of 2–4, temperature of 50°C–65°C, and retention times ranging from seconds to 10 min[32]. A retention tower is not necessary. The corrosive environment in the Z stage is described by the following:

- High redox-potential
- Low pH
- Moderate temperature
- Impurities such as chloride.

Ozone reacts with lignin breaking it into smaller molecules. The decomposition of ozone in the first stage results in gaseous oxygen and hydroxyl ions. Dissolved metal ions and high pH will increase oxygen decomposition rate. Hydroxyl ions will react with ozone to form perhydroxyl ions and oxygen. The perhydroxyl ion is a brightening agent. Treated and diluted pulp usually does not contain residual ozone. Addition of acid can destroy any ozone that is present. In ozone solutions, high molybdenum grades of stainless steels may perform worse than conventionally low-alloyed grades. Addition of ozone to a well-aerated solution with only oxygen dissolved can increase corrosion

CHAPTER 5

potential of a stainless steel with 500–900 mV[33]. Rubbers and other elastomers are usually not suitable for ozone use, but fluorinated special grades do have use.

The alkaline extraction (E stage) removes chlorinated lignin not washed away after the delignification. It generally uses sodium hydroxide at 45°C–60°C. Special grades use higher temperatures up to 100°C. Retention times can be 2 h. The pH of about 10 allows rather large amounts of chloride ions without risk of pitting corrosion of common stainless steels. The high amount of OH⁻ ions gives an inhibiting effect against chlorides. Even AISI 304L has had use, but molybdenum alloyed grades are usually selected[25]. In an aggressive atmosphere, external corrosion also requires consideration. Countercurrent washing makes the alkaline extraction stage more aggressive requiring use of type AISI 317L or even more alloyed grades. The extraction towers can use concrete with tile lining or steel towers lined with rubber. Oxygen alkali extraction (E$_O$) uses oxygen with alkali to lower consumption of other bleaching chemicals such as chlorine dioxide. The additional oxygen and peroxide increase the redox potential of the solution making it more corrosive compared with conventional extraction. The corrosive environment in the E stage is described by the following:

- High pH
- Usually moderate temperature
- Lower risk of chloride induced corrosion due to high pH
- High redox potential when using O_2 or H_2O_2.

The corrosion problems in a bleach plant are usually pitting and crevice corrosion and stress corrosion cracking of stainless steel. As Table 2 shows, the redox potential of bleaching solutions increases with increasing oxidant concentration. All chlorine compounds give chlorides as reaction products. Obviously, extensive use of bleaching chemicals can easily initiate localized corrosion of passivating metals. Earlier work notes that residual oxidants such as chlorine (Cl_2) and chlorine dioxide (ClO_2) are the primary cause of corrosion in the bleach plant[26]. As chlorine dioxide began to substitute for elemental chlorine in the 1970s to decrease chlorine release, corrosion problems became greater. Especially in the bleach plant, recycling of filtrates with high residual concentration has caused corrosion of existing materials[26]. Particularly in the first stages, the temperature and chloride concentration are so high that corrosion is possible even with highly alloyed stainless steels. Closing of the water circulation has increased operating temperatures. Modern closure concepts use countercurrent washing and extensive reuse of water in the bleach plant. This increases impurity content and redox potential of the previous process stages.

3 Corrosion in the paper mill

The main unit processes in a paper mill are stock preparation, stock cleaning, and the paper machine itself. The pulp is usually a mixture coming from various sources. All sources of pulp will add to the dissolved inorganic solids content of the white water. The ionic species in the pulp can include:

Corrosion in different environments

- Chemical pulp: Cl^-, SO_4^{2-}, SO_3^{2-}, and residual bleaching agents
- Mechanical pulp: Cl^-, SO_4^{2-}, $S_2O_3^{2-}$, HSO_3^-, S^{2-}.

In the paper machine, the stock is dewatered, pressed, and dried to form a finished sheet of paper. All paper machines have the same primary components: the wet section, the press section, and the dryer section. In the wet section, the stock is dewatered on one or more wires. The highly diluted stock flows onto the wire from a headbox and is dewatered. The smoothness of headbox surfaces is essential in maintaining smooth and uniform flow. Depending on the design of the wet section, paper machines are fourdrinier units where dewatering and sheet formation occur on an endless wire felt that stretches between a number of rolls, twin wire machines where the sheet forms between two wires and dewatering occurs in both directions, or hybrid machines featuring a fourdrinier table and a twin wire. Efficient operation requires rapid removal of water. Wires are cleaned by continuous or intermittent use of cleaning agents having considerably different pH values. These will affect the properties of the circulating white water. High rotation rates and stresses can cause cracking and fatigue in the moving parts. The stock leaves the wire as a web of paper with a dry solids content of about 20%.

Dewatering of the paper web continues in the press section where the web is pressed between press rolls. After the final press roll nip, the dry solids content is about 30%–50%. The presses usually have press felts that distribute the pressing pressure on the paper web and remove water. Several different types of presses are possible. The suction press consists of a perforated suction roll and a solid counter roll with a granite or rubber shell and a suction box fitted with a vacuum pump. The vented nip press has a vented (grooved) solid roll and a counter roll with a smooth surface. In the nip, water is pressed from the felt down into the grooves and is thrown out by centrifugal force. In a felt wire press, a coarse plastic wire is passed between the felt and the lower roll.

The felts also require use of cleaning agents. High rotating speeds, cyclic stresses, and pressure differences can cause cracking and fatigue. Localized corrosion can act as initiation sites for these damages.

In the dryer section, the paper web is dried to a dry solids content of about 95%. The dryer section is encased in a dryer hood to maintain good ventilation and moist air removal. Paper machine classification can use the way the dryer sec-

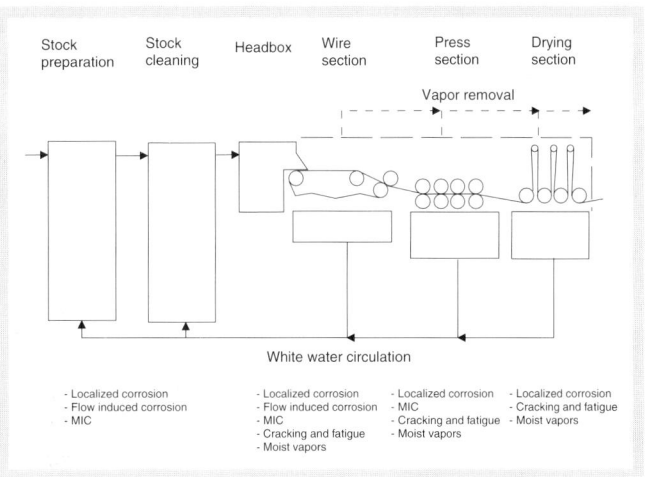

Figure 15. Components of a paper machine and some corrosion problems.

CHAPTER 5

tion uses multicylinder machines with a large number of steam-heated drying cylinders with diameters of 1.5–1.8 meters, Yankee machines use a large drying cylinder (Yankee cylinder) with a diameter of 4–6 meters, and combination machines use one Yankee cylinder and other ordinary drying cylinders. Corrosion problems may occur for example in ventilation ducting and cylinder surfaces. The schematic construction of a paper machine and some corrosion problems are shown in Fig. 15.

The feed material is a suspension blended from white water and pulp fibers. The corrosion problems in the paper mill are primarily in the wet end operations. The main electrolyte in papermaking is white water. Factors influencing white water corrosivity are pH, temperature, chlorides, sulfates, sulfites, and thiosulfates. The corrosivity also varies with degree of water recycling. Closed systems contain more dissolved salts and dissolved organic solids in water. Their pH is usually lower than in open systems. Closed systems are more corrosive due to dissolved salts and lower pH. White water pH is usually slightly acidic at about 4–6. The pH of standing white water may decrease by up to 2 units. The temperature of white water is about 40°C–50°C. If the water contains a few hundred mg/L of chlorides, this might be sufficiently high to initiate localized corrosion of conventional stainless steels. The main ions in white water are chlorides and sulfates with some thiosulfate and sulfite also possible. The sulfate to chloride weight ratio can be 2–5[34]. Higher content of sulfates will inhibit pitting and crevice corrosion. The amount of solid fibers varies and influences the solution flow. Fibers can accumulate on surfaces and create crevice sites if the flow rate is too low[23].

Papermaking environments are not as corrosive as in the pulping or bleaching processes. Construction materials have traditionally been carbon steel and stainless steel types 304 and 316, cast iron, and plastic. Austenitic and duplex stainless steels are major construction metals today for white water environments. Bronze has had use in forming wires and suction rolls, but it does not survive in modern white water systems. Premium materials for suction rolls are precipitation hardened and duplex stainless steels. For piping, type AISI 316L and duplex stainless steels are good. Other components use type 316L and 317L wrought stainless steels and cast stainless steel CF-8M (equivalent to AISI 316)[34]. The major corrosion problems are chloride-induced pitting and crevice corrosion, thiosulfate pitting, and microbiologically induced corrosion. The primary concern in white water service is exceeding critical concentration of chlorides, temperature, or both. Selecting stainless steels for chloride environment can use the molybdenum content. Table 6 gives a list of austenitic alloys representing increased resistance to chloride-induced localized corrosion. The order is by minimum molybdenum content. The type AISI 316L is usually resistant. Since it often has a molybdenum content just slightly above the 2% minimum, a safety margin is possible by selecting Swedish grade SIS 2353 or AISI 317L[26]. Unfortunately, the cost of these alloys increases with increasing molybdenum content.

Table 6. Austenitic alloys for white water environments[26].

Alloy	Molybdenum, % minimum
Type AISI 304L	0
Type AISI 316L	2
SIS 2353	2.5
Type AISI 317L	3
UNS N08904 (904L Alloy)	4
UNS S31254 (254 SMO)	6
UNS N10276 (C-276 Alloy)	15

Thiosulfate is an aggressive compound causing pitting corrosion especially on stainless steels that do not contain molybdenum. Unlike chloride pitting, thiosulfate pitting occurs below a critical potential that is the reduction potential of thiosulfate. The pits are stable and not spontaneously repassivated. Thiosulfate levels should be below 5 mg/L for equipment made from AISI 403L and 10 mg/L for AISI 316L[26].

Paper machines may suffer from microbiologically induced corrosion. The white water contains nutrients that can sustain bacterial growth. Design of the flow systems minimizes slime accumulation, but this does not always work. Slime collects in areas where flow velocity is low. Surface discontinuities may also allow slime to attach and remain on the surface. Wherever slime deposits begin to form, they also increase in thickness. The aerobic conditions in the slime layer may change to anaerobic as the layer thickens due to fiber accumulation. In an anaerobic environment, sulfate reducing bacteria grow and produce free sulfide ions. These will cause the environment beneath the biofilm to become reducing and result in depassivation of the stainless steel and pitting.

Paper machine headboxes are a special case. The headbox transfers pulp flow to uniform flat flow across the width of the paper machine. The slurry jet entering the sheet forming zone must have a high degree of uniformity locally and chronologically to obtain a paper sheet that is uniform in the micro and macro scale. Headbox construction uses AISI 316 steel or even a more alloyed steel. The surface finish of the headbox is very important for the machine operation since corrosion damages will assist scaling and hang-up of stock with concomitant influence on pulp flow. The internal surfaces of the headbox are pickled or sometimes electropolished to obtain a superior surface finish and prevent weak points that may start to corrode during operation.

Two major items involved in closing a white water system are filtration and reuse of the water. Removal of dissolved and colloidal material may be necessary to avoid slime and microbial corrosion problems. Peroxide and SO_3 from bleaching stages can cause changes in the white water. Their effect is difficult to estimate. Sulfur trioxide can cause problems in dryer hoods. Peroxide will assist chloride-induced localized corrosion but may inhibit thiosulfate-induced pitting. Total hardness, total alkalinity, and pH require estimation of scaling tendency. Scaling is usually a problem, but carbonate scales may also protect carbon steel from corrosion. Chloride concentrations of greater than 200 mg/L are common in recycled white water. Sulfate will remain the dominant white water anion[34]. Increased conductivity due to higher ion content may increase corrosivity and influence corrosion morphology.

4 Corrosion in chemical recovery

Kraft pulp mills normally run very closed. About 99% of the incoming organic material is made into products or burned for energy. Approximately 95% of the inorganic material is recycled into the process leaving a small solid waste stream. A mean value for aqueous effluent is about 90 m^3/ton pulp with some as high as 200 m^3/ton. Lower than 30 m^3/ton is also possible. The economy of modern mills depends heavily on the continuous effective operation of the chemical recovery stages. Different pulp making processes have different recovery systems. In the kraft process, the chemicals are recovered from the black liquor. The spent black liquor has a pH of about 12, and the residual active alkali is usually less than 30 g/L[26].

The black liquor from brownstock washing goes through a series of process steps known as cooking chemical recovery. The chemical recovery includes aqueous and high temperature processes. The black liquor is first concentrated by evaporation. Evaporation is a separation process that leaves a pure condensate and a concentrate containing the nonvolatile components of the feed.

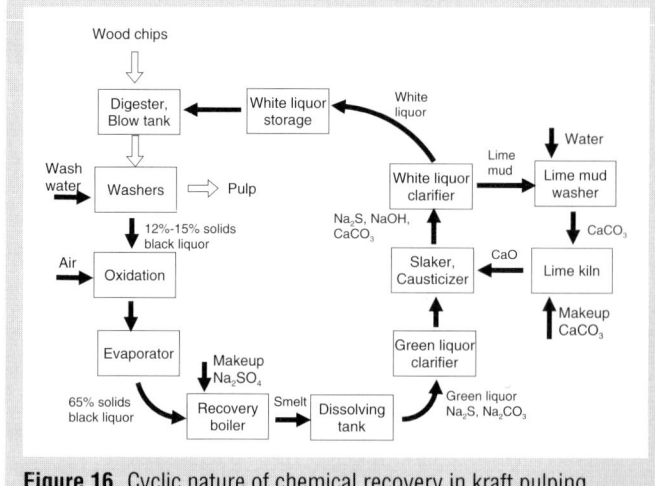

Figure 16. Cyclic nature of chemical recovery in kraft pulping.

Evaporation uses several steps each of which results in more concentrated liquor. Black liquor is oxidized for odor control if necessary. The concentrated black liquor is burned with sodium sulfate in a recovery boiler to obtain a smelt that contains sodium carbonate and sodium sulfide. This smelt dissolved in water produces the green liquor. Treatment of green liquor with calcium hydroxide converts sodium carbonate into sodium hydroxide and produces new white liquor. The conversion causes precipitation of calcium carbonate. This is subsequently recovered and calcined in a kiln to calcium oxide that is dissolved with water to regenerate the used calcium hydroxide. Figure 16 shows the cyclic nature of chemical recovery in kraft pulping.

After the separation of pulp in brownstock washers, the spent black liquor with less than 20% solids is concentrated in a multiple-stage evaporator train to make the

Corrosion in different environments

liquor combustible. On combustion, the dry solids content must be as high as possible since this results in better thermal economy, smaller emissions of sulfurous gases, and more efficient conversion of the chemicals into active cooking chemicals. During evaporation, the water in the liquor is boiled off using steam as the heating medium. If evaporation occurs in one stage, an equivalent amount of steam is consumed to the amount of water evaporated. Because of this, evaporation is usually divided into five or six stages to reduce the steam requirement to approximately one-fifth. The evaporators can be vertical rising film or falling film tube and shell constructions or falling film lamellar constructions with black liquor outside. The series of evaporators operate at different pressures so that the vapor of one evaporator becomes steam supply for the next. The flows of steam and black liquor in the entire system are countercurrent, but in every evaporator stage the flow is usually in the same direction. The first evaporator effect is the hottest. As the black liquor moves from one evaporator to the next, the pressure, boiling point, and solids concentration increase. A higher solids content and alkali content require a higher temperature to drive off the remaining water. The solids concentration of the strong black liquor discharged from the last evaporator effect is usually 50%–55%. This requires further concentration with a concentrator or a direct contact evaporator. A direct contact evaporator uses alternating contact between liquor and hot gases or a cyclone. A concentrator is a special, directly heated evaporator with large solution volume flow[35]. The final solids concentration of the heavy black liquor for burning is 60%–65% to more than 70%[26,35]. Carbon steel is not always resistant in modern mills. Stainless steels such as AISI 304 and AISI 316 are the main materials, but concentrated alkali solutions near and above the boiling point can be also be corrosive to stainless steels. Black liquor evaporation uses both alloys, but the acidic sulfite process red liquor requires use of molybdenum alloyed grades such as AISI 316. The corrosivity of environments in the evaporation stage depends on temperature and liquor composition and especially alkali and chloride concentration.

Hazardous gases such as H_2S and mercaptans release during cooking and require treatment. Black liquor oxidation in the chemical recovery process lowers total reduced sulfur emissions that contribute to the characteristic kraft mill odor. Oxidation of black liquor eliminates malodorous sulfide compounds. The process converts residual sulfite (Na_2S and $NaHS$) to thiosulfate. This eliminates the stripping of hydrogen sulfide (H_2S) in the direct contact evaporator. In newer recovery boiler designs, the direct contact evaporator is eliminated, and black liquor oxidation is not necessary[35]. The high concentration of thiosulfate does not cause heavy corrosion of carbon steel. In oxidized black liquor, the alkali has been consumed during digestion and sulfides have been oxidized. From the corrosion prevention viewpoint, oxidizing lean black liquor is desirable[26]. From the operational and economical viewpoint, oxidizing concentrated liquor is better[35]. Carbon steel is useful for oxidation equipment and storage of oxidized liquor[26], but stainless steel of type AISI 304 has had use for gas lines due to safety reasons[25].

The recovery boiler or recovery furnace is the most important unit in the recovery of spent cooking chemicals. In the recovery boiler, organic constituents of the black

CHAPTER 5

liquor are burned and oxidized sulfur compounds are reduced to sulfide and recovered in molten form. The recovery boiler also supplies steam to the mill. The heavy black liquor is sprayed to the lower part of the furnace. It is pyrolyzed before falling to the bottom. At the bottom of the furnace, air flow remains low so the atmosphere is reducing. Carbon and carbon monoxide act as reducing agents converting sulfur compounds back to sulfide. Sodium carbonate forms simultaneously. Operation of the recovery boiler ensures a continuous supply of chemicals for green liquor. Energy efficiency and steam generation have secondary importance. A separate section discusses corrosion problems associated with the kraft recovery boiler. The main problem is due to the sulfur compounds in the furnace atmosphere since these do not form a protective layer on iron. Reducing conditions in the lower part and oxidizing conditions higher up the furnace with varying sulfur and oxygen compounds make corrosion predictions difficult. In the lower part of the furnace, water wall tubes are actually protected by a layer of solidified smelt.

The molten sulfur and carbonate from the recovery boiler dissolve in water to produce green liquor. This liquor production can be a very corrosive environment since hot molten salt is disintegrated in a steam jet followed by particle dissolution in an agitated vessel. Flowing particles can cause erosion corrosion. The composition of green liquor is Na_2S and Na_2CO_3 with pH about 10–11. Added water and wash solutions control the strength of green liquor. Green liquor equipment has previously used ceramic coated carbon steel. Today, low alloyed austenitic stainless steels are common. The green liquor also contains unburned organic material, sodium and calcium carbonates, iron compounds, etc. These are removed in clarifiers and washers before the liquor is mixed with quick lime (CaO) to obtain white liquor. The purpose of this recausticizing is to convert sodium carbonate to sodium hydroxide and to remove impurities. Causticizing occurs in two stages as Eq. 16 shows: slaking and causticizing.

$$CaO + H_2O \rightarrow Ca(OH)_2$$
$$Ca(OH)_2 + Na_2CO_3 \rightarrow 2\ NaOH + CaCO_3$$
(16)

The first step is exothermic, i.e., it produces heat. High temperature and vigorous agitation promote the first step. Causticizing efficiency depends on the second stage, and a higher percentage conversion to sodium hydroxide is preferable to minimize sodium carbonate recycling. Causticizing efficiency increases with decreasing alkalinity and sulfidity[35], but such solutions are low in active chemicals for cooking. After slaking, the liquor overflow goes to a series of agitated tanks to obtain sufficient retention time for the slow causticizing step to complete. White liquor must also be clarified to remove solid $CaCO_3$ called "lime mud." The lime mud converts back to lime by burning. This conversion process is calcining. It occurs in a rotary kiln or fluidized bed. The calcination reaction requires temperature of about 800°C to cause decomposition of calcium carbonate into quick lime and carbon dioxide as Eq. 17 shows. The heat requirement is about 8000 MJ/ton of lime[35].

$$CaCO_3 \rightarrow CaO + CO_2 \tag{17}$$

The recausticization process does not influence sodium sulfide. The recovered white liquor is stored in a tank and used in digestion. The alkaline conditions of recausticizing are suitable for austenitic stainless steel. The main problem may be erosion due to solids in green liquor and white liquor clarifiers.

The chemical recovery system and especially the recovery boilers sometimes have use to treat various wastes. This can increase the impurity levels in the cooking liquor cycle. The increased impurity levels will affect the entire chemical recovery system. For example, higher chloride concentration will not necessarily cause increased corrosion of the digester, but it may lead to formation of low-melting compounds in the recovery boiler causing fireside corrosion. Sodium and potassium hydroxides are highly volatile, and they may concentrate on the coldest parts of the furnace, e.g., close to the primary air ports. Molten hydroxides are more corrosive to stainless steels than carbon steel. Excessive formation of hydrogen sulfide and other reduced sulfur compounds with excessively low air flow to oxidize properly may cause attack by sulfur in the superheaters[26]. The ash deposits on the superheater and boiler tubes are usually due to the carryover of black liquor particles or condensation of sodium compounds. These deposits can be enriched with chlorides, adsorb sulfur trioxide, etc. Spent acid additions from regeneration of bleaching chemicals and excessive use of firing oil with high vanadium content may lead to very corrosive environment under the ash deposits.

The chemical recovery systems of sulfite mills vary much more than that of kraft mills. Sometimes, the purpose of treating spent sulfite cooking liquor is not to recover chemicals but to fulfill environmental requirements for aqueous discharge. The calcium base in the original acid sulfite process was never recovered for economical reasons. More expensive bases were recovered almost from the beginning. Most sulfite recovery processes begin with evaporation and burning of the spent cooking liquor. The choice of disposal system depends on base; requirements to recover base, sulfur, or both; heat recovery; and economical factors. Only heat recovery is practical with calcium base due to formation of gypsum, $CaSO_4$. Burning of ammonium-based liquors with and without sulfur recovery is possible. No base can be recovered since ammonia (NH_4) decomposes into nitrogen and hydrogen that react with water. Complete chemical recovery is possible with magnesium base solutions, and the process is uncomplicated. This process differs from kraft recovery since no smelt is produced and chemicals are removed with flue gases such as MgO and SO_2[35]. Processes for chemical recovery of sodium base sulfite liquors use the burning of spent liquor in a kraft type furnace with recovery of the chemicals as a smelt. The smelt is higher in sulfidity, and flue gases contain more SO_2 than those of kraft recovery boilers. Different approaches can regenerate cooking liquor from the smelt chemicals. Carbonation of the smelt followed by decomposition of sodium bicarbonate to recover sodium carbonate is a common technique[35]. The hydrogen sulfide present in these reactions may cause corrosion problems.

CHAPTER 5

Sulfite liquors are more corrosive than kraft black liquor. Storage of spent liquor requires use of molybdenum containing stainless steel grades. If free sulfur dioxide can oxidize, dew point corrosion by sulfuric acid is possible, and UNS N08904 (904L Alloy) type alloys with 4%–4.5% minimum Mo may be necessary[36]. The evaporation train commonly uses AISI 316L or 317L stainless steel. Pitting and under-deposit corrosion are the most commonly reported problems[26]. Magnesium based solutions are more corrosive than calcium or sodium based solutions. Ammonia based solutions are least corrosive[36]. Sulfite recovery boilers are primarily carbon steel construction with corrosion allowances. Thermally sprayed stainless steel has sometimes prevented corrosion of carbon steel water walls. Composite tubing is becoming more popular. Flue gases contain more sulfur trioxide, hydrogen sulfide, and possibly hydrochloric acid. The AISI 316L and 317L stainless steels and glass reinforced plastics have use for scrubbers. Pitting and crevice corrosion may require use of higher alloyed stainless steels or nickel alloys[26].

5 Corrosion under thermal insulation

Most processes require use of thermal insulation to reduce energy loss and associated costs in heating and cooling. Insulation covers the metal surface allowing a corrosive environment and degradation to develop and proceed unnoticed. The critical factors for corrosion under insulation are availability of oxygen, high temperature, and concentration of dissolved species. As the temperature increases, the solubility of oxygen normally decreases resulting in reduced corrosion rates. Under insulation, the moisture remains in a closed system. Figure 17 shows the schematic behavior of corrosion caused by dissolved oxygen in open and closed vessels. When the temperature increases, the corrosion rate increases. In an open system, the corrosion rate begins to decrease above the temperature where solution vapor pressure increases rapidly reducing oxygen partial pressure. In a closed system, the oxygen cannot escape, and corrosion rate continues to increase as temperature increases. Field data has confirmed that corrosion of steel under insulation increases steadily with increasing temperature[37]. The corrosion of steel under insulation is identical to corrosion in a closed hot water system. The thermal insulation does not corrode steel but provides sites for moisture and impurity collection. It prevents the escape of water and water vapor.

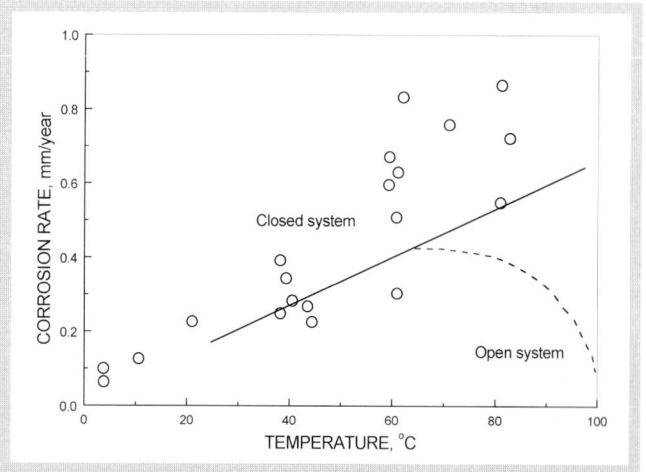

Figure 17. Effect of retaining dissolved oxygen in open and closed systems[37].

Thermal cycling below the dew point and above ambient temperature will cause condensation of water on the surfaces, drying, and impurity enrichment. A similar case occurs when a pipe runs through a vessel wall and environment changes from cold and wet to hot and dry or vice versa. The outer cladding that protects the insulation material must remain intact to prevent water from reaching the coated metal surface. The ingress of water and saturation of the insulation material create the corrosive environment. If the insulation becomes wet, drying under field conditions is almost impossible. In cases where precipitation becomes trapped on the metal surface by insulation, chlorides and sulfuric acid can concentrate and accelerate corrosion. Sometimes, chlorides are already present in the insulation material to cause corrosion when becoming wet. The measured corrosion rates under insulation follow trends to higher corrosion rates commonly associated with pressurized systems. Austenitic stainless steels present a greater problem since they are susceptible to stress corrosion cracking by chlorides. Chlorides can cause external cracking because the temperatures under insulation can be very high and bending of pipes causes stresses.

Thermal insulation generally consists of the following groups:

- Rigid insulation such as boards made from polystyrene, fiberglass, or polyisocyanurate
- Batt insulation such as blankets made from cotton, mineral wool, or fiberglass
- Loose-fill insulation blown in and foamed in-place using cellulose or cementitious isocyanurate.

Cellulose insulation is primarily recycled paper. The remainder of the material is a fire retardant product. Fire resistance has been a concern with cellulose insulation. Some products use boron, but borates are soluble in water. Ammonium sulfate fire retardants will improve fire retardancy performance. Concern exists about corrosion of metals in contact with the insulation. The insulation can be dry blown or poured in a loose fill application into enclosed cavities. A more common application is the wet spray method in which it is mixed with water upon spraying into the wall or ceiling cavity. Cotton insulation is batt and loose fill insulation made from recycled textile fiber with polyester fiber for improving tear strength and recoil characteristics. It is treated with fire retardants. Cementitious insulation uses magnesium silicate. It is foamed in place and expanded with compressed air. Mineral wool insulation is a by-product of metal production. Rock wool uses natural rocks such as basalt and diabase. It is noncombustible and noncorrosive, exudes no odor, and does not support the growth of fungus or bacteria. Some mineral wool materials are bonded with a thermoset resin. Fiberglass insulation uses sand and limestone as raw materials. Conventional fiberglass insulation uses phenol-formaldehyde resin as binder. Foam insulation uses petrochemicals in the production such as polyurethane, polyisocyanurate, extruded and expanded polystyrene, and phenolic foams. Extruded polystyrene is a thermoplastic. This means it can be heated and remolded into a new product. Polyurethane and phenolic products cannot be remolded but can be recycled as a filler. Polystyrene and phenolic foams are available as boards. Polyurethane and polyisocyanurate may be boards or foamed in-place.

CHAPTER 5

Mineral wool and polymer foams are the most common types of insulation materials. Corrosion can occur under every insulating material. Polyurethane foams are useful for cold and anti-condensation service. They are permeable to water vapor. The maximum service temperature is 80°C. If used in continuously cold conditions, it will not corrode unprotected metals. At high temperatures, these foams may release chlorides. Polyisocyanurate foam is fire resistant. It is permeable to water vapor in cold service. The maximum temperature is 120°C. When heated, the cell structure breaks down to release chlorides from the fire retardant and blowing agents. Flexible foamed elastomers do not absorb water and have a maximum service temperature of 80°C. Cellular glass is a rigid glass foam whose blowing agent contains CO_2 and H_2S. It does not absorb water, and the maximum use temperature is almost 500°C. When water is present and cells are damaged, the harmful compounds of the blowing agents can release. Glass fiber insulation is usually pure glass fibers with some binding material. It will absorb water, but excess moisture can drain out. The maximum service temperature is 230°C. With special grades, it is more than 450°C. Mineral wool is basically impure glass. It will also absorb water. Maximum temperatures are 650°C–1000°C. Mineral wools will usually hold water well. This makes them potentially dangerous for unprotected metal surfaces. Calcium silicates are cementitious materials that can absorb water up to 400% of their own weight. Water extracts from calcium silicate based materials, fiberglass, cellular glass, and ceramic fiber are usually neutral to alkaline. The pH values are 7–11. Cellular glass is free from chloride, but the other materials may contain some chlorides. Mineral wool gives a neutral environment with pH 6–7 and low chloride content of 2–3 mg/L. Water extracts from organic foams can be very acidic with pH values of 2–3. If halogenated fire retardants are components of the foam, the concentration of free halides (Cl^-, F^-, and Br^-) can be high depending on the hydrolysis reaction[38].

The most important factor in eliminating corrosion under insulation is to keep the insulation material and metal surfaces dry. Metal jacketing and various sealants and mastics have use to keep insulation dry, but they will not prevent water vapor and air from contacting the metal surface. Inhibition of the insulation material has been tried. The inhibitors have usually been inorganic salts. The insulation itself will not protect the underlying substrate. Careful selection of insulation materials to exclude high levels of corrosive impurities is critical in reducing corrosion under insulation. Selection of insulation material, inhibiting the insulation, and use of waterproof coatings are not considered effective methods to prevent corrosion under insulation[37].

Protective coating systems have use to prevent corrosion under insulation. Under insulation, the conditions are more severe than in atmospheric service. The environment and actual temperatures under the insulation especially in the operating temperature range of ambient to more than 100°C vary to make the selection of protective systems difficult. The epoxy phenolics have use for corrosion protection of carbon steel under thermal insulation with operating temperatures to 100°C–120°C. Other operations such as steam cleaning could force temperatures above the heat resistance capacity of an epoxy resulting in premature coating decomposition. The protection of steel structures under insulation with operating temperatures above 120°C can use zinc silicates and flame sprayed aluminum or Al-5Mg alloy.

A particular problem with corrosion under insulation is that the metal structures are concealed. Corrosion can therefore proceed unnoticed until failure occurs. Corrosion under insulation is local so spot checking will not reveal all corroding areas. Visible wet areas are easier to notice. These can be investigated after removing the insulation. Complicated geometries especially in the piping systems restrict the use of most nondestructive inspection methods. A checklist of potential trouble areas includes the following[37]:

- All surfaces exposed to frequent hot and cold temperature cycling
- Cold-temperature equipment where nozzles, clips, and brackets extend through the insulation
- Hot-to-cold interface areas of cold-temperature distillation columns with a hot base temperature
- Horizontal piping especially at joints or piping branches and on the bottom of the pipe
- Insulation weather barrier that has been mechanically damaged or removed
- Insulation that has changed shape or started to swell indicating a possible corrosion product accumulation.

6 Atmospheric corrosion

Atmospheric corrosion accounts for more failures on a tonnage and cost basis than any other type of environment. About half the costs of corrosion prevention are due to galvanizing, painting, etc., to prevent atmospheric corrosion. Classification of atmospheric environments includes indoor, rural, marine, and industrial. The aggressiveness of the environments increases in the order mentioned. Corrosion severity increases when salt, sulfur compounds, and other atmospheric contaminants are present. Industrial environments contain sulfur compounds, nitrogen compounds, and other acidic agents that can promote the corrosion of metals. In addition, industrial environments contain airborne particles that also contribute to corrosion. A characteristic of marine environments is the presence of chloride. Rural environments are the least corrosive of the atmospheric environments. They have low levels of acidic compounds and other aggressive species.

The principal factors influencing atmospheric corrosion for a given metal are moisture, temperature, and the presence of contaminants in the environment. The mechanism of atmospheric corrosion is very similar to that of aqueous corrosion. Atmospheric corrosion is basically aqueous corrosion in a moisture film due to rain, mist, or condensation of water vapor. After formation of the moisture film, the rate of atmospheric corrosion depends on the oxygen supply and presence of impurities in the film. Most metals have a critical humidity that is the relative humidity of an atmosphere above which corrosion rate of the metals increases sharply.

CHAPTER 5

The cathodic reaction in atmospheric corrosion is usually reduction of oxygen. The thin moisture film absorbs oxygen from the surrounding atmosphere to such an extent that it may be considered as oxygen saturated. The corrosion resistance of metals depends on the formation of a dense adherent surface film. Cracked, porous, and soluble films do not give protection and may result in high local corrosion rates. Uniform corrosion is the typical form of atmospheric corrosion. A thin moisture film allows rapid diffusion of oxygen to the metal surface. Localized corrosion is rare in atmospheric corrosion, but galvanic corrosion is always possible. Atmospheric corrosion is often cyclic through the day and night as relative humidity and resulting time of wetness vary. Atmospheric corrosion is a discontinuous process whose effects occur when summarizing the corrosion losses during each separate corrosion period.

The formation of a moisture film depends on humidity. Humidity alone is not sufficient for atmospheric corrosion since corrosion of uncontaminated surfaces may be low even in very humid environments. In clean, unpolluted environments, protective oxide films may form. Pollutants and other impurities increase moisture film conductivity and change its chemical composition. Typical impurities and pollutants that result in formation of loose, porous, and nonprotective surface films are chlorides, sulfur dioxide, nitrogen oxides, and fluorine. Formation of hygroscopic salt films enhances atmospheric corrosion since they reduce the critical humidity and increase the time of wetness. Capillary action due to surface discontinuities such as cracks, solid particles, and porous corrosion products will decrease critical humidity. The effect of temperature is variable. Low ambient temperatures cause low corrosion rates but longer wet periods. Sunlight exposure may increase surface temperature and increase reaction rates but also dry the surface and reduce corrosion. Shaded areas often corrode more rapidly especially in rural and industrial environments. Since deposits in shaded areas are not removed, they remain moist longer.

Since atmospheric corrosion can happen only when the surface has an electrolyte film covering, the time of wetness is an important parameter. This depends not only on atmospheric parameters but also on the corrosion products on the surface. These will determine the critical humidity above which condensation occurs. The dew point temperature of a vapor is the temperature at which the rate of evaporation of the condensed phase from a clean surface equals the rate of condensation from the gaseous phase. The dew point of atmospheric water vapor is about 0°C–10°C, but in flue gases containing about 10% moisture it is about 40°C. The presence of sulfur dioxide or hydrogen chloride in the gaseous phase causes formation of acid condensate at temperatures well above those of water vapor alone. At normal atmospheric temperatures, the moisture in the air is sufficient to start corrosion. Salts and acids on metal surfaces increase the electrical conductivity of any moisture present and accelerate corrosion. Moisture collects on dirt particles.

Figure 18 shows examples of various metals on corrosion behavior. Some metals such as chromium rapidly form a protective film, and corrosion stops. Metals such as zinc are unable to do this. Their corrosion rate remains constant. The chromium-type behavior occurs with all chromium containing alloys, aluminum, and lead. The corrosion rate of copper alloys decreases slowly since the protective films grow thicker. The zinc-type behavior occurs with active metals such as zinc, magnesium, and cadmium and with nickel. Since the reaction product layer is constantly consumed, the corrosion rate will not decrease. The behavior also depends on the environment. For example, corrosion of steel in a rural atmosphere is copper-type, but in marine or industrial atmosphere it is zinc-type. Corrosion of zinc in marine atmosphere is copper-type.

Industrial environments are more severe than rural or urban environments. The macro climate is usually not as important as the micro climate for atmospheric corrosion. Local sources of impurities have more importance than general atmospheric conditions. Metals at areas where they become wet and retain moisture generally corrode more rapidly than metals exposed to rain. Rain washes away accumulated dirt that causes differential aeration corrosion. Acid rain due to local or global reasons is obviously an exception. Impurities in the atmosphere have a strong effect on corrosion rate as Fig. 19 shows. Sulfur dioxide accelerates corrosion, and charcoal particles extend the time of wetness and provide active sites for cathodic reaction.

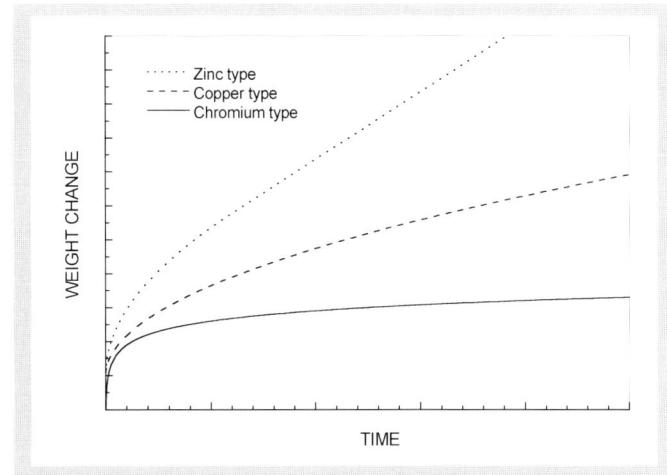

Figure 18. Corrosion behavior of various metals under atmospheric conditions.

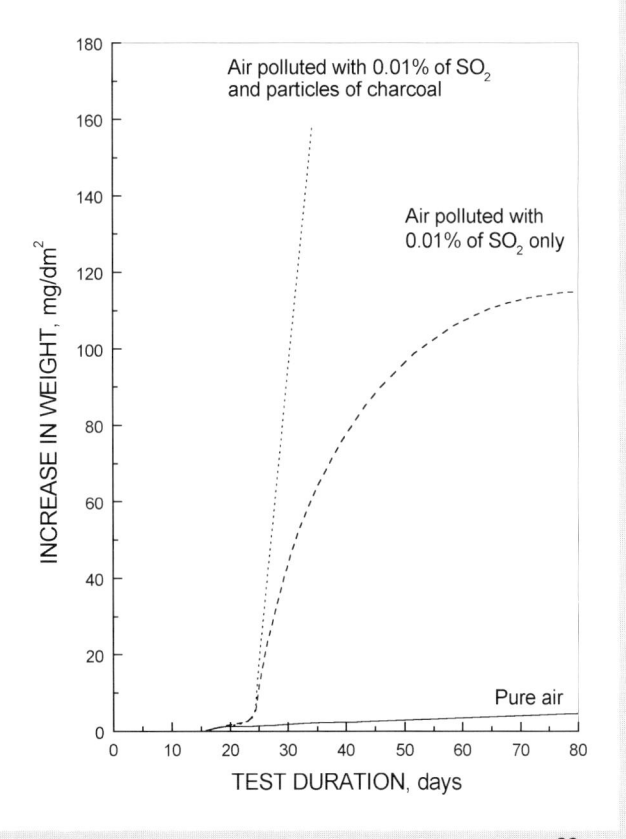

Figure 19. Effect of impurities on atmospheric corrosion rate[39].

The most common atmospheric contaminants are sulfur oxides, hydrogen sulfide, nitrogen compounds, saline particles, and other airborne particles[40]. Sulfur oxides of which sulfur dioxide is the most common will increase corrosion rates of steel and zinc. SO_2 in the atmosphere comes from combustion of sulfur-containing fuels and oxidation of natural H_2S. Sulfur dioxide adsorbs on metal surfaces and forms sulfuric acid under humid conditions. For nonferrous metals such as copper, lead, and nickel, sulfuric acid causes direct corrosion. For zinc and cadmium, it leaches the protective carbonate film and consumes the acid. In corrosion of ferrous metals, the resulting sulfates are oxidized, and sulfuric acid is regenerated. Hydrogen sulfide causes tarnishing of copper and silver and may have a significant effect in electrical equipment. Nitrogen compounds such as ammonia can increase the wetting of metals. Ammonia compounds can cause stress corrosion cracking of copper and copper alloys at very low concentrations. Saline compounds such as sodium chloride in marine atmospheres and ammonium sulfate in industrial atmospheres are usually hygroscopic. Some are acidic. Other particles will usually adsorb moisture and contaminants to increase the time of wetness and acidity. Higher moisture content and acidic conditions will increase corrosion rates of most metals.

Ventilation, exhaust, and gas scrubbing systems provide a special case of an atmospheric corrosion problem. The purpose of ventilation is to produce better operating conditions and a more controlled room environment. Exhaust and scrubbing systems remove gases and various components from the gas streams. In all these cases, moisture or evaporated water is removed from some system. As the gas phase cools, condensation occurs. Contrary to natural atmospheric conditions, cyclic wetting and drying is not happening, and the surfaces are almost constantly covered by a moisture film. Since the removed gases often contain corrosive compounds, the gas handling equipment may be subject to a very corrosive environment. For example, paper machine ventilation systems remove moisture and water vapor from the wet end of the machine. The removed gases therefore contain those compounds occurring in the white water. The micro climate inside the equipment and near exhaust valves and vents will determine atmospheric corrosivity. Process control systems, instrumentation, and electrical equipment are critical components in the operation of any industrial facility. They are also highly susceptible to corrosion from a wide range of contaminant gases produced in normal plant operation. This kind of corrosion attacks the process control systems with the least visual signs.

7 Microbiological corrosion

The terms microbiological corrosion, microbiologically-induced corrosion, and microbiologically-influenced corrosion (MIC) refer to a situation where corrosion begins or its rate increases due to microbiological activity. For MIC to occur in a system, microbes must be present, service temperature must be suitable for microbial metabolisms, nutrients must be present, and the metal must be susceptible to MIC. Microbiologically-influenced corrosion does not involve any new form of corrosion. Some doubt still exists if microbiological corrosion causes direct waste of metals although some microbes can metabolize metals. Identification of microbiological corrosion is essential when planning corrosion prevention measures. For example, organic inhibitors could be nutrients to bacteria. Microorganisms are usually associated with localized corrosion rather than uniform corrosion. Most cases of MIC involve deposits. Slimy films, discrete deposits, and tubercles are often associated with MIC. One possible indication of MIC is that the same kind of corrosion happens all the time even when using more resistant materials. Microbiologically-influenced corrosion usually occurs in nearly neutral solutions at pH of 5–9 at about 20°C–40°C. It may occur even at 75°C. The mechanism of microbiological corrosion varies. In some cases, the cathodic reaction, e.g., sulfate-reducing bacteria, enhances the microbiological effect. In other cases, the metabolic reactions in microbiological films may cause local acidification or local deoxygenation. Under the biofilm, pH may be several units lower and the redox potential up to 400 mV lower than in the bulk solution.

The most important reason for microbiologically-induced corrosion is formation of a biofilm. This is a microbial mass composed of bacteria, algae, fungi, etc. Biofouling describes undesirable biological growth and deposits on equipment surfaces. The biofilm changes the chemistry of the liquid beneath it in numerous ways. The biofilm itself is mostly water, and it does not completely isolate the metal to liquid interface from the bulk solution. It causes large concentration gradients perpendicular to the metal surface. The composition of a biofilm may vary in three dimensions and with time.

Microorganisms found in the biofilm are bacteria, fungi, and algae. Bacteria represent the largest group of troublesome organisms. Slime forming bacteria produce a dense, sticky mass that can cause fouling of heat exchangers. Impeding water flows can result in loss of heat transfer. Additional microbiological growth occurs when system water flow decreases. Spore forming bacteria are difficult to control if a complete kill is necessary. The organism becomes inert if its environment becomes hostile and begins to propagate once the environment becomes suitable again. Spore formers do not influence most processes when the organism is in the spore form. Sulfate-reducing bacteria generate sulfides from sulfates and can cause serious localized corrosion. They convert sulfur compounds to acidic hydrogen sulfide. This process usually occurs at the center of large reddish-black deposits and results in deep pitting under the deposit. Iron-reducing bacteria occur in waters with a high ferrous iron content. The ferrous iron converts to insoluble ferric hydroxide. This leads to tubercular attack that results in increased flow resistance and restricted carrying capacity. The bacteria associated with corrosion can be aerobic or anaerobic and autotrophic or heterotrophic. Aerobic bacteria require oxygen, and anaerobic bacteria require the absence of oxygen. Autotrophs do not depend on the environment for a source of organic substances as nutrients, but heterotrophs need energy and carbon from organic sources. Bacteria may exist as single cells or multicellular colonies. Categorization can use physical and metabolical characteristics. Some of these categories are as follows[41]:

- Shape

    Coccus: spherical

    Bacillus: rod shaped

    Spirillum: curved or comma shaped

CHAPTER 5

- Temperature requirements

    Psychrophilic: cold liking, 0°C–25°C

    Mesophilic: moderate temperature liking, 20°C–45°C

    Thermophilic: heat liking, 45°C–70°C

- Oxygen requirements

    Aerobic: require oxygen to live

    Facultative: can live with or without oxygen but growth rate increases in the presence of oxygen

    Indifferent: can live with or without oxygen but growth rate increases in the absence of oxygen

    Anaerobic: require absence of oxygen to live

- Nutritional requirements

    Autotrophic organisms: derive energy from the oxidation of inorganic materials

    Heterotrophic organisms: derive energy from the oxidation of organic and inorganic materials

    Paratrophic organisms: parasites that feed on living organic matter

    Saprotrophic organisms: subsist on dead or decaying matter.

Since algae require sunlight to grow, they occur in the open, exposed areas of a cooling tower. Algae grow in dense, fibrous mats that can plug piping. Algae growths also provide an ideal growth medium for anaerobic bacteria. Algae are not sensitive to pH, and some species can live at 50°C–60°C or higher. Most algae belong to three subdivisions:

- Cyanophyta: blue-green algae
- Bacillariophyta: diatoms
- Chlorophyta: green algae.

Blue-green algae prefer a neutral or alkaline environment. A few genera of blue-green algae can grow in an environment that provides only nitrogen, carbon dioxide, water, sunlight, and a few essential minerals such as magnesium and potassium. Green algae are indifferent to pH, and they can thrive at pH 4.5–9.3. Sunlight, carbon dioxide, and a few minerals are essential for their growth. The optimum temperature for growth of most fungi is 25°C–30°C. Many species found in cooling waters do prefer slightly higher temperatures. All the fungi grow optimally near pH 6. Changes in pH and the presence of dissolved heavy metals easily disturbs their metabolism.

7.1 Biofilm formation

The formation of biofilm occurs in several stages: adsorption and multiplication of primary colonization bacteria, multilayers of cells becoming embedded in their own polymer material, and development of mature biofilm. When a metal surface is exposed to natural or certain industrial waters, it becomes colonized with microbial species. The first step is adsorption of macromolecular organic film, the "conditioning film." The initial bacteria also begin to attach to the surface. Bacteria will attach to any metal in contact with an aqueous solution. Colonization is easiest at locations where the surface is rough or a stagnant solution pocket exists. Micro colonies of different species develop and eventually merge to form a biofilm. The micro colonies have random distribution. Some surface areas are less heavily colonized than others. This easily leads to various concentration cells. The second step in biofilm formation consists of cells on a substrate becoming embedded in slime. Slime is an organic polymer matrix excreted by microorganisms. This action creates a barrier between the metal surface and bulk electrolyte. More organic and inorganic material is then added to the film. Flowing water can detach some film to produce an equilibrium film thickness as Fig. 20 shows.

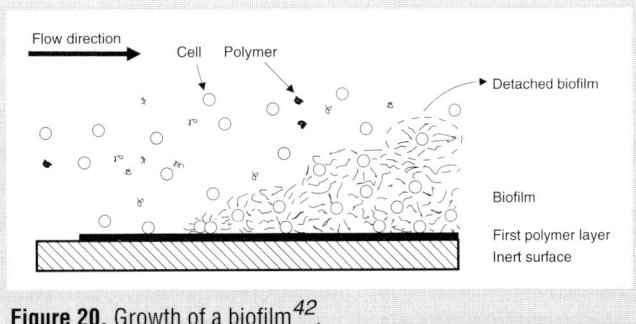

Figure 20. Growth of a biofilm[42].

A fully developed biofilm has cells, slime, fungi, and algae. Fungi and algae adhere to the metal surface and form differential aeration and concentration cells. Fungi require oxygen to grow, and they obtain energy by anaerobic fermentation or aerobic oxidation of organic materials. The algae are chlorophyll containing plants. They obtain energy from light or by oxidation of inorganic material and carbon for growth by assimilation of carbon dioxide. Even if the nutrients are low in flowing water, they may become concentrated at the metal surface.

When microorganisms form colonies on the surface of a metal, they do not form uniform layers but local sites. The sites of initial colonization may relate to such metallurgical features as roughness, existing corroded areas, or inclusions. After the colony forms, it attracts and includes other biological and nonbiological species. Depending on available oxygen, iron, manganese, etc., film formation and metabolism of species result in more severe corrosive environment inside and under the colonies compared with those on the surrounding metal. This leads to the formation of crevices and oxygen and ion concentration cells. Most microbiological colonies usually remain fixed to the initial colonization site. This causes the anodic site to become permanent. Hydrated slimes cover the surface, create differential aeration cells, and finally form an anaerobic environment below. In an aerobic environment, microbes can create tubercles that have inside oxygen concentration lower than that of the bulk solution.

7.2 MIC mechanisms

Microbiologically-influenced corrosion is primarily the result of different bacteria. The role of other microorganisms is to produce differential aeration and concentration cells. Most bacteria produce organic acids. Heterotrophic bacteria break organic compounds into organic acids, and autotrophic bacteria oxidize them to acids. Some bacteria produce carbon dioxide that can react to form carbonic acid. The main types of bacteria associated with corrosion are sulfate-reducing, acid-producing, and metal-oxidizing bacteria. Sulfate-reducing bacteria (SRB) are anaerobic bacteria. They reduce sulfate ions to sulfide ions and hydrogen sulfide. They occur in many industrial waters and are very resistant to biocides. Some species of sulfate-reducing bacteria promote formation of sulfide films that can catalyze a hydrogen reduction reaction. Aerobic bacteria can oxidize sulfur and sulfides or oxidize metal ions to higher valency states, generate acids through metabolic reactions, and form slime. Sulfur and sulfide oxidizing bacteria produce elemental sulfur or sulfate ions that can later form sulfuric acid. Short chain organic acids such as acetic acid are common products of microbiological activity. Acetic acid is very aggressive to carbon steel when concentrated under a colony or other deposit. Iron and manganese oxidizing bacteria derive energy by oxidizing ferrous ion to ferric ion. The ferric ion is a powerful oxidizer and can therefore accelerate cathodic reaction rate and corrosion rate. The most dangerous biofilms probably contain several strains of bacteria simultaneously changing many environmental factors to become more corrosive. Figure 21 shows some schematic mechanisms of MIC.

The first known cases of MIC occurred in the 1800s when traces of sulfidic reaction products were observed on iron objects. The classic explanation for microbiologically-

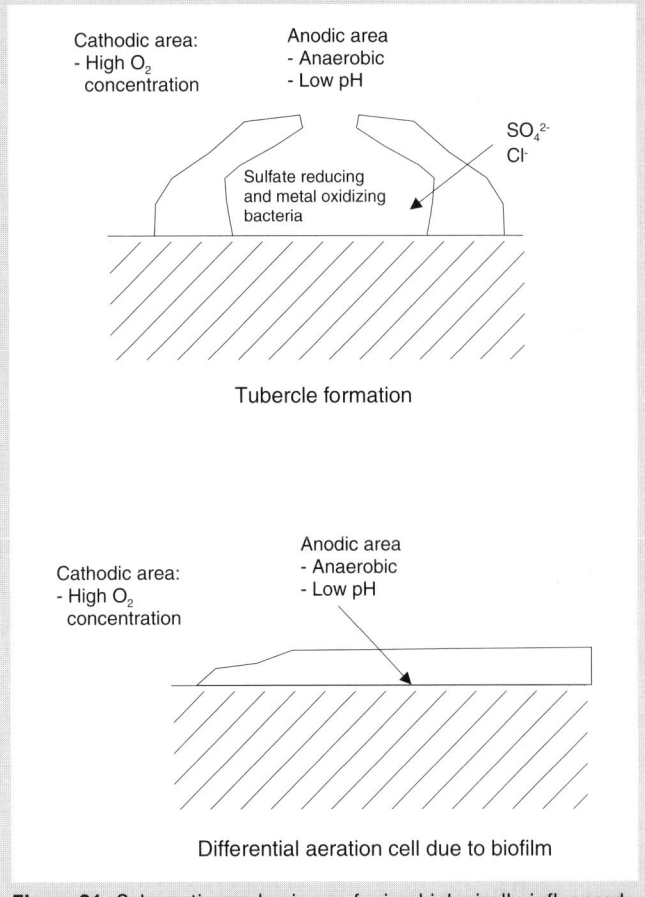

Figure 21. Schematic mechanisms of microbiologically-influenced corrosion.

influenced corrosion of steel and iron caused by sulfate-reducing bacteria is cathodic depolarization proposed by von Wolzogen Kühr in 1934[43]. This mechanism uses the idea that the rate limiting step in corrosion is dissociation of hydrogen from the cathodic site. Accumulation of hydrogen in a closed site increases activation overpotential and decreases the cathodic reaction rate. Sulfate-reducing bacteria probably consume hydrogen through the action of their hydrogenase enzymes and "depolarize" the cathode. This increases the corrosion rate back to the original level. Some investigators still believe that this mechanism is the important one for MIC of iron and steels. Later studies have shown that the depolarization mechanism of consuming hydrogen cannot explain all cases. The mechanism does not consider other important factors such as the effects of sulfide, bisulfide, and hydrogen sulfide produced from sulfate reduction, the catalyzing effect of sulfides on hydrogen evolution reaction, production of other corrosive compounds, etc. Although the mechanism may be unclear, the principal effect of sulfate-reducing bacteria is to accelerate the cathodic reaction rate. Most MIC research focused on sulfate-reducing bacteria since they were probably the primary cause of MIC problems[43].

Occluded areas on a metal surface form because microbes make local colony centers on the surface of a metal rather than uniform layers. The colony attracts other biological species, metals, and chloride ions resulting in conditions within and under the colonies different from those on the surrounding metal. The formation of crevices, oxygen, and ion concentration cells allows corrosion to continue. Most microbiological communities usually remain fixed to the primary colonization site causing this area to become permanently anodic. Depending on the species in the biofilm, the conditions under it may become acidic, concentration of oxidized metal ions and aggressive anions may increase, or oxygen concentration may decrease. Sulfur-oxidizing bacteria are aerobic and can produce sulfuric acid up to 10% concentration[44]. Iron bacteria are a group of microorganisms found in industrial waters, streams, lakes, wells, and potable water supplies. Iron-oxidizing bacteria produce ferric ion that causes migration of chloride ions to restore electrical equilibrium such as in crevice corrosion mechanisms. Some bacteria can also oxidize manganese with the same results. The microbiologically-induced corrosion sometimes has an autocatalytic nature. Acidic environment with a high concentration of aggressive ions will definitely cause corrosion. Iron bacteria can deposit iron in the form of hydrated ferric hydroxide. Water will turn "brick-red" leading to the common reference of "red water." In addition to discoloring the water, this group of micro organisms produces undesirable accumulations in pipes, nozzles, ponds, etc. These deposits will eventually slough off, plug lines, and foul pumps and valves. They can also influence the quality of finished products. While these microorganisms are aerobic, they grow in waters with very low oxygen content. Slime-forming organisms produce layers that create differences in oxygen concentration. On top of the biofilm, oxygen concentration is high decreasing to virtually anaerobic conditions in extreme cases. A high concentration of oxidants in the biofilm combined with absence of oxygen under the biofilm will create a strong concentration difference cell. The biofilm and species in it do not directly cause corrosion but create an environment where corrosion is very likely to begin.

CHAPTER 5

The important variables in microbiologically-influenced corrosion are difficult to specify. Besides the variables usually associated with the form of corrosion under study, the variables that affect the life cycle of the microbiological species become important. Such variables are dissolved oxygen, pH, temperature, and available nutrients. Many variables may also influence corrosion behavior without microbes. Usually, the organisms and phenomena that affect corrosion are inside the biofilm or tubercle. Bulk electrolyte properties may therefore have little relevance to the corrosion problem. Direct counts of micro organisms present in the bulk solution may also have little use when predicting MIC. Sulfate-reducing bacteria can live at a wide pH range from slightly acid to slightly alkaline. Most bacteria live between some minimum and maximum temperatures but also have a narrow optimum temperature range at which their activity is highest. These temperatures are not definite and may change as other environmental factors vary.

7.3 MIC problems

MIC is a common problem in industrial processes due to the presence of microbes, adequate nutrients, and corrosive by-products. For carbon steels, microbiologically-influenced corrosion is often pitting under tubercles. On austenitic stainless steels, sulfate-reducing bacteria usually produce black deposits, and iron bacteria give brown or reddish-brown mounds. Construction materials in the pulp and paper industry are the same as the total chemical industry, but the environments are often very suitable for biological activity. In a typical pulp mill, the systems potentially susceptible to MIC are the following[45]:

- Raw water systems
- Treated process water
- Economizers
- White water systems
- Paper machine headboxes and suction rolls
- Cooling water systems
- Fire water piping.

Raw ground water and surface water can contain iron-oxidizing bacteria and sulfate-reducing bacteria. The problems with carbon steel piping are usually tubercle formation and associated pitting corrosion, reduced flow, and contamination of water by corrosion products such as dissolved iron, iron compounds, and hydrogen sulfide.

Treated process waters have use for showers, washing, and dilution. Filters, clarifiers, and storage tanks are suitable environments for metal-oxidizing and slime-forming bacteria. These can transfer to every location using process water. Cooling water systems and their problems are similar to other industrial processes. Open recirculating cooling systems are especially suitable for microbiological activity. The biofilm can cause losses in heat transfer and corrosion by aerobic and anaerobic bacteria. The same applies for economizers and heat exchangers that use condensate or treated process water and operate with lukewarm water containing dissolved solids.

The most common case of microbiological activity in pulp and paper manufacture is the white water environment of paper machines. The white water is warm and contains dissolved organic nutrients and sulfate ions for the metabolism of sulfate-reducing bacteria. Unclarified white water also contains suspended solids that help to build biofilms. Although the white water usually contains oxygen to support aerobic microbial activity, thick sludges cause anaerobic activity to be more common. Microbiologically-influenced corrosion causes general corrosion and pitting corrosion in several places and is not restricted to a particular material class, flow rate, temperature range, etc. Besides corrosion problems in white water systems, MIC can cause deterioration of product and odor problems[45]. Microbiologically-influenced corrosion is a particular problem in alkaline papermaking. Paper machine headboxes are vulnerable, and even slight corrosion can cause problems in production. Pitting of stainless steel and tubercular attack of carbon steel can be due to sulfate-reducing and iron-oxidizing bacterial. Tubercular action is a corrosion process that produces hard mounds (tubercles) of corrosion products on the metal surface. This result of corrosion increases friction and reduces flow in a water distribution system. Since corrosion results in rougher surfaces, the biofilm can attach more effectively, deposits can become more extensive, and control of corrosion and MIC can become more difficult. Suction and couch rolls are subject to MIC since fine suspended solids are in close contact with the rolls. Slime-forming bacteria can adhere tightly to the surfaces and create differential aeration cells. Sulfate-reducing bacteria can be active under slime and fiber deposits. Since the rolls are under constant load and suction rolls also have numerous holes acting as stress raisers, minute corrosion under deposits can initiate cracking[45].

Prevention of microbiologically-influenced corrosion can use material selection or chemical treatment. Material selection has the same guidelines as for most aqueous environments. Higher corrosion resistance will result with a higher alloyed metal. The mechanism for microbiologically-influenced corrosion is different for carbon steel and stainless steel. Replacement of carbon steel with types AISI 304 or AISI 316 stainless steels may therefore be ineffective and result in other MIC problems. Stainless steels often corrode due to differential aeration cells formed under tubercles created by metal-depositing bacteria. The corrosion mechanism is analogous to the deoxygenation-acidification mechanism crevice corrosion. More alloyed stainless steels and especially those with high amounts of molybdenum are therefore more resistant than the common Types AISI 304 or AISI 316. Chapter 6 of this volume discusses the chemical treatment to prevent MIC under "Control of biofouling."

8 High temperature corrosion

A "high temperature" for a material is any temperature where it begins to creep or starts to react with low humidity atmosphere, molten salt, or liquid metal. The rate of attack often increases with increasing temperature. High temperature corrosion is direct oxida-

CHAPTER 5

tion of metals by dry gases, i.e., without any moisture film. It has some similarities to aqueous corrosion. Most metals react with oxygen to form an oxide film. Oxidation can occur at the gas to scale interface or at the metal to scale interface. The reaction proceeds due to anion or cation transfer through the scale that behaves as a solid electrolyte. As the oxide film grows thicker, it restricts mass transfer between the metal and the surrounding atmosphere to decrease the corrosion rate. The rate determining step becomes metal or oxygen diffusion through the film. In an aqueous solution, a concentration of dissolved compounds in the solution describes the environment. In high temperature systems, the partial pressures of gases describe the environment. The partial pressures, i.e., the relative concentrations of gases, will determine which reaction products will form. Contrary to aqueous corrosion, high temperature corrosion depends more on thermodynamic factors than kinetic factors. One can estimate the various compounds that can form corrosion products with theoretical calculations knowing the partial pressures of gases.

An Ellingham diagram can estimate the stability of a metal under a gaseous atmosphere. Such diagrams show the change of Gibbs free energy for compound formation as a function of temperature. If the partial pressure is sufficiently high, the compound will form. If the partial pressure is sufficiently low, the metal will not react, and existing compounds will reduce back to the metallic state. This partial pressure is the dissociation pressure. The data has practical value in predicting which alloy components will react with various compounds. The partial pressures of gases required for compound formation are usually low at high temperatures. Low partial pressures occur in gas mixtures rather than in vacuum systems. Figure 22 shows a simplified Ellingham/Richardson diagram for oxide formation. The auxiliary scales show partial pressure of oxygen as gaseous oxygen or partial pressure of oxygen in gas mixtures such as H_2+H_2O and $CO+CO_2$. The lines describing oxide formation have positive slopes. This means that oxide stability decreases with increasing temperature. The diagram in Fig. 22 shows that oxide formation is more probable at high partial pressures of oxygen. Corresponding diagrams are constructed for sulfides, carbides, etc.[46].

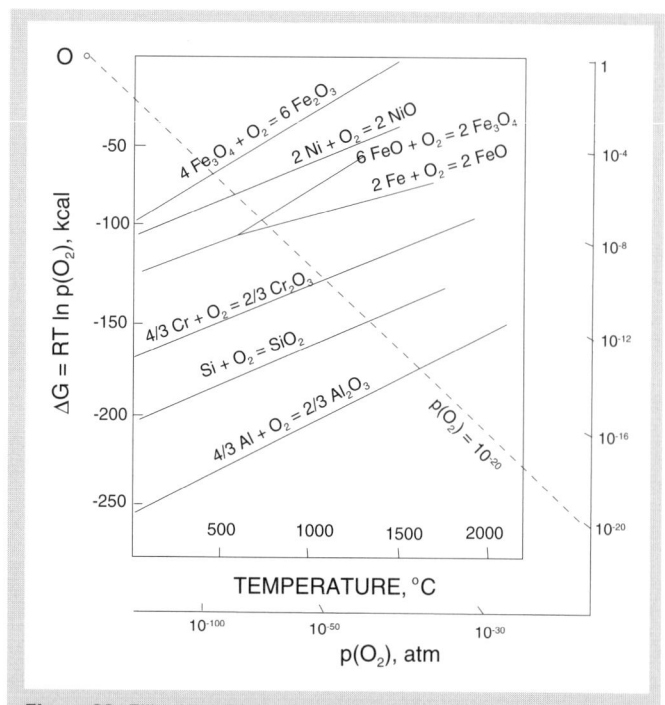

Figure 22. Ellingham/Richardson diagram for oxide formation.

A phase stability diagram is more convenient to estimate the reaction products in gas mixtures. These diagrams show the stability boundaries of metals and compounds as a function of partial pressures of the gaseous compound at constant temperature. The diagrams are high temperature analogues of the Pourbaix diagrams in aqueous solutions. If the partial pressures of gaseous reactants are sufficiently low, the metal exists as a pure metal. Increase of partial pressure of a gaseous compound will shift the equilibrium first to the stability domain of the most metal-rich compound and then to those of less metal-containing compounds. Figure 23 shows the stabilities of nickel and its compounds in an Ni-O-S system at 700°C. The sulfur compounds are pure sulfur in Fig. 23(a) and sulfur trioxide in Fig. 23(b). The resulting phases are essentially the same, but their stability areas are different.

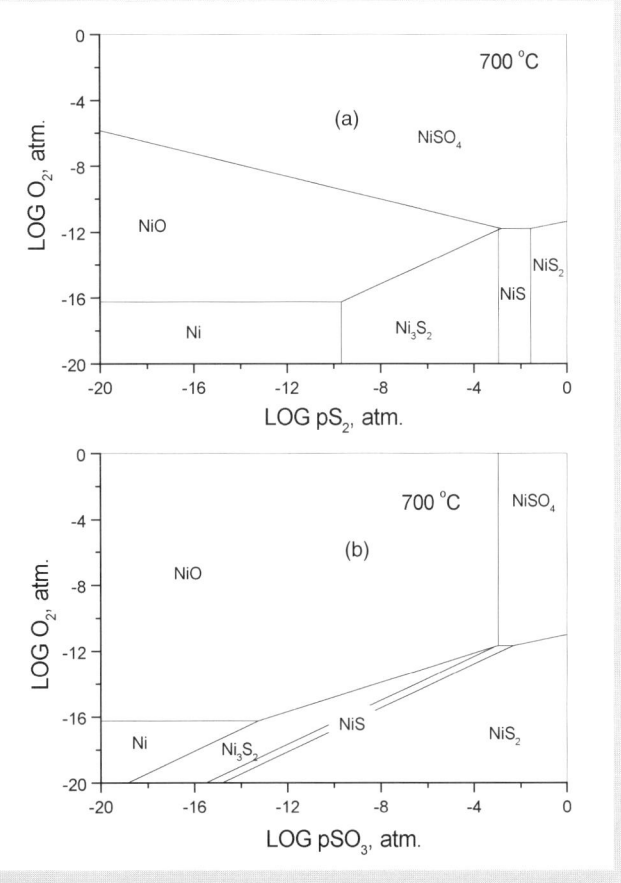

Figure 23. Phase stability diagram for an Ni-O-S system at 700°C.

Oxidation in clean air and oxygen gives a thin oxide film that may grow thicker. In polluted gases, oxide films are more often porous. The reaction product is not necessarily an oxide. It can be sulfide, carbide, or other compound. Protective scales are usually oxides, but they can also be sulfides, carbides, or mixtures of these materials. Usually, only oxides are considered protective. The rate of oxidation increases with increasing temperature. If the scale is continuous and nonporous, ionic transport through the film will be the rate determining step. Properties of the oxide scale will determine the high temperature corrosion resistance of the metal. The desired properties for a protective oxide scale are high thermodynamic stability, high melting temperature, good adherence to the surface, and low conductivity and diffusion coefficients of reactive species so the scale grows slowly. The important environmental factors influencing scale formation are temperature, thermal cycles with respect to time, concentrations of corrosive agents and their chemisorption and diffusion characteristics, and static and cyclic stresses.

CHAPTER 5

The rate of oxidation depends on the properties of the reaction products. The reaction products are usually brittle and lack ductility. Initiation of cracks on the reaction product scale depends on whether the surface scale is under tensile or compressive stress and the magnitude of the stress. The structure and formation of the scale also influence the rate law. One can theoretically estimate the protectiveness of the scale to some degree by using the Pilling-Bedworth ratio. This is the molar volume ratio of oxide formed to metal consumed in producing the oxide as Eq. 18 shows.

$$Pilling-Bedworth\ ratio = \frac{V_{oxide\ produced}}{V_{metal\ consumed}} = \frac{M_{oxide} \cdot \rho_{metal}}{z \cdot M_{metal} \cdot \rho_{oxide}} \quad (18)$$

If the molar volume of oxide is smaller than that of the metal, the ratio has a value lower than one. This means not enough oxide can cover the entire surface. The result is a porous, nonprotective film. Since the film is porous, the oxidation will follow a linear rate law. If the oxide molar volume is larger than that of the metal, the ratio has a value larger than one. In this case, the oxide film should be continuous and protective. This is true for some metals such as aluminum. Complications usually arise if the ratio is too high. Compressive internal stresses developed in the growing oxide film may lead to blistering and cracking of oxide. Usually, metals that have Pilling-Bedworth ratios of approximately 1–2 should form protective oxides. In practice, this is seldom valid.

Metal oxides are ionic compounds. Oxides are usually nonstoichiometric and contain vacancies. Oxides with excess metal ions in interstitial sites or with anion vacancies are n-type oxides. Oxides deficient on metal ions are p-type oxides. In n-type oxides, the film grows by simultaneous diffusion of excess metal to the surface and electron diffusion in the opposite direction. On the surface, metal ions will react to oxide. In p-type oxides, the growth is due to a vacancy concentration gradient that then causes metal ions to migrate toward the outer surface. For a metal forming a p-type oxide, alloying with a metal with lower valency will strengthen the protective film. The oxidation rate decreases due to a lower number of vacancies and an increased number of parent metal ions. For a metal forming the n-type oxide, the alloying metal must have a higher valency. The n-type oxides grow by anion diffusion or by interstitial diffusion of cations. In the former case, anion vacancies decrease. In the latter case, the number of interstitial metal ions decreases[47].

Most high temperature alloys contain chromium. Oxidation rates of ferrous alloys decrease sharply as chromium is increased to 10%. Nickel alone also has an inhibiting effect on oxidation of iron. With chromium, nickel strongly increases the oxidation resistance of steels. Eventually, nickel will diffuse and concentrate on the FeO at the substrate interface and suppress FeO formation. This stabilizes the protective Fe_3O_4 and Fe_2O_3 phases. It also stabilizes the creep resistant austenite phase and improves resistance to thermal cycling. Silicon forms an adherent resistant film alone or with chromium but decreases creep resistance. Aluminum also forms protective oxides, but it may cause embrittling. Aluminum therefore requires use at the lowest possible concentration[48].

8.1 Mixed atmospheres

Oxygen normally forms resistant barrier scales on heat resistant alloys. In actual service, the environments contain water vapor, sulfur compounds, and carbon monoxide and dioxide. Oxygen containing environments are oxidizing with excess free oxygen or reducing without free oxygen at equilibrium for the gas mixture. Mixed environments with several oxidizers and low oxygen activity are usually more corrosive. Thermal cycling is also frequent. This has a detrimental effect on alloy resistance. Many high temperature gas streams are actually oxygen deficient. This decreases thermodynamic stability of oxides and increases the probability of formation of other compounds. Thermal cycling may cause cracking of oxide films. Oxides that are ductile at high temperatures may be brittle at low temperatures. Different thermal expansion coefficients with metals cause stresses when temperature changes. The growth of a protective film may abruptly change to linear oxidation due to physical or chemical transformation. At first, the oxide film seems to be protective, and mass gain decreases with time. If the protective film begins to crack at a continuously increasing number of sites, the cracking leads to breakaway corrosion.

Catastrophic oxidation and hot corrosion are specific terms separate from general high temperature corrosion. Hot corrosion is due to impurities such as sulfur, vanadium, and sodium that may form low melting compounds. These compounds can form molten slags on surfaces and flux the protective oxide layers. Then the metal is again vulnerable to corrosion. Hot corrosion is an accelerating oxidation process. For example, sulfide scales can convert to more stable oxides, but the released sulfide ions penetrate to the metal matrix to form more sulfides that then undergo oxidation. Sequential oxidation and sulfidation cause more rapid corrosion than either reaction alone. Hot corrosion limits the use of high nickel alloys especially in environments with high sulfur partial pressures.

Several gaseous components can cause more severe corrosion than mere oxidation. If the partial pressure, i.e., concentration, of gaseous sulfur is sufficiently high, sulfide phases can form instead of oxides. Dangerous sulfur compounds are sulfur vapor, hydrogen sulfide, and the less aggressive sulfur dioxide. Sulfide scales are almost always less protective than the corresponding oxides since they are more porous. Sulfides will also melt at lower temperatures than oxides. The extent of corrosion depends on the partial pressures of sulfur and oxygen and relative stabilities of sulfides and oxides. Oxygen is a stronger oxidizer than sulfur. Usually, oxides will form instead of sulfides on alloys containing chromium and aluminum. In many combustion product gases, the oxygen has been consumed so that sulfur compounds present in the gas stream can react more easily with the metal. Especially in damaged oxide scales, sulfur compounds can react with alloyed Cr and Al, hinder oxide film repair, and allow sulfidation of the matrix metals such as iron and nickel. Carburization is possible at very low oxygen partial pressures, i.e., under reducing conditions. Carbon monoxide and dioxide can penetrate oxide layers through pores and cracks. The metal will oxidize. As carbon content increases, carbides may form. Carburization decreases mechanical strength, oxidation, and sulfidation resistance and may cause internal stresses due to carbide precipitation. It differs from other high temperature reactions because the carbon must diffuse into the metal before reacting.

Sulfidation describes the gaseous attack of metals by H_2S, SO_2, gaseous sulfur, and other gaseous sulfur species. Hydrogen sulfide is present in reducing atmospheres, and sulfur dioxide is present in oxidizing atmospheres. The amount of air blown in the furnace can change the sulfur compounds from one to another. In mixtures of hydrogen and hydrogen sulfide, the partial pressure of oxygen is so low that protective oxides are not stable. Sulfides are the stable phases. Besides sulfur compounds, more complex mixed gas environments contain carbon monoxide, carbon dioxide, and water vapor. The partial pressures of oxygen and sulfur are so high that oxides and sulfides can form. In sulfur dioxide environments, the flue gases contain sulfur dioxide or a mixture of sulfur dioxide and oxygen. The partial pressure of oxygen is sufficiently high for oxides to form[51]. Hydrogen sulfide is perhaps the most corrosive sulfur compound. It can act as an acid in low temperature aqueous corrosion and react directly with iron at higher temperature to form a brittle and nonprotective sulfide scale. The rate of attack on iron by hydrogen sulfide does not decrease with time. Low alloy additions to steel do not solve the problem. Chromium contents over 12% may be necessary to achieve any benefits under operating conditions containing high temperature sulfide. Austenitic stainless steels and straight chromium steels (12%–16% Cr) have been resistant[49]. In mixed environments, the alloys are protected if the environment is oxidizing and breakaway corrosion does not occur. In this case, rapid sulfidation attack may follow. High nickel alloys are usually very susceptible to sulfidation. Chromium will improve resistance of iron-, nickel-, and cobalt-based alloys. In sulfur dioxide environments, chromium-alloyed metals are resistant when oxygen is present. In atmospheres with low partial pressure of SO_2, the attack may be stronger. Using excess oxygen can prevent this[49].

Halogens are also harmful impurities in high temperature applications. Reaction mechanisms are often the same as for oxidation or sulfidation. Halogens do not usually form scales since the reaction products are too volatile and will vaporize. Halogen species are very mobile. They diffuse and dissolve in a metal matrix. This can lead to serious internal oxidation damage. Chlorides disrupt protective scales since their vaporization leaves vacancies and forms voids in the scales. A common guideline for material selection in halogen service is to keep temperature below where the halogen vapor pressure reaches 10^{-4} atm[49]. Mixed pollution environments where composition of the flue gases vary are particularly difficult for high temperature service. Attempts to burn different wastes in a single furnace may lead to rapid corrosion since various impurities attack the protective oxide scales with different mechanisms.

8.2 Recovery boiler corrosion

The recovery boiler is the main unit in the chemical recovery process of a pulp mill. It regenerates the spent cooking chemicals and simultaneously uses the heat of the black liquor for steam production. The kraft recovery boiler is essentially an ordinary steam boiler burning spent cooking liquor containing organic material. The main reason for the recovery boiler is conversion of the spent cooking liquor back to cooking chemicals. The energy produced is a byproduct. The recovery boiler has two sections. The lower section uses reducing conditions, and the upper section uses oxidizing conditions to complete combustion. Concentrated black liquor and sodium sulfate (Na_2SO_4) are burned under reducing conditions. Organic elements of the black liquor convert to carbon dioxide and water vapor. Sodium carbonate forms when carbon dioxide reacts with sodium. Addition of sodium sulfate replenishes elements lost in cooking. The oxidized sulfur compounds such as sulfate and thiosulfate reduce to sulfide under the reducing conditions in the lower part of the boiler. Sulfur release from burning black liquor droplets is a decomposition reaction that starts when pyrolysis begins. The sulfur in black liquor releases in the lower furnace primarily as hydrogen sulfide gas. Reactions in the atmosphere and smelt result in sulfur leaving the boiler as sodium sulfide and sulfate in the smelt and hydrogen sulfide and sulfur dioxide in flue gases. The degree of reduction is a measure of how much sodium sulfate converts into sulfide. This is usually 90%–92%. It is a measure of the efficiency of the recovery boiler.

The recovery boiler operates with a molten salt bed. The main components of this smelt are sodium carbonate and sodium sulfide. Boiler tubes with circulating water surround the molten salt. This is the water wall construction. Steam is produced in the array of tubes that form the heat exchangers and walls of the recovery boiler. A layer of smelt freezes on the fireside of the water wall tubes providing a barrier between tubes and boiler atmosphere. The external surfaces of boiler tubes may corrode by the high temperature processes mentioned earlier. The internal surfaces may corrode due to poor water treatment. Stress corrosion cracking, thermal fatigue cracking, and dew point corrosion are also possible. The lower part of the water wall is the most critical area. If a tube ruptures, high pressure water can release into the boiler. When water contacts molten chemicals, it instantly vaporizes. A leak in the tubes is intolerable since the reaction between smelt and water can cause an explosion.

Before the 1960s, heavy corrosion damage in boiler furnace tubes made of carbon steel was uncommon. Corrosion damages were usually due to local overheating when deposits hindered water flow inside the tubes. With the trend to construct larger boilers and operate at higher pressures, the service life of carbon steel tubes decreased to a few years. Recovery boiler corrosion became a serious problem in the beginning of the 1960s. Severe gas-side corrosion occurred in Finland and Sweden. This was possibly due to the higher temperatures and higher sulfidity of the pulping process in Scandinavia than in Canada and the United States. High pressure recovery boilers provide more efficient power generation, but the associated higher-temperature flue gases are also more corrosive. The corrosion was due to hydrogen sulfide in the flue gases. Protection of water wall tubes was sought by using thermally sprayed coatings or chromium oxide based refractory materials on studded tubes. Since the late 1960s, tubes with stainless steel outer cladding have had use[50]. The type of outer stainless steel cladding has gradually increased from AISI 304 to 6% Mo grades and nickel-based alloys. The cost of highly alloyed cladded tubes has caused a return to use of studded tubes and ferritic tubes.

CHAPTER 5

Figure 24 shows the schematic structure of a recovery boiler. The boiler has tubes joined together. The boiler contains three zones: reduction zone at the bottom, drying and combustion zone in the middle, and complete combustion zone at the top. Black liquor is fed to the middle drying and combustion zone. The dry organic compounds fall to the boiler bottom where they build a heap of solidified smelt with a molten layer on top. All the reactions occur in the active molten layer. The amount of primary air fed to the reduction zone is only 40% of the oxygen necessary for combustion to maintain reducing conditions. The reducing atmosphere is due to carbon, carbon monoxide, and hydrogen. These reduce the sodium sulfate back to sodium sulfide. The smelt temperature is about 800°C, but the wall temperature of the tubes is below 300°C in low pressure boilers or slightly above that in high pressure boilers. The solid smelt itself is not corrosive and will protect the walls from corrosive gases and molten smelt. The flue gases then pass through a superheater, steam generating bank, and economizer sections before entering precipitation and desulfurization.

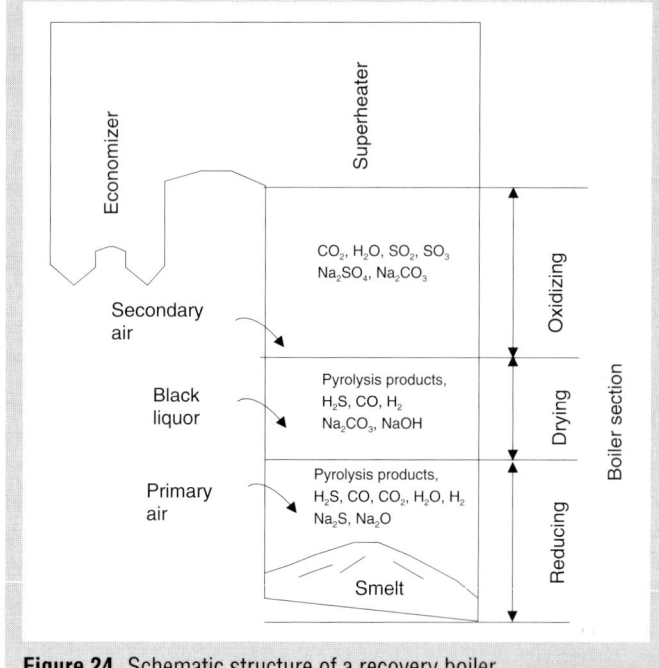

Figure 24. Schematic structure of a recovery boiler.

The atmosphere in the bottom of a recovery boiler is reducing. In principle, materials that survive in such conditions should have high chromium and nickel contents and contain some aluminum. Many original boilers used ferritic, low alloyed CrMo-steels. Corrosion problems with those alloys resulted in use of austenitic stainless steels as cladding. Tubes cannot use solid stainless steel since a risk of stress corrosion cracking exists. Protection by thermally sprayed iron- and nickel-based alloys has had use. The tubes sometimes have protruding studs to assist adhesion of frozen smelt on the tubes, but these may cause thermal gradients. The corrosivity in the furnace hearth and smelt increases as operating pressure and consequently surface temperature of the tubes increase. Molten smelt in contact with the tubes can cause rapid corrosion. The melting point decreases with increasing chloride and potassium content[50].

The tubes around black liquor and air ports sometimes suffer from localized corrosion. This is probably the result of condensation of molten alkalis at stagnant areas around the ports. The corrosion can even extend outside the boiler[50,51]. A solid hydroxide deposit will not necessarily cause corrosion, but molten hydroxides will flux protective films. Furnace design is important in eliminating stagnant gas flow areas. The flue gases contain carbon dioxide and water from burning of organic material and sulfur dioxide from sulfur species. Hydrogen sulfide is also present in the flue gases.

Corrosion has recently become worse by increased solids content of black liquor and increased temperature. The dry solids content of the burning liquor has risen from 65% to about 70%. Efforts are underway to reach dry solids content as high as 75% by modifying the conventional evaporator and boiler arrangement. Attempts to increase solids content to improve efficiency and reduce environmental load are common. The dry solids content of the concentrated waste liquor has importance in operation of the recovery boiler. A higher dry solids content requires a higher temperature in the furnace. This increases the degree of reduction and lowers the content of environmentally harmful sulfur dioxide and hydrogen sulfide in the flue gas. A high dry solids content also increases steam production. Higher solids firing has caused a more corrosive environment inside the recovery boiler, and existing units can suffer from unexpected problems.

Thermal efficiency for a kraft recovery boiler is the percentage of black liquor fuel value actually realized in the generated steam. A modern design may achieve an efficiency as high as 70%. Two major factors affecting efficiency are liquor solids concentration and exit gas temperature. The capacity limit of a kraft recovery boiler depends on solids loading. When the loading increases, temperature of the gas and entrained solids entering the superheater section also increase. The optimum capacity of the furnace has been exceeded when the gas temperature entering the superheater section is so high that the ash particles in suspension are sticky. Mechanical methods can no longer control fouling of the surfaces. Melting of fireside deposits determines the upper limit for the superheated steam temperature. Above this temperature, the corrosion rate of superheater tubes increases. The deposits consist primarily of sodium sulfate and sodium carbonate with smaller amounts of chloride, sulfide, and potassium salts. The chlorides and potassium compounds decrease the melting point of deposits.

When the flue gases reach the economizer, their temperature has decreased. The metal temperature can be as low as 150°C–200°C. In boilers with high sulfur dioxide or trioxide levels, sulfuric acid can condense on the surfaces causing dew point corrosion. The damages are pits or vertical grooves. To avoid dew point corrosion, gas and metal temperatures must be sufficiently high.

8.3 Corrosion in high temperature water

High temperature aqueous environments differ from the aqueous environments at ambient temperatures. The water at high temperatures is usually conditioned, i.e., dissolved salts and gases have been removed. The corrosion problems in high temperature waters are often due to faulty water treatment. The boiler system uses heat to produce steam, and the water contains impurities and pollutants. Water impurities can be in the gaseous or solid form. Dissolved gases include oxygen, carbon dioxide, and nitrogen. Solids occur in two forms: suspended or dissolved. Suspended solids include any impurities in the water that are undissolved including sand, silt, or particles of organic matter.

CHAPTER 5

If undisturbed, the suspended solids will settle to the bottom of a vessel or container. Dissolved solids include any impurities dissolved in the water. Total dissolved solids is a measure of all impurities dissolved in a given water supply. Most dissolved mineral impurities found in water are present as ions. The dissolved impurities increase the conductivity of the boiler water. A temperature compensated conductivity meter is a useful instrument for checking boiler water and feed water total dissolved solids (TDS) levels.

Cycles of concentration refer to the accumulation of the impurities in a water supply. As water evaporates, water impurities remain behind. When a boiler is replenished with fresh water, more impurities enter the system. If no compensating action occurs, the impurities in a boiler system increase indefinitely until the water cannot dissolve its own impurities or hold them in solution. Boiler water solids remain in the drum and tubes as steam is generated. Unless removed, these solids will continue to accumulate and eventually cause unsatisfactory boiler operation. Every boiler has a limit for total solids. This limit decreases as boiler pressure increases. Some concentrated boiler water must be removed from the circulation system to control solids content. The removal of a portion of concentrated water is boiler blowdown. Replacement of concentrated water with feedwater lowers the general solids concentrations in the boiler.

The corrosion process on the steam side between iron and oxygen-free water results in the formation of a protective magnetite film according to the overall reaction of Eq. 19.

$$3\ Fe + 4\ H_2O = Fe_3O_4 + 4\ H_2 \tag{19}$$

This reaction proceeds in three main steps: iron dissolution, precipitation of iron hydroxides, and conversion of these hydroxides to protective magnetite scale. This last step is the Schikorr reaction of Eq. 20. At temperatures below 100°C, the rate of the Schikorr reaction is very slow, but at temperatures above 200°C it happens so rapidly that virtually no ferrous hydroxide exists[52].

$$3\ Fe(OH)_2 \rightarrow Fe_3O_4 + H_2 + 2\ H_2O \tag{20}$$

The stability of the magnetite layer depends on water alkalinity. Excessively acid or alkaline conditions will dissolve the magnetite resulting in corrosion of steel. The optimum pH is about 8.5–12.7. Figure 25 shows the effect of pH on the corrosion rate of steel in water at 310°C[53]. Only a sound and continuous magnetite film can protect from corrosion and permit use of carbon steels. The thickness of the protective layer is only about 5–20 µm. The objective of primary water treatment is to preserve the magnetite layer.

Special problems with boiler water are corrosion by dissolved oxygen and carbon dioxide, caustic corrosion, hydrogen embrittling, and scaling. Water treatment primarily solves these problems. All parts of the steam and water system are interdependent. Before applying any water treatment method, one must consider its effects on the entire system. The presence of oxygen causes pitting in boilers and steam and condensate systems. Oxygen concentration must remain as low as economically possible. This becomes more important when operating pressure increases. Mechanical deaeration is usually not sufficient. This requires the use of chemical oxygen scavengers. Common scavengers are sodium sulfite and hydrazine. Sodium sulfite has use in low and medium pressure boilers. In high pressure boilers, it will cause excessive carryover of dissolved solids. Oxygen scavenging in high pressure boilers almost solely uses hydrazine or organically catalyzed hydrazines. Decomposition of sulfite into sulfur dioxide and hydrogen sulfide can cause corrosion problems in the after boiler sections. Typically, a maximum residual oxygen level of 5 ppb (5 µg/L) is safe[54].

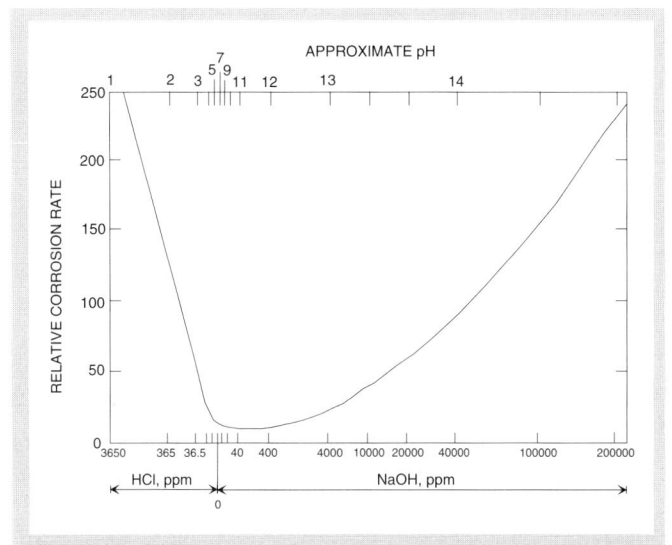

Figure 25. Effect of pH on the corrosion rate of steel in water at 310°C[53].

Corrosion in the after boiler section is a common problem in steam generating systems. The after boiler section includes the superheater, equipment using the steam, and condensate lines. Deposits may form in these areas because of dissolved solids carryover, contamination, and migration of corrosion products. The most common causes of after boiler corrosion are oxygen corrosion and low pH caused by carbon dioxide. Carbon dioxide related corrosion is almost always a problem of condensate lines. Carbon dioxide forms by the decomposition of bicarbonate by heat as Eq.21 shows[55]:

$$2\ HCO_3^- \rightarrow CO_3^{2-} + CO_2\uparrow + H_2O$$
$$CO_3^{2-} + H_2O \rightarrow 2\ OH^- + CO_2\uparrow$$

(21)

In the condensate, carbon dioxide will react with water to form carbonic acid. Carbon dioxide lowers the pH and causes corrosion. In addition, carbonic acid leads to formation of bicarbonate ions. These can react with dissolved iron and prevent formation of protective magnetite scale. When pressure drops or carbon dioxide content in the vapor decreases, the ferrous bicarbonate will precipitate as hydroxides and hydrated oxides causing scaling. When oxygen and carbon dioxide are both present, the corrosion can be rapid. Carbon dioxide disturbs passivation, and oxygen increases cathodic reaction rate and the corrosion rate on the unprotected surfaces. Corrosion caused by carbonic acid usually results in uniform thinning of the metal.

CHAPTER 5

Corrosion resulting from excess sodium hydroxide will cause general corrosion and cracking of steel. Sodium hydroxide has use in the treatment of boiler water to maintain optimum pH for protective magnetite scales and to keep dissolved solids as nonadhering sludges instead of solid scales. In high pressure boilers, excessive hydroxide will result in corrosion of magnetite and iron and form nonprotective Na_2FeO_2. The formation of porous deposit layers due to any reason creates another problem since sodium hydroxide can concentrate beneath these. Boiler water

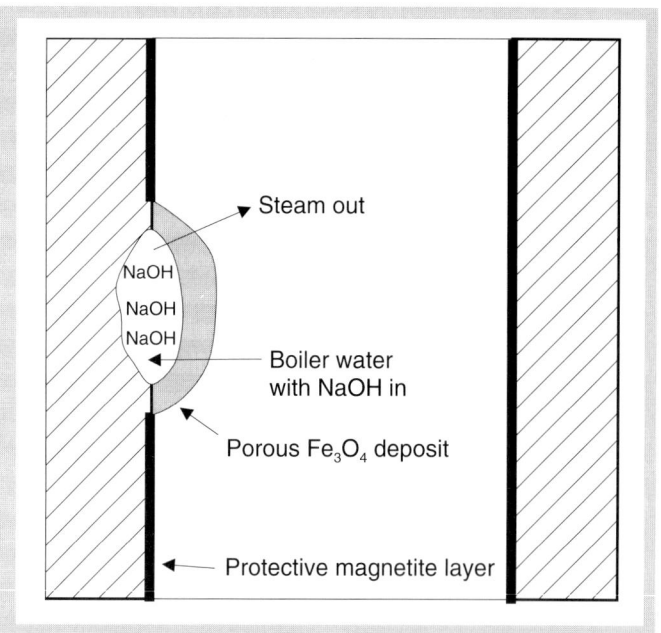

Figure 26. Concentration of boiler water solids such as sodium hydroxide under porous deposits.

enters below the deposit where it converts to steam. Steam escapes through the deposits. Any dissolved compounds will remain below the deposit as Fig. 26 shows. The sodium hydroxide concentration in boiler water is not high at about 5–10 mg/L. Under porous deposits, it can increase up to 100 000 mg/L. Such high caustic concentrations are on the level of kraft cooking liquors.

Hydrogen damage may occur if the internal surfaces corrode rapidly. This will release high amounts of nascent hydrogen in a short time. Some atomic hydrogen can migrate through the surface of tube steel and react with carbides. This reaction results in formation of methane, CH_4. This is a large molecule that creates internal pressure in the metal[53,55]. Hydrogen damage is possible in boilers with a low pH water chemistry, and deposits cover the surfaces.

Scale deposits interfere with heat exchange by insulating the heat transfer surfaces. The most common form of scale is calcium carbonate, $CaCO_3$. In natural waters, calcium ions combine with bicarbonate to form calcium bicarbonate as Eq. 22 shows. Calcium bicarbonate has a solubility of 300–400 mg/L at room temperature.

$$Ca^{2+} + 2\ HCO_3^- = Ca(HCO_3)_2 \tag{22}$$

As the temperature of the system increases, calcium bicarbonate converts to calcium carbonate by decomposition as Eq. 23 shows.

$$Ca(HCO_3)_2 = CaCO_3 + H_2O + CO_2 \tag{23}$$

The Langelier index can estimate the danger of calcium carbonate precipitation. Precipitation is possible when too much calcium is present or alkalinity is too high. Other sources of scale deposits are contaminated condensate and corrosion products. Localized scale deposits will sometimes result in locally higher wall temperatures on the furnace side. This can cause unexpectedly high temperature corrosion. Scale deposition can also result in metal softening, expansion, and failure since boiler tubes are overheated. This usually occurs in the areas of highest heat transfer.

9 Summary

The key variables for characterizing corrosive environments are solvent, pH, redox-potential, temperature, dissolved salts and gases, and flow conditions. Some of these factors will affect electrochemical reaction rates while others will affect stabilities of protective surface films. Local variations in solution chemistry often happen. Variations can occur due to uneven flow, uneven heating, chemical feed, deposits, biofilms, etc.
pH will affect the corrosion mechanism and corrosion rate. In most neutral and alkaline environments, corrosion rate is limited by mass transfer of cathodic reactant such as oxygen. In acid environments, hydrogen evolution rate often controls corrosion rate. Most metals can form protective surface films at some pH range. Outside this range, they will corrode. Redox-potential is a measure of the oxidizing or reducing potential of the solution. It often determines the suitability of various alloys. For example, passivating alloys are useful in oxidizing environments, but copper and nickel alloys are useful in reducing environments. Highly oxidizing solutions can easily initiate localized corrosion. Temperature will influence reaction rates. Corrosion is usually initiated or accelerated by increase in temperature.

Dissolved salts in the solution will often affect formation and stability of protective surface films. For example, increase in chloride concentration can initiate localized corrosion of stainless steels. Salts can also affect electrochemical reaction rates. Dissolved gases are often reactants for cathodic reactions and can affect formation and stability of protective surface films and scales. A minimum supply of oxidizing compound is necessary for passivation and maintaining passivity.

Fluid flow conditions can affect reaction rates and formation and stability of surface films. Increased flow rate can accelerate general corrosion rate and cause erosion corrosion, but stagnant conditions can cause pitting, differential aeration corrosion, and crevice corrosion.

CHAPTER 5

References

1. Pourbaix, M., *Atlas of Electrochemical Equilibria in Aqueous Solutions*, Pergamon Press, Oxford, 1966, Ch. 1.1.

2. Katz, W., in *Korrosion und Korrosionsschutz* (F. Tödt, Ed.), Walter de Gruyter, Berlin, 1955, p. 118.

3. Wensley, D. A. and Charlton, R. S., *Corrosion* 38(8):385(1980).

4. Macdonald, D. D. and Owen, D., *Journal of the Electrochemical Society* 120(3):317(1973).

5. Henrikson, S. and Kucera, V., *Effect of Environmental Parameters on the Corrosion of Metallic Materials in Swedish Pulp Bleaching Plants*, Pulp & Paper Industry Corrosion Problems, vol. 3., NACE, Houston, 1982, p. 137.

6. Heidersbach, R. H., in *ASM Metals Handbook*, Vol. 13, Corrosion, ASM International, Metals Park, 1987, p. 893.

7. Szlarska-Smialowska, Z., *Pitting Corrosion of Metals*, NACE, Houston, 1986, Chap. 12.

8. Crowe, D. C., *Corrosion on acid cleaning solutions for Kraft digesters*, 1992 Proceedings of the 7th International Symposium on Corrosion in the Pulp and Paper Industry, TAPPI PRESS, Atlanta, p. 33.

9. Anon., *Principles of Industrial Water Treatment*, Drew Chemical Corporation, Boonton, 1977, Chap. 4.

10. Hasenberg, L., in *DECHEMA Corrosion Handbook* (G. Jänsch-Kaiser, Ed.), vol. 11, VCH, Weinheim, 1992, pp. 1–64.

11. Fredette. M. C., in *Pulp Bleaching: Principles and Practice* (C. W. Dence and D. W. Reeve Eds.), TAPPI PRESS, Atlanta, 1996, Chap. II:2.

12. Garner, A. and Newman, R. C., *Thiosulfate Pitting of Stainless Steels*, 1991 NACE Corrosion Conference, NACE, Houston, Paper no. 186.

13. Fontana, M. G., *Corrosion Engineering*, 3rd edn., McGraw-Hill, Singapore, 1987, Chap. 2.

14. Smook, G. A., *Handbook for Pulp & Paper Technologists*, TAPPI PRESS, Atlanta, 1989, Chap. 4.

15. Smook, G. A., *Handbook for Pulp & Paper Technologists*, TAPPI PRESS, Atlanta, 1989, Chap. 7.

16. Laliberte, L. H., Corrosion Problems in the Pulp and Paper Industry, Pulp & Paper Industry Corrosion Problems, vol. 2., NACE, Houston, 1977, p. 1.

17. Troselius, L., Field exposure of carbon steel and stainless steel in digesters, 1995 Proceedings of 8th International Symposium on Corrosion in the Pulp and Paper Industry, Swedish Corrosion Institute, Stockholm, p. 46.

18. Mueller, W. A., Mechanism and Prevention of Corrosion of Steels Exposed to Kraft Liquors, Pulp & Paper Industry Corrosion Problems, vol. 1., NACE, Houston, 1974, p. 109.

19. Ahlers, P. E., Polysulfide in kraft cooking and its effect on corrosion of carbon steel, 1983 Proceedings of 4th International Symposium on Corrosion in the Pulp and Paper Industry, Swedish Corrosion Institute, Stockholm, p. 53.

20. Stockmann, L. and Ruus, L., Svensk Papperstidning 57(22):831(1954).

21. Roald, B, Norsk Skogsindustri 10(8):285(1956).

22. Maspers, E., General Corrosion in Continuous Digesters, 1995 Proceedings of 8th International Symposium on Corrosion in the Pulp and Paper Industry, Swedish Corrosion Institute, Stockholm, p. 1.

23. Kiesling, L., A study of the influence of modified continuous cooking processes on the corrosion of continuous digester shells, 1995 Proceedings of 8th International Symposium on Corrosion in the Pulp and Paper Industry, Swedish Corrosion Institute, Stockholm, p. 12.

24. Jonsson, K. -E., in Handbook of Stainless Steels (D. Peckner, I. M. Bernstein, Eds.), McGraw-Hill, New York, 1977, Chap. 43.

25. Anon., Stainless steels for pulp and paper manufacturing. Nickel Development Institute Report No 9009, American Iron and Steel Institute, Toronto, 1982, pp. 5–47.

26. Garner, A. et al., in ASM Metals Handbook, Vol. 13, Corrosion, ASM International, Metals Park, 1987, pp. 1187–1220.

27 Smook, G. A., Handbook for Pulp & Paper Technologists, TAPPI PRESS, Atlanta, 1989, Chap. 6.

28. Smook, G. A., Handbook for Pulp & Paper Technologists, TAPPI PRESS, Atlanta, 1989, Chap. 5.

29. Smook, G. A., Handbook for Pulp & Paper Technologists, TAPPI PRESS, Atlanta, 1989, Chap. 11.

30. Yeske, R. and Garner, A., Processing changes and materials engineering challenges in the pulp and paper industry, 1992 Proceedings of 7th International Symposium on Corrosion in the Pulp and Paper Industry, TAPPI PRESS, Atlanta, p. 1.

31. Andreasson, P. and Troselius, L., The corrosion properties of stainless steel and titanium in bleach plants, Swedish Corrosion Institute, Stockholm, 1995, Chap. 4. (In Swedish).

CHAPTER 5

32. Rounsaville, J. and Rice, R. G., Ozone Science and Engineering 18:549(1997).

33. Klarin, A. and Pehkonen, A., Materials in Ozone Bleaching, 1995 Proceedings of 8th International Symposium on Corrosion in the Pulp and Paper Industry, Swedish Corrosion Institute, Stockholm, p. 96.

34. Thompson, C. B. and Garner, A., Paper machine corrosion and progressive closure of the white water system, 1995 Proceedings of 8th International Symposium on Corrosion in the Pulp and Paper Industry, Swedish Corrosion Institute, Stockholm, p. 207.

35. Smook, G. A., Handbook for Pulp & Paper Technologists, TAPPI PRESS, Atlanta, 1989, Chap. 10.

36. Jonsson, K. -E., in Handbook of Stainless Steels (D. Peckner, I. M. Bernstein, Eds.), McGraw-Hill, New York, 1977, Chap. 43.

37. Asbaugh, W. A., in ASM Metals Handbook, Vol. 13, Corrosion, ASM International, Metals Park, 1987, pp. 1144–1147.

38. Valentine, R. J., in ASM Metals Handbook, Vol. 13, Corrosion, ASM International, Metals Park, 1987, pp. 1226–1231.

39. Chandler, K. A. and Hudson, J. C., in Corrosion, Vol. 1 (L. L. Shreir, R. A. Jarman, G. T. Burstein, Eds.) 3rd edn., Butterworth-Heinemann, Oxford, 1994, Chap. 3.1.

40. Fyfe, D., in Corrosion, Vol. 1 (L. L. Shreir, R. A. Jarman, G. T. Burstein, Eds.) 3rd edn., Butterworth-Heinemann, Oxford, 1994, Chap. 2.2.

41. Anon., Principles of Industrial Water Treatment, Drew Chemical Corporation, Boonton, 1977, Chap. 5.

42. Boffardi, B. B., in ASM Metals Handbook, Vol. 13, Corrosion, ASM International, Metals Park, 1987, pp. 487–497.

43. Borenstein, S.W., Microbiologically influenced corrosion handbook, Industrial Press, New York, 1994, Chap. 2.

44. Stein, A. A., in A Practical Manual on Microbiologically Influenced Corrosion (G. Kobrin, Ed.), NACE International, Houston, 1993, Chap. 10.

45. Lutey, R. W., in A Practical Manual on Microbiologically Influenced Corrosion (G. Kobrin, Ed.), NACE International, Houston, 1993, Chap. 4.

46. Kofstad, P., High Temperature Corrosion, Elsevier Applied Science, London, 1988, Chap. 1.

47. Bradford, S. A., in ASM Metals Handbook, Vol. 13, Corrosion, ASM International, Metals Park, 1987, pp. 61–76.

48. Pinder, L. W., in Corrosion, Vol. 2 (L. L. Shreir, R. A. Jarman, G. T. Burstein, Eds.) 3rd edn., Butterworth-Heinemann, Oxford, 1994, Chap. 7.2.

49. Lai, G. Y., Journal of Metals (11):54(1991).

50. Sharp, W. B. A., Overview of recovery boiler corrosion, 1992 Proceedings of 7th International Symposium on Corrosion in the Pulp and Paper Industry, TAPPI PRESS, Atlanta, p. 23.

51. Bruno, F., Primary air register corrosion in kraft recovery boilers, 1983 Proceedings of 4th International Symposium on Corrosion in the Pulp and Paper Industry, Swedish Corrosion Institute, Stockholm, p. 68.

52. Hömig, H. E., Metall und Wasser, 4th ed., Vulkan Verlag, Essen, 1978, Chap. 2.

53. Dooley, R. B., in ASM Metals Handbook, Vol. 13, Corrosion, ASM International, Metals Park, 1987, pp. 990–993.

54. Kingerley, D. G., in Corrosion, Vol. 2 (L. L. Shreir, R. A. Jarman, G. T. Burstein, Eds.) 3rd edn., Butterworth-Heinemann, Oxford, 1994, Chap. 17.4.

55. Anon., Principles of Industrial Water Treatment, Drew Chemical Corporation, Boonton, 1977, Chap. 11.

CHAPTER 6

Corrosion prevention

1	**Design**	**250**
1.1	Vessels	251
1.2	Joining	254
1.3	Piping	256
1.4	Outdoor constructions	259
2	**Material selection**	**261**
2.1	Carbon steels	267
2.2	Stainless steels	269
2.3	Nickel alloys	274
2.4	Titanium	276
3	**Modification of environment**	**279**
3.1	Removal of dissolved gases	280
3.2	Inhibitors	281
3.3	Control of biofouling	283
4	**Protective coatings**	**286**
4.1	Corrosion protection paints	286
4.2	Paint film protection mechanisms	296
4.3	Organic coatings	297
4.4	Metallic coatings	301
	4.4.1 Hot dipping	302
	4.4.2 Cladding, lining, and welding	304
	4.4.3 Thermal spraying	306
	4.4.4 Electroplating	310
4.5	Inorganic coatings	312
	4.5.1 Ceramic coatings and linings	312
	4.5.2 Conversion coatings	314
5	**Electrochemical protection**	**315**
5.1	Anodic protection	316
5.2	Cathodic protection	320
6	**Summary**	**326**
	References	327

CHAPTER 6

Jari Aromaa

Corrosion prevention

Most engineering materials are thermodynamically unstable and corrode. Totally stopping corrosion is usually not practical. A common practice is to decrease corrosion rate to an acceptable level. The first step in preventing corrosion is understanding the specific mechanism. The second and more difficult step is designing the protection method. Several corrosion prevention methods are available, and economical factors determine the method that is most feasible to solve a problem. The purpose of this chapter is to give an overview of different corrosion prevention methods. The chapter gives general guidelines on design, materials selection, coatings, water treatment, and electrochemical protection.

Using the principle of a corrosion cell, four methods are possible to halt corrosion:

- Stop the cathodic reaction
- Stop the anodic reaction
- Remove the electrolyte
- Remove the electrical connection between the electrodes.

Removing the oxidant such as oxygen in a closed system can stop cathodic reaction. Using inhibitors or coatings to hinder oxidant transfer to cathodic areas will slow the cathodic reaction. Selecting an inherently more resistant material, enhancing the passive film formation by anodic protection, or using inhibitors or coatings can slow or stop an anodic reaction. Electrolyte removal is primarily a design problem to minimize accumulation of stagnant solution and dirt. Removal of an electrical connection is a useful method in bimetallic joints where insulating gaskets, sleeves, and washers have use. The use of cathodic protection does not follow the closed electrical circuit principle of a corrosion cell as presented above. The corrosion cell can also explain the effect of cathodic protection. The cathodic protection system delivers additional anodic current. With this current, replacing some or all the corrosion current that otherwise would be produced by anodic dissolution reactions is possible.

The corrosion prevention process begins before the drawing board stage. Structural planning and process selection should include a corrosion risk analysis. The design of process vessels and equipment is very complex and must often meet official requirements. The principles for a sound structure from the corrosion viewpoint are very simple but sometimes difficult to realize. Process selection and development require that the corrosive environment will not become unnecessarily severe. Coating and electrochemical protection methods are useful when other methods are uneconomical. A good quality coating is often the best solution. The main task of corrosion prevention today is to ensure continuous operation with good product quality without costly,

CHAPTER 6

unplanned shutdowns. Corrosion prevention tasks should be part of the entire life of an installation. Corrosion prevention is definitely not applying a coat of paint at the last minute.

1 Design

The purpose of a sound design is to ensure a system lifetime with adequate safety margins but without overestimating material thicknesses. The rate, extent, and type of corrosion damage that are tolerable in a structure vary depending on the specific application. Structures and components have exposure to various environments that are usually not constant. For example, changes in relative humidity, temperature, concentration of dissolved species, pH, oxygen concentration, solid and dissolved impurities, and flow rate influence corrosion rates. These variations and their effects must be part of the design stage. While corrosion control during the operation might receive consideration, corrosion during transit, storage, or starting and shutdown are often forgotten. The benefits of sound design can also be lost with poor installation practice. Starting problems are typically excessively high temperatures, varying concentrations, inadequate inhibitor supply to all parts of the system, and incomplete oxygen removal. Proper cleaning before starting is an important step in the construction or manufacture of new facilities and equipment. It can avoid serious problems that are costly to correct. Cleaning can include removal of temporary corrosion preventive compounds, greases, dirt, sand, cutting oils, mill scale, and other corrosion products and passivation of the freshly cleaned metal surfaces. Downtime problems also often relate to improper cleaning followed by concentration of corrosive agents at certain locations.

    Different metal and solution heterogeneities and geometrical factors will cause potential differences leading to corrosion as Fig. 1 shows. Some factors leading to potential differences are unavoidable, but proper manufacturing and installation procedures might remove others. Differences in metallurgical structure will always be present in any commercial alloy. A designer can do little to compensate for these. Grain boundaries, inclusions, etc., will usually become anodic to the rest of the alloy as will cold worked areas and stressed areas. Stress relieving heat treatment and proper manufacturing and installation practices to avoid external stresses require consideration. The surface must be smooth, clean, and free from corrosion products. Holes and damages in protective films, scales, or coatings will usually become anodic. Proper manufacturing and installation practices are again necessary. For passive films, material selection is the key to corrosion prevention. Crevices, deposits on the surface, and other geometric details can result in concentration differences of the cathodic reactant. In these cases, the area in contact with the lower cathodic reactant concentration will become anodic. Material selection to ensure corrosion resistance or proper design and operation to avoid formation of concentration differences can help. Avoiding dissimilar metals in contact is necessary since the corrosion rate of the less noble metal will increase. Proper design or selection of materials close to each other in the galvanic series will help. The designer should also consider the effect of differences in temperature, velocity, pH, and dissolved salt concentration. The area that contacts the higher temperature, higher velocity, lower pH, or lower salt concentration will usually become anodic. The part of the structure with higher stress also becomes anodic, and replacements with new, uncorroded surface and higher initial corrosion rates become anodic.

A smoother surface often improves corrosion resistance by preventing formation of deposits. Better surface quality through polishing will also remove unalloyed surface layers and weak points from the metal. An extremely important factor is removal of surface scales before use. Stress corrosion cracking resistance especially depends on the metallurgical state. Thermal treatment in fabrication and cold working may have a strong effect on corrosion resistance. Prevention of loss of corrosion resistance during installation is often a problem. Heat treatment of welds and stress relieving by annealing are often necessary, but conducting them in the field requires expertise.

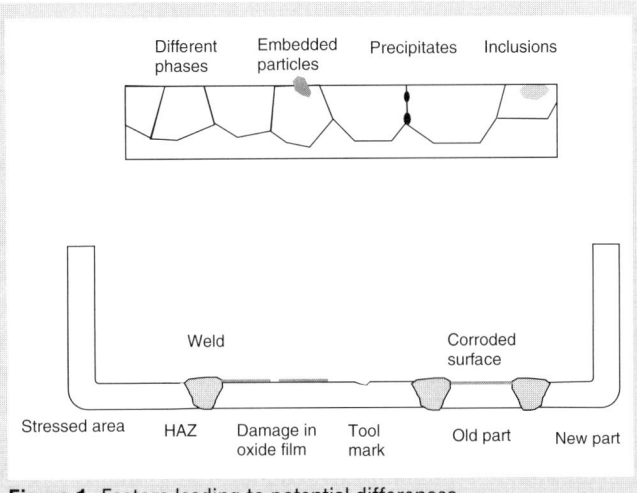

Figure 1. Factors leading to potential differences.

1.1 Vessels

Basic rules exist for good design details to prevent vessel corrosion. Again, the most fundamental rule is to avoid heterogeneities. This includes metal, process solution, heat, stress, and flow. The inside and outside of the process equipment both require attention. Storage containers and vessels should allow complete drainage so corrosive agents do not remain on the bottom and concentrate with time. Complete drainage is also important when using highly concentrated chemicals. Any remaining solution can absorb moisture, dilute, and start to corrode the metal. An example is sulfuric acid with carbon steel. Debris may also accumulate by not considering smooth and adequate flow from the vessel. The bottom of the vessel should slope to the center with drainage from the lowest point of the bottom. A vessel should also be totally filled since corrosion is often most rapid at the liquid and gas interface. If the vessel sometimes remains empty, good ventilation is necessary to avoid condensation and concentration of corrosive agents. The inlet lines should extend to the middle of the vessel to prevent local concentration of corrodents. Heaters should also have a central location. Solid particles and gases or steam require removal from the liquid before it leaves the vessel. Stiffeners should be on the outside of the vessel or tank and have good drainage. Outlet lines should have flanges rather than threads. Corners inside the vessel should be round. Figure 2 shows examples of the vessel design points.

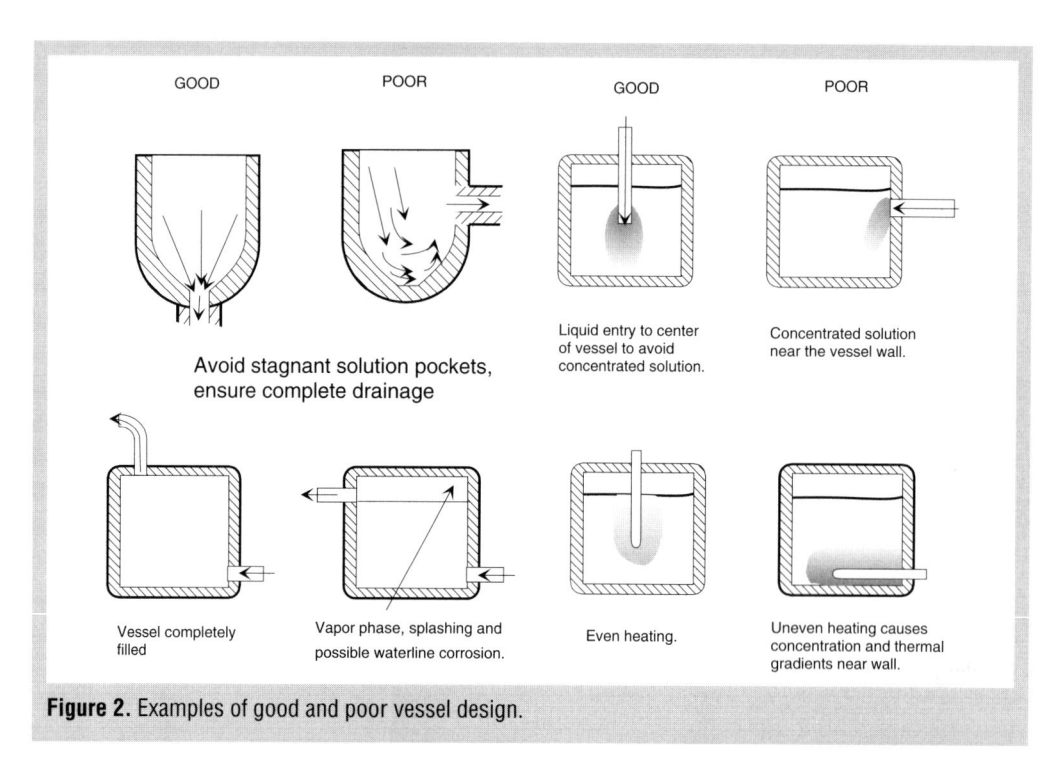

CHAPTER 6

GOOD   **POOR**   **GOOD**   **POOR**

Avoid stagnant solution pockets, ensure complete drainage

Liquid entry to center of vessel to avoid concentrated solution.

Concentrated solution near the vessel wall.

Vessel completely filled

Vapor phase, splashing and possible waterline corrosion.

Even heating.

Uneven heating causes concentration and thermal gradients near wall.

Figure 2. Examples of good and poor vessel design.

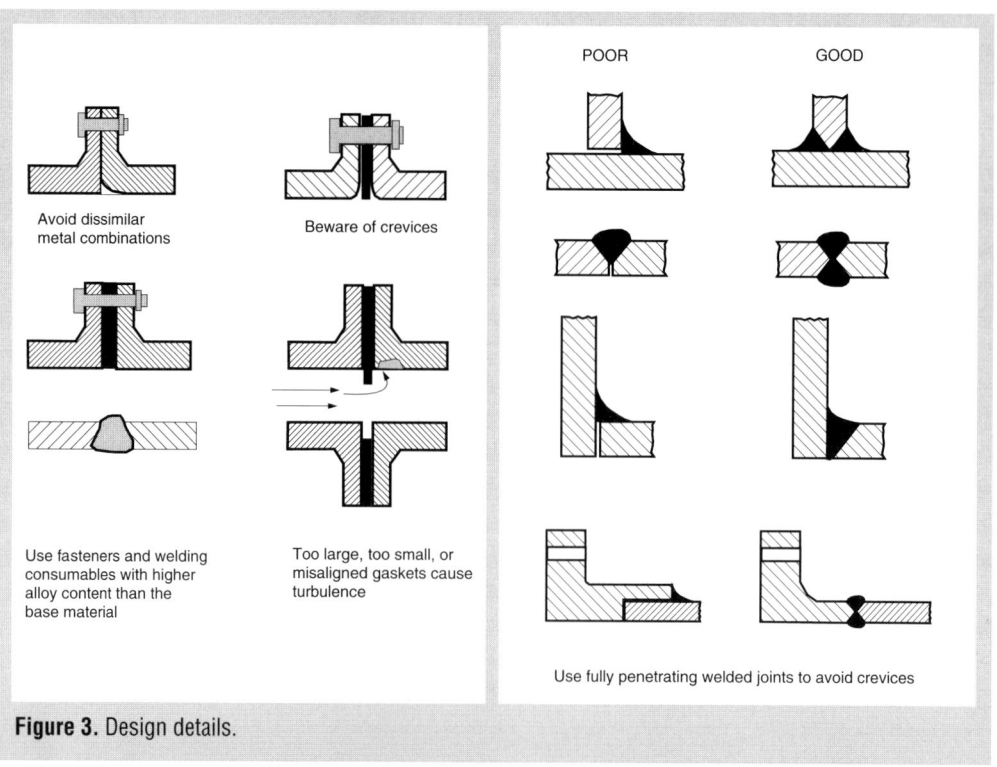

Avoid dissimilar metal combinations

Beware of crevices

POOR   **GOOD**

Use fasteners and welding consumables with higher alloy content than the base material

Too large, too small, or misaligned gaskets cause turbulence

Use fully penetrating welded joints to avoid crevices

Figure 3. Design details.

252

Small details to avoid in a sound design are numerous: crevices, dissimilar metal joints, tack welds, round-edged joints and other potential water traps, sharp corners, etc. Figure 3 shows examples of small design details. Different kinds of crevices are perhaps the weakest points in every process flow. Crevices typically occur in joints. Dangerous crevices are usually 0.025–0.1 mm wide. They are possible in bolted, riveted, and welded joints. A tighter crevice requires a more corrosion resistant material. Excessively large or small diameters for seals and gaskets can cause crevice sites and turbulence leading to erosion. Gasket materials that absorb water can also cause crevice corrosion. Joints should always use more noble bolts, rivets, screws, and welding fillers than the structure material. If two components made of dissimilar metals require joining, they should be electrically insulated from each other. Small critical components should be made from a noble corrosion resistant material so that the less noble massive structures can give cathodic protection. When using coatings, one should never coat the less noble parts only. Coating of the more noble material will decrease the total cathodic area and decrease galvanic corrosion. In the opposite case, small damages in the coating can cause extremely rapid corrosion of the less noble material.

Local temperature differences may result due to several reasons. These will usually lead to local increase in corrosivity as Fig. 4 shows. Cold surfaces will condense liquids from the atmosphere. Condensate droplets may contain various amounts of corrodents. If condensation occurs continuously on the same sites and droplets flow on the surface, corrosion may result. The same may also happen if the condensed droplets remain on downward facing surfaces over long periods. Condensation on copper piping can cause corrosion of structures located below these since the droplets contain copper ions causing corrosion by displacement reaction, i.e., cementation reaction. External stiffeners and other structures may lead to cold bridges. These will cause condensation from the gas phase inside the vessel to areas that are colder than the vessel wall. Such cases require insulation not only for the vessel but also for all metallic structures directly in contact with it. Hot surfaces and hot spots can also cause corrosion. Increasing temperatures usually increase reaction rates, and temperature may be sufficiently high to cause initiation of localized corrosion such as pitting. Hot spots in contact with solution may enhance bubble formation in surface inhomogeneities resulting in locally hot concentrated solutions.

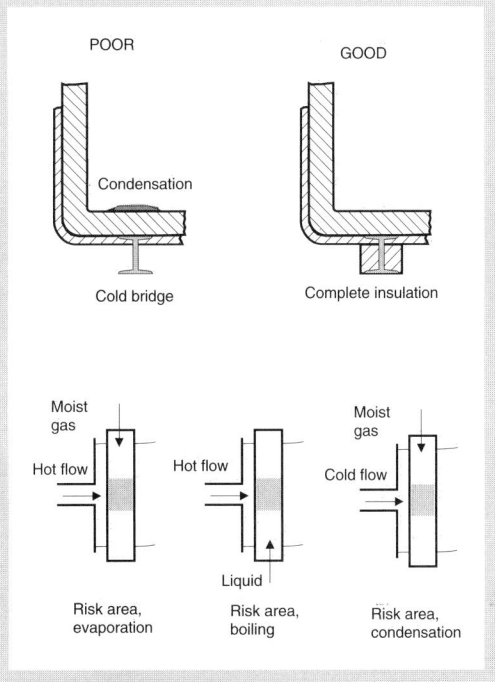

Figure 4. Effect of cold sites, insulation, etc., on localized corrosion.

CHAPTER 6

1.2 Joining

Reliable bolt joints require a firm, strong connection, materials that are stable in the environment, stable geometry, and appropriate stresses in the parts including a correct clamping force. Bolted joints should prevent slip, separation, vibration, misalignment, and wear of parts. Rigid joints result from specifying sufficient bolts of adequate diameters, sufficiently strong bolt and joint materials, sturdy joint members, and correct installation procedures. The bolt size should be such that tensile stress is no greater than 60% of bolt yield strength depending on the application. If stress corrosion cracking is a danger, 20% loads may be specified. Bolt and joint materials should have similar coefficients of thermal expansion. Temperature also influences stress corrosion cracking, hydrogen embrittling, and fatigue. Gasketed joints may start leaking after thermal cycles due to creep of gaskets and metal parts, change in gasket material properties, and loss of clamping force on the gasket caused by thermal expansion and contraction.

Fasteners come in a variety of materials. Selecting a material should use such considerations as corrosive environment, temperature, weight, properties, stresses, reusability, and expected life. To keep costs low, standard materials should be used. Most fasteners use steel. Stainless steel fasteners are useful where corrosion, temperature, and strength are problems. Common martensitic fastener alloys are AISI Types 410, 416, and 431. Ferritic steels are useful for economic reasons when corrosion resistance requirements are not too severe. A common fastener alloy is AISI 430. Austenitic alloys offer the greatest corrosion resistance. Typical alloys are AISI 304 and AISI 316. Nickel fasteners can be made from commercially pure metal, Ni-Cu alloy, or Ni-Cr-Fe alloy. They have use where toughness, corrosion resistance, and strength at high temperatures are desirable. Pure nickel is suitable for applications involving contamination and strength retention at high and sub zero temperatures. Ni-Cu alloy is an economical selection with adaptability to forming and machining. Ni-Cr-Fe alloys are excellent for fasteners that must retain high strength and oxidation resistance at temperatures up to 850°C.

The finish on a fastener is an important and integral part of the entire joint attachment. Coatings or finishes improve appearance, increase corrosion resistance, and provide lubricity. Electrodeposited zinc, cadmium, nickel, and chromium coatings are very common for fasteners. Zinc is preferable for industrial atmospheres. Cadmium is best in marine atmospheres, but it is more expensive than zinc. Cadmium is also a health risk. Corrosion products form a protective film on zinc and cadmium surfaces. Chromate treatment after zinc coating increases corrosion resistance. This is particularly effective in drying moisture and preventing rusting between mated parts. Chromate coatings are not very abrasion resistant. Zinc as a fastener finish can be alloyed with nickel, cobalt, or iron. These alloys provide more corrosion resistance than pure zinc of the same thickness:

- Zinc-cobalt alloys provide optimum corrosion resistance at 0.4%–1.0% cobalt. Lower or higher cobalt content does not improve corrosion resistance.

- The nickel content in zinc-nickel alloys is 5%–20% Ni. Higher percentages of nickel give corrosion resistance that exceed cobalt alloyed zinc of the same thickness.
- Iron content in zinc-iron is in the range of 0.4%–1.0% Fe. These deposits can be chromated.

Electroplating a threaded fastener can change the thread fit requiring allowances. Hot-dip coated fasteners are either aluminized or galvanized. They provide low cost protective coatings for inexpensive, high strength, ferrous fasteners. Chemical conversion coatings such as a phosphate coating give a corrosion resistant deposit. Usually, fasteners are dipped in a solution of zinc or manganese phosphate. Phosphate coating is less expensive than electroplating especially when heavy deposits are necessary. Organic coatings provide a tight film barrier against corrosion. They can further extend corrosion resistance. Organic finishes previously used alkyd and phenolic paints, but new coatings use fluorocarbon and other polymers.

Welding is often a cost-effective fabrication method. Since it does not require overlapping materials, it eliminates excess weight. Welded joints distribute operating stresses evenly. The design of the joint affects the selection of welding processes. Five basic types of joints are possible:

- Butt joints
- Corner joints
- T joints
- Lap joints
- Edge joints.

Each welding process has its own characteristics and capabilities. Joint design must be suitable for the desired welding process. Joint design will also affect access to the weld. Ideal welding procedures are those that produce acceptable quality welds at the lowest total cost. Joint design must consider the heating produced by welding and accessibility for the various processes involved. The filler materials should always be equal or more alloyed than the construction materials to be welded. Welding can induce distortions and residual stresses in structures. The joints should be located away from high stress areas whenever possible. Heating can relieve stress in most welded materials. The size and number of welds should be minimum, and welds should match but not exceed the strength of the base metal. When using full penetration joints through the thickness of the material, weld to metal strength must equal or exceed that of the construction metals.

Welding uses highly localized heating to fuse the material. This causes nonuniform stresses in the component because of expansion and contraction of the heated material. Compressive stresses form in the cold parent metal when the weld pool forms, but tensile stresses occur on cooling when the contraction of the weld metal and the immediate heat-affected zone are resisted by the bulk of the cold parent metal. If the

CHAPTER 6

stresses generated exceed the yield strength of the parent metal, localized plastic deformation of the metal occurs. Plastic deformation causes a permanent reduction in the component dimensions and distorts the structure. The main types of distortion are longitudinal and transverse shrinkage, angular distortion, bowing, dishing, buckling, and twisting. Contraction of the weld area on cooling results in transverse and longitudinal shrinkage. Nonuniform contraction through the thickness produces angular distortion in addition to shrinkage. Longitudinal bowing in welded plates occurs when longitudinal shrinkage in the welds bends the section into a curved shape. Clad plate can bow in two directions due to longitudinal and transverse shrinkage of the cladding. This produces a dished shape. Dishing also occurs in stiffened plating. Long range compressive stresses can cause elastic buckling in thin plates resulting in dishing, bowing, or rippling. Twisting in a box section is due to unequal longitudinal thermal expansion. Increasing the number of tack welds can prevent shear deformation and reduce the amount of twisting, but the use of tack welds is not a good selection for corrosion prevention.

Parent material properties that influence distortion are coefficient of thermal expansion and specific heat per unit volume. For example, a stainless steel has a higher coefficient of expansion than plain carbon steel. It will therefore more likely suffer from distortion. If a component is welded without any external restraint, it distorts to relieve the welding stresses. A restraint produces higher levels of residual stress to give a greater risk of cracking in weld metal and heat-affected zone (HAZ). A suitable joint type can balance the thermal stresses through the plate thickness. This could mean a double-sided weld in preference to a single-sided one. Part fitting should be uniform to produce predictable and consistent shrinkage. Excessive joint gap can also increase the degree of distortion by increasing the amount of weld metal needed to fill the joint. Welding procedure influences the degree of distortion primarily through its effect on the heat input. Since selection of a welding procedure usually considers reasons of quality and productivity, the welder has limited scope for reducing distortion. As a general rule, weld volume should be minimum.

1.3 Piping

The amount of piping in a process plant is enormous. The components can include straight pipe, elbow, tee section and branch, reducer to change pipe diameter, end plate, and flange and flanged joints. Usually, the bends, reducers, and branches use the same material as straight piping, but joints often use other materials. Piping systems frequently use welding, but flanged joints are common when reopening is necessary, the materials are not compatible to weld, welding is otherwise impossible, or when piping connects to vessels and other equipment. Most common corrosion causes are excessively high or low flow rate, different discontinuities in the flow channel causing turbulence, and solid particles.

To avoid corrosion of piping, several guidelines are available. Velocity is the most important single factor influencing design and corrosion in piping systems. Velocity of the medium through the system influences pressure losses and therefore pumping costs. The design velocity chosen controls the dimensions of many components such as pumps and valves. The costs of these components increase rapidly with pipe diameter. Velocity also influences the corrosion behavior of some construction materials such as carbon steel and copper-based alloys. For stainless steel systems, corrosion under flow conditions is not a problem. The flow rate must be sufficiently high to prevent settling of particles and possible under-deposit corrosion. Nickel-based alloys and titanium are not subject to impingement attack even at high velocity, but high price limits their use to special applications. Local velocities may vary considerably from design velocity. This is particularly important where features of the system that can generate turbulence such as small radius bends, orifices, partly throttled valves, misaligned flanges, etc., cause locally high velocities. Especially in a system with a low nominal flow velocity, the valve may cause turbulence and much higher local velocities particularly when the valves are used for throttling. Design and fabrication of the system should try to minimize turbulence raisers.

Solid impurities such as sand, fibers, corrosion products, etc., require removal by filtering or other methods. Gas, air, and steam lines should be free from moisture, and air should be removed from liquid lines. The piping design should be such that liquid lines are always completely filled with no air space. The lines must also be adaptable for total emptying if necessary. Piping system design should have a minimum number of bends and joints. Smooth flow is necessary to avoid turbulence. This is the result of proper design of joints, diameter changes, and flow direction changes. The radius of a bend should be as high as possible. For steel and copper, it should be at least three times the pipe diameter. If the material is prone to erosion or stress corrosion cracking, even larger bend radii are necessary. A higher flow rate requires a higher bend radius. Erosion corrosion and cavitation will occur if solution flow will hit the metal surface too hard. Discontinuous flow will cause turbulence that initiates stresses on pipe walls and damages protective films. Pipe diameter changes should be smooth, and the length must be at least twice the diameter of the smaller pipe. Control of flow rate should use a straight length of the pipe. It should never occur immediately before a bend, pump, or other component. If the pressure in the line remains sufficiently high, it will prevent bubble nucleation and cavitation.

When using dissimilar metals, the more noble metal must always be after the more active one. Wherever dissimilar metals join in piping systems, an evaluation of corrosivity is necessary. If galvanic corrosion is likely to occur, the dissimilar materials should be electrically isolated or the more noble material internally coated close to the coupling to reduce the cathodic area. For example, the length of the coated section could be ten times the pipe diameter. The necessary corrosion allowance could be specified in transition areas, or some less corrosion resistant material could be used as a sacrificial waste piece. At dissimilar metal connections, one can assume that the local corrosion rate near the joint can be anything above the average corrosion rate. Depending on the solution conductivity, corrosion rate decreases exponentially away from the joint within a length of up to five pipe diameters. This can establish magnitude of corrosion allowance or waste piece length. An important point with electrical isolation of dissimilar metals is to ensure removal of the connection from the external side. Process

CHAPTER 6

piping can be fixed to metal structures far from the actual dissimilar joint, but electrical connection still exists. Fixing of pipes can also result in crevices and crevice corrosion on the external side.

Figure 5 provides some design guidelines for piping. The main task is to avoid formation of turbulence and risk of erosion corrosion. This results by making all changes in flow rate or flow direction smooth. In principle, a piping system with low pressure loss is a sound design from the corrosion viewpoint. All bends, transitions to smaller diameter pipes, and joints must be as smooth as possible. The diameter of the gasket in a flanged joint must match the inner diameter of the pipe.

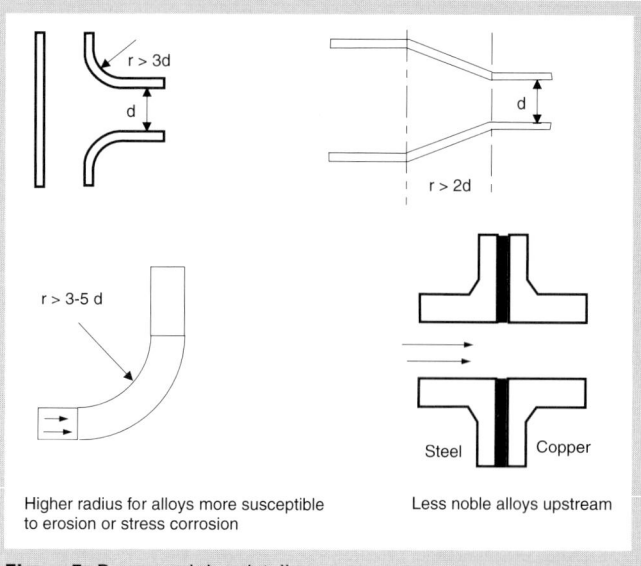

Figure 5. Process piping details.

Welds must not create obstructions to flow. Weld beads should be smooth, and the penetration must be complete. If the weld does not fill the root fully, a situation similar to a flanged joint with an excessively small gasket occurs.

Design of pumps and valves primarily considers their mechanical performance. They usually can withstand common corrosive environments. For more corrosive solutions, special designs or materials may be necessary. Pumps and valves have exposure to corrosive environments internally and externally (ambient conditions). Since either source of attack can cause problems, each deserves serious attention. Hostile external environments can destroy a valve as surely as a corrosive solution. Metal valves often fail due to corrosion of the stem at the packing box.

The basic low-cost valve used in ferrous pipe systems has a cast iron body with more corrosion resistant internal materials. The cast iron body cathodically protects the internal materials unless a layer of graphitic corrosion product forms. Coatings on valve bodies are often useful, but their success depends primarily on the valve design. Any break in a coating can result in intense corrosion and perforation of the valve body. Upgrading of valve body materials to give higher reliability requires the use of alloys such as nickel containing austenitic cast irons. Stainless steels have high resistance to flow, and few problems will arise due to erosion corrosion when using them for valve bodies. Cavitation that often occurs when flow is disturbed within and downstream from valves limits flow rate. Since pitting and crevice corrosion resistance are important in valves, alloys with similar resistance to that of the piping are necessary for valve bodies. Cast versions of several high alloy stainless steels are available.

Valve seats especially used in throttling service experience high water velocities. Valve seats can be machined areas on the valve body, disc, gate, or ball depending on the valve design. Some cases use separate seats for easy replacement in case of damage. In such cases, the seat material can be different from the body or disc material. Stems in stainless steel valves use materials with similar corrosion resistance to the body. In ball and butterfly valves, one seat may be nonmetallic such as an elastomer. Valves are an expensive part of a piping system, but the cost of a valve depends on its design. A ball valve will be more expensive than a butterfly valve because of its greater weight. When use of a ball valve is desirable such as for its good flow control characteristics, the additional cost of more corrosion resistant materials must be accepted. Several sharp changes of flow direction occur inside the ball valve. This causes severe turbulence. Materials of high impingement resistance are recommended. For stainless steel systems, turbulence in a ball valve presents no problem, and alloys similar to the pipe materials are suitable.

The pump impeller contacts fast flowing, highly turbulent flow. Circulating pumps used frequently should have a material with high resistance to these conditions. Some alloys do not suffer from impingement attack but may pit when the pump is stationary and filled. For cast iron or alloyed cast steel pump casings, stainless steel impellers are preferable. Cathodic protection by the casing reduces the risk of pitting on the impeller when the pump is stationary. For stainless steel pumps, the impeller can use the same material as the casing. Rapid pump wear is usually due to abrasives, excessively low viscosity, or excessively high pressure or temperature. A common practice is to use an anodic material in the pump body compared with the critical components such as impeller shafts and seals. The same principle is useful with valves where the valve trim is more cathodic.

1.4 Outdoor constructions

Design of structures subjected to atmospheric corrosion also has some simple rules. One must first evaluate the aggressiveness of the micro climate. In plant areas especially, the aggressive condensates and attack directions are important. The surface area subjected to corrosive environment should be minimum. Simple box structures are therefore preferable over complex diagonal profile structures. The flow of corrosive solutions away from the surfaces is necessary. In most cases, solutions and dust that collect and remain on surfaces will cause the most severe attack. When selecting profiles for the structure, those with the least sides and corners are preferable. Tube and inverted L-profile are better than U-profile or I-profile as Fig. 6 shows. Crevices and joints should be minimum, and the joining method requires careful consideration. Continuous welding is preferable. The structures require isolation from soil if possible. Steel bars driven in soil will corrode near the surface without a good coating or concrete basement.

Protection with paints is the most common method of corrosion prevention. To ensure proper application of paint, some design details require consideration. The use of paint requires thorough pretreatment, paint application, and quality checking. Paint films must be applied to very pure surfaces with a proper surface roughness. To ensure

CHAPTER 6

removal of contaminants, the constructions should not contain crevices. To ease paint application and minimize voids, the accessibility to all parts of the structure must be good as Fig. 7 shows. Most equipment uses paint coatings. This requires some constructional details. Crevices that cannot be cleaned and coated properly should be avoided. Full-seam welds are preferable to tack welds, and round corners are better than sharp ones. Mating surfaces of bolted and skip-welded joints require sealing with caulk.

Hot-dip galvanizing is a common corrosion protection method against atmospheric corrosion. Several guidelines exist for design and fabrication of articles for galvanization. The first item to consider is that the articles will undergo immersion in molten zinc with a temperature of at least 460°C. All structures must have ventilation and drainage holes since large closed areas could cause explosions if trapped pickling liquids evaporate. Good zinc coating deposition only occurs when the surface is free from organic contaminants. Contaminants not removed by pickling require sand blasting or cleaning with power tools. Factors that improve the quality of galvanizing are the following:

- Design structures so that they will be assembled after galvanizing preferably by using fasteners. Provide lifting points and ensure complete drainage of vessels.

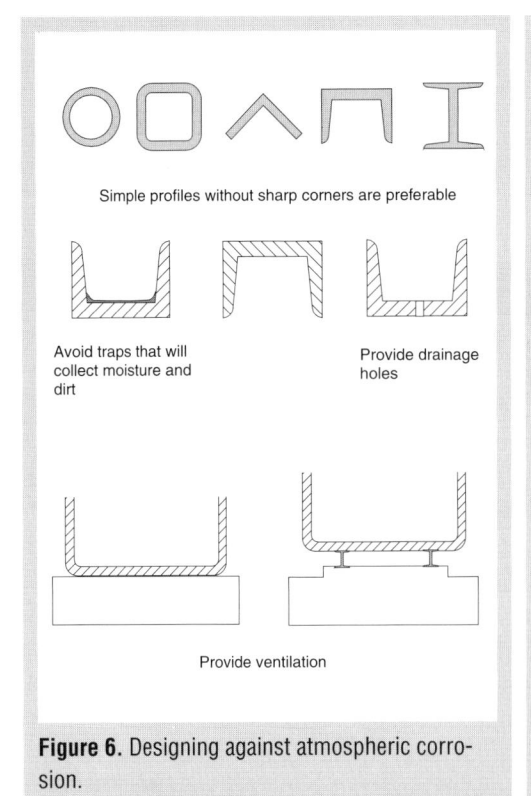

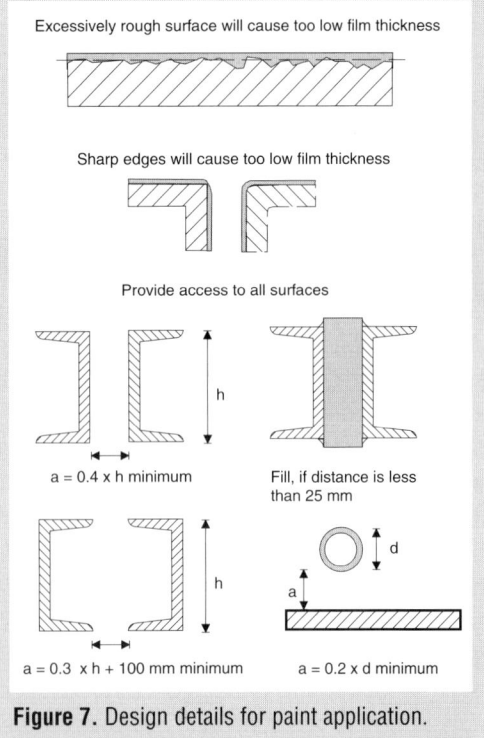

Figure 6. Designing against atmospheric corrosion.

Figure 7. Design details for paint application.

- Install parts that are supposed to move in relation to each other after galvanizing or have at least 1 mm clearance between surfaces.
- Avoid using materials with highly varying thicknesses. Uneven heating during immersion can cause warping, buckling, etc. Avoid long, slender structures. Large areas with less than 3–4 mm material thickness usually are not suitable.
- Avoid different materials. As-rolled steel, corroded steel, and cast iron require different galvanizing procedures. The coating will be uneven if different materials are welded together in the assembly. Steels produced with different deoxidation methods should not be welded.
- Avoid acid traps such as tight crevices, tack welds, and lap joints. If covered areas are larger than about 70 cm^2, ventilation holes are necessary. If acid penetrates into a closed area or tight crevice, it cannot be removed or even diluted. The molten zinc will cover the access. The trapped acid will eventually corrode steel. Cast structures to be galvanized must have smooth surfaces.

2 Material selection

The most cost-effective method for corrosion protection is often proper material selection that includes a coating. The material selection procedure includes four steps:

- Defining the material requirements by analyzing the environment, construction, etc.
- Identification of possible materials
- Evaluation of selected candidate materials
- Selection of the proper material.

Economical material selection in the chemical process industries is complex because the process-side environment is usually more aggressive than the external environment. Thermal and mass transfer due to flow of the process-side environment change the corrosive environment. Corrosion resistance estimates by available data may be too optimistic. Material selection is further complicated by the fact that standardized grades often have large compositional tolerances. For example, the molybdenum content of AISI 316L stainless steel is standardized as 2%–3%. The minimum value for good performance may be somewhere between this range. An example of this is the "papermakers grade," i.e., type AISI 316 stainless steel with 2.75% Mo minimum[1]. Commonly used terms such as "stainless" or "acid proof" steel can be misleading since steels assumed to be equivalent may have different compositional ranges as Table 1 shows for selected AISI types, German grades, Swedish SIS, and Finnish SFS grades. The differences are not large but may influence the corrosion resistance.

The decisive factor for material selection should be the total cost during the system designed operation time. The total cost should include initial material and installation costs, maintenance and replacement costs, and as far as possible costs by

CHAPTER 6

interrupted production. Over-specified material is initially costly, but under-specified materials increase production cost per amount of product through downtime and maintenance costs during the entire operation time. Material selection requires optimization considering investment and operational costs. Life cycle costs (LCC) should be minimum while providing acceptable safety and reliability. Materials with good market availability and documented fabrication and service performance are preferable. The number of different material types should be minimum considering costs, interchangeability, and availability of relevant spare parts. Specifying highly resistant and often expensive materials requires caution because it can give false security and require high capital investment.

Table 1. Austenitic stainless and acid-resistant steels usually considered equivalent with analyses in weight percent.

"Stainless"	"Acid proof"	C_{max}	Cr	Ni	Mo
AISI 304 (UNS S30400)		0.1	18.0–20.0	8.0–10.5	
W.Nr 1.4301		0.1	17.0–19.5	8.0–10.5	
SIS 2333		0.1	17.0–19.0	8.0–11.0	
SFS 725		0.1	17.0–19.0	8.0–11.0	
	AISI 316 (UNS S31600)	0.1	16.0–18.0	10.0–14.0	2.0–3.0
	W.Nr 1.4436	0.1	16.5–18.5	10.5–14.0	2.5–3.0
	SIS 2343	0.1	16.5–18.5	10.5–14.0	2.5–3.0
	SFS 757	0.1	16.0–18.5	10.5–14.0	2.5–3.0

Corrosion data for material selection is available from various sources. Some data is quantitative, and other data is qualitative. Manufacturers publish data that sometimes supports their materials. These publications often contain results from laboratory tests and case histories. For example, a comparison of corrosion performance of commercial alloys in some standard test can be very helpful. Some reference works contain huge amounts of corrosion data. For simple system design, charts may be available, but actual processes are often too complex. Another problem is that large tables compiled from several sources do not recognize minor variations that can have a large influence on corrosion resistance. Corrosion data for various materials is available in the following publications:

- Dechema Corrosion Handbook (12 volumes) presents corrosion performance of different materials in a particular media. It has wide coverage of metals and nonmetals in certain environments. Available from Dechema, Frankfurt am Main, Germany.

- Corrosion Data Survey, metals and nonmetals sections, gives corrosion rates of a particular material and material groups ranked in four categories in a particular environment as a function of chemical concentration and temperature. The book is a collection of reference points. Available from NACE International, Houston, Texas.

- ASM Handbook of Corrosion Data is organized using environment. A general overview of specific alloy systems is also available. Numerical values and diagrams are given. Available from ASM International, Metals Park, Ohio.

- ASM Handbook, volume 13, Corrosion, gives information about metals, corrosion forms, and environments. The information is often the same as in the ASM Handbook of Corrosion Data. Available from ASM International, Metals Park, Ohio.

Journals and conference proceedings may give solutions to particular problems. Performance data from identical or similar installations can be very valuable. Using such data requires caution until one verifies that process conditions, operation, and materials are comparable. Because of system closure causing more corrosive conditions, past experience has lost much of its value in the pulp and paper industry. Corrosion testing is the last source for material performance data. The problem with laboratory experiments is how to reproduce the actual process conditions and its many small details. The usefulness of material performance information from various sources increases in the order of laboratory tests, pilot plant tests, field tests, and prior service under actual, comparable conditions.

Material selection usually uses physical or mechanical properties. Corrosion resistance seldom receives consideration as carefully as other properties. All the published data will only give a starting point for the material selection process. Knowing the environment and material recommendations from various sources gives a list of candidate materials. When an alloy with adequate resistance to uniform corrosion has been found, its resistance to various forms of localized corrosion must be ensured. The properties of any candidate materials also require evaluation outside the normal plant operating conditions. At startup or during shutdown, the conditions can be very different from those during normal operation. For example, chloride concentration or temperature may increase to a level that initiates pitting or stress corrosion cracking of stainless steel that would be resistant under normal operation. Selecting a material that has corrosion resistance under more severe conditions is always safer, but this might be too costly. Excluding nonmetallic materials, selection uses the following eight groups:

1. Carbon and low-alloyed steels
2. Stainless steels
3. Aluminum and its alloys
4. Copper and its alloys
5. Nickel and its alloys
6. Lead, zinc, and tin
7. Valve metals and refractory metals, i.e., titanium, niobium (columbium), tantalum, and zirconium
8. Noble metals.

CHAPTER 6

Under reducing or nonoxidizing conditions, nickel, copper, and their alloys are resistant in principle. With oxidizing conditions, passivating alloys such as chromium containing alloys are useful. In very strong oxidizing media, chromium is not sufficiently resistant, and titanium or other valve metals are preferable. Selection of other corrosion resistant materials such as plastics and nonmetallic inorganic materials is also possible. These are often extremely suitable in severe environments. Most ceramics are more corrosion and wear resistant than metals, but their mechanical properties are not as good.

For the pulp and paper industry, the main metallic material groups are carbon steels, stainless steels, nickel alloys, and valve metals. The corrosion resistance of alloys in different groups and within every group can vary considerably. The carbon and low alloy steels have almost equal corrosion resistance, and compositional differences do not have much influence. Separate instances do exist where the concentration of minor constituents has a remarkable effect. In kraft digesters and associated white liquor service, carbon steels with very low silicon and higher copper, nickel, and chromium concentrations have been superior[2]. The corrosion resistance of different grades of stainless steels varies considerably, but corresponding grades from different manufacturers are usually similar. Different grades of nickel alloys can behave totally differently. This presents a higher risk of selecting an unsuitable alloy. The titanium grades are not usually problematic since their corrosion resistance values are approximately equal.

The corrosion resistance evaluation procedure begins with the decision of system operation time, determination of operational environment and operational parameter value ranges, and preliminary selection of candidate materials. The selection of candidate materials considers their uniform corrosion rate vs. operational time. The corrosion rate in the major compound of the environment is usually used. Many reference sources contain isocorrosion diagrams to assist this task. An isocorrosion diagram may contain recommended alloys for a particular corrosion rate or different corrosion rates for a single or a few alloys. The diagrams may also show risk to some specific form of corrosion such as stress corrosion cracking (SCC). Figure 8 gives examples from isocorrosion diagrams:

- Figure 8(a) shows the corrosion rate of austenitic stainless steels, AISI 304 and AISI 316, in sodium hydroxide solutions. The corrosion rate depends on caustic concentration and temperature.

- Figure 8(b) shows special information regarding the caustic stress corrosion cracking of carbon steel. This kind of diagram is more simple than the one in Fig. 8(a) since it does not include different corrosion rates. The information shows whether SCC occurs.

- Figure 8(c) shows areas of application of several materials. In this kind of diagram, the resistance criterion is usually a predetermined corrosion rate. In Fig. 8(c), the diagram shows areas where steel and nickel are suitable in sodium hydroxide service.

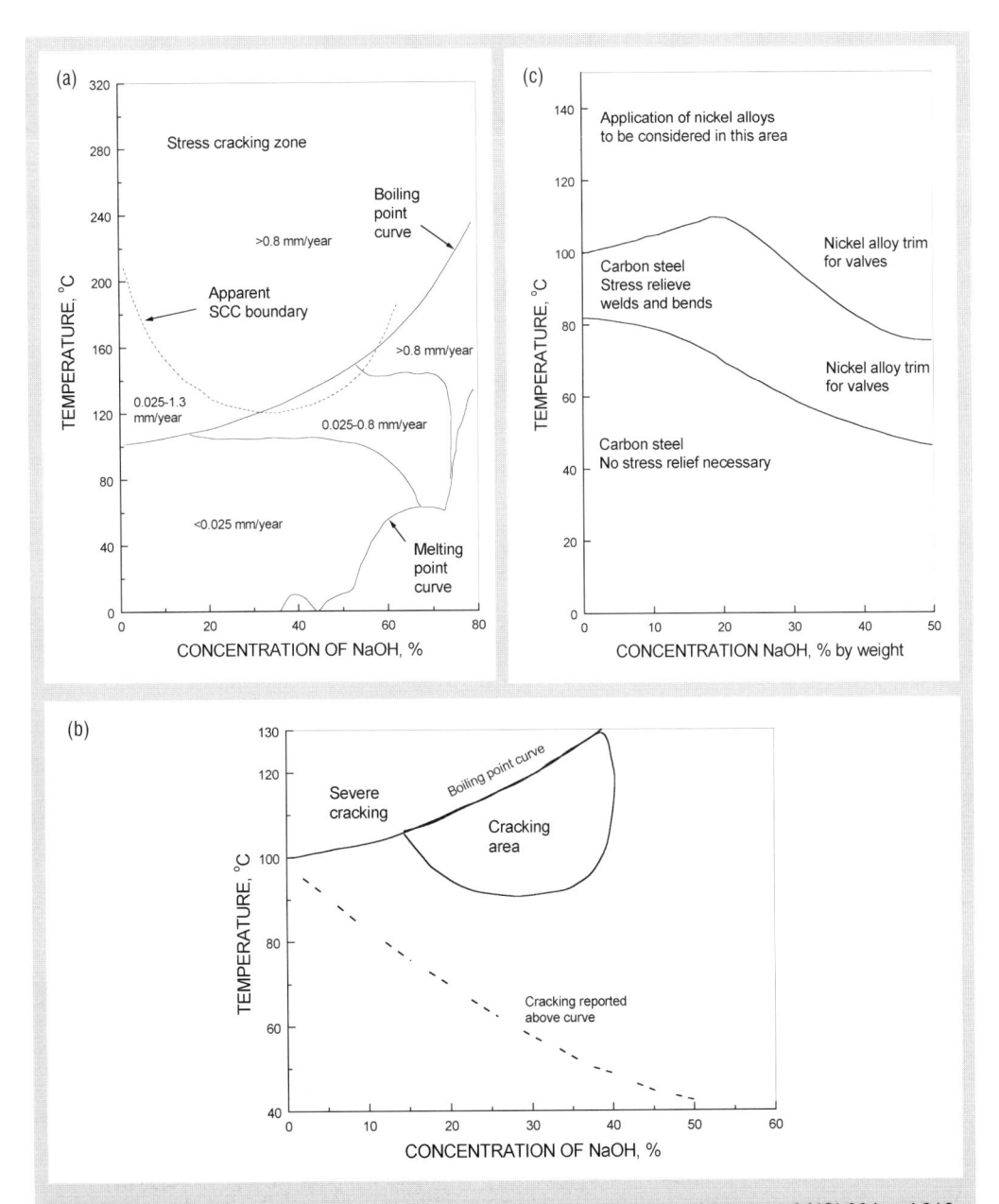

Figure 8. Different types of isocorrosion diagrams with (a) showing corrosion rates of AISI 304 and 316 stainless steels[3], (b) indicating risk of stress corrosion cracking of carbon steel[3], and (c) giving application ranges of steel and nickel[4].

CHAPTER 6

The next step in corrosion resistance evaluation is to estimate the effect of process variables. Special attention is necessary for those variations that may cause change in the corrosion mechanisms. Typical variables are temperature, pressure, flow rate, and variations in pH or ionic concentration. Abnormal conditions during startup or shutdown also require consideration. For example, decrease of pH can shift corrosion of carbon steel from the mass transfer controlled oxygen reduction type to a more rapid activation controlled hydrogen evolution type. More complex dependencies occur when several solution factors are present. For example, temporary increase in chloride concentration or temperature can induce pitting or crevice corrosion of stainless steel. Different kinds of diagrams are available for this purpose. Figure 9 shows some schematic examples for stainless steels:

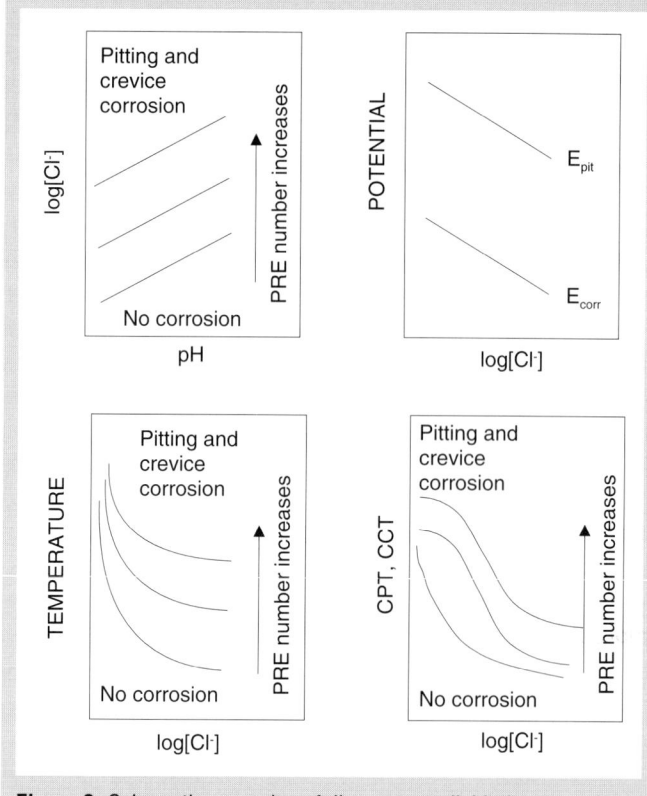

Figure 9. Schematic examples of diagrams available for corrosion resistance estimation.

- Increasing pH permits higher chloride concentration. The dependence of pH and log [Cl⁻] is nearly linear.

- The effect of chloride concentration on corrosion potential and pitting potential is often also linear. A higher concentration of chlorides or other aggressive anions decreases corrosion resistance. This is evident in lower corrosion and pitting potential values.

- Clear borderlines between corrosion and corrosion-free situations often occur. These depend on solution composition and alloy. Increasing the alloy content of the stainless steel will give more security against unexpected corrosivity changes.

- The concentration of aggressive anions and temperature is very complex. Higher chloride concentration usually allows operation only at low temperatures. Operation at high temperatures is possible only with low chloride contents. Critical pitting temperatures (CPT) and critical crevice temperatures (CCT) are not always simple functions of solution composition.

2.1 Carbon steels

Iron and steel are very common construction materials. They corrode in many environments including the outdoor atmosphere. Structural steels are seldom selected for their corrosion resistance but for their strength, ease of fabrication, and cost. Coatings, inhibitors, and electrochemical protection are useful to protect steel. All steels corrode in moist, impure atmospheres. Addition of small amounts of copper can decrease the corrosion rate by 50% since the rust scale becomes more dense and adherent. Use of weathering steels is possible without paint coating. The formation of a protective surface layer requires cyclic wetting and drying. If the surface is constantly wet, the protective layer will not form. If the corrosion resistance of structural steel is not sufficient, a logical step is to consider a stainless steel. The corrosion rate of carbon steel in the atmosphere depends on the location. In a dry rural atmosphere, it is a few µm/year. In a marine atmosphere, it is 50–100 µm/year. It can be more than 100 µm/year in an industrial atmosphere. Local salt or chemical loading can increase corrosion rate to several hundred 100 µm/year. Initially, the corrosion rate of carbon steel in the atmosphere is high, but it decreases because of formation of surface films. Figure 10 compares corrosion rates of unalloyed carbon steel and a commercially available weathering steel. The results show that the corrosion rate of more alloyed steels becomes constant very rapidly.

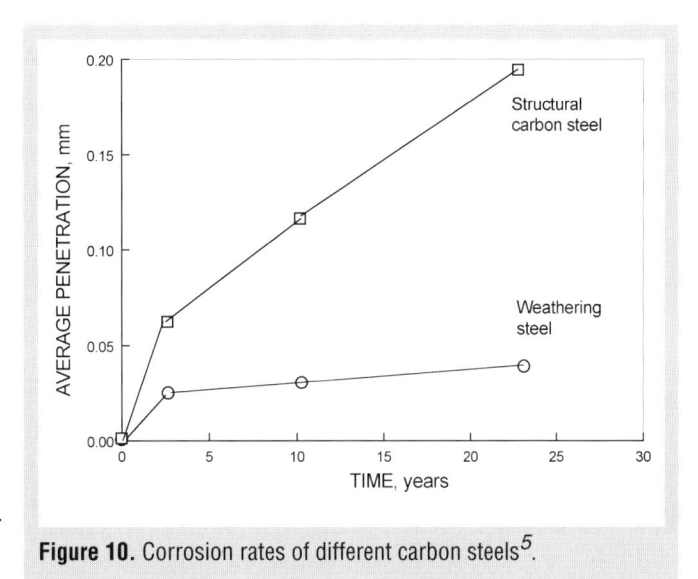

Figure 10. Corrosion rates of different carbon steels[5].

Structural steels have very good corrosion resistance to alkalis, strong oxidizing acids, and many organic compounds. Dilute alkali solutions at normal temperatures are not aggressive and can act as inhibitors to steel. Corrosion of carbon steel at room temperature ceases above a concentration of 1 g/L sodium hydroxide or 0.7 g/L potassium hydroxide[6]. As alkali concentration or temperature increase above some critical level, severe corrosion begins. Carbon steel can handle sodium hydroxide up to 50% and 55°C. Aeration, carbon dioxide, and the presence of chlorides will increase corrosion rate. Corrosion is often local because protective films form in alkaline environments on cathodic areas that concentrate the corrosion at active anodic areas. On the anodic

CHAPTER 6

areas, the film is weaker or has been damaged. Figures 11 and 12 show corrosion rates for carbon steel in sodium hydroxide and sulfuric acid[7, 8].

In most aqueous solutions, corrosion is due to dissolved oxygen. In process liquors, other oxidants may also be present. The corrosion rate of carbon steel in batch kraft digesters varies from less than 15 mpy to almost 150 mpy (approximately 0.38–3.8 mm/year)[9]. These values are very high. Changes in operation have increased them further leading to use of stainless steel overlay welding and cladding. Figure 3 of Chapter 5 shows the effect of pH and oxidants on the corrosion rate of carbon steel. The same tendency is obvious in Fig. 13 showing instantaneous corrosion rates from laboratory experiments of carbon steel in simulated sewage[10]. The redox potential is not very high because the solution contains only dissolved oxygen and no dissolved chlorine. Corrosion rate increases rapidly when the solution becomes too acidic.

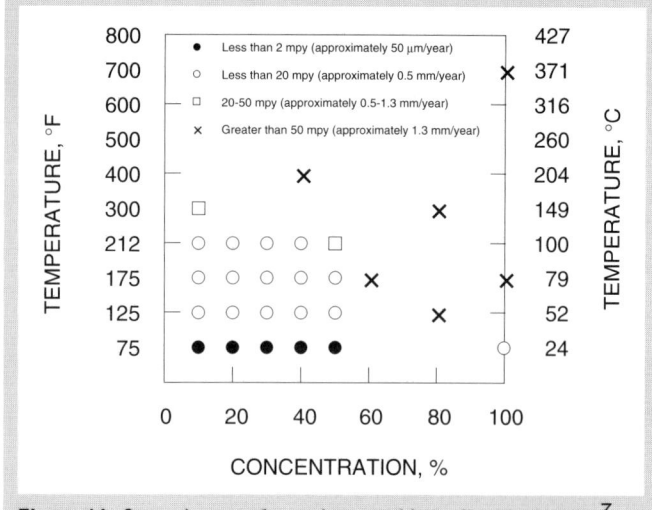

Figure 11. Corrosion rate for carbon steel in sodium hydroxide[7].

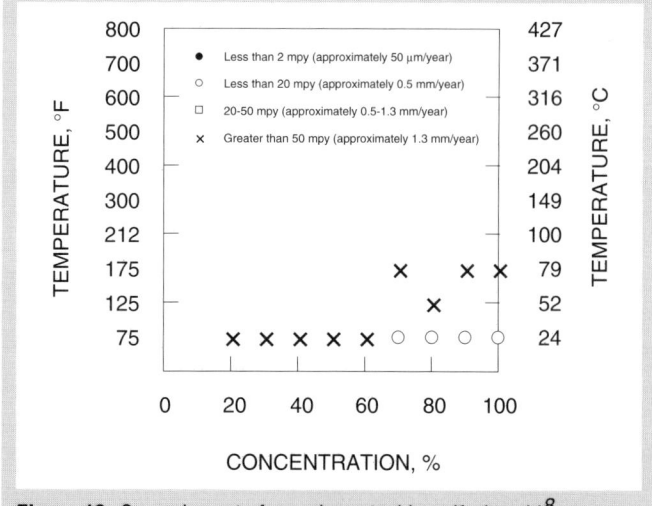

Figure 12. Corrosion rate for carbon steel in sulfuric acid[8].

Mild steels can be susceptible to stress corrosion cracking in nitrate, hydroxide, ammonia, and hydrogen sulfide containing media. Hydrogen evolution may cause embrittling and blistering. Figure 14 shows some environments that cause stress corrosion cracking on carbon steels[11]. For some environments, the dangerous corrosion potential range is large. For hydroxide for example, two narrow potential ranges as discussed in Chapter 4 are present. Large corrosion potential ranges correspond to more probable SCC.

Low alloy steels have better strength, and their corrosion resistance is similar to structural steels. The corrosion rates of mild steel, low alloy steel, and wrought iron are approximately equal when immersed or buried. High-strength, low-alloy steels can be very susceptible to SCC even in a mildly corroding environment when hardened to their maximum hardness.

2.2 Stainless steels

Stainless steels are the second important material group. They are particularly resistant to oxidizing conditions. Since the corrosion resistance of stainless steel varies depending on steel microstructure and composition, their selection is not easy. Many stainless steel types possess similar corrosion resistance but are designed for high strength or machinability. This can make material selection confusing to those not familiar with modifications and their reasons. The service life of stainless steels depends on understanding their properties. Misunderstanding often leads to premature failure. Historically, the microstructure of stainless steels has provided classification. This does not indicate corrosion resistance completely. The

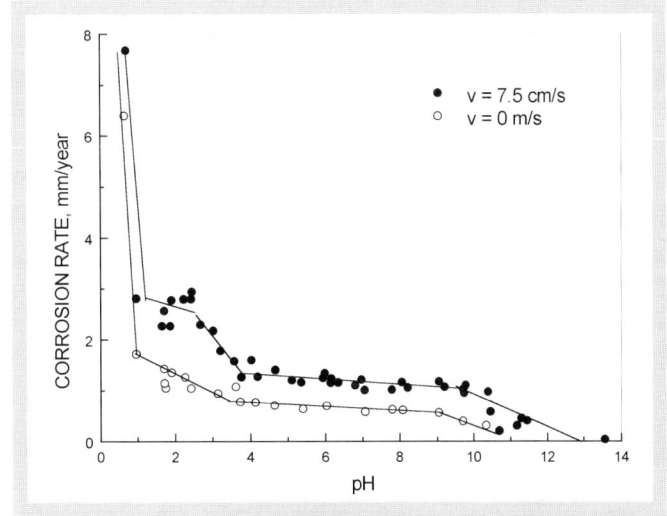

Figure 13. Instantaneous corrosion rates from laboratory experiments of carbon steel in simulated sewage[10].

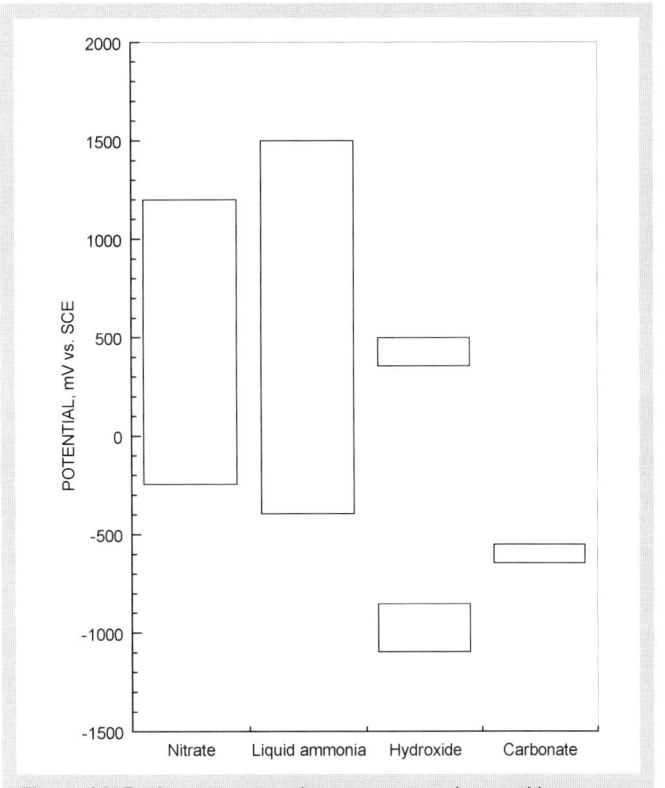

Figure 14. Environments causing stress corrosion cracking on carbon steels.

CHAPTER 6

microstructure has particular importance in chloride-induced stress corrosion cracking that is a problem for austenitic steels but not for ferritic or duplex steels. Stainless steel grade must be selected by its UNS number or standard designation. Common terms and trade names can be misleading. For example, "6-Mo" grades are super austenitic stainless steels with high chromium and molybdenum content (more than 24% Cr and about 6% Mo), but "7-Mo PLUS" is a duplex stainless steel with 26.5% Cr and 1.5% Mo.

Some general guidelines for stainless steel corrosion resistance comparison are available. A higher chromium content gives a more resistant steel to oxidizing media and high temperature corrosion. Nickel, molybdenum, and copper improve resistance to nonoxidizing media. Molybdenum is especially beneficial in reducing localized corrosion caused by chlorides. Figure 15 shows the principal effect of alloying elements on the corrosion of stainless steels. Alloying elements that decrease passivation potential or passivation current density will make passivation easier. Decrease of passive current density by alloying with chromium, nickel, and copper will decrease the reaction rate in the passive state. This can be beneficial when crevice corrosion is a problem.

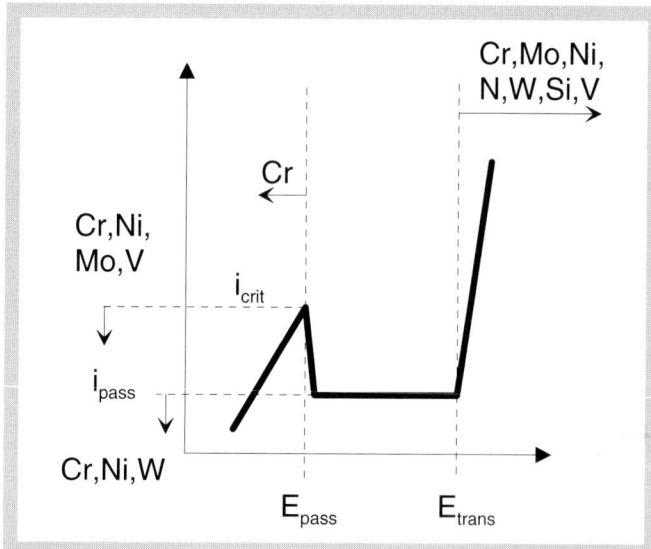

Figure 15. Effect of alloying elements on corrosion and passivation of stainless steels.

The pitting resistance equivalent (PRE) calculated from nominal composition gives an estimate of the resistance to localized corrosion. These values are best to give an approximate ranking between different steels. Calculation of the PRE numbers uses various equations. The most commonly used expressions consider the effect of chromium, molybdenum, and nitrogen. With duplex grades, the effect of nitrogen can be stronger than with austenitic grades. Introduction of tungsten as an active alloying element in duplex grades results in another expression:

- Austenitic stainless steels, PRE = %Cr + 3.3·%Mo + x·%N, factor x for the effect of nitrogen has had values of 12.8, 13, 16, 27, and 30

- Austenitic stainless steels in hydrogen sulfide environment,
  PRE = %Cr + 3.3·%Mo + 11·%N + 1.5·(%W+%Nb)

- Duplex stainless steels PRE = %Cr + 3.3·(%Mo + 0.5·%W) + 16·%N.

Corrosion prevention

For duplex alloys in particular, one must consider the pitting resistance of the two phases separately due to alloying element partitioning. An alloy with high PRE value may have a low resistance due to preferential corrosion of a phase with low actual PRE value. Figure 16 shows the effect of microstructure and composition on pitting and stress corrosion cracking resistance of stainless steels. In all alloys, increase of chromium, molybdenum, and nitrogen content increases resistance to pitting. In austenitic and duplex stainless steels, nickel and molybdenum will increase resistance to stress corrosion cracking. In ferritic steels, they have an adverse effect.

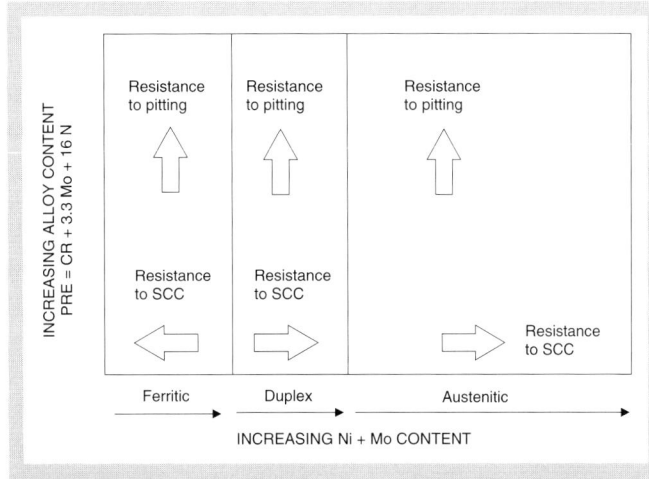

Figure 16. Effect of microstructure and composition on pitting and stress corrosion cracking resistance of stainless steels.

Various parameters can describe the resistance of stainless steel to pitting or crevice corrosion. Since critical temperature, chloride concentration, and temperature must be exceeded for pitting or crevice corrosion initiation, plotting any of these parameters or their combinations vs. another parameter or steel composition has use for material comparisons. As a general rule, the crevice corrosion resistance of stainless steels relates to pitting corrosion resistance. Figure 17 shows results of laboratory tests made for determining critical pitting temperature (CPT) with anodic polarization curves[12]. The experiments were done in 1M NaCl solution. CPT values naturally depend on solution composition and potential. If these remain the same, the results of Fig. 17 are useful when the corrosion resistance of some alloys is known and used as a benchmark.

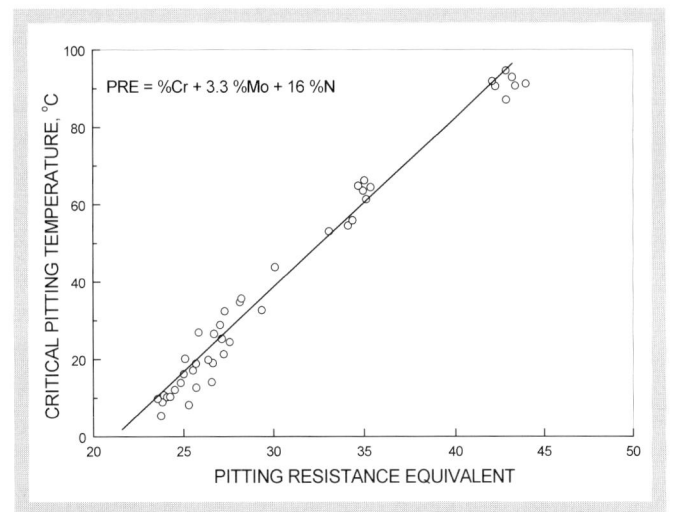

Figure 17. Critical pitting temperature as a function of stainless steel composition[12].

CHAPTER 6

Another possibility is to create charts that show the corrosion dependence on solution temperature and concentration. The data in the charts of Figs. 18 and 19 results from several tests. The solution was ammonium chloride at pH 6. In these figures, the samples were tested with cyclic polarization curves at various temperatures and chloride concentrations. If the curve showed no hysteresis and the sample was free from visible corrosion, the test was a pass result[13]. The results in Fig. 18 show the maximum temperatures for pitting corrosion, and Fig. 19 shows the corresponding results for crevice corrosion. As the figures show, the critical temperatures are much lower for crevice corrosion than for pitting corrosion.

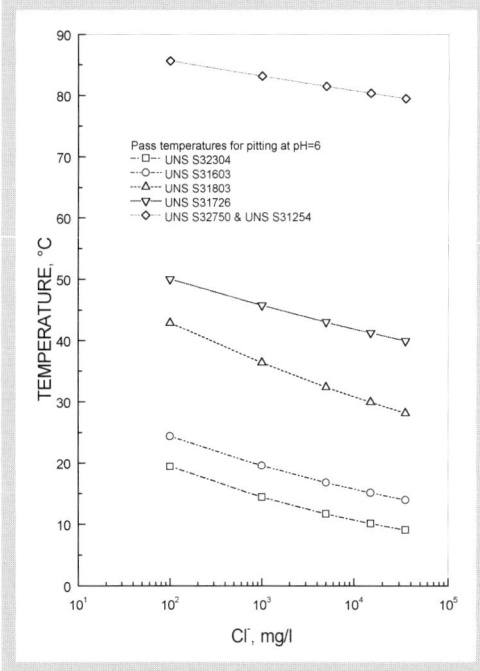

Figure 18. Critical temperatures for pitting corrosion of austenitic stainless steels in NH_4Cl solution.

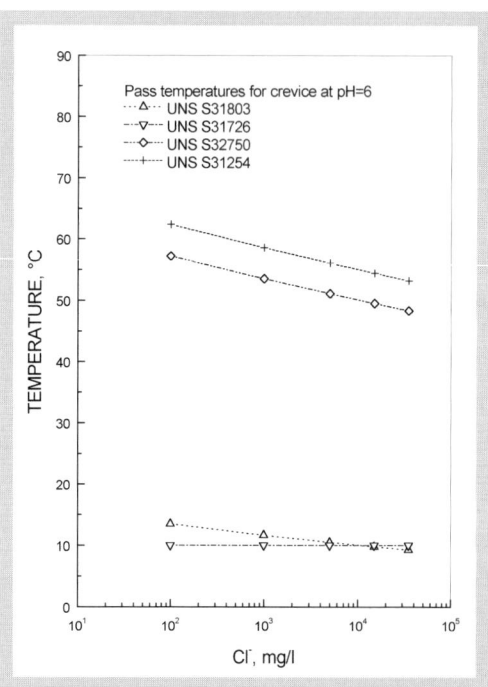

Figure 19. Critical temperatures for crevice corrosion of austenitic stainless steels in NH_4Cl solution.

The effect of chloride concentration is also lower for crevice corrosion than for pitting corrosion. For material selection, the safe area is located below these curves.

The resistance of a stainless steel to mineral acids depends on the hydrogen ion concentration and oxidizing power of the acid. Stainless steels are applicable in oxidizing nitric acid. Most austenitic AISI 300 series stainless steels in the annealed state have good resistance to HNO_3 at concentrations of 0%–65% up to the boiling point. In nitric acid service, chromium content is important for corrosion resistance of stainless steels and nickel alloys. The most widely used grades of stainless steel for nitric acid are the low-carbon or stabilized austenitic grades: Type 304L (S30403), Type 321 (S32100), and Type 347 (S34700). Stainless steels with higher levels of carbon such as Type 304

(S30400) are subject to intergranular attack at the heat-affected zone near welds as a result of sensitization. Molybdenum additions are generally considered to improve the resistance of stainless steels to acids. In nitric acid, Type 304L generally performs better than the molybdenum grades because the molybdenum promotes the sigma phase that is less resistant to nitric acid. With mixed acids or contaminated acid containing halides, Type 316L is preferable.

In sulfuric acid, stainless steels may be in the active or passive state. The conventional austenitic grades resist very dilute or very concentrated H_2SO_4. Intermediate concentrations are more aggressive, and types AISI 304 and 316 or 317 are not suitable as Fig. 20 shows[14]. Good aeration or addition of oxidizing species will increase the redox potential of the acid. Changing the acid from reducing to oxidizing assists passivation of stainless steels. Steel grades with high nickel and copper contents are more resistant. Conventional ferritic grades are not suitable for sulfuric acid[15].

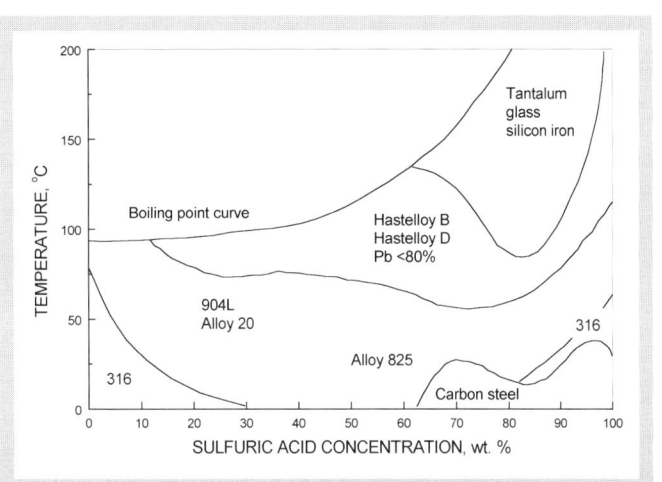

Figure 20. Stainless steels and other alloys in sulfuric acid using as a basis 20 mpy (0.5 mm/year) maximum corrosion rate with dotted lines showing the inhibiting effect of metal sulfate salts[14].

Sulfurous acid, H_2SO_3, is reducing, but many stainless steels are still suitable in it. For example, types AISI 316 and 317 have had use in sulfite digesters. Types AISI 316 and 317 and 20Cb-3 alloy have had use in wet sulfur dioxide and sulfurous acid services.

Commonly used stainless steels such as AISI 304 and 316 are not suitable for use in hydrochloric acids. Corrosion rates are high even at ambient temperatures. High alloyed grades may have possible use in very dilute acids at room temperature. Nickel, molybdenum, and copper may improve corrosion resistance in dilute acids, but pitting and stress corrosion cracking may result. The standard ferritic grades such as types AISI 410 and 430 are not suitable for hydrochloric acid. The poor resistance of stainless steels to hydrochloric acid may cause problems if used in descaling of equipment. Hydrochloric acid even if inhibited is not suitable for cleaning of overlay welded or cladded digesters.

Stainless steels are generally resistant to alkali solutions. In sodium hydroxide, conventional grades AISI 304 and AISI 316 are resistant to uniform corrosion at all concentrations up to about 65°C and in a 20% solution up to the boiling point. They may show stress corrosion cracking at 100°C as Fig. 8(a) showed. In ammonia and ammonium hydroxide solutions, stainless steels are resistant at all concentrations up to the boiling point.

Stainless steels are resistant to neutral or alkaline nonhalide salts. Halide solutions cause various forms of localized corrosion because they can penetrate and damage the passive film. Pitting is more probable in aerated or mildly acidic solutions. In very oxidizing conditions such as those in bleaching, stainless steels can suffer from various forms of localized corrosion. The corrosion results from excessively high chloride concentration, temperature, and oxidizing power of the solution. More alloyed grades are more resistant. The common stainless steels for bleach plant equipment have been the austenitic types AISI 316L and 317L. If they show problems, the usual replacement is a stainless steel with more molybdenum such as 904L (4%–5% Mo) or 254 SMO (6% Mo). The conventional alloys are usually unsuitable for most washer environments, and even the 6% Mo alloys may show crevice corrosion[16]. Common ferritic grades in the AISI 400 series are not suitable for bleach plant use, but superferritic grades of type 29-4-2 can be suitable. Molybdenum containing duplex stainless steel such as 2205 can also be suitable.

2.3 Nickel alloys

Nickel alloys are the next material group when stainless steels are not suitable. Nickel alloys are expensive. They have use only when no alternatives are possible. Nickel alloys often find use for chloride containing and reducing media where localized corrosion attacks stainless steel. Nickel and its alloys are very suitable in alkalis. Commercial nickel alloys can handle most types of acids. They are also often resistant to acid, neutral, and alkaline salt solutions. Nickel alloys have two groups. Depending on the composition, they are intrinsically resistant, or their corrosion resistance results from chromium alloying. The alloys in the first group are suitable for nonoxidizing conditions and in the second group for oxidizing conditions. The resistance of nickel to reducing media improves with molybdenum and copper additions. Addition of chromium to nickel gives resistance to several oxidizing and reducing media. Ni-Cr-Mo alloys are among the few metals that resist oxidizing conditions and halide ions. The high nickel alloys are generally resistant to transgranular SCC in chloride solutions at elevated temperatures where common austenitic stainless steels fail. The resistance to stress corrosion cracking increases rapidly when nickel content is above 10%.

The nickel alloys are an even more heterogenous group than stainless steels. Since the nickel alloys are useful in many corrosive environments, giving application areas and corrosion rates is also more difficult than with stainless steels. The main uses of pure nickel are in strong alkaline solutions. They resist sodium hydroxide at concentrations higher than 50%. The corrosion resistance of nickel and nickel containing alloys to caustic solutions is approximately proportional to the nickel content. Figure 21 shows an isocorrosion diagram for nickel alloys in sodium hydroxide[17]. Only at high concentration does the corrosion rate exceed 25 µm/year (1 mil/year). When higher strength or resistance to other corrosive agents is necessary, alloys such as Inconel 600 (N06600), Inconel 800 (N08800), and Hastelloy C-276 (N10276) are useful.

Nickel-chromium alloys are suitable in high temperature water, steam, and warm flue gases. Sulfide containing atmospheres at high temperatures may cause corrosion and embrittling of nickel containing alloys. Alloys with about 50% Ni and 50% Cr are resistant to hot corrosion caused by sodium sulfate and vanadium pentoxide. They are also suitable for alkali solutions. Nickel-copper alloys are suitable for saline waters and less than 50% alkali solutions. Nickel-molybdenum alloys such as Hastelloy B find use in nonoxidizing acid solutions. Typical applications are HCl, H_2SO_4, and H_3PO_4. The high molybdenum content of Ni-Mo alloys is beneficial in reducing solutions but harmful in oxidizing solutions.

Nickel-chromium-molybdenum alloys are suitable for oxidizing and nonoxidizing acids and their mixtures. They have frequent use for chloride containing solutions where stainless steels fail. They also have use in moist chlorine gas and chlorine and chlorine dioxide containing bleach solutions. Addition of copper to Ni-Cr-Mo alloys improves corrosion resistance to impure, halogen containing nonoxidizing acids and saline waters. The typical bleach plant environment is oxidizing with chlorides. Such environments often require use of chromium containing high molybdenum alloys.

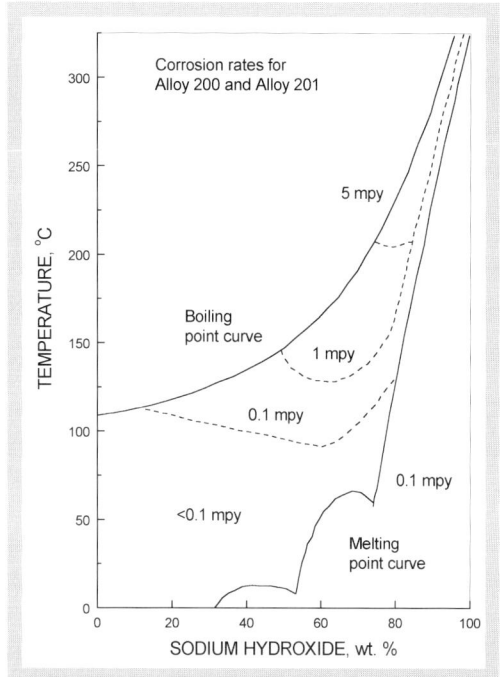

Figure 21. Isocorrosion diagram for nickel alloys in sodium hydroxide[17].

In pulp and paper manufacture, nickel alloys find use in pulping processes, bleaching operations, dryer rolls, and flue gas desulfurization such as in SO_2 scrubbers. High nickel Ni-Cr-Fe alloys with approximately 70% Ni have use in alkaline pulping processes. Alloys belonging to this group are Inconel 600 and its derivatives. Inconel 600 and 800 alloys have had use in digester liquor heater tubing because the high nickel content prevents stress corrosion cracking. Using Alloy 600 and 625 filler metals and sheet liners over weld areas in cladded equipment can prevent cracking in the digester weld area. High nickel Ni-Cr-Fe alloys also have use in steam heated dryers. Molybdenum alloyed grades such as Ni-Cr-Fe-Mo alloys are useful in digesters. In the bleach plant, nickel alloys are typically Ni-Cr-Mo alloys that resist hot, acidic, and oxidizing liquors with chlorine, chlorine dioxide, hypochlorites, chlorides, etc. The suitable alloys have high molybdenum content such as Inconel 625 (N06625) with 9% Mo, Hastelloy C-22 (N06022) with 13% Mo, and Hastelloy C-276 (N10276) with 16% Mo. An old application of nickel alloys is precipitation hardening alloy K-500 (N05500, 63% Ni, 29% Cu, and 3% Al) doctor blades. The alloy provides abrasive wear and corrosion resistance[18]. Table 2 gives a list of nickel alloys and applications in the pulp and paper industry.

Table 2. Nickel alloys and their applications in the pulp and paper industry[18].

Alloy	Type	Application
N08026 Carpenter 20Mo-6	Ni-Cr-Fe-Mo	Bleach plant
N08825 Incoloy 825	Ni-Cr-Fe-Mo	Bleach plant
N06007 Hastelloy G	Ni-Cr-Fe-Mo	Bleach plant
N06985 Hastelloy G-3	Ni-Cr-Fe-Mo	Bleach plant
N06625 Inconel 625	Ni-Cr-Mo	Bleach plant
N10276 Hastelloy C-276	Ni-Cr-Mo	Bleach plant and kraft blow tank plate
N06022 Hastelloy C-22	Ni-Cr-Mo	Bleach plant

2.4 Titanium

The corrosion resistance of valve metals and refractory metals is second only to that of the noble metals. Valve metals are characterized by their dielectric oxide layer that acts as a current rectifier. The oxide film will pass cathodic currents, but it is very resistive to anodic currents giving corrosion resistance in oxidizing conditions. The best known valve metals are titanium, tantalum, and niobium. Refractory metals are characterized by very high melting points compared with iron and steel. They are inherently reactive but will form a strong protective layer. In high temperature applications, they require a suitable coating for protection. The group of refractory metals includes niobium, molybdenum, tantalum, tungsten, and zirconium as examples. Some but not all these metals are in both groups. The most common engineering metals are titanium, niobium, tantalum, and zirconium. Except titanium, they have use only in special applications. Table 3 gives a material selection guide for common titanium alloys.

Table 3. Material selection guide for common titanium alloys where 1 is lowest and 5 is highest performance.

Corrosive environment	Oxidizing					Oxidizing, temporarily reducing	
Titanium grade	Grade 1	Grade 2	Grade 3	Grade 4	Grade 5	Grade 11	Grade 12
Strength, N/mm^2	200–300	200–300	300–400	300–400	400–800	200–300	300–400
Formability	5	4	3	2	1	5	3
Weldability	5	5	5	4	2	5	5

Tantalum has the highest corrosion resistance in this group. It has had use for many years. It is resistant to acids, dry and moist chlorine, and chloride solutions. Tantalum is also resistant to biofouling. Fluorine, hydrofluoric acid, strong sulfuric acid, sulfur trioxide, concentrated strong alkalis, and certain molten salts attack tantalum. It easily adsorbs hydrogen so corrosion in reducing media can therefore lead to embrittling. Tantalum is expensive, and it lacks strength. Its primary use is as a lining. Niobium or columbium can be a less expensive alternative to tantalum. Niobium is resistant to dry

Corrosion prevention

and moist chlorine, chloride solutions, oxidizing acids, and reducing sulfuric and hydrochloric acid within certain temperature and concentration limits. The mechanical strength of niobium is less than that of tantalum, but it also has use as a lining. Hydrofluoric acid and strong hot sulfuric and hydrochloric acids will attack niobium. The corrosion resistance to alkalis is poor. Zirconium has good corrosion resistance to acids and alkalis except hydrofluoric and hot, concentrated sulfuric and hydrochloric acids. Highly oxidizing salts such as cupric and ferric chlorides can cause pitting. The main applications of zirconium and its alloys are in high temperature water and steam[19].

Table 4. Resistance of titanium to some chemical compounds[20,21].

Chemical	Concentration, % by weight	Temperature, °C	Corrosion rate[20]	Corrosion rate[21]
Calcium hypochlorite	2 and 6	100	<0.125 mm/year	0.001 mm/year
Calcium hypochlorite	18–20	21–24	<0.125 mm/year	
Calcium hypochlorite	18	25		nil
Chlorine gas, dry less than 0.005% H_2O		30	>1.25 mm/year	
Chlorine gas, wet with more than 0.013% H_2O		75	<0.125 mm/year	
Chlorine-saturated H_2O		room	<0.125 mm/year	
Chlorine-saturated H_2O		75		0.003 mm/year
Chlorine-saturated H_2O		97		0.07 mm/year
Hydrogen peroxide	3, 6, and 30	room	<0.125–1.25 mm/year	
Hydrogen peroxide	5, pH = 4.3	66		0.061 mm/year
Hydrogen peroxide and 500 ppm Ca^{2+}	5, pH = 1	66		nil
Hydrogen peroxide	20, pH = 1	66		0.686 mm/year
Hydrogen peroxide and 500 ppm Ca^{2+}	20, pH = 1	66		nil
Sodium hydroxide	5–10	21	<0.125 mm/year	
Sodium hydroxide	10	boiling	<0.125 mm/year	
Sodium hydroxide	50	38–57	<0.125 mm/year	
Sodium hydroxide	73	113–129	<0.125–1.25 mm/year	
Sodium hydroxide	73	110		0.05 mm/year
Sodium hydroxide	50–73	188		>1.1 mm/year
Sodium hypochlorite	6	25		nil
Sodium hypochlorite		boiling	<0.125 mm/year	
Sodium hypochlorite 10% w/v chlorine	40	80	<0.125 mm/year	
Sodium sulfide	saturated	room, 60	<0.125 mm/year	
Sodium sulfite	saturated	room	<0.125 mm/year	
Sulfurous acid	6	room	<0.125 mm/year	

CHAPTER 6

The most common valve metal in the pulp and paper industry is titanium. The use of titanium as a construction material began in the 1950s. Its strength and light weight made its first applications in aircraft and missiles. Chemical industries have used titanium for many years in oxidizing and chloride containing environments. The high strength to weight ratio gives an advantage over steels when weight or excessive wall thickness would cause problems. Table 4 shows the resistance of titanium to some chemical compounds. Note that essentially the same information is given as performance classes or as corrosion rates in different sources. Comparing these sources without checking the specific form of information used can cause misunderstanding.

The main applications of titanium in the pulp and paper industry are in bleach plants. Titanium is resistant to oxidizing conditions, and chlorides or acidity do not heavily influence it. Titanium has become a standard material for drum washers, diffusion bleach washers, pumps, piping systems, and heat exchangers especially for the equipment developed for chlorine dioxide bleaching systems. Titanium resists solutions of chlorites, hypochlorites, chlorates, perchlorates, and chlorine dioxide. For example, titanium is immune in a chlorine and chlorine dioxide washer atmosphere where all stainless steels corrode. Unalloyed Grade 2 titanium has been a traditional material for construction of chlorine and chlorine dioxide bleach plants. In addition, titanium has been used in the associated storage tanks, transfer piping, mixers, and washers. Titanium has use in chloride solutions especially at higher temperatures. Very low corrosion rates occur in chloride solutions over the pH range 3–11. Oxidizing metal chlorides such as $FeCl_3$ or $CuCl_2$ and other oxidizing impurities extend titanium's passivity to lower pH levels. A limiting factor of titanium alloy application in aqueous chloride solutions can be crevice corrosion in metal to metal and gasket to metal joints or under deposits. Depending on pH and temperature, localized corrosion of unalloyed titanium and other alloys may occur in hot chloride containing media.

Titanium is generally highly resistant to alkaline media. In highly concentrated sodium or potassium hydroxide solutions, the useful application of titanium may be limited to temperatures below 80°C. In hot, strongly alkaline media, titanium alloys may embrittle due to hydrogen adsorption. Titanium is often the choice for alkaline media containing chlorides, oxidizing chloride species, or both. Even at higher temperatures, titanium resists pitting and stress corrosion cracking causing problems for stainless steels.

The most difficult problem to date has been use of titanium equipment for the alkaline peroxide stage. The alkali stage after chlorination has been reinforced with hydrogen peroxide. In some cases, peroxide has completely replaced chlorine dioxide. Existing equipment intended for the chlorine dioxide stage has frequent use. Such equipment is corroding due to hydrogen peroxide. The corrosive agent is HO_2^-. Titanium alloys shift from passive to active state when pH, hydrogen peroxide concentration, and temperature are too high. Salts causing natural hardness in water such as Ca^{2+} and Mg^{2+}, silicate, and lignin can inhibit corrosion. Transition metal ions and complexing agents can accelerate corrosion[22,23]. For titanium grade 2, the critical levels are pH > 11, H_2O_2 concentration more than 3 g/L, and temperature above 80°C. Using equilibrium calculations, a higher pH can be compensated by lower H_2O_2 concentration or temperature[23]. When using existing titanium equipment, the location of hydrogen peroxide feed is important to avoid local concentrated solutions. For example, hydrogen peroxide could be fed only after titanium equipment[24].

3 Modification of environment

Removing corrosive agents or enhancing formation of protective surface films can lower the corrosivity of a system. Methods of environmental modification may include oxidizer removal, inhibition, and pH or temperature change. Some cases combine several methods. An example is boiler feed water treatment with removal of dissolved solids and gases, adjustment of pH, and use of additives to combine residual ions. These corrosion prevention methods are less permanent than the use of a more corrosion resistant material. Modification of environment to prevent corrosion requires continuous monitoring to maintain correct chemical feed.

The methods of environmental modification fall into two groups. The first group contains methods to assist metal passivation. The second consists of methods that decrease corrosion rate in the active corrosion range. Figure 22 shows the principal methods to passivate a metal. Depending on the environmental factors of the active corrosion range, it is possible to move to the passive area by increasing potential or changing pH in either direction. The methods to decrease corrosion rate on the active corrosion area often use the removal of the cathodic reactant from the system.

Simple modifications of the environment include temperature change, pH change, and removal of dissolved salts to lower conductivity and prevent scale deposition. Since the rate of uniform corrosion and probabilities of various forms of localized corrosion usually increase with increasing temperature, lowering the temperature is an obvious corrosion prevention method. The effect of temperature becomes more critical when trying to increase production rate. In most cases, this requires higher temperatures to increase reaction rates and eventually also corrosion rates.

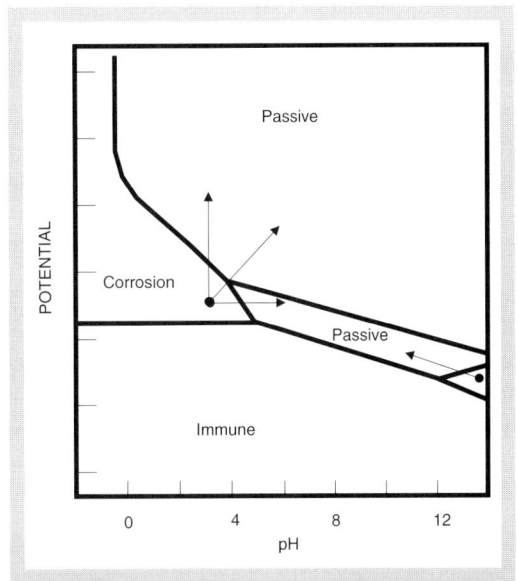

Figure 22. Principal methods to passivate a metal. Depending on the environmental factors, increase of potential or change of pH can move corrosion potential to the passive area.

The effect of pH on corrosion of metals is complex and depends on the nature of the metal. Most metals will corrode more rapidly in acid solutions due to rapid hydrogen evolution as a cathodic reaction. This usually combines with lack of protective surface films. Very high pH values can also prevent formation of protective surface films. Most metals are resistant to neutral or slightly alkaline solutions. Control of pH can also ensure precipitation of protective carbonate scales in municipal water supplies as an example.

Lowering the concentration of dissolved solids can prevent many corrosion and scaling problems. Especially in the case of boiler feed water treatment, removal of calcium and magnesium salts is essential. Higher operational pressure of the boiler requires more complete removal of dissolved solids. Inorganic scale deposits can form at any location where the solubility of a particular anion and cation pair is exceeded. Some common minerals found in the deposits are carbonates, sulfates, oxalates, and silicates of calcium; barium sulfate; iron and aluminum oxides; and hydroxides, sulfates, and carbonates of sodium. Scale problems also occur in various places in the paper manufacturing process. In continuous kraft digesters, deposits will interfere with the smooth flow of chips and liquor. Deposits on strainers can be particularly troublesome since these affect liquor circulation capacity. In the bleach plant, calcium carbonate is the most common scale mineral. Calcium oxalate deposits occur primarily in hypochlorite and chlorine dioxide equipment. The first way to prevent scaling is by depositing the scale forming minerals away from raw water. In some cases, treatment of raw water is not possible. This requires use of inhibitors. Inhibitors also have use to control the dissolved solids in treated water. In this case, their function is to keep dissolved ions in soluble complexes to prevent deposition.

3.1 Removal of dissolved gases

In many systems, dissolved oxygen is the main corrosive agent. Oxygen removal can therefore be an effective method for corrosion prevention. Oxygen removal is suitable for closed or semi-closed systems where constant oxygen replenishment is hindered. Typical applications are heating and cooling systems, boiler feedwater, etc. Oxygen removal is usually done physically with heating or gas purging, chemically with oxygen scavengers, or by using a combination of these techniques.

Mechanical oxygen removal or mechanical deaeration is the removal of oxygen and other aggressive dissolved gases such as carbon dioxide and ammonia by raising the solution to the boiling temperature corresponding to the process operating pressure. Removal of the dissolved gas is easier by decreasing its partial pressure in the surrounding atmosphere. A technique is bubbling another gas through the solution or vacuum treatment. Depending on the application, steam bubbling can remove oxygen and air bubbling can remove free carbon dioxide. Mechanical deaeration to remove the dissolved gases typically occurs before the addition of chemical oxygen scavengers. Efficient mechanical deaeration can reduce dissolved oxygen to as low as 6.5 ppb[25].

Oxygen scavengers such as sodium sulfite and hydrazine react with dissolved oxygen binding it into some compound. Sodium sulfite reacts with dissolved oxygen and forms sodium sulfate that is a very soluble compound. To assure complete oxygen removal, maintaining continuous sulfite feed and residual sulfite in the solution is necessary. The problem with sulfite addition is increase of dissolved solids that may cause deposit formation. In high pressure boilers, sulfite can decompose to sulfur dioxide and hydrogen sulfide and cause acid corrosion in the return condensate system. Hydrazine, N_2H_4, reacts with oxygen and forms water and nitrogen. It is a reducing agent that does not increase dissolved solids content because the reaction products are gases. Hydrazine is a very effective oxygen scavenger. Due to environmental and occupational hazards, alternatives for hydrazine are continuously sought. Theoretically, 7.88 ppm of sulfite as Na_2SO_3 or 1 ppm of hydrazine will scavenge 1 ppm of dissolved oxygen. In practice, about 10 ppm sulfite and 1.5–2 ppm hydrazine are necessary to scavenge 1 ppm dissolved oxygen[25,26]. Equation 1 shows the reaction of sodium sulfite with oxygen, and Eq. 2 shows the reaction of hydrazine.

$$2\ Na_2SO_3 + O_2 = 2\ Na_2SO_4 \tag{1}$$

$$N_2H_4 + O_2 = 2\ H_2O + N_2 \tag{2}$$

Feeding of oxygen scavengers to the feedwater line should be as far back from the boiler as possible to allow sufficient time for them to react. Catalyzed scavengers operate much more rapidly than pure chemicals. In low temperature systems especially, catalyzed scavengers are preferable. Analyzing the residual level of sulfite or hydrazine controls the oxygen level. High pressure boilers use lower residual levels than in low pressure units.

A popular misconception is that the system will become stable and no more corrosion will occur after consuming all the oxygen in a closed system. Usually, this is not true because most so-called closed systems are not truly closed. Provisions are necessary to compensate automatically for any pressure changes or water losses in the system. To accomplish this, the system will normally use an expansion tank and relief valve. Expansion tanks usually have a volume of air trapped above the water. When makeup water enters the system, it brings a fresh supply of oxygen. The net result of this is the introduction of a small but continuous supply of oxygen.

3.2 Inhibitors

Inhibitors are compounds that will effectively decrease the corrosion rate when used in small amounts. The inhibitors can be anodic, cathodic, or film-forming. Anodic inhibitors will decrease the rate of the anodic reaction, and cathodic inhibitors will decrease that of the cathodic reaction. Film-forming inhibitors are essentially a barrier between the material and environment. Contrary to anodic and cathodic inhibitors, film-forming inhibitors adsorb over the entire surface rather than at specific anodic and cathodic sites. This can retard both reactions. Anodic inhibitors are often more effective than cathodic inhibitors, but they are also dangerous inhibitors since an excessively low inhibitor concentration will lead to local anodic reactions where corrosion rates are high. Cathodic inhibitors are generally not as efficient as anodic inhibitors, but they are safer. An excessively low cathodic inhibitor concentration will increase cathodic reaction rate, but the resulting corrosion covers the entire area. The risks associated with anodic and cathodic inhibitors are therefore analogous to galvanic corrosion. In many cases, several different types of compounds have use simultaneously to provide a synergistic effect. The use of inhibitor mixtures is necessary when several alloys are present. Table 5 gives a listing of traditional inhibitors and their uses.

Table 5. Traditional inhibitor types and their uses in the nearly neutral pH range[27].

Metal	Chromates	Nitrites	Benzoates	Borates	Phosphates	Silicates	Tannins
Mild steel	Effective	Effective	Effective	Effective	Effective	Reasonably effective	Reasonably effective
Cast iron	Effective	Effective	Ineffective	Variable	Effective	Reasonably effective	Reasonably effective
Zinc and zinc alloys	Effective	Ineffective	Ineffective	Effective	–	Reasonably effective	Reasonably effective
Copper and copper alloys	Effective	Partially effective	Partially effective	Effective	Effective	Reasonably effective	Reasonably effective
Aluminum and aluminum alloys	Effective	Partially effective	Partially effective	Variable	Variable	Reasonably effective	Reasonably effective
Lead-tin soldered joints	–	Aggressive	Effective	–	–	Reasonably effective	Reasonably effective

Anodic inhibitors are usually inorganic oxidizing substances. They will shift the corrosion potential to the anodic direction and form a protective layer on the metal. Anodic inhibitors in contact with a metal surface will first initiate high anodic current densities at the remaining anodic sites allowing them to passivate. After formation of the passive film, the anodic current densities on the entire surface are low. Anodic inhibitors are effective only when present at sufficient concentration and when the metal can undergo active-passive transition. Cathodic inhibitors will usually increase hydrogen evolution overpotential or form a surface film that decreases cathodic reactant diffusion rate. Corrosion inhibition is a reversible phenomenon. A minimum inhibitor concentration must therefore be maintained by constant supply. Thorough circulation and absence of stagnant solution pockets are essential.

The inhibition efficiency is a percentage calculated from Eq. 3. A closer corrosion rate with inhibitor to that without inhibitor gives lower inhibitor efficiency. Usually, inhibitor efficiency of at least 95% is desirable.

$$\eta = \frac{rate\ without\ inhibitor - rate\ with\ inhibitor}{rate\ without\ inhibitor} \cdot 100\% \qquad (3)$$

By calculating the inhibitor efficiency with different inhibitor concentrations preferably using a logarithmic scale, one can estimate a minimum inhibitor concentration. Clean and smooth metal surfaces usually require a lower concentration of inhibitor than rough or corroded surfaces. The presence of oil, grease, or other film-forming compounds on the metal surface will affect the required inhibitor concentration.

Inhibition of acid cleaning solutions used in digesters is in principle the same as inhibition of metal pickling solutions. The advantages of using inhibited solutions are to save metal and pickling acid and reduce acid fumes due to hydrogen evolution. The cleaning solution corrosivity increases with increasing acid concentration and temperature. This requires higher inhibitor concentrations, but inhibitors are usually less effective at elevated temperatures. Being organic compounds, they may break down above their thermal limit. Most pickling acid inhibitors are film forming. They will effectively block hydrogen evolution and metal dissolution. Some inhibitors preferentially decrease the rate of one reaction. Adsorption is usually uniform across the entire surface. Pickling inhibitors are often mixtures of organic compounds added to the acid at a typical level of 0.01%–0.1%.

3.3 Control of biofouling

Control of biofouling is often necessary because deposits may retard heat transfer, cause corrosion, and decrease product quality. Prevention of MIC requires frequent mechanical cleaning of the surface and use of biocides to control bacterial growth. Chemical agents usually control biofouling, but without mechanical cleaning the chemicals may not reach organisms protected by thick biofilms. Cleaning can be more effective by using acids and chelating compounds. Cleaning acids require thorough washing, and biocides may be necessary in the water. For a given agent, a minimum concentration above which it will kill microorganisms exists. Below this level, the agent can only inhibit their growth. The microorganisms become immune with constant use of the same biocide. Batch dosing and use of different biocides gives better results. Common toxicants are chlorine and its compounds, bromine, chlorinated phenols, quaternary ammonium compounds, and copper and tin compounds. Some of these biocides are being replaced by environmentally less harmful ones such as DBNPA (2,2-dibromo-3-nitrilopropionamide), carbamate, and glutaraldehyde. Surface active agents and dispersants are often added to standard biocides for deposition control. Most biocide formulations are not secrets, because regulations require listing of biocide ingredients.

The effective control of microbiologically-influenced corrosion requires correct identification of the bacteria and verification that the bacteria actually are causing corrosion. Before trying to attack the biofilm, excluding the more common corrosion forms is essential to avoid mistakes. An effective micro biocide program to control the growth of microorganisms involves three steps:

- Identification of the types and concentrations of microorganisms present in the system
- Selection of proper biocides using system design, discharge restrictions, and types of microorganisms (Selection of biocide treatment depends on the type of application, construction materials, and water supply.)
- Proper application, dosage, and control of the selected biocides.

CHAPTER 6

General classification of biocides are oxidizing and nonoxidizing. Oxidizing biocides will irreversibly oxidize protein groups. This results in loss of normal enzyme activity and subsequently rapid death of the cell[28]. Common oxidizing biocides that use the oxidizing effect of chlorine and its compounds are chlorine, chlorine dioxide, hypochlorous acid, and hypochlorites. Chlorine has had common use for disinfection, and it is effective against bacteria. The pH, temperature of water, and chlorine demand determine the amount of chlorine compound necessary. Chlorine demand refers to the exact amount of chlorine that will react with biocidal mass and contaminants. Any additional chlorine will pass through the system as free residual chlorine. When chlorine gas dissolves in water, it dissociates into hypochlorous and hydrochloric acids. Hypochlorous acid will hydrolyze to hydrogen ions and hypochlorite ions. The ratio of hypochlorous acid to hypochlorite ion determines the biocidal efficacy. Hypochlorous acid, HOCl, is 20–80 times as effective as a biocide as the hypochlorite ion, OCl^-[28,29]. Addition of hypochlorites such as NaOCl or $Ca(OCl)_2$ that are salts of hypochlorous acid to water also results in formation of hypochlorous acid. Chlorine gas can therefore be substituted by hypochlorous acid and hypochlorites that will eventually dissociate to produce hypochlorite ions.

Chlorine oxidizes the active sites on certain coenzyme groups that constitute intermediate steps in the production of adenosine triphosphate that is essential to respiration[28]. Hypochlorous acid is a very powerful oxidant. The solution is highly oxidizing and destroys bacterial cells and slime by reacting with cellular components producing stable nitrogen to chlorine bonds with the cell proteins[28,30]. Hypochlorous acid treatment is most effective at nearly neutral pH of 6.7–7.0. At pH values above 9.5, hypochlorous acid completely dissociates into hypochlorite ions and is no longer effective. At low pH values, the high oxidizing power of the solution may make biocide solutions too corrosive. The chlorine demand is difficult to estimate. The free residual chlorine is therefore used to determine the amount of chlorine fed to the system. The free residual chlorine for proper biocidal use is a factor of pH. At pH of 6.0–8.0, 0.2 ppm free residual chlorine is usually sufficient. At higher pH, the concentration of free residual chlorine doubles for every unit in the pH scale, e.g., 0.4 ppm free chlorine for pH values up to 9.0[28]. Chlorine and its compounds require addition to the system immediately before the biofilm containing areas to delay chlorine breakdown. The concentration of free residual chlorine must be measured at the end of the system to assure the presence of adequate chlorine level throughout the system.

Chlorine dioxide is more oxidizing than chlorine, and its biocidal effect also uses its high oxidation power. Chlorine dioxide does not form hypochlorous acid. It exists solely as dissolved chlorine dioxide. It is generally less effective as a biocide than the other chlorine compounds, but it is more effective than chlorine at high pH ranges.

Ozone is a new oxidizing biocide. When dissolved, it retains its oxidizing character and works very similar to chlorine biocides. Ozone combines with proteins and inactivates enzymes essential to respiration. Bacterial cells subject to ozone treatment appear to be ruptured with the loss of cytoplasm. The pH, temperature, organic material, and dissolved metal ions also influence ozone. Ozone has an ozone demand like chlorine compounds have their chlorine demand. Common ozone treatment residuals are about 0.5 ppm[28].

Most oxidizing biocides have use with continuous or batch feeding. The latter is preferable because of lower chemical cost and better effectiveness. Many microbes adapt to the environment. Continuous addition of one biocide at constant concentration may therefore make them immune to that one. Table 6 compares some chlorination treatment programs. Free residual refers to the portion of total chlorine that will react chemically and biologically as hypochlorous acid. Combined residual refers to the portion of total chlorine that will react chemically and biologically as inorganic or organic chloramines. In the form of chloramines, chlorine is a mild bactericide and oxidizing agent.

Table 6. General comparison of chlorinating programs[30].

Program	Remarks
Continuous chlorination - free residual	Most effective Most costly Not always technically or economically feasible due to high chlorine demand
Continuous chlorination - combined residual	Less effective Less costly Inadequate for severe problems
Intermittent chlorination - free residual	Usually effective Less costly than continuous chlorination
Intermittent chlorination - combined residual	Least effective Least costly

Nonoxidizing biocides are often more effective than the oxidizing ones, and they have use in conjunction with them. A common practice in cooling water treatment is to use an oxidizing biocide intermittently with additional nonoxidizing treatment. The nonoxidizing biocides range from metals to organic metal compounds and organic compounds. The simplest nonoxidizing biocide is dissolved copper. Copper is effective against bacteria and algae, but molds and fungi are generally resistant. They have been effective when used in the 1–2 ppm range of copper sulfate. Copper ions are more noble than steel, and in excessively high concentration they can therefore cause corrosion of iron by the cementation reaction. Organic tin compounds are usually not very toxic, but they are effective against algae and molds. Organo-tin compounds work best at alkaline pH values.

Chlorinated phenolics have common use as nonoxidizing biocides[28]. They function by adsorbing onto the microorganism cell wall. After adsorption, they diffuse into the cell and precipitate protein to influence respiration. Quaternary ammonium salts are surface active chemicals. They are often the most effective materials against bacteria and algae in alkaline solutions. Quaternary ammonium salts create stresses in the cell wall by electrostatic bonds causing cell death. They can also distort the permeability of the cell wall. Dirt and oil limit their activity because of surface activity. A variety of

CHAPTER 6

organic sulfur compounds are available for biocides. They have very similar operating mechanisms but different optimum pH ranges[28]. They function by inhibiting cell growth.

4 Protective coatings

Coatings used for metal protection can be metallic, inorganic, or organic. They protect the metal in three possible ways:

- By forming a barrier between the metal and corrosive environment
- By sacrificial dissolution
- By inhibition.

All coatings will separate the metal from its environment to some degree. For a barrier coating to be effective, it must cover the entire surface, be sound, and resist mechanical damage. Sacrificial coatings act in two ways. If the coating is sound, it forms a barrier. If the coating is damaged, it will dissolve. Being anodic to the substrate metal, it will give protection analogous to cathodic protection with sacrificial anodes. Inhibited coatings include many paints, oils, and greases used for temporary protection. They work in a dual role by forming a barrier and also acting as inhibitors.

The life expectancy of a metal structure and the durability of its appearance depend on the quality of surface preparation before coating and coating performance. Surface cleaning usually has two stages. First, the organic impurities such as oil, grease, and paint are removed with organic solvents, strongly alkaline solutions, emulsion cleaning, or steam cleaning. It is then possible to remove inorganic impurities like mill scale, rust, and other corrosion products. The common methods are mechanical brushing, grinding, sandblasting, flame cleaning, and chemical pickling with strong acids.

Selection of a corrosion-resistant coating system consists of identifying the type of corrosive environment, surface preparation, and service life requirements. Determining these parameters is essential since certain finishes have formulations more resistant to given environments.

4.1 Corrosion protection paints

Protection of metals by organic paint coatings is the most common corrosion prevention method. Paints are liquids or powders that form a film on a surface and harden to a solid adhering coating. The purpose of a paint coating is to give corrosion protection and desired appearance. Selection of a paint or paint system depends on the type of protected structure, environment, application method, and surface preparation. The coating must have physical and chemical protective properties. Most paints are barrier coatings, but paint films are still permeable and will therefore allow water and oxygen diffusion to the metal surface. The included pigments, impermeability, and adhesion give paint film its protective properties. Painting often serves multiple functions that include corrosion protection, erosion resistance, safety, identification, and appearance. (Rusted equipment looks unreliable.) This leads to different requirements for applications such as flooring, electrical apparatus, process equipment, piping, buildings, control rooms, supports, and hangers. No universal coating can handle all these surfaces. Changes in regulations affecting the use of coatings have narrowed the field of available paint industrial types. Stricter environmental regulations have limited volatile organic solvents permitted in a paint. New paints have been developed that have a higher percentage of solids content compared with older paints. Some generic classes of coatings can no longer be formulated in compliance with regulations.

The first steps in correct paint application are to specify the proper coatings, surface preparation, and application technique. The generic type is the most useful paint classification principle because coatings of the same generic type have similar handling and performance properties. The name for most generic types of coatings uses the binder in the formulation. The binder is only broadly classified by its general chemistry, i.e., inorganic or organic. For example, vinyl and epoxy are generic coating types with names using the binder. Generic coating classifications are general and broad and can be subdivided into narrower, more descriptive classes. A secondary generic classification is by curing mechanism or some other compositional element. For example, the epoxies include different resins and hardeners with different specific properties. Generic coating types can have even broader classifications such as inorganic zinc-rich and organic zinc-rich coatings. Zinc-rich indicates that high loadings of zinc dust are part of the formulation.

A paint film consists of a binding medium or binder, pigments, and additional compounds dispersed into it. The binder is dissolved in an organic solvent or emulsified in water to decrease viscosity. After application of the paint on the surface, the solvent or water evaporates. The binder is the primary compound of the paint. It forms the film that adheres on the surface. The dry film then contains all the pigments and other components bonded together. Most paint film properties depend on the binder. This includes mode of drying, adhesion strength, resistance to water and chemicals, and durability. The binders are often large molecule organic polymers or reactive small molecular resins that form polymers when paint dries. Most binders are synthetic resins. Natural resins and modified natural products such as rubber and oils have also use. Solvents and diluents dissolve the binder and decrease its viscosity. Solvents and binders affect paint film drying speed, ease of application, adhesion, and durability. The application method determines the selection of solvent and diluent. For example, spraying requires more rapidly evaporating compounds than brushing. A pigment is essentially any compound that gives the paint some desired property. The main pigment groups in corrosion protection paints are color and corrosion prevention pigments and various fillers. The pigments affect color, gloss, coverage, viscosity, and durability. With colored pigments, a certain color and coverage is sought. Additional filling pigments provide gloss and strength. In corrosion protection paints, aluminum and iron oxide flakes decrease permeability. Corrosion prevention pigments such as zinc dust and zinc phosphate will decrease corrosion of the metal substrate. The paint also contains small quantities of additives. Common additives are thickeners, antisettling agents, surface active agents, antioxidants to prevent skinning, etc.[31].

CHAPTER 6

A single paint will seldom have all the required properties. Paint systems with a primer and a top coat or finish are therefore necessary. The system can also contain several intermediate layers. The purpose of the primer is to protect the substrate metal by pigments, high electrical resistance, and impermeability and give the entire paint film good adhesion. The intermediate layers increase film thickness and impermeability. They often contain inhibiting or flaked pigments. The intermediate layer must have good adhesion to the primer. The finish must be resistant to the environment and protect the intermediate and primer layers. It must also have the required color and gloss. A shop primer is a thin protective coating for protection during transport and storage applied at the steel supplier facility. A stripe coat is a supplementary coat applied to ensure adequate protection of critical areas such as edges, welds, etc. Before the application of each coat, a stripe coat is applied by brush to all welds, corners, behind angles, edges of beams, and areas not fully reachable by spray to obtain the specified coverage and thickness. Primers have the highest relative amount of pigments, and topcoats have the lowest. A typical paint coating is about 100–300 µm thick. In very aggressive environments, paint systems may have several layers and be over 500 µm total thickness. The uniform corrosion rate of the unprotected metal should not be very high when using paint or damages in paint film may cause rapid and localized corrosion. If corrosion rate of the unprotected metal is more than about 1 mm/year, durable thick coatings or linings are more useful.

The successful use of paint coatings requires considerable information about the conditions of application and environmental exposure. The drying conditions will determine suitable binders, and binder selection will determine suitable solvents or solvent mixtures. Pigments and additives give the additional desired properties. The binder may dry by one of three mechanism:

- Physical drying involves evaporation of solvent constituents.

- Chemical drying involves some chemical reaction such as oxidation that results in cross-linking of the binder. In oxidation, the paint dries from the outer surface inward. A number of thin coatings are necessary to build a thick layer.

- Polymerization drying requires a chemical reaction between the binder and a curing agent (hardener) mixed into the paint before application. The paint polymerizes to create a cross-linked structure. Binders that dry by solvent evaporation in air can be acrylic, bitumen, chlorinated polymers such as chlorinated rubber, vinyl resins, and vinyl tar.

Aqueous emulsion paints such as acrylic latex also dry with physical evaporation. Chemical drying with oxidation occurs with oil-based alkyds (enamels) and epoxy esters. Drying with polymerization occurs with epoxies, some modified alkyds, acetates, phenols, polyesters, and polyurethanes. These are two-component coatings prepared by mixing resin and curing agent immediately before application. Many coatings applied as liquid or powder cure further by baking in an oven to crosslink the resin or remove residual solvent. Binders for these paints are alkyds, polyester, acrylics, and epoxies. Figure 23 shows the classification of paint types by their drying method.

Corrosion prevention

```
                                    Binder
                                      |
          ┌───────────────────────────┴───────────────────────────┐
    Chemical drying                                         Physical drying
          |                                                       |
    ┌─────┴─────┐                       ┌───────────────────┬─────┴─────┐
    |           |                       |                   |           |
Evaporation   Evaporation of     Evaporation of       Evaporation of  Unsolvated
of solvent    water              solvent or water     solvent or water compounds
No chemical   No chemical        Oxidation of         Components of    of binder react
reaction      reaction           binder               binder react
    |           |                       |                   |           |
Bitumen      Latex paints        Oil-based            One-component   One-component
Vinyl                            Alkyd                - heat cured     - powder coatings
Chlorinated rubber               Epoxy ester            alkyd,        Two-component
                                 Urethane alkyd         acrylic and    - solvent-free
                                                        polyester        epoxy
                                                      Two-component
                                                      - epoxy
                                                      - coal tar epoxy
                                                      - polyurethane
```

Figure 23. Classification of paint types by drying method.

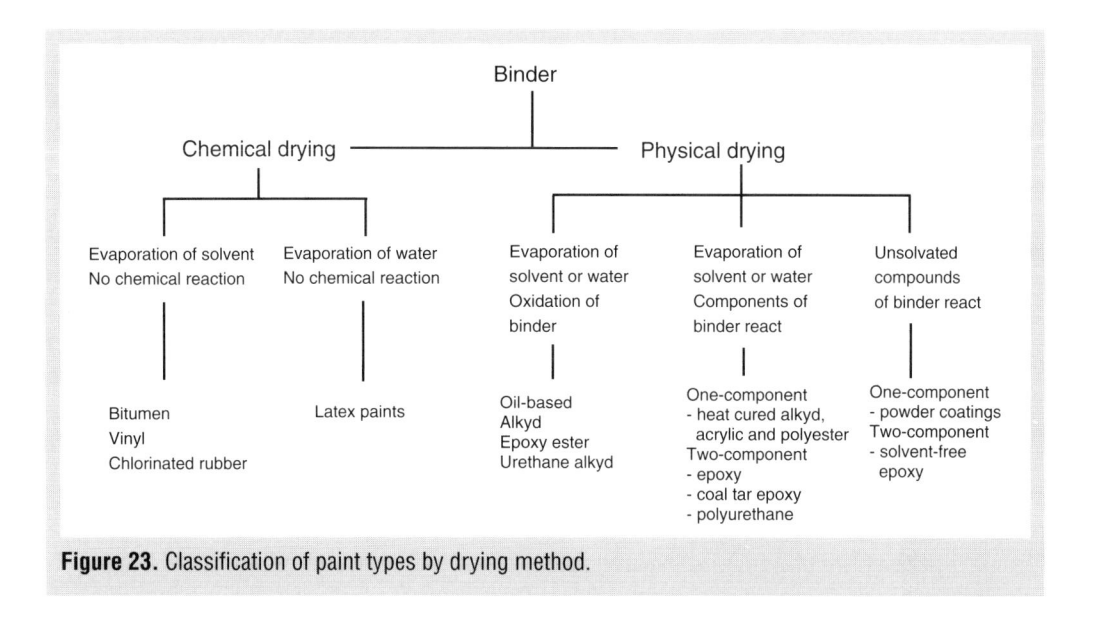

Anticorrosion coatings primarily use epoxy, urethane, ethyl silicate, vinyl, and chlorinated rubber binders. Paint systems for industrial applications vary depending on the exposure to atmosphere, water, soil, or chemicals. Oil-based alkyds are probably the most common corrosion protection paints. They have use in outdoor and indoor atmospheric exposure if chemicals are not present. They are resistant only to occasional condensation or immersion. Modification of alkyds can improve different properties depending on the modifier group. Modified alkyds are more resistant to weather and wear. They are also heat and oil resistant but not suitable for acids or alkalis. Alkyd paints are economical coatings easily applied over most paints without damaging them. Peeling may occur when applying alkyd paints on a previous film of fully hardened alkyd paint. Application should not be too thick.

Paints using chlorinated rubber and vinyl resins have use in chemical applications. They provide good corrosion protection with easy application. Chlorinated rubber paints dry physically with no temperature dependence. Repair and maintenance of chlorinated rubber films is easy since no danger exists for peeling between successive coats. Application of chlorinated rubber over other types of paint requires special precautions because solvents might lift previous coats. Chlorinated rubber paints require good pretreatment and special solvents. The lowest temperature for chlorinated rubber is -10°C. Chlorinated rubber is suitable for field work. Its resistance to water moisture and mechanical strain is high. Vinyl coatings are very resistant to water, moisture, and mechanical strain. Good pretreatment is necessary, but drying is independent of temperature and does not take long. Repainting of vinyls is easy with no risk of peeling between successive coats of paint. As with chlorinated rubbers, use of vinyls on top of other paints requires special precautions to avoid the risk of lifting previous coats. The

CHAPTER 6

lowest application temperature for vinyls is 0°C. Vinyl tar is a mixture of vinyl and coal tar suitable for immersion in water. The lowest application temperature is -10°C.

Epoxies are notable for very good chemical and abrasion resistance and for excellent adhesion. Two-component epoxies are resistant to solvents and chemicals, and the coating is hard, ductile, and wear resistant. Epoxy coatings have a high solids content and are very resistant to water, chemicals, solvents, oils, and mechanical strain. They may be unsuitable in strongly oxidizing conditions. Epoxies require a minimum temperature of +10°C to polymerize. Solvent-free epoxies made from liquid resin are suitable for immersion service. They are also wear and chemical resistant. Epoxy tar coatings are mixtures of epoxy and coal tar. These black coatings are resistant to water and chemicals and used in water and soils. Epoxy-resin coatings modified with oils are epoxy esters. They are similar to the alkyds.

Polyurethanes have different properties depending on the components. They may be weather or chemical resistant, hard or soft, etc. Polyurethanes modified with coal tar are also suitable for water or soil. Polyurethanes have excellent color, gloss retention, good abrasion resistance, and flexibility. Aqueous paints that dry by evaporation are useful in atmospheric exposure indoors and outdoors. They may require weeks to dry depending on temperature and relative humidity. Drying the paint in an oven gives good results. Solvent-free polyvinyl chloride plastisols are useful for thin sheet coating. Powders using thermosetting polymers have application for various equipment.

Numerous criteria exist for selecting coating systems. The resistance of a generic paint type to the environment is obviously the most important criterion. Close to that are ease of application, extent of surface preparation, appearance, drying time, and cost. Table 7 shows the resistance of some paint types. The maximum rating in each category is 10.

Table 7. Resistance of some paint types to environmental conditions[32,33].

Condition	Vinyl	Epoxy	Phenolic	Alkyd	Oil-based	Urethane	Inorganic zinc
Sunlight and water	10	9	9	10	10	8	10
Stress and impact	8	3	2	4	4		
Abrasion	7	6	5	6	4	10	10
Heat	7	9	10	8	7		
Water	10	10	10	8	7	10	5
Salts	10	10	10	8	6	10	5
Solvents	5	8	10	4	2	9	10
Alkalies	10	9	2	6	1	10	1
Acids	10	10	10	6	1	9	1
Oxidation	10	6	7	3	1	9	10

Table 8. Examples from paint systems for industrial applications. For surface preparation levels see Table 9.

Application	Paint type	Surface preparation level	Primer and intermediate film	Top coat	Film thickness	Notes
Structural steel in industrial atmosphere	Alkyd	St2	2 x 40 μm red lead brush	2 x 40 μm alkyd brush	160 μm	suitable for field use
	Alkyd	St2	2 x 40 μm red lead brush	1 x 80 μm alkyd high-pressure spray	160 μm	
	Alkyd	Sa2	1 x 80 μm alkyd brush or HP spray	1 x 80 μm alkyd high-pressure spray	160 μm	thixotropic and rapid drying, for shop and field use
Structures, machinery, vessels, etc., in atmosphere with chemical spills or corrosive gases	Two-component Epoxy	Sa2½	primer 1 x 60 μm epoxy inter. 1 x 80 μm epoxy high-pressure spray	1 x 40 μm epoxy high-pressure spray	180 μm	
Process industry structures and equipment	Two-component Epoxy	Sa2½	primer 1 x 80 μm epoxy inter. 1 x 80 μm epoxy high-pressure spray	1 x 50 μm epoxy high-pressure spray	210 μm	
Internal surfaces of vessels and tanks, steel under immersion	Two-component Epoxy	Sa2½	primer 1 x 100 μm epoxy inter. 1 x 100 μm epoxy high-pressure spray	2 x 50 μm epoxy high-pressure spray	300 μm	
Wet end of a paper machine	Two-component Epoxy	Sa2½	primer 1 x 40 μm Zn epoxy inter. 1 x 100 μm epoxy high-pressure spray	1 x 90 μm epoxy high-pressure spray	230 μm	
Galvanized structures	Chlorinated rubber	Brushing, washing	1 x 60 μm chlor. rubber high-pressure spray	1 x 40 μm chlor. rubber high-pressure spray	100 μm	suitable for field use
Steel structures subject to spills, dust and corrosive gases	Chlorinated rubber	Sa2½	2 x 80 μm chlor. rubber high-pressure spray	1 x 40 μm chlor. rubber high-pressure spray	200 μm	suitable for field use
As above, very corrosive environments	Chlorinated rubber	Sa2½	2 x 80 μm chlor. rubber high-pressure spray	2 x 40 μm chlor. rubber high-pressure spray	240 μm	suitable for field use
Steel structures in pulp mill atmosphere	Chlorinated rubber	Sa2½	primer 1 x 40 μm Zn dust 2 x 80 μm chlor. rubber high-pressure spray	2 x 40 μm chlor. rubber high-pressure spray	280 μm	
Galvanized structures	Vinyl	Brushing, washing	1 x 60 μm vinyl high-pressure spray	1 x 40 μm vinyl brushing, HP spray	100 μm	suitable for field use
Steel structures subject to spills, dust and corrosive gases	Vinyl	Sa2½	2 x 60 μm vinyl high-pressure spray	2 x 40 μm vinyl brushing, HP spray	200 μm	suitable for field use
Galvanized structures	two-component polyurethane	Brushing, washing	2 x 70 μm epoxy high pressure spray	1 x 40 μm polyurethane brush, spray	180 μm	gloss and color very durable
Steel surfaces under very corrosive atmospheric conditions	two-component polyurethane	Sa2½	1 x 40 μm Zn epoxy primer 1 x 100 μm epoxy high pressure spray	2 x 40 μm polyurethane brush, spray	220 μm	gloss and color very durable
Steel surfaces in soil or immersed, complex structures	two-component polyurethane tar	Sa2½	1 x 100 μm urethane tar high pressure spray	3 x 100 μm urethane tar high pressure spray	400 μm	thick black or brown films. Can be applied at -10 °C

Corrosion prevention

CHAPTER 6

Previously, lead-based alkyd materials bonded to less prepared surfaces with few problems. New high performance coating systems require a higher degree of surface preparation for proper adhesion. To decrease the high costs in preparation of the surfaces, manufacturers are constantly developing new paint systems that may not last as long as the traditional zinc-based systems but require a lower quality surface. Table 8 shows some recommended paint systems for industrial applications. The examples come from product sheets of a major Finnish paint manufacturer. A zinc primer on a properly cleaned surface is a minimum. In zinc-rich primers, zinc powder and a liquid base are mixed immediately before application. Inorganic or two-component epoxy based primers are suitable. They can be applied with normal industrial paint equipment. The resultant film is equal to hot-dip galvanizing for weathering protection and superior to the electroplated zinc in its chemical resistance. Unlike hot-dip galvanizing, inorganic zinc can be topcoated with a wide variety of coatings to improve lifetime and appearance. A compatible top coat resistant to acid or alkaline conditions will increase coating life. The highly corrosive atmosphere and constantly changing environmental conditions typically found at pulp and paper mills requires detailed adherence to coating specifications.

Correct paint application is important. The importance of proper surface preparation never can have too much stress. Good surface preparation is more important than good paint application. The Steel Structures Painting Council (SSPC), NACE International, ASTM, and various European standardization bodies have standards for surface preparation. Table 9 lists some of the surface preparation levels. Each has its advantages and disadvantages in terms of effects on surface preparation requirements considering coating type, environment, and operator. Certain surface preparations are more effective in removing some contaminants, and most coating systems will perform better on a surface treated at a high level.

Table 9. Descriptions for surface preparation levels.

Method	SSPC	NACE	ASTM	European	Description
Solvent cleaning	Sp1				removes oil, grease, wax, dirt
Hand tool cleaning	Sp2		St2	St2	removes loose rust, mill scale, and coating
Power tool cleaning	Sp3		St3	St3	removes loose rust, mill scale, and coating
Brush-off blast	Sp7	No4	Sa1	Sa1	does not remove tightly adhering mill scale, rust, or old coating
Commercial blast	Sp6	No3	Sa2	Sa2	66% of surface area free of all visible residue
Near-white metal blast	Sp10	No2	Sa2$^{1}/_{2}$	Sa2$^{1}/_{2}$	95% free of all visible residue
White-metal blast	Sp5	No1	Sa3	Sa3	complete removal of all visible residue

Most coating failures are due to improper surface preparation and application. The surface must be clean from dirt, grease, mill scale, and dust before paint application. Various grades of surface preparation range from solvent cleaning of contaminants to pickling or white-metal blast cleaning to remove all mill scales, rust, paint, and other foreign matter. Before blasting, sharp edges, fillets, corners, and welds require rounding or smoothing by grinding. Hard surface layers resulting from flame cutting need removal. The surfaces must be free from any foreign matter such as weld flux, residue, slivers, oil, grease, salt, etc. Solvent or alkali cleaning must remove any oil and grease contamination. Major surface defects and especially surface laminations or scabs detrimental to the protective coating system must be removed. All welds require inspection. If necessary, they must be repaired before final blast cleaning of the area. Blasting abrasives must be dry, clean, and free from contaminants. The size of abrasive particles for blast cleaning should be such that the prepared surface profile height meets the requirements for the applicable coating system.

The final surface preparation and primer application should occur on the same day. The surface to be coated should be clean, dry, free from oil and grease, and have the specified roughness and cleanliness. Dust, blast abrasives, etc., should be removed from the surface after blast cleaning. Proper paint application requires rather stringent environmental conditions. The temperature and relative humidity of the atmosphere are the most significant factors. Typically, the metal surface should be 5°C–50°C and a few degrees above the dew point. The relative humidity should be below 85%. Paint films may show different kinds of defects due to poor application workmanship. Most failures occur in applications where surface preparation or coating application specifications are not followed.

Incomplete cleaning can lead to incomplete adhesion of the coating. Inadequate curing between coating layers and improper mixing of multicomponent systems will destroy coating properties. Blistering is formation of broken or unbroken bubbles of various sizes under or within the coating. Most common blistering results from improper solvent, oil, or moisture contamination, surface contamination with salts (osmotic blistering), or excessive cathodic protection. Peeling and flaking are most often due to inadequate surface cleaning or painting over an incompletely dried coating. Orange peeling forms a morphology on the coated surface resembling an orange peel. This problem typically results from excessively high viscosity during application or improper solvent evaporation rate. Undercutting is a form of blistering and peeling where exposed substrate corrodes. Propagation of corrosion under the coating causes debonding. Enhancing coating adhesion and using an inhibited primer before top coating can reduce this. Cracking may result from excessively thick paint and unclean, cold, or hot surfaces. Chalking forms a powdery material on the surface of the coating. It often results from inadequate resistance of the coating to ultraviolet light[34]. Table 10 gives an overview of painting problems, possible causes, and remedies.

CHAPTER 6

Table 10. Painting problems, possible causes, and remedies.

Problem	Effects	Possible cause	Remedy
Skinning	Some paint is wasted Poor appearance if skins remain in paint film	Unsealed bucket Storage too warm	Storage in cool, full buckets Apply solvent before closing Filter before use
Sediment due to insufficient mixing	Poor distribution of pigments resulting in uneven gloss, stripes, and film formation	Excessive storage or storage at excessively high temperature	Store in cool place Mix properly before application
Drying troubles	Sticky film, adhesion of dirt, poor durability	Insufficient hardener Film too thick Painting over fresh film Moist or cold conditions Dirty or greasy surface	Follow directives given by paint manufacturer Ensure complete drying of primer films Clean surfaces
Wrinkling	Poor film formation Adhesion of dirt	Painting over wet primer Film too thick	Ensure complete drying of primer films Correct film thickness
Lifting	Paint film delaminates	Primer is incompatible with solvents of top coat	Use compatible paint types Avoid strong thinners
Orange peel	Wrinkled paint film	Wrong thinner, method, or viscosity during spraying	Proper selection of thinner and viscosity
Grains	Poor appearance	Dirty paint or equipment Dirty surface Incompatible thinner Blast cleaning dust	Filter and mix paint Use clean equipment Clean surfaces and atmosphere Correct thinner
Unequal gloss	Stripes in the film	Incompatible thinner Uneven surface Uneven application Absorbing surface	Correct thinner Spot paint over absorbing areas Apply evenly
Porosity	Pores decrease protection level and film easily becomes dirty	Incompatible thinner Air in paint or moisture in spraying air Film too thin Drying too fast	Correct amount of correct thinner
Chalking	Pigments become dislodged as the binder decomposes	Excessive thinner Wear too heavy	Correct paint system Follow paint manufacturer directions

Table 10. Painting problems, possible causes, and remedies.

Problem	Effects	Possible cause	Remedy
Blistering	Paint film delaminates as round blisters	Porous surface Air in paint Incompatible thinner Moist atmosphere Paint film too thick Surface too hot Moisture under paint film Corrosion under paint film Cathodic overprotection	Thin primer on porous surface Careful mixing of paint to avoid air entrapment Paint under appropriate conditions only Good surface cleaning Use paints resistant to corrosive environments, e.g., alkali resistant paints with cathodic protection
Discoloration	Lighter or darker shade of paint	Bright pigments do not tolerate weathering Chalking	Use recommended paint systems only Allow primer films to dry completely Avoid different paint types over another
Flaking, scaling, peeling	Paint film or parts of it become loose and protective effect is lost	Painting on moist or greasy surface Painting on mill scale or rust Painting under poor conditions Paint incorrectly mixed or thinned Insufficient time between paint film applications Paint system incompatible with surface	Clean surface properly Paint on dry surfaces at temperatures sufficiently high Follow paint manufacturers directions on mixing, time intervals, and film thicknesses Use a suitable paint system
Chipping	Paint film becomes separated from the film below	Previous film is too dirty or too hard Incompatible top coat	Clean painted surfaces before applying next layer Hard or glossy films must be ground Use compatible paints

CHAPTER 6

Since the life of the coating is usually less than the life of the structure it must protect, some form of maintenance is necessary. Three maintenance strategies are common:
- Spot repair
- Overcoating
- Complete recoating.

In spot repair, only rusted or delaminated areas are removed from the surface, and a new coating is applied. In overcoating, all defective areas are removed, and the entire structure is recoated. Complete recoating requires removal of all existing paint followed by appropriate surface treatment and application of paint.

4.2 Paint film protection mechanisms

The paint film works by stopping anodic or cathodic reactions or by creating a high resistance. The cathodic reaction stops when oxygen and water diffusion to the metal surface is hindered. The permeability of the paint film is important in this regard. Paint films typically contain numerous microscopic pinholes. Paint films as barrier coatings are therefore not completely impermeable to moisture and will eventually break down to allow the corrosion process to begin. The barrier coatings will only protect as long as the coating is intact. If a barrier coating is scratched or damaged in some way to expose the underlying metal, corrosion begins. Proper selection of binder and pigments can diminish the permeability. Moisture penetrates these pin holes to reach the unprotected substrate. When this occurs, corrosion results in pits, and blisters lift away the coating. This is the primary reason to use inhibited primers. The corrosion prevention methods of paints are high electric resistance pigments, anodic passivating pigments, and sacrificial pigments (such as zinc).

Most organic paint films have a large electrical resistance. Although paint films cannot totally exclude cathodic reactants from the metal surface, the film will resist charge transfer by ions through the film. Any remaining salts due to poor surface preparation will dissolve in penetrating water and form ions to decrease paint film resistance. High resistance films use very water resistant binders such as epoxy, tar epoxy, chlorinated rubber, and vinyl. Solvent free epoxies have use as the sole paint in structures immersed in water.

Passivating corrosion protection pigments form a protective reaction product layer with water on local anodic or cathodic sites. These pigments have use in primers for atmospheric exposure. Phosphates, borates, and due to environmental reasons restricted lead and chromium compounds are passivating pigments. Alkaline compounds such as lead and zinc oxides or hydroxides and many carbonates can act as neutralizers against local pH changes.

Zinc paints contain metallic zinc dust as a sacrificial pigment in an organic or inorganic alkali resistant binder. They differ from the other paints because they can give galvanic protection. Zinc paints can be used without a topcoat, but a topcoat will naturally increase protection level. Paints using organic binder such as epoxy or chlorinated rubber require less surface preparation and are easier to paint with a topcoat. Inorganic binders such as ethyl silicate are more heat resistant and not as flammable. The zinc in paints acts similarly to hot dip or sprayed zinc coatings preventing corrosion propagation in paint film defects. Zinc paints have use as primers on steel construction in severe corrosive environments such as occur in the chemical process industry. Containing metallic zinc, they are effective primarily in neutral or slightly alkaline solutions. Zinc paints are a suitable alternative for large structures that cannot be hot-dip galvanized.

4.3 Organic coatings

Polymer coatings have wide applications in the pulp and paper industry. Most polymers used as functional coatings are fluorocarbons. Polymer coatings designed for corrosion protection are usually tougher and applied in heavier films than paint films. Requirements of such coatings are much more stringent. They must adhere well to the substrate and must not chip easily or degrade from heat, moisture, salt, or chemicals. Damaged organic coatings can lead to penetration of corrosive solution between coating and substrate metal. This can lead to serious corrosion that is very difficult to detect before the system fails.

Powder coatings are applied to metal substrates to form highly durable and attractive finishes. They are manufactured and applied without the use of organic solvents. Several methods can apply organic coatings. Some methods require heating the part. In fluid bed coating, the heated part is dipped into a fluidized bed of cold powder. In hot flocking, a cold powder is sprayed on the heated part. In both methods, the temperature of the part must be higher than the melting temperature of the powder. The powder sticks and melts on the substrate with cure by residual heat of the part or by baking in an oven. Heating of the part can be avoided by using an electrostatic fluid bed. The particles in the bed are electrically charged and will be attracted to the part. The powder is subsequently cured. This method is most suitable for simple, uniform geometries. Process components such as pipe, fittings, filter housings, and vessels can be coated with polymer powder. New, high-build powder coatings are replacing rubber and polymeric sheet-lined vessels.

Thermal spraying as a method for coating with polymer powders has had use since the 1950s. The powder is fed into a plasma arc where it melts and is propelled to the substrate. Almost any material can be thermally sprayed provided it does not decompose before melting. As the molten particles reach the surface, they solidify and form a uniform coating layer. Plasma spraying does not require curing, and the size of the coated part is not restricted. The coatings can also be applied as liquids by using conventional spray equipment or dipping. Solvent and aqueous systems have use. Liquid applied coatings require curing after evaporation of the solvent. Thick polymer coatings have use as liners by using adhesives and mechanical fasteners or as loose liners.

CHAPTER 6

Figure 24 shows the schematic coating process for polymer coating. The part is first degreased by solvent cleaning or alkaline cleaning. Sometimes, heating at 300°C–400°C will burn organic contaminants. The part is then blast cleaned to Sa 2 1/2 –Sa 3 by using sufficiently coarse grit to produce a rough surface for good adhesion. Immediately after blast cleaning, the fresh surface requires treatment with a primer. The coating is then applied to the part using a method mentioned above. The coating procedure is repeated to achieve the specified thickness. If curing is necessary, it occurs after each coating layer. Finally, the thickness, adhesion, and porosity are tested.

The interesting properties of polymers are chemical resistance, impact resistance, and abrasion resistance. The operating temperatures depend on the polymer type and vary between 75°C–250°C. Chemical resistance depends on polymer type. Hardness and wear resistance are strongly dependent on polymer type and fillers. The following conditions will effect the suitability of a specific resin laminate:

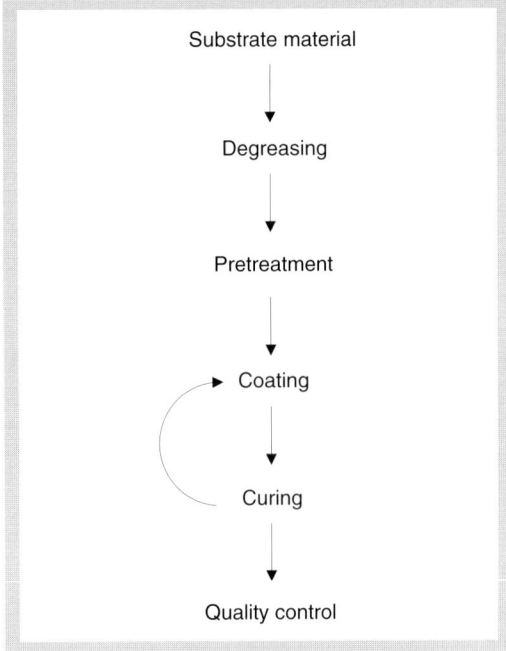

Figure 24. Schematic flow of a polymer coating process.

- Periodic changes in temperature
- Temperature spikes
- Changes in chemical concentrations
- Combinations of chemicals
- Exposure to vapors only
- Exposure to frequent splashes and spills
- Exposure to intermittent splashes and spills
- Frequency of maintenance wash down
- Load bearing or nonload bearing requirements.

The organic coatings can be thermosetting or thermoplastic polymers. The properties of the polymer coating depend on the chemical nature of the polymer and the influence of coating process conditions. Mechanical properties can be improved by using reinforcements. Impermeability can be improved by using inorganic or organic fillers. This is identical to pigments in paints. The resistance of various polymers to processes and chemicals is well documented by manufacturers of chemical-resistant resins such as vinyl ester, polyester, epoxy resins, and thermoplastic polymers. Table 11 gives some approximate guidelines to the chemical resistance of certain thermoplastic polymers. Materials with good resistance are recommended in most cases, materials with limited resistance require testing, and materials with poor resistance are not recommended.

Table 11. Chemical resistance of some thermoplastic polymers.

Compound	Soft PVC	Hard PVC	PE	PB	PP	Fluoropolymers
Dilute acid	Good	Limited	Good	Good	Good	Good
Concentrated acid	Limited	Limited	Limited	Limited	Limited	Good
Oxidizing acid	Poor	Poor	Poor	Poor	Poor	Good
Organic acid	Good	Good	Good	Good	Good	Good
Dilute alkali	Good	Good	Good	Good	Good	Good
Concentrated alkali	Good	Limited	Good	Good	Good	Good
Acid salt solution	Good	Good	Good	Good	Good	Good
Neutral salt solution	Good	Good	Good	Good	Good	Good
Basic salt solution	Good	Good	Good	Good	Good	Good
Oxidizing salt solution	Limited	Limited	Good	Good	Good	Good

Phenolic resin coatings are available in baked and air-dry formulations and are popular for their extremely hard finish. Phenolic coatings possess excellent resistance to moisture, solvents, and a wide variety of concentrated acids at temperatures reaching 65°C with air-dry formulations and to 200°C when baked.

Epoxy resin coatings are normally provided in a catalyzed formulation. Epoxies have excellent chemical resistance to various acids, alkalis, and salts. Most have a temperature limitation of 100°C–150°C. Epoxies are thermoset materials, and films are created by a chemical reaction when resin is mixed with a hardener immediately before application. Epoxies normally serve as barrier coatings. They are typically tough and have excellent abrasion and chemical resistance. Two types of epoxies previously existed: amines and polyamides. Novel resins and curing agents combine the best properties of all the earlier epoxies. Epoxies are preferable for their overall physical properties. Epoxy resin coatings are also suitable for floor coatings. Epoxy-phenolic resin coatings have use primarily for alkali resistance in moderate temperatures up to 150°C. They are baked or catalyzed. Different epoxy types include the following:

- Bisphenol-A is a cost-effective, general-purpose resin that provides excellent alkali resistance, good acid resistance, and fair-to-good solvent resistance.
- Bisphenol-F is a low viscosity material with excellent alkali resistance and improved acid and solvent resistance compared with Bisphenol-A.
- Novolac epoxy is fast curing and offers excellent resistance to strong alkalis, acids, and solvents.

CHAPTER 6

Vinyl ester coatings have excellent adhesion to steel. They are available in baked or air-dry formulations. In corrosion prevention applications, both Bisphenol-A and novolac epoxies have use as base materials for vinyl esters. Vinyl ester coatings possess very good chemical resistance to mixed corrosive environments, but they are not useful in solvent containing environments. They offer satisfactory results for most corrosive fume applications below 70°C. Vinyl ester resins are the most common thermoset materials used in paper mill fiberglass reinforced plastic (FRP) applications followed closely by polyester resins. Vinyl ester coatings and fiber reinforced pipes have had use in white water systems to replace corroding stainless steel.

Polyester resin coatings cure at room temperature after catalysis. They are resistant to mild acid, alkalis, and solvents. Temperature limitations vary according to each specific coating manufacturer's formulation. Polyesters exhibit good outdoor weathering properties. Polyester resins have use for similar but less severe services than vinyl ester resins.

Silicone coatings have use for medium to high temperature service where temperatures seldom fall below 100°C–150°C. These coatings normally exhibit good fume resistance from acids, alkalis, solvents, and salt solutions, but they are not suitable for splash, spillage, or immersion service.

The most effective barrier coatings among the polymers for a variety of corrosive environments are fluorocarbons. Common fluorocarbon coating materials are perfluoroalkoxy (PFA), polytetrafluoroethylene (PTFE), ethylene chlorotrifluoroethylene (ECTFE), fluorinated ethylene propylene (FEP), and polyvinylidene fluoride (PVDF). PTFE is the oldest and most commonly used fluoropolymer. Very few materials will stick to PTFE. Production interruptions for cleaning and maintenance can be fewer by elimination of paper machine rolls accumulating sticking material. A heat shrinkable PTFE resin can provide a surface that resists sticking and protects against corrosion. Thermally sprayed wear resistant coatings such as tungsten carbide can be sealed with PTFE. These coatings have good adhesion combined with wear resistance and a nonadhering surface.

PFA coating is elastic and ductile and will therefore withstand mechanical wear. With its low coefficient of friction, it finds use in vessels, piping, and valves. In paper manufacture, PFA is useful as a coating for rolls, cylinders, and supporting structures. FEP has properties similar to PFA, but its ductility and heat resistance are lower. For applications under mechanical load, PVDF and ECTFE coatings are suitable. The other fluorocarbons are also acceptable, but they creep. ECTFE and PVDF have use for the same applications as PFA except supports. ECTFE powder coatings provide excellent resistance to strong acids and strong bases at elevated temperatures where other plastics cannot be used and where corrosion resistant metals would be too expensive. PVDF is the best selection for abrasive conditions since it has the highest compressive strength among the fluorocarbons. PVDF has use in scrubber blades, agitator blades, fan wheels, and exhaust fan systems.

The rubber lining materials are broadly classified into natural rubbers and synthetic rubbers. In practice, compounds are generally mixtures of synthetic and natural materials. The rubber lining materials are available in thicknesses up to 12–13 mm.

Higher thicknesses result by applying multiple layers. The rubber lining application uses the following steps:

- Pretreatment of the surface to remove contaminants and provide a suitable surface roughness
- Application of adhesive to the cleaned surface
- Application of rubber sheet
- Vulcanizing or curing of the lining using heat
- Inspection of the lining for pinholes using a high voltage spark tester.

Rubber linings that have use in the pulp and paper industry involve wet chlorine gas, hypochlorite and ozone, and flue gas scrubbers. Paper machine rolls can use rubber coatings. Problems with rubber linings often relate to high concentrations of organic contaminants and excessively high temperatures.

4.4 Metallic coatings

Metals and alloys can be applied to almost all other metals using several techniques. The most common methods are hot dipping, spraying, cladding, and electroplating. Metallic coatings give protection by forming a barrier or by sacrificial dissolution. Barrier coatings are more noble and sacrificial or active coatings less noble than the substrate material. A more noble metallic coating must be absolutely nonporous or it will create galvanic couples leading to substrate corrosion on pore areas. With a noble coating, care is necessary in operation to avoid damage. On the other hand, a damaged sacrificial coating corrodes and corrosion products may fill the damaged areas. The corrosion products of an active coating may contaminate a product or spoil appearance. Active coatings are often less expensive than more noble ones.

An active coating has a greater tendency to dissolve than the protected metal. When both are present, the coating becomes the anode protecting the more noble metal. This is similar to galvanic corrosion. Figure 25 shows an example of corrosion protection with sacrificial and noble coatings. The sacrificial zinc coating dissolves giving up electrons and becomes corroded, but the carbon steel remains undamaged. The zinc coating will protect carbon steel until depletion of the zinc. Corrosion products of the dissolving coating

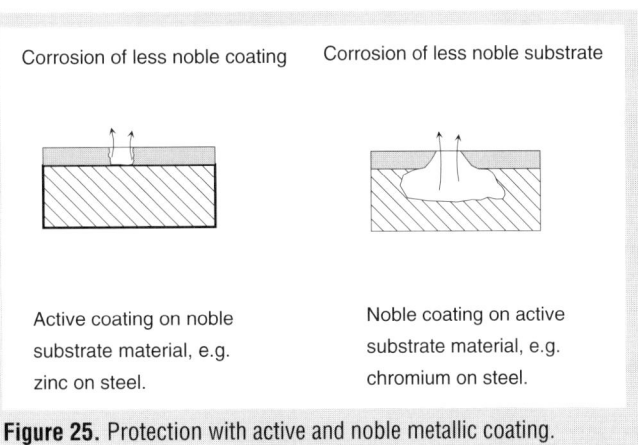

Figure 25. Protection with active and noble metallic coating.

CHAPTER 6

material can precipitate and block damages in the coating to resume the barrier properties. When using a noble coating, a risk of coating defects and heavy localized corrosion always exists. Most common noble coatings in the process industries are nickel and various cladded stainless steels and nickel alloys.

Active coatings are usually produced by hot dipping or thermal spraying. Methods to make noble metal coatings include cladding, overlay welding, thermal spraying, and electrodeposition. When using a noble metal, the coating must be free from pores or galvanic corrosion can result. The coating methods used for noble metals usually have some specific limitations that may cause weak points in the coating. Cladding and overlay welding have welded joints that may corrode. Using proper techniques and materials can avoid problems. Thermally sprayed coatings are inherently porous. Some techniques produce more dense coatings than others. A higher velocity of the coating particles typically gives a more dense coating. Using sealants can improve the corrosion resistance of thermally sprayed coatings. Electrodeposited coatings are usually free from pores, but their use is limited to small components and simple geometries. The most common noble metal coatings in the pulp and paper industry are various stainless steels and nickel-base alloys. These have had use for corrosion resistance in aqueous and high temperature environments.

4.4.1 Hot dipping

Hot dipping is simple and often the cheapest method to make a metallic coating. In hot dipping, the cleaned component is immersed in a molten metal bath. The coating metal must have a lower melting temperature than the substrate metal. The method has frequent use to produce thick coatings when the coated part is not geometrically complex and the uniformity of coating thickness is not a prime consideration. The substrate metal must withstand the temperature of the molten metal bath, and fabricated structures require special attention for venting and draining. The coating metal must wet and alloy with the substrate. Before hot dipping, the parts are first cleaned from rust and mill scale in a caustic cleaning bath and activated in an acid pickling bath. They may then be immersed in a flux to remove oxides and prevent oxide formation before dipping into the molten metal bath. The reaction between the molten metal and the substrate metal will not occur unless the substrate surface is chemically clean. The parts should remain in the bath until they reach the bath temperature. The articles are then slowly withdrawn from the molten metal bath and excess metal is removed. Coatings formed by hot dipping are usually very ductile. Steel is the most important substrate metal. The most common coating is hot-dip galvanizing, but zinc-aluminum alloys, tin, aluminum and their alloys also have use. On a tonnage basis, most hot dip coatings are sacrificial such as zinc and aluminum on steel.

Hot-dip galvanizing is a common method to protect steel from atmospheric corrosion. The use of zinc and galvanizing has a long history. The early patents for hot-dip galvanizing were issued in France and England in 1836 and 1837. This technology was quickly adopted and widely used in the late 1800s. The zinc used for hot dipping must be at least 98.5% pure. The most harmful impurities are iron, lead, copper, and cadmium. The quality of the zinc layer depends on many factors including steel composition, wall thickness, immersion time, and bath temperature. With excessively short immersion times, a thick and poorly adhered layer will form. With long times, a layer of brittle zinc-iron alloy forms. If the silicon content is about 0.05%–0.12%, the zinc layer can become too thick and crack during exposure. The generic composition of a steel for galvanizing that will produce an acceptable coating will be carbon less than 0.25%, phosphorous less than 0.05%, and manganese less than 1.3%. Rough surfaces will usually result in thicker coatings due to the increased surface area, but these coatings will be rough and have a poor appearance.

The structure of the hot-dip galvanized coating is inhomogeneous containing layers with varying iron content. The outermost layer is pure zinc. Iron content increases toward the substrate from less than 0.2% to 21%–28% Fe. Typically, the intermediate zinc-iron layers are harder than the base steel, and the outermost layer is very ductile. As the zinc/iron alloys form, they will grow perpendicular to the steel surface. This causes coating in corners and edges to be thicker than the surrounding coating. This contrasts with other types of protective coatings that are thin at the edges and corners of a material. Hot-dip galvanizing can cause run-offs of solidified zinc that affect the appearance of the structure.

Zinc is an amphoteric metal meaning it will dissolve in acids and alkalis but not in neutral solutions. It is also a very active metal that can corrode in neutral solutions unless protected by a zinc hydroxide layer. The zinc hydroxide layer forms rapidly in the atmosphere and will protect for a long time. When in contact with a sulfur containing atmosphere, it will slowly react to zinc sulfate that in turn will dissolve in water. Corrosion resistance is then lost. Zinc does not corrode at pH 6–12. The rate of zinc depletion is slow when the pH of the electrolyte is 4–13. This range covers many industrial environments. Zinc is much more resistant to hard water than soft water. A protective hydroxide layer will not form in water containing no carbon dioxide. Sodium chloride, sodium sulfate, and calcium sulfate will dissolve the hydroxide layer. Zinc has poor corrosion resistance at temperatures above 42°C in all waters, but corrosion rate decreases above about 55°C. The relative nobility of iron and zinc changes at about 60°C. Hot dip galvanized parts are therefore not useful at elevated temperatures. Dissolved noble metal cations will deposit on zinc by cementation causing dissolution. This may lead to localized corrosion of zinc-coated parts[35]. Figure 26 shows the service life of zinc coatings in different atmospheres. The

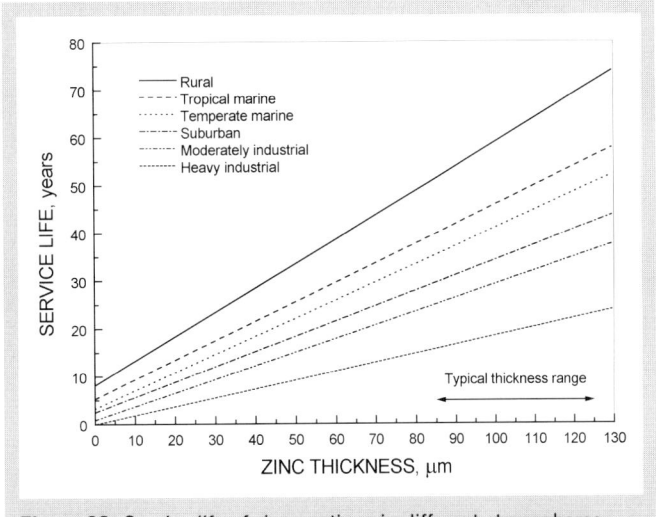

Figure 26. Service life of zinc coatings in different atmospheres.

CHAPTER 6

service life is the point when 5% of the area corrodes. In a moderate atmosphere, a rule of thumb is that 100 µm Zn corresponds to 50 years.

Aluminum coatings made by hot dipping are new. Aluminum baths for hot dipping usually contain about 10% silicon to retard growth of a brittle Al-Fe intermetallic layer. Another type of aluminum bath contains over 97% Al. Aluminum coatings have a layered structure with pure aluminum outermost and iron content increasing toward the steel substrate. Aluminum is also an amphoteric metal. Compared with zinc, it is stable at a much narrower pH range of 5.5–8. In the atmosphere, an aluminum hydroxide layer will form. Since this layer will also withstand slightly acid conditions, aluminum coatings are better in industrial atmospheres than zinc coatings. Aluminum will withstand many dilute acid solutions. Aluminum coatings may show pitting in chloride containing solutions particularly in crevices or stagnant solution areas. In soft water, aluminum coatings may be more noble than steel.

Most sacrificial coatings can also use thermal spraying methods. The metals and their application systems vary, but in most applications thin coatings are applied to surfaces to improve corrosion or abrasion resistance. The deposits are porous, but this is not a problem when using a sealant or paint coating. Flame spray and arc spray methods are the most common methods. Thermally sprayed metal coatings are excellent corrosion resistant undercoatings for paint to replace primers. The porosity of a sprayed coating may be a benefit over a hot dip coating because it gives better adhesion for the paint film. Paint films on iron and steel usually fail due to corrosion under the paint resulting in lifting of the coatings. Thin zinc and aluminum coatings prevent corrosion of the substrate and offer better bond to the organic coating. Aluminum and zinc are usually suitable for protection of iron and steel in atmospheric conditions and also in natural water immersion. Zinc is usually 99.9% pure and is not contaminated in the spraying process. Sprayed zinc coatings are purer than those applied by hot-dip galvanizing. Sprayed zinc has very poor resistance to almost all organic or inorganic acids. Zinc spraying may be an alternative to zinc paints. Aluminum is usually 99.0% pure, but an alloy with 5% magnesium can also be specified. Sprayed aluminum may be effective against corrosion under wet insulation. Very dilute nitric or sulfuric acid solutions and many organic acids have little effect on sprayed aluminum coatings treated with a suitable sealer.

4.4.2 Cladding, lining, and welding

Cladding means coating one or both sides of a less corrosion resistant structural material with a thin sheet of more resistant metal. A typical combination is steel clad with stainless steel or nickel alloy. This is a common corrosion control technique in the process industries. The base material satisfies the strength requirements, and the clad layer gives the required corrosion resistance. The metals bond together by hot rolling or explosive forming. Both methods create a metallurgical bond. Cladding is a cost-effective alternative for simple geometries. It has primary use in plate constructions such as storage tanks, reactors, and pressure vessels. The cladding thickness varies between 5%–50% of the total thickness of the composite plate. It is usually 10%–20%. Corrosion problems relating to cladding are diffusion of unwanted metals to the clad metal during bonding or welding and metallurgical changes due to stress relieving heat treatments.

Corrosion prevention

Welding of clad metals requires special procedures to ensure mechanical properties and corrosion resistance. The cladding is stripped for some distance around the joint. First, the carbon steel part is welded. The cladding is then repaired with suitable, more alloyed electrodes.

Figure 27 shows two methods for producing a welded joint for stainless steel clad plate. The first method is the most common and is the most economical. The plate edges are first beveled for welding. The bevel must end 1/16 in. (1.6 mm) minimum above the cladding. The carbon or low-alloy steel weld is made first. This weld should not penetrate closer than 1.6 mm to the cladding. Using a low hydrogen method for initial passes is a good practice. The joint is then gouged from the clad side. Material removal must be at a minimum, but the sound carbon steel weld metal must be reached. Stainless steel is deposited in a minimum of two layers. At least the first layer must be a more alloyed grade than the clad material. In severely corroding environments, an additional stainless steel strip can be welded on top of the joint. The second method in Figure 27 is more expensive since more material removal is necessary. It allows use of conventional welding methods since the cladding does not contaminate the carbon steel weld. The first step includes beveling and removal of clad layer. The strip must extend to 3/8 in. (4.8 mm) minimum from the joint. Carbon steel is deposited with any conventional method. The root is then ground flush with the carbon steel plate. The area where cladding was removed is overlay welded with at least two layers of weld metal. Again, the first layer must be made with a sufficiently high alloy grade to prevent dilution. An alternative to overlay welding is to use a strip of wrought stainless steel welded in place[36].

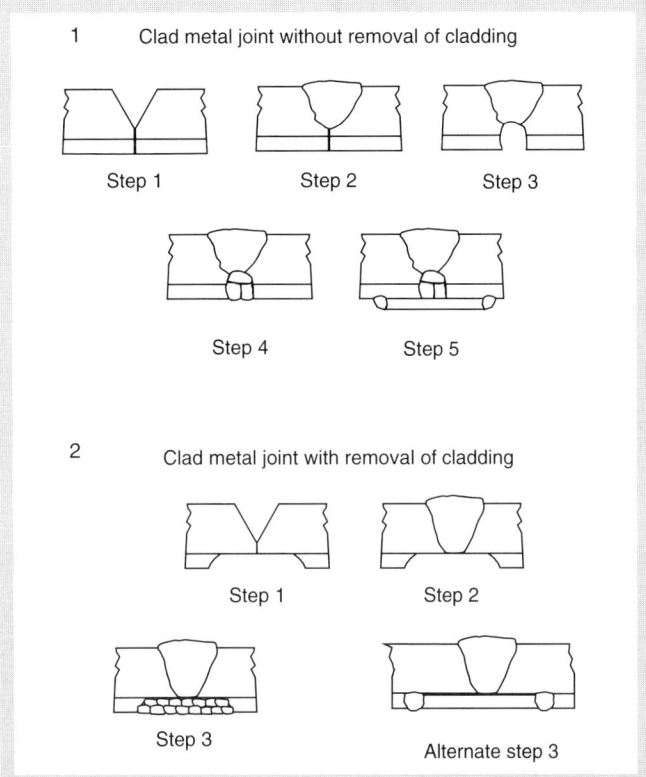

Figure 27. Welding of clad metals[36]. In method 1, the steps are beveling, welding of carbon steel, back gouging, welding of stainless steel, and optional welding of stainless steel strip for severely corroding environments. In method 2, the steps are beveling, removal of clad layer, welding of carbon steel, grinding of root, and overlay welding or welding of a stainless steel strip to repair the cladding.

Wallpaper lining or sheet lining is another method to coat with a more corrosion resistant metal. This has been available since the late 1920s and had first use in the chemical industry[37]. Wallpapering, strip lining, and plug lining are different terms for this method. Figure 28 shows the principle of wallpaper lining. Thin sheets of the more noble metal are welded so that their edges overlap. Wallpaper lining has had success in continuous digester or flue gas desulfurization systems. The lining should use controlled conditions with new structures. If the structure has had exposure to corrosive liquors and the lining will occur during a shutdown period, thorough cleaning is necessary, and welding can be difficult.

Common materials used in wallpaper lining are stainless steel Types AISI 304L and 316L. Because welding must use field conditions, heat treatment can be impossible. Selection of low carbon grades therefore avoids sensitization.

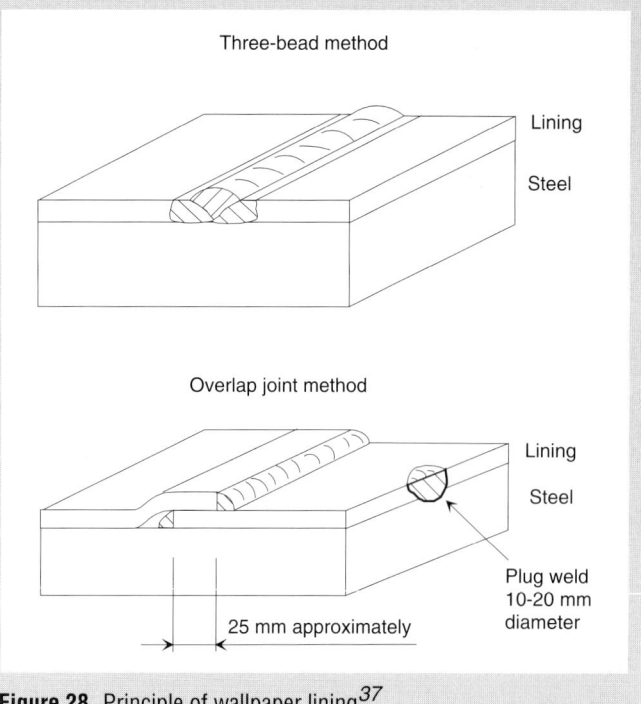

Figure 28. Principle of wallpaper lining[37].

Overlay welding is also a common method for application of a corrosion resistant coating. Typical application is a stainless steel on carbon steel sheet or cast iron. Weld overlaying by placing a continuous layer of overlapping weld beads is also a type of cladding. When choosing the welding method, penetration into the base metal must be minimum. This reduces the dilution of weld metal by base metal and keeps the corrosion resistance high. The first electrode applied should have a high alloy content to permit a considerable dilution. Avoid a hard martensitic structure. Lower alloy electrodes have use in depositing over the first layer. Typical weld electrodes for the first layer are AISI 309, AISI 309-Mo, and AISI 312. Type 309 contains 22%–24% Cr, 12%–15% Ni, and 2% Mo. Type 312 weld electrode contains 28%–32% Cr and 8%–10.5% Ni. The second layer uses lower alloyed grades. AISI types 308, 308L, and 347 have common use. Alloy AISI 308 has a nominal composition of 19%–21% Cr and 10%–12% Ni, and alloy AISI 347 has 17%–19% Cr and 9%–13% Ni with stabilization by niobium and tantalum[36].

4.4.3 Thermal spraying

In many applications, corrosion or wear resistant alloys are applied on the surface using some thermal spraying method. "Thermal spray" is a general term for an entire group of processes for depositing metallic and nonmetallic coatings. The main thermal spraying

Corrosion prevention

processes are flame, arc wire, plasma, and high velocity oxy-fuel (HVOF). In these methods, the feed material is a wire or powder. The material melts using combusting gases, electric arc, or plasma and is blown to the surface. The coating that forms is therefore composed of metal droplets that deform and solidify on the surface as Fig. 29 shows. The temperature of the substrate does not increase significantly during the spraying process. Metallizing is another term for thermally sprayed metal coatings. It usually refers to flame or arc wire spraying. Fused alloys are usually hard-facing, self-fluxing alloys with a high chromium-nickel or cobalt base and hard carbides. These coatings fuse after their application, and they require heating of the coating and substrate to over 1 000°C. Fused coatings are essentially nonporous due to the fusing procedure. Unfused coatings are not heated after spraying, and the coating is used in the as-sprayed condition but usually machined or ground. These coatings are always porous and have a lamellar structure. Unless the coating is sealed, it does not protect from corrosion.

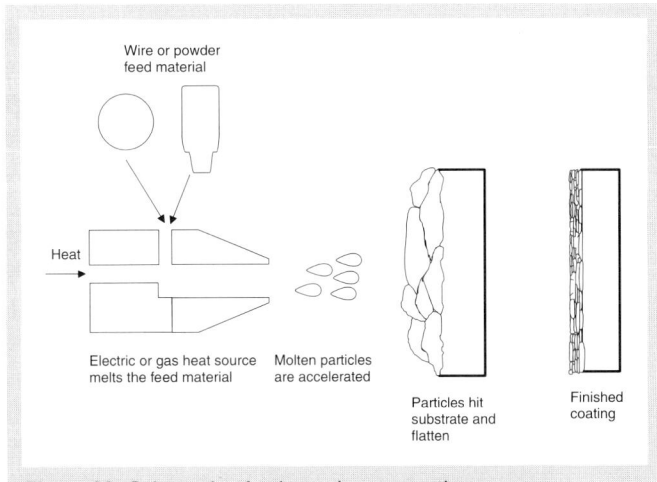

Figure 29. Schematic of a thermal spray coating process.

In common thermal spraying methods, the molten coating particles react with the atmosphere before they reach the workpiece. This leads to inclusions of various compounds having an adverse effect on coating properties. Figure 30 shows the structure of a thermal spray coating with various kinds of defects. The coating has an inhomogeneous structure, and it contains oxides and other impurities unless using a special coating chamber and equipment. Coating under controlled atmosphere or vacuum has use for small workpieces. The resulting metallic coatings are free of oxides and other contamination, and ceramics sprayed using inert atmospheres are very pure.

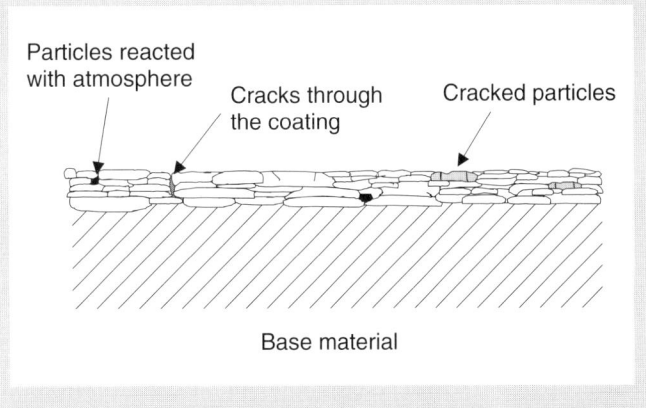

Figure 30. Structure of a thermal spray coating with various kinds of defects.

CHAPTER 6

Flame spraying uses wire or powder as the coating material. Wire flame spray was the first thermal spray process developed. It is still useful in certain applications today. The spray material in wire form feeds continuously into a fuel gas-oxygen flame where it melts. Compressed air surrounds the flame and atomizes the molten wire. The spray of molten particles is accelerated and blown toward the workpiece. Suitable fuel gases are acetylene, propane, and hydrogen. The powder flame spray process is similar to the wire process except it uses powdered feed materials. This offers a wider range of coating materials. The spray material is fed continually into a fuel gas-oxygen flame where it melts. A carrier gas transports the powder into the flame, and the mixed gases transport the material toward the workpiece. In flame spraying of wire or powder, the temperature range can reach 3 000°C. Particle velocities impact at 150–200 m/s for wire and 100–150 m/s for powder. The wire flame spray method is suitable for large structures and high production rates. The simple feeding system provides good atomization. The choice of coating materials is limited to alloys in wire form. Powder flame spray offers a wider range of materials including ceramics and cements. Both systems have use to make aqueous, atmospheric, and high temperature corrosion resistant coatings and to salvage worn or damaged parts. Powder coatings with hard ceramic particles can also provide wear resistance. Both systems are portable and therefore suitable for field operations.

Arc wire spray uses two metallic wires as feed material. The wires are fed into the arc gun at controlled speed. At the nozzle of the gun, an electric arc creates sufficient heat to melt the wires continuously. Compressed air atomizes the molten material and blows it onto the workpiece. The arc wire spray systems do not require a combustible gas supply. The temperature ranges up to 5 000°C, and particle velocities at impact are 150–200 m/s. Arc wire spray has the same applications as flame wire spraying. The equipment is also portable. The arc spray is a newer method, and it usually gives faster output and better adhesion. Arc and flame spray both have use for fabricated structures. Access to difficult areas favors flame spraying, and large areas favor arc spraying.

Plasma spraying is a process by which an inert gas is ionized by an electric arc to produce a hot gas stream with temperatures over 16 000°C and particle velocities at impact of 300–600 m/s. Injecting the coating material into the gas melts and propels it toward the target. By varying the process parameters, powders of most metals and their alloys, tungsten and chromium carbides, ceramics, and other materials can be successfully applied. Plasma spray is perhaps the most flexible of all the thermal spray processes since it can develop sufficient energy to melt any material and uses powder feed. Typical plasma gases are argon and nitrogen with hydrogen and helium as additional gases. Gas mixtures and applied current to the electrode control the amount of energy produced by the plasma system. The distance of the plasma gun from the target, gun and workpiece relative speeds to each other, and workpiece cooling keep the workpiece temperature usually below 250°C. The resulting coatings are dense. Coating production is predictable and repeatable. Typical uses are repair of worn parts, wear resistance, corrosion resistance, and thermal or electrical insulation.

Detonation coating and high velocity oxy-fuel coating have similar backgrounds. Detonation coating uses a mixture of combustible and inert gases such as oxygen, propane, nitrogen, and argon and a powder coating material fed to a tubular gun. The gas mixture is then ignited to give an explosion that melts and propels the coating material. The cycle can repeat a few times per second. The method is very noisy and requires special housing or separate buildings. High velocity oxy-fuel (HVOF) spraying involves injecting powder materials into a high velocity jet stream of oxygen and hydrogen with a flame temperature of about 2 700°C. The powdered coating material is fed through the gun using nitrogen as a carrier gas. This technique uses special torch designs in which a compressed flame undergoes free expansion upon exiting the torch and gas speed increases to supersonic values. The high thermal and kinetic energies of these methods produce dense coatings suitable for the most demanding applications. The coatings have high bond strengths due to the high speed of the molten particles. Residual internal stresses are usually low, and coatings can be sprayed to higher thicknesses than with other methods. As a result of the high kinetic energy, the coating material generally does not require full melting. Instead, only the particle surfaces are molten, and they deform as they hit the workpiece. Typical applications are repair of worn parts, coatings for wear resistance, corrosion resistance, and thermal or electrical insulation. Detonation spraying is difficult to control making HVOF preferable.

The first step in reducing abrasion, erosion, fretting, adhesion, cavitation, or oxidation/corrosion is defining the type of expected wear or type of resulted damage. Distinct patterns, substrate material properties, and design aspects of the engineered part allow determining the cause or causes of damage. Analysis of the component damage assists in determining the most appropriate coating material and process for repair. Coating materials include metals, ceramics, carbides, composites, and plastics used individually or in combination. Some overlays for wear and corrosion resistance are the following:

- Cobalt-tungsten carbide has outstanding abrasion resistance up to 260°C. Typical applications are rolls, shafts, and mixing equipment.
- Nickel-chromium-silicon-boron is a spray and fused coating having outstanding galling resistance and excellent corrosion resistance. Typical applications are pump sleeves, wear rings, and valve trim.
- Nickel-chromium-iron produces machinable "stainless steel" coatings that are useful for salvage and build-up applications on ferrous or nickel alloy substrates where high hardness is not necessary.
- Nickel-iron-chromium-silicon coatings are heat and oxidation resistant.
- Chromium oxide has excellent resistance to sliding wear. It is chemically inert. Typical applications are shafts and compressor and pump parts.
- Chromium oxide with SiO_2 and TiO_2 has high wear and corrosion resistance. It resists mechanical shock better than other ceramics and has low friction.

CHAPTER 6

- Cobalt-chromium-tungsten-carbon has good impact strength and corrosion resistance. It has extensive use in the valve industry and in many pump and chemical applications.

- Iron-molybdenum-carbon with high molybdenum is suitable as an alternative to hard chrome plating and used to protect against abrasive wears, wear from hard bearing surfaces, and fretting. It has a low coefficient of friction.

- Iron-aluminum-molybdenum-carbon has use for salvage and build-up of ferrous base substrates.

Thermal spraying with stainless steels is a common method for repair of various equipment. Restoration of Yankee dryer cylinders with martensitic stainless steel and press rolls with austenitic Mn-containing grades has been done. The part is first coated and then machined to restore dimensions and surface finish. Thermal spraying is used for protection of the recovery boiler water wall using metallic or ceramic coatings. Metal coatings are iron or nickel-based. Hastelloy C-276 has also had use as a coating material[18].

4.4.4 Electroplating

Electroplating means depositing a metal or alloy in an adherent form on an object serving as a cathode. The metal is deposited by passing an electrical current into an electrochemical cell. The processes in the cell are exactly opposite to those in corrosion. Electroplating is done for corrosion resistance, friction reduction, heat tolerance, and decoration. Electroplating is a common method for depositing metals and alloys as thin coatings for small parts. Almost every metal can be deposited with some method from an aqueous or non-aqueous bath. Electroplated coatings must be applied to conductive substrates. Polishing, pre-treatment, and post-treatments are often more critical than the electroplating step itself. Electroplated coatings require metallurgical bonds. The surface metal must also be free of oils, grease, wax, rust, tarnish, and passive films before coating. Table 12 shows some common electroplated metals, finishes, and their uses.

Nickel is an important electroplated metal having three primary types of solutions: bright nickel, Watts nickel, and sulfamate nickel also categorized as bright, semi-bright, and high-sulfur strike. Bright nickel is lustrous with a yellowish-brown color to the metal. Semi-bright has a grey haze. Bright nickel and Watts nickel (nickel deposited from Watts bath) offer corrosion protection. Watts nickel has matte to semi-bright appearance, and the deposits have high ductility. High sulfur nickel is also grey and hazy with matte to semi-bright appearance. It is the most ductile of all nickels and also provides corrosion protection.

Chromium deposits in decorative applications are applied as thin coatings over thicker nickel surfaces. Hard chromium is deposited from hexavalent chromium baths to achieve corrosion protection and wear resistance properties. Chromium provides bright surfaces. Hard chromium is usually deposited on steel, copper, or any alloy of these metals. Hardness, corrosion resistance, and wear resistance make chromium an outstanding surfacing material for cylinders and pins.

Corrosion prevention

Table 12. Electroplated metallic coatings and their uses.

Coating	For use on	Corrosion resistance	Characteristics and uses
Cadmium	Most metals	Excellent	Bright silver-gray, dull gray, or black finish. For decoration and corrosion protection and especially marine applications.
Chromium	Most metals	Good, improves with increased copper and nickel undercoats	Bright blue-white, lustrous finish with hard surface. Used for decorative purposes, corrosion, and wear resistance.
Copper	Most metals	Fair	Electroplated finish. Used for nickel and chromium-plate undercoat. Can be treated to obtain various finishes.
Bright nickel	Steel, die-cast zinc, copper	Indoor excellent Outdoor good	Silver finish used for appliances, hardware, etc.
Watts nickel	Steel, die-cast zinc, copper	Same as bright nickel	Dull finish if no brighteners in the bath.
Electroplated tin	All metals	Excellent	Silver-gray color. Excellent corrosion protection for parts in contact with food.
Electroplated zinc	All metals	Very good	Bright-blue-white gray coating. For corrosion protection of steel parts.
Black chromate finish	Zinc plated steel	Added corrosion protection	Black, semilustrous. Used for outdoor purposes.
Clear chromate finish	Zinc plated parts	Very good to excellent	Clear bright or iridescent chemical conversion coating for added corrosion protection, coloring, and paint bonding.
Olive drab, gold, or bronze chromate finish	Zinc plated parts	Very good to excellent	Green, gold, or bronze tones same as clear chromate. Colored coatings usually have greater corrosion resistance than clear.
Dichromate finish	Zinc plated parts	Very good to excellent	Yellow, brown, green, or iridescent colored coating. Same as clear chromate.

Zinc is a common electroplated metal in industry because it gives corrosion protection for steel by sacrificial action. Even higher corrosion resistance is possible through the use of chromates. Zinc and zinc-alloy electroplates require application of chromate conversion coatings to improve corrosion resistance. These treatments delay the formation of white corrosion products that form as zinc begins to oxidize. A chromate is a gel-like coating of a hexavalent chromium complex. Different color chromates have differing corrosion resistance: blue bright, yellow, bronze, green, and black in the order of increasing resistance. In applying zinc electroplates, finishers say the yellow chromate is the most common followed by black, green, and clear coatings. While alloy plates can be chromated, the appearance may be different from what would be expected in chromating pure zinc plates. This does not mean it is less protective. It is simply different in color. Chromated zinc parts have wide use. Rust does not form until the zinc has penetrated to the base metal. This is a function of the thickness of the zinc plate and its ability to provide sacrificial protection of the steel.

Brush plating or selective plating is a portable electrodeposition process. The process uses high currents providing a very dense deposit. The equipment consists of a lightweight dc power pack with a flexible lead to the workpiece and another to an anode handle. Anodes of suitable size and shape attach to the handle and have a wrapping with an absorbent material such as cotton. With the anode and handle positive and the workpiece negative, touching the workpiece with the covered anode causes current to reduce metal electrolytically from the solution in the wrapping and deposit the metal onto the base metal. Brush plating has use for repair or resizing and mechanical or corrosion resistant properties. The equipment used in brush plating can be taken into the work area to avoid disassembly and reassembly. This can reduce downtime. Brush plating offers some benefits over other metal repair processes. The system is portable, easy to operate, can coat to size, and has excellent adhesion of dense coatings.

Immersion plating means depositing a metallic coating on a metal immersed in a liquid solution without an external electric current. Other terms are dip plating or electroless plating. The metal ions in a dilute aqueous solution are plated onto a substrate using an autocatalytic chemical reduction. The electroless process is suitable for nonmetallic parts or metallic substrates when a uniform coating is necessary on irregularly shaped articles. A typical example is electroless nickel. It is really not nickel but a nickel-phosphorous alloy with several unique advantages. The deposition process does not depend on current distribution as in electroplating. The coating can therefore be applied to inside diameters and even blind holes. The coating is also extremely uniform in thickness and can restore undersized threads without requiring subsequent grinding. Heat treating can provide high hardness. Electroless nickel deposits are usually harder, more brittle, and more uniform than electroplated deposits.

4.5 Inorganic coatings

Inorganic coatings are typically ceramics, but enamels, brick and tile linings, and conversion coatings can also fall in this group. Ceramic coatings are often corrosion resistant at low and high temperatures, but they have use primarily to improve wear resistance. The pulp and paper industry uses high amounts of advanced ceramic materials with excellent wear and chemical resistance. Ceramics are prone to cracking when subjected to thermal shock conditions. The machining and grinding costs for ceramics are also considerable to obtain the precise tolerances and smooth surface finishes required for paper machine components. Ceramic brick linings with resistant mortars and enamels are often very competitive with metals.

4.5.1 Ceramic coatings and linings

Ceramics are inorganic nonmetallic materials. Typical ceramics are hard, high-temperature resistant materials that do not deform plastically. Typical properties are good corrosion resistance, low electrical conductivity, and low thermal conductivity. Thermal spraying methods often apply ceramic coatings. The coating thickness may vary from

0.1 mm to several millimeters depending on the coating material and its actual function. Thermally sprayed coatings improve corrosion and wear resistance or repair worn parts. Parts coated with ceramics are seal rings, pump sleeves, shafts, and other small parts that require corrosion protection. Chromium oxide is a common ceramic coating. Since thermally sprayed ceramic coatings are porous and often more noble than the base material, they require impregnation with a sealing compound.

Enamel coatings are made on a treated substrate by addition of a paste that is a melt at high temperatures. The glassy coating results as the part cools. The enamels consist primarily of inorganic oxides such as silica, feldspar, borax, and titanium oxide. The primary glass-forming component is silica, SiO_2. Increasing the amount of silica or alkali oxides tailors the enamels to acid or alkaline conditions. Addition of boron and alumina increases the chemical resistance. Enamel coatings use two different layers. A primer enamel provides adhesion and corrosion protection, and a covering enamel protects the primer from mechanical damage. Enamels are glasses. They dry hard and wear resistant but may crack by impact or bending. The thermal expansion coefficient of the enamel must match that of the substrate material, and enamel coatings should preferably be under compressive stresses after cooling. Enamels are resistant to acids and alkalis, and they can be used in high temperature applications up to 1 000°C. They are also very resistant to thermal shocks. Enamel coatings have common use on cast iron and steel. The glossy surface of an enamel coating is easy to clean. Valves subjected to a very corrosive environment such as in a chlorine dioxide plant have been coated with enamel[38,39].

Chemical-resistant masonry includes brick and tile installed with an appropriate mortar or thick film cementitious systems. Brick and tile linings are among the most resistant protection methods in many large constructions. Brick liners with organic membranes have use against prolonged or continuous chemical exposure. Often the exposure combines with high temperature and abrasive service. Bricks have good resistance, but they are permeable. Eventually, chemicals will penetrate the brick. A properly selected organic membrane will provide a barrier to prevent chemicals from attacking the substrate material. Tile has frequent use on surfaces subject to intermittent or short-term chemical exposure combined with abrasion or wear. Organic membranes do not have common use with tile in this service.

The bricks consist of silica and alumina with oxides of iron, calcium, and magnesium. Brick made from Portland cement and alkali resistant aggregates has use in alkaline conditions. Graphite bricks are useful in acid and alkaline environments. The mortars used for joining include synthetic resins, Portland cement, and inorganic silicates. Synthetic furan resins are resistant to nonoxidizing acids and alkalis. Polyester resin mortars are resistant to oxidizing media found in a bleach plant with the exemption of hot hypochlorite solutions. Epoxy resins resist nonoxidizing alkalis and dilute acids. Epoxies generally have very good adhesion. Phenolic resin mortars resist nonoxidizing acids but will not withstand alkalis. Some modified phenolic resins are resistant to acids and alkalis. Mortars using Portland cement with sand or silica as aggregates are suitable for less severe alkaline solutions and in slightly acidic solutions. Grades with alkali resistant aggregates have use in more aggressive alkaline media under oxidizing and nonoxidizing conditions. Sodium or potassium silicate based mortars are very suitable in strong acids. Hydrofluoric acid is an exception. Silicate based mortars are not suitable for alkalis. When using chemical resistant masonry, a membrane is almost always used behind the masonry as a barrier. The membrane material can be fiberglass reinforced plastic or an elastomer such as polyurethane, rubber latex, or rubber sheet[40,41]. A brick lining often protects floors and foundations. Monolithic flooring systems use various types of synthetic resin and mineral fillers. Corrosion resistance of the monolithic system depends on the resin. Fillers improve mechanical resistance.

Brick and membrane linings find use in couch pits, bleach tanks, causticizers, and all pulp vats and storage vessels. Refractory applications include lime sludge kilns, boilers, and burners. Acid brick linings have use in the sulfite pulping industry in acid storage units, digesters, and tanks. Acid brick with polyester resin mortar is suitable for chlorine dioxide service. Ceramic tiles joined with Portland cement are useful in hypochlorite and peroxide bleach and in caustic extraction towers. Alkali resistant Portland cement bricks and mortar find use in many places in kraft process chemical recovery[40].

4.5.2 Conversion coatings

Conversion coatings are inorganic compounds formed by controlled dissolution of the metal. Most conversion coating processes use immersion. The dissolved metal atoms convert to solid phosphates, chromates, oxides, etc., forming a protective layer on the metal surface. Typical phosphate and chromate applications prepare metal for painting by modifying the metal with another inorganic coating. Conversion coatings can improve corrosion resistance, act as an adhesion layer for paint coatings, improve abrasion resistance, and improve appearance. Conventional conversion coatings applied to steel and zinc-coated steels inhibit corrosion and provide an effective bond with paints, lacquers, and other topcoats. Phosphate coatings are useful for ferrous and nonferrous metals. The most popular are iron, zinc, and manganese phosphatizing. Immersing a metal in an acid bath containing phosphate salts causes corrosion, but the resulting pH increase in the metal surface causes phosphate precipitation. The resulting films are about 5–15 μm thick, fine or coarsely crystalline, and have about 0.5% open pores. The phosphates are generally soluble in acids but not in neutral or alkaline solutions. Phosphate coatings themselves are not so resistant that their use alone would be sensible. They are almost always used as base layers for paint film or to absorb corrosion-preventive oils and waxes. Phosphate coatings can also protect complex parts since the coating forms by a chemical reaction. Iron phosphate treatment is sometimes applied to rusted surfaces. This combines the cleaning and phosphatizing steps.

Chromate coatings are usually applied to various nonferrous metals. Iron and steel treated with phosphate can also be treated to improve corrosion resistance. The chromate coatings have primary use as a paint base or final coating. Passivation treatments for zinc or zinc coated parts are often chromate conversion processes. Chromate films are usually thin (less than 1 μm thick). The coating forms in an acid solution containing hexavalent chromium ions. The metal corrodes, and chromium ions are partially reduced to form a chromate coating. Chromates are generally more resistant than phosphates. They are also less porous than phosphates or oxides and have use as sealants to them.

Hot alkali solutions, thermal oxidation, or anodizing applies oxide coatings to various metals. Many oxide coatings are not sufficiently resistant themselves but act as a layer for impregnation of oil, waxes, or lacquer. A common oxide coating is black oxide on steel. The process is blackening. The coating and coating method is also called black oxide or caustic black. The unalloyed steel is immersed in oxidizing concentrated alkaline solution at 135°C–145°C for 5–20 min. The resulting oxide coating is 1–2 μm thick and consists of iron (II) and iron (III) oxides. The most common oxide coating is the anodized layer on aluminum. In this case, aluminum is dissolved in an electrolytic cell. The surface layer consists of hydrated aluminum oxide. The natural oxide layer with a thickness of about 10 nm increases to a maximum of about 60–80 μm. The anodized layer on aluminum consists of a dense and thin barrier layer and a thick and porous layer. The thickness of the barrier layer is not much more than that of the natural oxide film. Titanium alloys can also be anodized. The film on a freshly pickled tube is a few nanometers. This natural oxide film grows in thickness over time and in a neutral or oxidizing medium generally results in increasing corrosion resistance of the metal. Any surface treatment that will thicken the oxide film will improve the corrosion resistance of the metal. During anodizing, the TiO_2 film formed is the anatase type. The custom has been to specify anodizing for titanium tubes operating in a severe environment especially where hydrogen uptake was a concern. A thermal oxide treatment will produce a rutile type film of TiO_2 that can be more resistant than the anodized film. The anodized films can usually be removed by acid pickle, but thermally produced films generally require sandblasting or processing in a caustic descaling bath for removal.

5 Electrochemical protection

Electrochemical protection includes anodic protection and cathodic protection. Anodic protection is maintaining the passivity with an external current source. The protected structure is deliberately polarized to the anodic direction. Cathodic protection is production of surplus anodic current by sacrificial anodes or external current source to satisfy the needs of cathodic reaction. This results in a decrease in corrosion current density and potential shift to the cathodic direction. Cathodic protection is usually used with coatings. The coating protects the structure from the corrosive environment, and cathodic protection ensures that faults in the coating have electrochemical protection. Anodic protection and cathodic protection with external current source require much equipment: power source, auxiliary electrodes, reference electrodes, cables, and control system. Table 13 compares the main properties of anodic and cathodic protection.

CHAPTER 6

Table 13. Comparison of anodic and cathodic protection[33].

	Anodic protection	**Cathodic protection**
Applicability to metals	Passivating metals only	All metals
Applicability to corrodents	Weak to aggressive	Weak to moderate
Relative installation cost	High	Low
Relative operation cost	Very low	Medium to high
Throwing power	High	Low
Significance of applied current	Often a direct measure of protected corrosion rate	Complex: does not indicate corrosion rate
Operating conditions	Can be accurately and rapidly determined by electrochemical measurements	Must usually be determined by empirical testing

When using any electrochemical protection method, the protected structures must have proper isolation from other structures and equipment. Stray current corrosion can cause rapid failure if protection systems are in electric contact with nonprotected equipment. Insulating gaskets must be used, and cables and connections must be secured.

5.1 Anodic protection

Anodic protection is a method to assist passivation of a metal. It is usually applied to steel or stainless steel. Protection by anodic polarization was considered impossible until the beginning of the 1960s. Despite suppression of microscopic corrosion cells, the thinking was that dissolution of metal would occur due to applied current. For application of anodic protection, the metal should show borderline passivity. In this case, passivation is theoretically possible but will not happen under normal operating conditions because the corrosion rate is lower than the critical current density required for passivation. During anodic protection, the metal is polarized to the anodic direction, and the resulting higher dissolution rate enables passivation. Anodic protection is a suitable protection method only if the metal has a broad passive potential range and low passive current density in

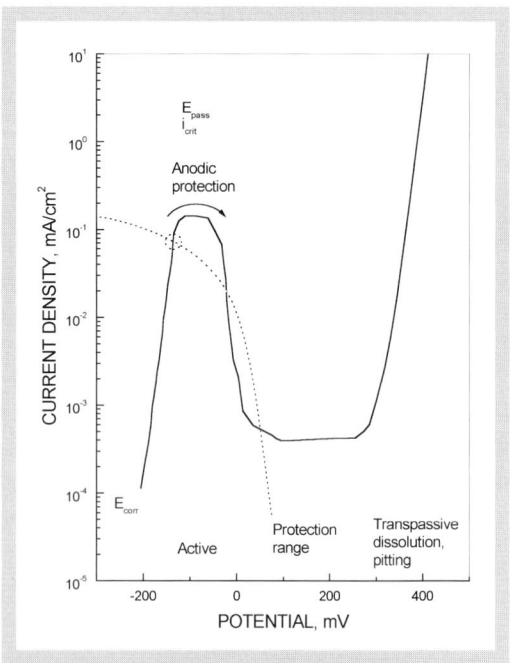

Figure 31. Electrochemical principle of anodic protection.

the process solution. The environment should not cause localized corrosion, and anodic protection is therefore usually not useful in halide solutions. In strong solutions of halogen acids such as hydrochloric acid, anodic protection is clearly an unsuitable method since the metals may not show any active to passive transition. Polarization to higher potentials would only increase general dissolution rate.

Figure 31 shows the electrochemical principle of anodic protection. A spontaneous passivation does not happen because of the excessively low cathodic reaction rate compared with the critical current density of passivation. Passivation results by increasing the dissolution rate with an external current source. When the dissolution rate exceeds the passivation current density, passivation will begin and spread from the initial point that is usually the area of the anode connection. Care is necessary not to polarize the structure to excessively high potentials since this can initiate pitting corrosion or transpassive dissolution in extreme cases.

Suitable combinations of metal and environment are steel in sulfuric acid, stainless steel in phosphoric acid, and steel in caustic solutions as in kraft digesters and liquor tanks. The method has some limitations. The electrical equipment is expensive to install and maintain. If protection is temporarily lost, corrosion rate may be very high since the corrosion potential shifts to the active area. Difficulties in maintaining a constant potential throughout the entire structure and increased corrosion with excessively low or high potentials are the principal problems of anodic protection. Variations of minor component concentrations may change the polarization behavior causing localized corrosion or preventing passive film formation.

Design of an anodic protection system begins by determining the polarization behavior, passive potential range, and current requirements for the specified metal in an actual process solution or representative simulated solution. Determination of the actual metal temperature is essential when protecting heat transfer equipment. The optimum protection potential corresponds to the minimum current density on the passive range. The current necessary to achieve and maintain protection depends on the critical current density for passivation and passive current density. These current values will vary in the field depending on solution parameters. Current values determined by a laboratory experiment may be too high. They therefore give some safety factor but may cause excessive equipment and installation costs. If the system has a high corrosion rate, passivation should occur in a short time. Long passivation times should be allowed only in systems where passive film is infrequently damaged and unprotected corrosion rate is low.

Design of power supply capacity uses the current required to passivate the equipment rather than maintain the passive state. To start anodic protection, the metal must be polarized from the active range over its passivation potential where the current density has its maximum value, i_{crit}. The critical current density, protected area, and solution conductivity determine the required output current and voltage of the power source. The output current must be sufficiently high to passivate an area around the anode cable connection that is sufficiently large for a stable passive film to spread further. Passivation of a large system containing a very corrosive solution may require such large power units that anodic protection is not feasible. In this case, the initial current requirement can be decreased by filling the system first with diluted, less corrosive, or lower temperature solution for passivation or by partial filling of a vessel and raising solution level as the system passivates. The deliberate dissolving of metal to form a passive film is always a compromise between equipment cost and metal waste since rapid polarization would require larger and more expensive hardware but decrease the metal loss. Since the anodic film needs time to develop on the entire surface, frequent localized film breakdowns or variations in the solution corrosivity may preclude the application of anodic protection[42].

The anodic protection requires cathodes for current transfer and reference electrodes for protection monitoring and control. The cathodes should be stable in the environment and inert or able to withstand cathodic polarization. The power requirement for the protection system depends heavily on the cathode area. A higher cathode area equates to lower cathode resistance and system cell voltage. Using cathodes that are physically as large as possible is therefore economically feasible. For the cathode materials, platinum cladding is a good candidate although it is an expensive one. Corrosion resistant iron- or nickel-based alloys are other possibilities. Nonmetallic materials such as magnetite or ferrites may also be suitable. Figure 32 shows the principle of an anodic protection system installation. The electrodes are inserted through the top of the storage tank. The polarization is usually made potentiostatically. The potential of the protected surface is measured with respect to a reference electrode and maintained at the set value by changing the current passing from anode to cathode. Since the reference electrodes monitor the protected structure potential and control the output current, they must be compatible with the environment. The control unit changes output current of the dc power supply to maintain the measured potential difference between tank wall and reference electrode within set limit values. Potentiostatic systems can be dangerous without special safety precautions. If the system cannot sense a potential difference between protected structure and reference electrode, it may start to supply maximum available current and cause transpassive dissolution or strong hydrogen evolution.

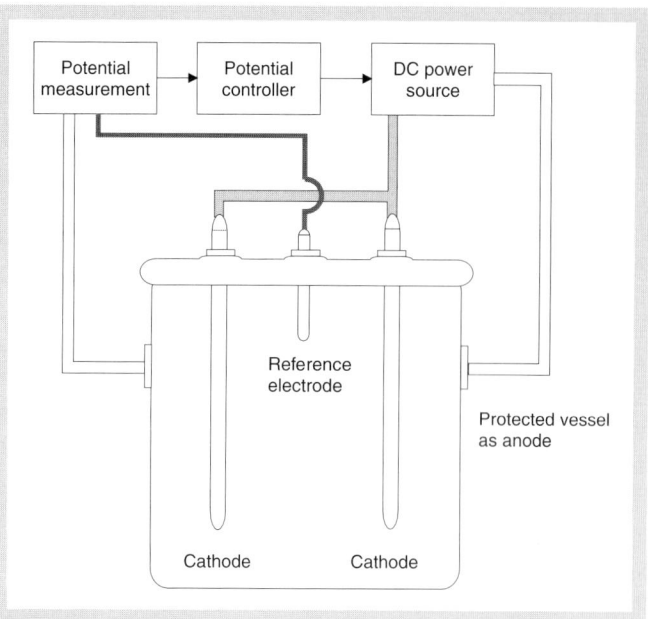

Figure 32. Principle of anodic protection system installation.

Depending on the system geometry, the electrodes can be installed in various ways. In some early installations for protection of vessels, the electrodes were inserted through walls. Inserting the electrodes through roofs to prevent leaks and minimize assembly downtime soon replaced this practice. The electrodes are usually inserted through welded flanged joints and electrically isolated from the equipment. The insulating gaskets should be resistant to vapors and condensates of the solution. The cathodes should not be too close to the vessel walls or bottom. The current following Ohm's law adopts the path of lowest resistance. Excessively short anode to cathode distances may therefore result in local transpassive dissolution of the protected material. Reference electrodes should be installed close to the protected metal to minimize solution potential drops. The arrangement of the electrodes must ensure their solution contact in all cases. The protection system will not pass current through the system if cathodes are out of solution and will usually try to pass maximum current if the reference electrode cannot sense any potential.

The current distribution around the protected structure will ultimately determine the success of anodic protection. The protected areas are large compared with auxiliary cathode areas. Maintaining low current densities and high solution conductivities is necessary. This ensures good potential distribution and long protection distances in simple geometries such as vessels and tubing. Complex structures including heat exchangers require more cathodes around the structure to ensure protection. Dividing complex process systems into smaller independent and isolated subsystems may be the only sensible solution. Crevices are difficult to protect. The bottom of a crevice especially may not have protection although the crevice mouth does have protection. If the anodic current density inside the crevice is below the critical current density, it will accelerate dissolution and cause severe local corrosion[43].

Anodic protection has had use in batch digesters, continuous digesters, storage tanks, clarifiers, and liquor heat exchangers. The construction material for large vessels has traditionally been carbon steel because carbon steel is cost-effective and reasonably corrosion resistant in alkaline pulping liquor. The corrosion resistance of carbon steel in alkaline environments relies on a protective oxide film. Process development for improved productivity has changed chemical dosing and process temperatures. Increased chemical circulation has increased the concentration of sulfur compounds in pulping liquors. The natural oxide film is often not resistant to flowing liquor since mechanical stresses break the surface layers and the inability to repassivate causes corrosion. The purpose of anodic protection is to enhance the passivation capability. Anodic protection of a kraft digester was suggested in the 1950s with the first application in 1958. Continuous digesters have been protected since the beginning of the 1980s. Anodic protection in digesters can prevent general corrosion, erosion corrosion, and stress corrosion cracking[44]. Prevention of general corrosion and erosion corrosion results from passivation of the steel surface if the solution is not sufficiently oxidizing for spontaneous passivation. The effectiveness of anodic protection against stress corrosion cracking may relate to moving away from the dangerous electrode potential range. Stress corrosion cracking usually happens on the active to passive transition range that can be very narrow (about 100 mV). The main problem with anodic protection in digesters is determination of the actual passive range since the complex sulfur chemistry of the cooking liquor will affect any electrochemical measurements.

CHAPTER 6

5.2 Cathodic protection

The cathodic protection of metal against corrosion was first demonstrated in the early nineteenth century. It has been used to stop corrosion of metallic structures and components in a wide variety of environments. Essentially, cathodic protection is the intentional application of a direct electric current in opposition to the naturally occurring electrochemical corrosion of metal. Cathodic protection is a common method for corrosion prevention. Theoretically, it can prevent corrosion of any alloy in any environment since it works by compensating the harmful cathodic reaction currents with surplus anodic current. This results in decrease in corrosion current and

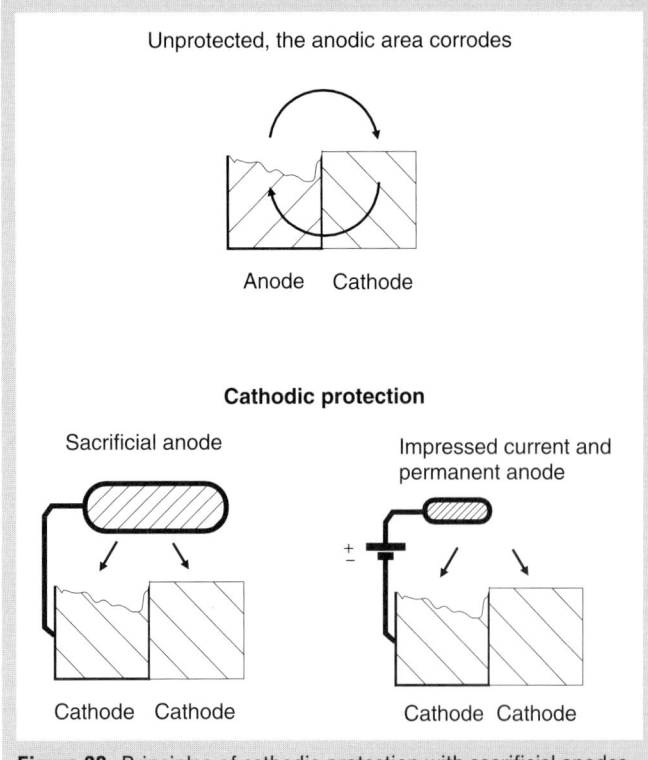

Figure 33. Principles of cathodic protection with sacrificial anodes and impressed current.

cathodic polarization of the protected metal. Cathodic protection can stop corrosion totally, but in most cases it is easier merely to decrease corrosion rate to a tolerable level. Cathodic protection uses sacrificial anodes or an external current source known as impressed current cathodic protection (ICCP). Protection with sacrificial anodes is a situation analogous to galvanic corrosion. In principle, protection with impressed current is electrolysis. It eliminates micro scale corrosion cell currents by forcing currents in macro scale in a certain direction. Figure 33 shows the principles of cathodic protection with sacrificial anodes and impressed current.

Figure 34 shows the electrochemical theory of cathodic protection. Without cathodic protection, the system stabilizes to a steady-state with known corrosion potential and corrosion current density. The anodic corrosion current drawn from the anode equals the cathodic current. By adding sacrificial anodes to the system as Fig. 34(a) shows, most anodic current comes from preferred, spontaneous dissolution of these. The cathodic current remains constant. Depending on the current delivering capacity or polarizing capacity of the anodes, the system adopts a new, lower corrosion potential. The sacrificial anodes must have lower equilibrium potential than the protected metal. By adding an external current source and insoluble anode, the potential of the protected system can be a selected value as Fig. 34(b) shows. If the potential of the protected

Figure 34. Mixed potential theory of sacrificial anode protection (a) and impressed current protection (b).

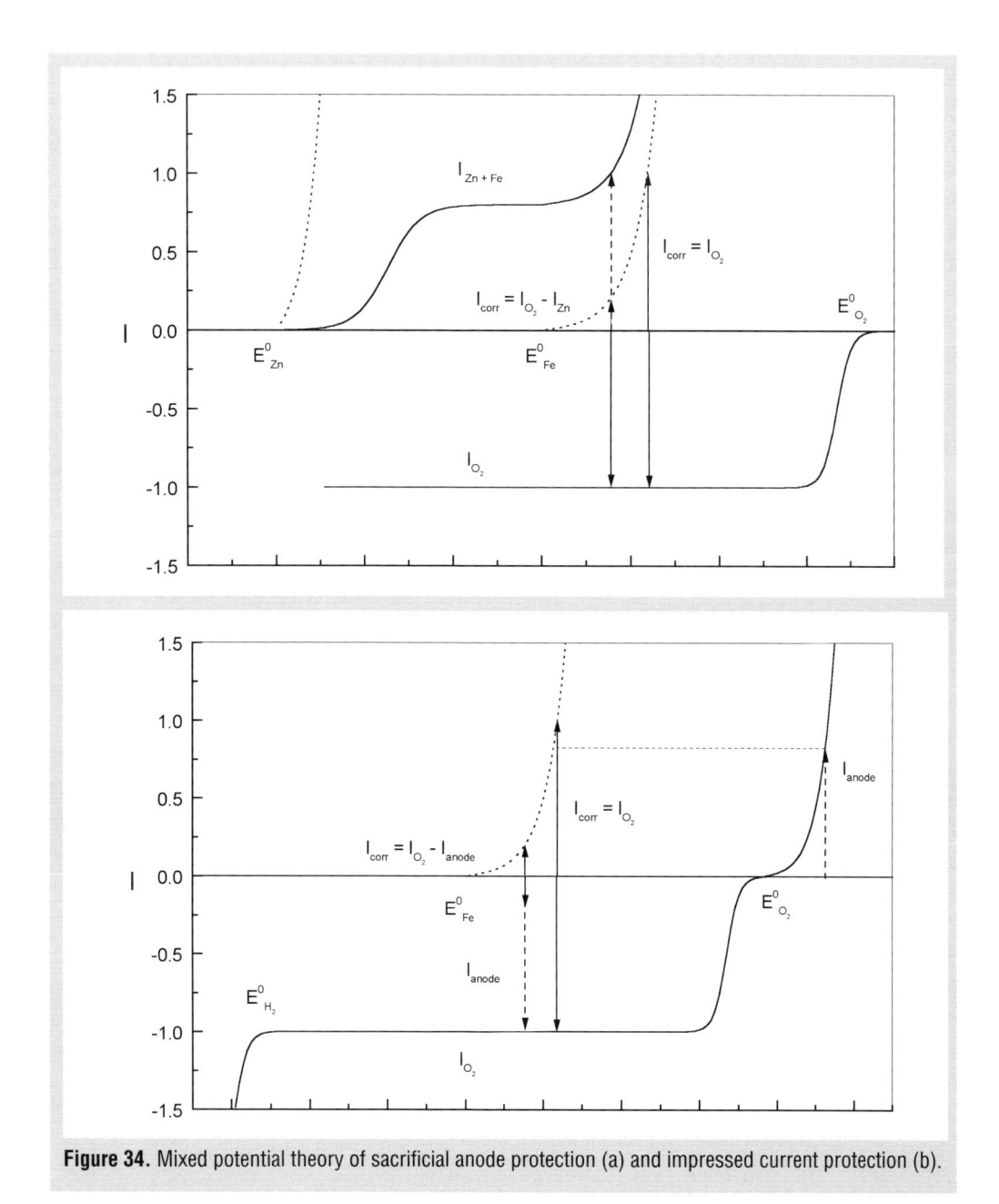

structure is set sufficiently low, it operates primarily as a cathode, and anodic corrosion is minimized or totally stopped. Thus cathodic protection mitigates corrosion mainly by eliminating anodic areas on the corroding metal surface. The magnitude of the original cathodic current affects the magnitude of potential change but not the absolute value of the selected potential.

CHAPTER 6

Usually, cathodic protection has use against uniform corrosion, and it results in a potential shift toward the corrosion potential or below it. The corrosion rate reduction depends on the magnitude of the potential shift. A lower structure potential under protection gives a lower corrosion rate. If the potential of the protected structure is too low, overprotection results. Overprotection is often associated with hydrogen evolution. This can cause damages in protective coatings and hydrogen induced damages in high-strength materials. In some special cases, impressed current cathodic protection has use against localized corrosion or even transpassive dissolution. In these cases, the structure is polarized back to a passive state. Since the transpassive dissolution is often associated with pitting corrosion, polarizing the metal to potentials between the beginning of the passive range and the protection potential of pitting is necessary. A risk of stress corrosion cracking might exist if stainless steel is protected to the active-passive transition range. Uniform corrosion will occur if the steel is polarized too low to the active range as Fig. 35 shows.

Although the principles of cathodic protection are simple, design for a real system is usually inexact and depends on the experience of the designer. The hardware is the same as in anodic protection: power source, power controller, and potential measuring device as Fig. 36 shows.

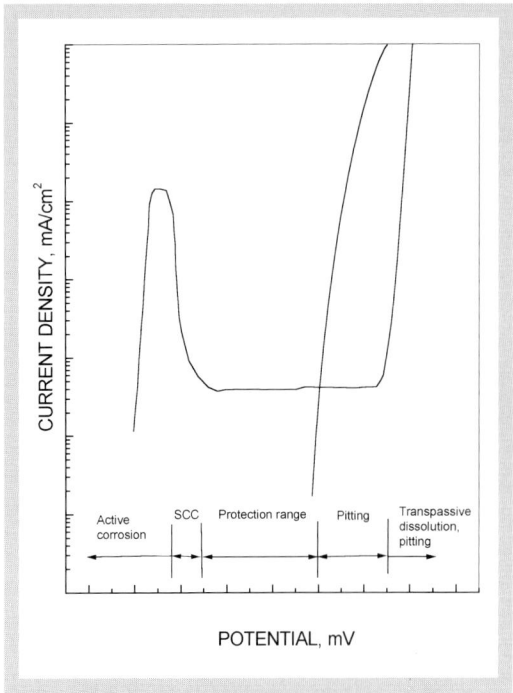

Figure 35. Cathodic protection of stainless steel.

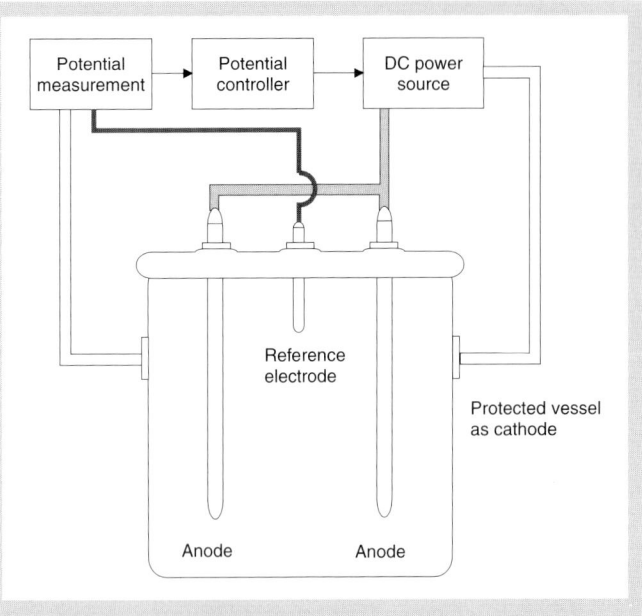

Figure 36. Principle of cathodic protection system installation.

The main problem in cathodic protection design is to assure uniform current distribution along the protected structure because the throwing power of cathodic pro-

Corrosion prevention

tection is low. The current between anode and cathode flows along the path with lowest resistance. This is usually the shortest distance between anode and cathode. Areas close to the anode therefore draw more current and may become overprotected while areas far from the anode have no protection. Spread of protection into corners and down a pipe are common problems. Cathodic protection often has very little protective effect upstream in solution flow. Overprotection means polarizing the protected structure to excessively low potentials. This can also cause material degradation. To obtain protection in areas far from the anode, the protected structure is coated. The coating must have high electrical resistivity. Cathodic protection has severe limitations in a process plant due to complex geometries and different materials handling different corroding substances close together. In cases with simple geometries and constant corrosivity of the solution, cathodic protection can give long protection with little additional cost.

The design of a cathodic protection system requires information on the materials, system geometry, and corrosivity of the system. Cathodic protection works by supplying external current to the system requiring protection. The necessary amount of external current is the protection current density. It varies depending on the metal and corrosive environment. Protection current density relates to the corrosion rate, i.e., the corrosion current

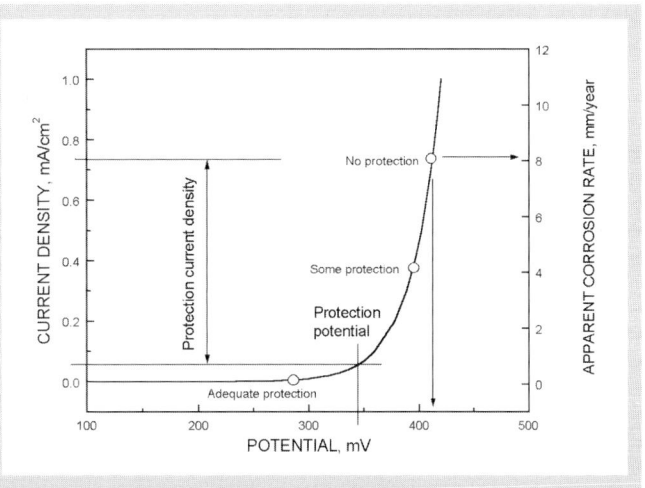

Figure 37. Corrosion rate, protection current, and protection potential in cathodic protection.

density. When protective current density is supplied to the protected surface, its potential will shift to the cathodic direction (lower potentials), and corrosion rate decreases. Measuring the potential of the protected structure and comparing it to a reference value known as protection potential determines the effectiveness of cathodic protection. Protection potential is a value at or below which the corrosion rate is sufficiently low as Fig. 37 shows.

When designing a cathodic protection system, ensuring that the entire protected surface receives sufficient protection current is essential, but no part of the surface should receive excessive current. Location of anodes, reference electrodes, and cathode connections will determine the primary current and potential distribution in the system. All surfaces of the metal structure must be cathodic to prevent corrosion. Inadequate current distribution can lead to shielded areas that remain unprotected. Figure 38 is a schematic presentation of current distribution along a protected pipe line. The maximum allowable potential drop is $U_0 = -0.6$ V, and the minimum potential drop corresponding to the protection potential is $U_L = -0.3$ V. The protection current density is highest close to the anodes.

Consequently, the structure potential is lowest at the vicinity of the anodes and increases when moving away from them. At some point, the current density on the surface is not sufficiently high to polarize the surface below the protection potential. The distance between two anodes marked as 2L must be sufficiently short to maintain adequate current density on the entire surface. Anode current, environment conductivity, coating quality, structure geometry, etc., control this distance.

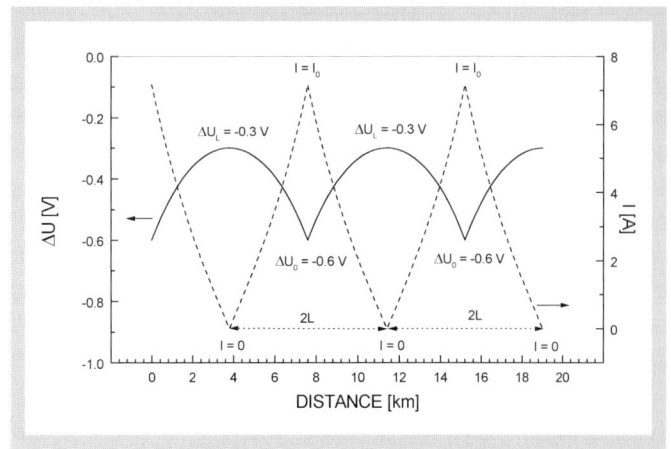

Figure 38. Current distribution along a cathodically protected pipe line.

Some design values for protection current density and protection potential are very universal. In natural, nearly neutral environments, the protection potential of steel is -850 mV vs. copper/copper sulfate electrode that is equivalent to -780 mV vs. saturated calomel electrode. This is valid for soil and sea water. The same criteria also applies for some process equipment. The current density for protection depends on the rate of cathodic reaction, flow rate, and condition of the coating. A higher rate of cathodic reaction means a higher external current is necessary to compensate for it. Increase of flow rate will increase the current requirement since the supply of cathodic reactant increases. The quality of the coating directly influences the protection current requirement. A good coating covers most structure surface, and only a small part of the structure that contacts the environment requires protection. A poor coating has more defects, and more area requires protection. Cathodic protection of an uncoated steel structure in natural water requires about 80–130 mA/m^2, but a good quality coating can decrease this by 99% or more. For large structures, the use of a coating is essential to keep current requirements at a sensible level.

Protection of large structures requires a coating to keep power consumption low and maintain proper current distribution. Bleach filters, pipelines, pumps, valves, and storage tanks are possible applications for cathodic protection. Cathodic protection is often useful in protection of existing equipment with known corrosion problems. Using cathodic protection allows extending the service life of the system without extensive coating or lining. A major benefit of cathodic protection systems is their short installation time. The systems will also adapt to varying corrosivity that is extremely useful during startup and shutdown periods. Some guidelines for cathodic protection are available in standards. The NACE standard RP0180-91, "Cathodic Protection of Pulp and Paper Mill Effluent Clarifiers," discusses structure design, cathodic protection requirements, protective criteria, and coatings. It also provides information about reference electrodes, selection and design of cathodic protection systems, installation and testing, and records concerning pulp and paper mill effluent clarifiers.

The pulp and paper industry has developed some special constructions for protecting rotating equipment and protection in very corrosive environments. Protection of paper machine Yankee dryer cylinders has been available since the 1960s. Pulp dries over a large cast iron or steel cylinder that is internally heated with steam. The Yankee cylinder not only dries the paper web but also imparts a smooth and glossy surface to the paper. This is due to the paper web pressing against the brightly polished Yankee cylinder. The web adheres to the cylinder until it is sufficiently dry to loosen. The paper produced is clear and smooth only if the cylinder itself is smooth, i.e., not corroded.

Protection of rotating drying cylinders has an interesting design problem. How does one provide electrolytic current to a structure not immersed in solution? Developing a special cell construction that fits the cylinder surface solved the problem. The cell contains a small chamber through which electrolyte constantly circulates with an impressed current anode (plated titanium) near the cylinder surface. The cell extends almost the entire length of the cylinder, but it is very narrow. Although the section that receives protective current from the external cell is very small compared with the entire area of the cylinder, the system can supply sufficient protection current[45,46]. By using high current densities (200 mA/cm^2), the system rapidly polarizes to sufficiently low potentials. The charge of the double layer decays slowly and keeps the section of the cylinder surface protected until it is again in the cell.

The corrosivity of bleached pulp washer filtrates increases with system closure. This can lead to pitting of stainless steels used in washer cylinders and associated equipment. The corrosion problems are typically crevice corrosion, pitting, and stress corrosion cracking. They become more severe with high residual chlorine, low pH, and high chloride concentration. Nickel alloys and titanium would obviously be more resistant, but their cost is often too high. Protection with glass reinforced plastic can also be expensive if the geometry is too difficult. One possible protection method is cathodic protection with impressed current. Protection of stainless steel equipment does not require polarization to below corrosion potential but to the passive range and potentials below pitting protection potential. Polarizing the structure to potentials corresponding to the immune region would require unacceptably high current densities. In acid chloride solutions, most stainless steels have pronounced active areas. Application of cathodic protection therefore requires stricter control than in neutral solutions such as sea water. The protection potential is often very close to the active region leaving little margin for potential variation due to risk of general corrosion and SCC.

A protection system for bleach plant washers made from stainless steels became available at the end of the 1970s. The idea came during development of a traditional cathodic protection system for bleach plant washers. The system consists of a power unit, an inert anode parallel to the drum axis, one or more reference electrodes, and slip rings for electrical connection to the drum. As the drum rotates, only the part submerged has protection. The basic idea in this system is not to decrease the corrosion potential of the metal but to reduce residual chlorine to chloride ions on the surface of the protected part[47,48]. This is equivalent to reducing residual chlorine with sulfur dioxide. The redox potential of the solution decreases by elimination of active chlorine. The risk of localized corrosion is therefore reduced. Excessive cathodic polarization

CHAPTER 6

increases the uniform corrosion rate. If the potential is set at excessively high values, crevice corrosion is not prevented. Optimization of the operating potential requires care and experience.

6 Summary

Totally stopping corrosion is usually not practical. A common practice is to decrease corrosion rate to an acceptable level. The first step in preventing corrosion is understanding the specific corrosion mechanism. The second step is designing the protection method.

The easiest way to eliminate corrosion problems is not to create them in the design stage. This means avoiding constructions that need or can cause metal or solution heterogeneities leading to corrosion. The purpose of a sound design is to ensure a system lifetime with adequate safety margins but without overestimating material thicknesses. The rate, extent, and type of corrosion damage that are tolerable vary depending on the specific application.

The most cost-effective method for corrosion protection is often proper material selection that includes a coating. Materials selection must not be limited to metals. The decisive factor for material selection should be the total cost during the system designed operation time. The total cost should include initial material and installation costs, maintenance and replacement costs, and as far as possible costs by interrupted production. Knowing the environment and material recommendations gives a list of candidate materials. When an alloy with adequate resistance to uniform corrosion has been found, its resistance to various forms of localized corrosion must be ensured. The properties of any candidate materials also require evaluation outside the normal plant operating conditions. At startup or during shutdown, the conditions can be very different from those during normal operation.

Corrosion prevention by modification of the environment is usually limited to closed or semi-closed systems. This can include removal of corrosive agents, pH control for maintaining the passive film, inhibition, and control of biofouling.

Protective coatings can be metallic, inorganic, or organic. They protect the metal by forming a barrier between the metal and a corrosive environment, by sacrificial dissolution, or by inhibition. All coatings will separate the metal from its environment to some degree. For a barrier coating to be effective, it must cover the entire surface, be sound, and resist mechanical damage. Paints are the most common protective coatings. Other common coatings in the pulp and paper industry are metal and polymer linings.

Electrochemical protection is usually the last protection method for consideration. Electrochemical protection includes anodic protection and cathodic protection. Anodic protection is maintaining the passivity with an external current source. The protected structure is deliberately polarized to the anodic direction. Cathodic protection is production of surplus anodic current by sacrificial anodes or external current source to satisfy the needs of cathodic reaction. This results in a decrease in corrosion current density and potential shift to the cathodic direction. Cathodic protection is usually used with coatings.

References

1. McGovern, D., A Review of Corrosion in the Sulfite Pulping Industry, Pulp & Paper Industry Corrosion Problems, vol. 3., NACE, Houston, 1982, p. 60.

2. Wensley, D. A. and Charlton, R. S., Corrosion studies in kraft white liquor (II) Effect of plain carbon steel composition, Pulp & Paper Industry Corrosion Problems, vol. 3., NACE, Houston, 1982, p. 20.

3. Nelson, J. K., in ASM Metals Handbook, vol. 13, Corrosion, ASM International, Metals Park, 1987, pp. 1174–1180.

4. Nelson, G .A., Corrosion data survey, 4th edn., NACE, Houston, 1967, p. 4.

5. Bryson, J. H., et al., in ASM Metals Handbook, vol. 13, Corrosion, ASM International, Metals Park, 1987, pp. 509–530.

6. Uhlig, H. H, in Corrosion Handbook (H. H. Uhlig, Ed.), John Wiley & Sons, New York, 1948, p. 125–143.

7. Nelson, G .A., Corrosion data survey, 4th edn., NACE, Houston, 1967, p. S-6.

8. Nelson, G .A., Corrosion data survey, 4th edn., NACE, Houston, 1967, p. S-11.

9. Mueller, W. A., Mechanism and prevention of corrosion of steels exposed to Kraft liquors, Pulp & Paper Industry Corrosion Problems, NACE, Houston, 1974, p. 109.

10. Charlton, R. S. and Tromans, D., Corrosion in Kraft mill combined outfall sewers - pH and velocity effects, Pulp & Paper Industry Corrosion Problems, vol. 2., NACE, Houston, 1977, p. 76.

11. Poulson, Corrosion Science 15(8):469(1975).

12. Alfonsson, E., and Qvarfort, R., Materials Science Forum 111–112:483(1992).

13. Forsén, O., Aromaa, J., Tavi, M., et al., Materials Performance 36(5):59(1997).

14. Schillmoller, C. M., Selection and performance of stainless steels and other nickel-bearing alloys in sulfuric acid, NiDI technical series report No 10057, Nickel Development Institute, Toronto, 1990, pp. 1–9.

15. Craig, B. D., Handbook of corrosion data, ASM International, Metals Park, 1989, pp. 581– 642.

16. Willis, J. D. and Johnson, R. S., Corrosion resistance of stainless steels used in bleach washing environments, 1992 Proceedings of the 7th International Symposium on Corrosion in the Pulp and Paper Industry, TAPPI PRESS, Atlanta, p. 41.

CHAPTER 6

17. Schillmoller, C. M., Alloy selection for caustic soda service, NiDI technical series report No 10019, Nickel Development Institute, Toronto, 1988, pp. 1–9.

18. Asphahani, A. I., et al., in ASM Metals Handbook, Vol. 13, Corrosion, ASM International, Metals Park, 1987, pp. 641–657.

19. Fontana, M. G., Corrosion Engineering, 3rd edn., McGraw-Hill, Singapore, 1987, Chap. 5.

20. Cotton, J. B. and Hanson, B. H., in Corrosion (L. L. Shreir, R. A. Jarman, and G. T. Burstein, Eds.), vol 1, 3rd edn., Butterworth-Heinemann, Oxford, 1994, Chap. 5.4.

21. Schutz, R. W. and Thomas, D. E., in ASM Metals Handbook, vol. 13, Corrosion, ASM International, Metals Park, 1987, pp. 669–706.

22. Macdiarmid, J. A., Charlton, R. J., and Reichert, D. L., Corrosion and materials engineering considerations in hydrogen peroxide bleaching, 1992 Proceedings of the 7th International Symposium on Corrosion in the Pulp and Paper Industry, TAPPI PRESS, Atlanta, p. 97.

23. Andreasson, P., The corrosion of titanium in hydrogen peroxide bleaching solutions, 1995 Proceedings of 8th International Symposium on Corrosion in the Pulp and Paper Industry, Swedish Corrosion Institute, Stockholm, p. 119.

24. Bardsley, D. E., Compatible metallurgies for today's new bleach washing processes, 1995 Proceedings of 8th International Symposium on Corrosion in the Pulp and Paper Industry, Swedish Corrosion Institute, Stockholm, p. 75.

25. Anon., BETZ Handbook of industrial water conditioning (J.J. Maguire, Ed.), 8th edn., BETZ Laboratories, Trevose, 1980, Ch. 10.

26. Anon., Principles of Industrial Water Treatment, Drew Chemical Corporation, Boonton, 1977, Chap. 11.

27. Mercer, A. D., in Corrosion (L. L. Shreir, R. A. Jarman, and G. T. Burstein, Eds.), vol 2, 3rd edn., Butterworth-Heinemann, Oxford, 1994, Chap. 17.2.

28. Anon., Principles of Industrial Water Treatment, Drew Chemical Corporation, Boonton, 1977, Chap. 5.

29. Anon., BETZ Handbook of industrial water conditioning (J. J. Maguire, Ed.), 8th edn., BETZ Laboratories, Trevose, 1980, Ch. 26.

30. Stein, A. A. and Mussalli, Y., in A Practical Manual on Microbiologically Influenced Corrosion (G. Kobrin, Ed.), NACE International, Houston, 1993, Chap. 11.

31. O'Reilly, M. W. and Pringle, J. T., in Corrosion (L. L. Shreir, R. A. Jarman, and G. T. Burstein, Eds.) vol 2, 3rd edn., Butterworth-Heinemann, Oxford, 1994, Chap. 14.2.

32. Weaver, P. E., in Paint handbook (G. E. Weismantel, Ed.), McGraw-Hill, New York, 1981, Chap 4.

33. Fontana, M. G., Corrosion Engineering, 3rd edn., McGraw-Hill, Singapore, 1987, Chap. 6.

34. Hess, M. and Bullett, T. R., in *Corrosion* (L. L. Shreir, R. A. Jarman, and G. T. Burstein, Eds.), vol 2, 3rd edn., Butterworth-Heinemann, Oxford, 1994, Chap. 14.4.

35. Chivers, A. R. L and Porter, F. C., in *Corrosion* (L. L. Shreir, R. A. Jarman, and G. T. Burstein, Eds.), vol 2, 3rd edn., Butterworth-Heinemann, Oxford, 1994, Chap. 13.4.

36. Anon., Welding of stainless steels and other joining methods, Nickel Development Institute report No 9002, American Iron and Steel Institute, Toronto, 1993, pp. 33–36

37. Avery, R. E., Harrington, J. D., and Mathay, W. L., Stainless steel sheet lining of steel tanks and pressure vessels, Nickel Development Institute Technical series report No 10039, Nickel Development Institute, Toronto, 1989, pp. 1–18.

38. Bakhvalov, G. T. and Turkovskaya, A. V., *Corrosion and protection of metals*, Pergamon Press, Oxford, 1965, Chap. 16.

39. Millar, N. S. C. and Wilson, C., in *Corrosion* (L. L. Shreir, R. A. Jarman, and G. T. Burstein, Eds.), vol 2, 3rd edn., Butterworth-Heinemann, Oxford, 1994, Chap. 16.1.

40. Anderson, T. F., Flynn, R. C., Oswald, K. J., et al., Use of nonmetal materials of construction in pulp mills, *Pulp & Paper Industry Corrosion Problems*, vol. 2., NACE, Houston, 1977, p. 53.

41. Sharp, W. B. A., Use of non-metals in the bleach plant, *1983 Proceedings of the Fourth International Symposium on Corrosion in the Pulp and Paper Industry*, Swedish Corrosion Institute, Stockholm, p. 179.

42. Riggs, Jr., O. R. and Locke, C. E., *Anodic protection*, Plenum Press, New York, 1981, Chap. 4.

43. France Jr., W. D. and Greene, Jr., N. D., *Corrosion* 24(8):247(1968).

44. Singbeil, D., Does anodic protection stop digester cracking, *1989 Proceedings of the Sixth International Symposium on Corrosion in the Pulp and Paper Industry*, The Finnish Pulp and Paper Institute, Helsinki, p. 109.

45. Almar-Nœss, A., Corrosion protection of drying cylinders in paper making machines, a new principle of cathodic protection, *1964 Proceedings of Fourth Scandinavian Corrosion Congress*, Kemian Keskusliitto, Helsinki, p. 201.

46. Nöglegaard, O., Service experience from a Yankee machine with cathodically protected cylinder, *1964 Proceedings of Fourth Scandinavian Corrosion Congress*, Kemian Keskusliitto, Helsinki, p. 231.

47. Garner, A., *Materials Performance* 21(5):43(1982).

48. Garner, A., "Electrochemical protection of bleached pulp washers", *1983 Proceedings of the Fourth International Symposium on Corrosion in the Pulp and Paper Industry*, Swedish Corrosion Institute, Stockholm, p. 160.ss

CHAPTER 7

Corrosion monitoring

1	**Physical monitoring techniques**	**332**
2	**Electrochemical methods**	**335**
2.1	Potential monitoring	336
2.2	Linear polarization technique	337
2.3	Galvanic current measurement	338
2.4	Tafel method	339
2.5	Electrochemical impedance spectroscopy	340
2.6	Electrochemical noise	341
3	**Other methods**	**341**
3.1	Hydrogen flux monitoring	342
3.2	Chemical analysis	342
4	**Data collection and analysis**	**343**
	References	344

CHAPTER 7

Jari Aromaa

Corrosion monitoring

Corrosion monitoring is a basic function required for safe and efficient industrial operations. It is the systematic and continuous measurement of corrosion rates to assist corrosion control. The various measurement methods available range from simple corrosion coupons to sophisticated electrochemical techniques. All systems should meet the specific needs of particular plant requirements to minimize downtime and equipment failures. The purposes of corrosion monitoring include supervising corrosion control methods, warning of system changes leading to corrosion damage, controlling the process, and estimating of service life with determination of inspection schedules, maintenance schedules, or both. Since corrosion problems are often complex, no single method will necessarily work in all applications. Combining multiple technologies may be necessary to provide reliable corrosion monitoring information. Most corrosion monitoring methods are essentially identical to the corrosion measurement methods discussed earlier. The monitoring methods fall into three main groups:

- Physical methods that measure the physical loss of material due to corrosion or erosion by physical or electrical means
- Electrochemical methods that measure the instantaneous corrosion rate not the actual material loss (These methods often relate to the corrosiveness of the environment.)
- Supporting analyses, such as iron counts, biofilm testing, pH monitoring, etc.

Industry prefers user-friendly methods and equipment to minimize the expertise necessary for operation and data interpretation. This will limit the suitability of many electrochemical techniques. An ideal monitoring system should have the following properties:

- Good resolution
- Short response time
- Reliability
- Maintenance free
- Simple data processing.

Corrosion monitoring methods have different classifications. Direct monitoring techniques directly measure the waste caused by corrosion or electrochemically measure the instantaneous corrosion rate. Corrosion coupons to determine mass loss and electrical resistance technique using the change in probe resistance produced by the change in cross-sectional area are direct physical measurement methods. Electrochem-

CHAPTER 7

ical test methods such as linear polarization techniques use the relationship of potential and current of a corroding electrode to determine corrosion current density or some other factor relating to the corrosion rate. Indirect corrosion monitoring techniques do not measure corrosion itself. They measure some result of the corrosion process instead. Two common indirect techniques are ultrasonic testing and radiography. Both techniques have use to determine the remaining wall thickness of a pipe, vessel, or other item. Direct methods usually use a sensor. Indirect methods apply to actual process components.

Another way to classify corrosion monitoring techniques is whether they require entry into the process stream. Almost all direct measurement methods require access to the inside of the equipment being monitored. These methods are therefore intrusive or invasive methods. Most indirect methods such as external hydrogen flux probes and analysis of water samples through an existing valve do not require access to the system. The efficient use of direct monitoring techniques requires the location of areas with the highest risk of corrosion. Differences in flow patterns, temperature, and other environmental variables can also give results that do not represent the actual process. Access to the process may limit locations available for inserting sensors. Sensor materials must match plant construction materials very closely in processing history and composition to obtain reliable results.

1 Physical monitoring techniques

The first corrosion monitoring method was probably the use of sentinel holes or sentry holes in piping. These holes were accurately drilled to a depth that corresponded to the corrosion allowance of the remaining material. They were at areas with anticipated high corrosion rates. When internal corrosion reduced the wall thickness to the drilled section, leaks developed. By using a series of holes drilled to various depths, estimation of corrosion rates was possible. This method has rare use today.

The use of corrosion coupons is a basic method of corrosion monitoring that is often overlooked. A sample of the material of interest is weighed and exposed to the corroding material for a period. Then it is removed, carefully cleaned, and weighed again. The change in weight can calculate the metal loss expressed as thinning or mass loss per area and time. Corrosion coupons are usually strips, discs, or cylindrical rods mounted in suitable racks. Corrosion coupons provide the most reliable monitoring method, but their response time is long. They give average uniform corrosion rate and estimate the extent of localized corrosion. They can also provide information on the nature of corrosion through analysis of their corrosion products. Coupon monitoring usually requires long times, and it only provides time-averaged data that has no use for on-line corrosion monitoring. Corrosion coupons provide periodic information. Their data has use as a basis for validation of other methods. The use of corrosion coupons is labor intensive, but the samples and their installation are not expensive.

Procedures for in-plant corrosion coupon testing are available in ASTM G4 "Standard Method for Conducting Corrosion Coupon Tests in Plant Equipment" and NACE RP0775 "Preparation and Installation of Corrosion Coupons and Interpretation of Test

Data in Oilfield Operations." These standards give methods for placement, installation, and exposure of in-plant corrosion coupons and their use in oilfield operation. ASTM G1 "Practice for Preparing, Cleaning, and Evaluating Corrosion Test Specimens" gives guidelines for preparation and use of corrosion coupons. The conversion of mass loss to a uniform corrosion rate uses the following:

$$Corrosion\ rate\ [mm/year] = \frac{87\ 600 \cdot M}{A \cdot \rho \cdot t} \tag{1}$$

where M is mass loss, g
  A coupon surface area, cm^2
  ρ material density, g/m^3
  t time of exposure, h.

Coupons are useful for long-term baseline corrosion rates. Long exposure times are necessary to obtain accurate results. Significant mass change detectable with an analytical balance is necessary. The corrosion rate of a new coupon is higher than the steady state corrosion rate, and loss of uncorroded material from the coupon occurs during the removal of corrosion products. A greater metal loss due to normal corrosion reduces the significance of these errors. Thirty days is usually the minimum exposure time. A good practice is to install the coupons on overlapping 30–90 day intervals or to use a planned interval test. The exposure interval in hours should be at least 50 divided by the expected corrosion rate expressed as mm/year. If the expected corrosion rate is 0.1 mm/year, the minimum exposure time is 500 hours or about three weeks.

Figure 1 shows schematically the differences that can originate between average corrosion rate and "instantaneous" corrosion rate in the beginning of a test or from an upset in the process system. The coupon method averages weight loss during exposure. It assumes that the corrosion of the metal is uniform. Evaluating coupons visually is important, and the density, size, and depth of localized attack require documentation. A useful parameter is the pitting factor, i.e., the ratio of maximum localized attack rate vs. general corrosion rate determined by mass loss. If this ratio is near one, the general corrosion rate can predict corrosion performance.

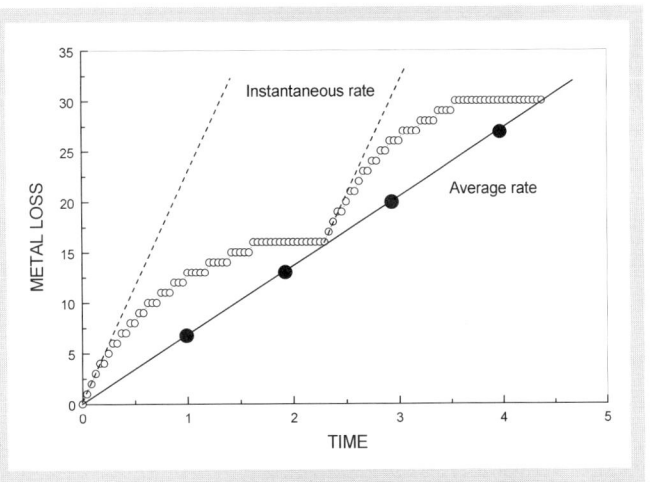

Figure 1. Average corrosion rate measured by corrosion coupons and instantaneous corrosion rate.

CHAPTER 7

Electrical resistance probes (ER probes) are sometimes described as on-line coupons because they also measure the loss of metal from a sample. An ER probe has a sensing element that is a loop of material made from a wire or strip used to conduct an electrical signal. When exposed to a corrosive environment, the cross section of the loop decreases. This increases the resistance of the sensing element and produces a change in the output of the electrical resistance meter. An ER probe provides the cumulative amount of corrosion. Differentiating this information with respect to time provides a calculation of corrosion rate. Figure 2 schematically shows a typical electrical resistance probe with the electric wiring principle.

The following expresses mathematically the electrical resistance of an element having a fixed length:

$$R = \rho \cdot \frac{L}{A} \qquad (2)$$

where ρ is specific resistivity of the element
L length
A cross-sectional area.

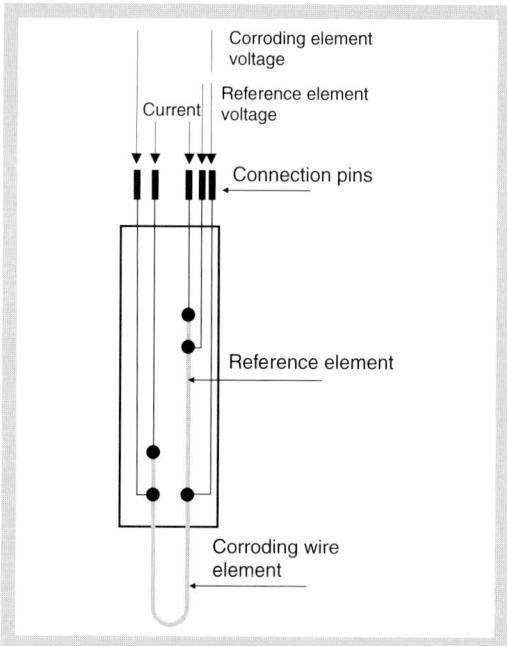

Figure 2. Schematic diagram of the electrical resistance probe with the principle of electric wiring.

Resistance change therefore provides a calculation of reduction in cross section. Since specific resistivity is very temperature dependant, a method of eliminating the effect of temperature changes is necessary. An element of identical material at the same temperature as the sensor element but protected from corrosion is used. Resistance readings of this element therefore always relate to the same specific resistivity as the sensor. When expressing the two resistance readings as a ratio, changes in this ratio represent changes in the sample cross section independent of temperature. Figure 3 shows typical electrical resistance data measured for stainless steel in a reducing process stream. The system behaves almost identically to paralinear, high-temperature oxidation. Short intervals of parabolic growth follow each other, but the passive film is not stable and breaks down. This lead to almost linear metal loss.

The main benefit of the electrical resistance technique is its usefulness in continuous, on-line process monitoring. The electrical resistance technique does not require a continuous electrolyte to make measurements. The system works in multiphase environments containing liquid hydrocarbons and has use for corrosion monitoring in nonaqueous and gaseous process environments. The limitation of the electrical resistance technique is that it only provides representative data for general corrosion. It does not accurately detect localized attack. Data is usually the rate of metal loss for the period of interest. ER probes typically require several days to determine a reliable corrosion rate trend. If the process is prone to rapid changes in corrosivity, ER probes typically may not provide accurate and reliable corrosion rate data. In some cases, they can give erroneous results due to the presence of conductive corrosion products on the sensing element. The results of ER probes require comparison with those obtained from coupon exposures during the same time period. While electrical resistance data may not give reliable indications of the absolute corrosion rate, they can give useful indications of trends and changes in plant corrosion activity.

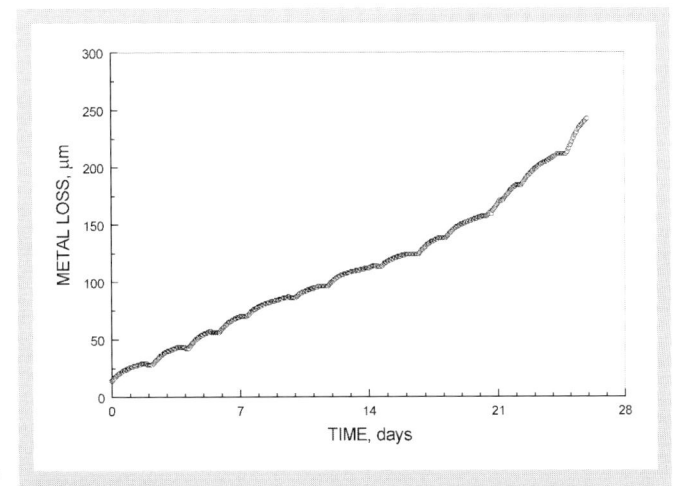

Figure 3. Typical electrical resistance data measured from a process stream.

A further development in the ER probe is the field signature method (FSM). This uses electric current and multiple connection points to create a unique "electrical fingerprint" for the system. The electrical fingerprint method is another term for this monitoring method. FSM can detect changes in the electrical field pattern resulting from corrosion, wear, and other phenomena. The current is fed between two electrical connection studs. As it spreads in the structure, the pattern is determined by the geometry and conductivity of the material. Small sensing pins are distributed in a selected pattern over the monitored area. Any changes will locally alter the conductivity of the material and change the electrical field pattern. Voltage measured between two selected sensing pins is compared with a measurement between a reference pair to compensate for temperature effects and relate to the original field pattern. Welds, bottom sections of pipes, joints, and elbows are typical areas for FSM monitoring[1].

2 Electrochemical methods

Electrochemical corrosion monitoring does not differ from electrochemical measurements for research purposes. The corrosion processes provide electrochemical potential that shows the thermodynamic driving force and a current that indicates the reaction rate of

CHAPTER 7

the process. Faraday's law converts the corrosion current to a corrosion rate. Accurate monitoring depends heavily on the ability to measure the current correctly. Electrochemical methods have limitations in many multiphase systems and have little use in nonaqueous and gaseous environments. The main difference between laboratory measurements and monitoring measurements is elimination of mass transfer and ohmic overpotentials. Although this can occur under controlled laboratory conditions using various procedures, actual process conditions are more difficult. Uncompensated resistance in the environment typically leads to errors in electrochemical measurement in low conductivity environments. Electrochemical techniques are discussed in more detail in Chapter 3.

Standard analyses of electrochemical data usually assume that the measured corrosion rate is the result of uniform corrosion. In many cases, only some surface of the specimen is corroding as in the case of pitting. In these cases, severe underestimation of the penetration rate is possible. In some environments, electrochemical reactions of dissolved species can lead to the measurement of current that does not contribute to corrosion. Soluble sulfur species can complicate electrochemical measurements because sulfur easily oxidizes and reduces especially at elevated potentials.

The main benefits of electrochemical methods are that they can provide an instantaneous corrosion rate in the system. They can therefore identify rapid changes in process corrosivity. Electrochemical techniques also have use to obtain more mechanistic information including identification of transitions between active and passive behavior and efficiency of inhibition. As with ER probes, electrochemically measured corrosion rates require comparison with corrosion coupon data. Again, the trends may be more useful than the absolute corrosion rates.

2.1 Potential monitoring

In some systems, knowing how the potential of a material changes with process variations is important. Typical examples are monitoring of passivity and electrochemical protection. For passivating alloys, potential measurements may also indicate the influences of process changes on the state of the metal or on the corrosion potential relative to the pitting protection potential. When cathodic or anodic protection prevents corrosion, variations in the potential can indicate if the proper levels of protection are being maintained or if local changes in corrosion behavior are occurring. Potential monitoring provides a useful way to differentiate active corrosion from passive conditions. If the potential ranges associated with active and passive states are sufficiently large, potential monitoring is possible. This method can assess risk of corrosion but not corrosion rate. It only determines whether the metal is corroding slowly or rapidly. Potential monitoring is applied to a plant instrument rather than a probe and will therefore describe the equipment state more reliably.

Potential monitoring under plant conditions requires a stable reference electrode. This obvious requirement is sometimes overlooked since accurate but fragile laboratory electrodes are used. The measurement accuracy does not need to be as high as in the laboratory, and a more rugged electrode with 10–20 mV reproducibility might be sufficient. For the evaluation of cathodic protection of buried pipelines and equipment, compensation for the resistance drop associated with soils of relatively low conductivity is also necessary. High solution resistance can also lead to similar problems when electrochemical measurements are made in high purity water environments used for cooling water in some systems.

The major applications of potential monitoring occur in electrochemical protection and monitoring of passivating metals. The use of potential monitoring for actively corroding metals is often not sufficiently accurate. Assuming that an actively corroding metal obeys Tafelian behavior, a change of one order of magnitude in corrosion current density would shift the corrosion potential 40 and 60 mV for three-electron and two-electron mechanisms, respectively. Such changes are very high in laboratory experiments but are often only normal fluctuations under field conditions. For metals with active and passive transition, the corrosion potential may shift several hundred millivolts when changing from one state to another. Figure 4 shows this schematically. The metal has two stable corrosion potential values marked with P in the passive range and A in the active range. For reliable use of corrosion potential monitoring, the active and passive potential ranges require determination beforehand.

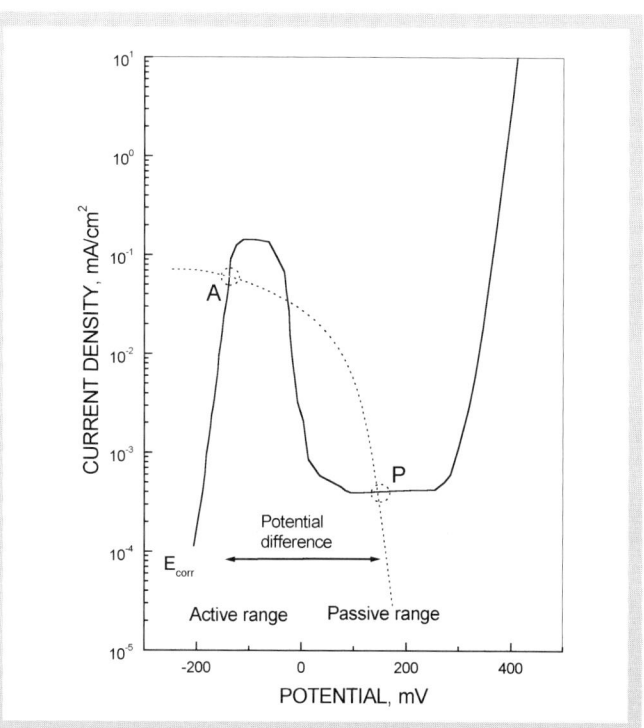

Figure 4. Corrosion potential monitoring for a metal with active and passive transition.

2.2 Linear polarization technique

The most popular electrochemical technique used for corrosion monitoring is the linear polarization (LPR) technique. As described in ASTM G59 "Standard Practice for Conducting Potentiodynamic Polarization Resistance Measurements," it uses a measurement of the slope of the potential vs. current plot ±20 mV around the corrosion potential to define a parameter called the polarization resistance. Automated corrosion monitoring equipment very often uses a constant Tafel slope, i.e., 120 mV/decade, for the anodic and cathodic polarization. The corrosion current density is calculated from Eq. 3 where b_a and b_c are the anodic and cathodic Tafel slopes, respectively. B values in different systems are available[2].

$$i_{corr} = \frac{b_a \cdot b_c}{2.303 \cdot (b_a + b_c)} \cdot \frac{1}{R_p} = \frac{B}{R_p} \qquad (3)$$

The corrosion rate obtained from LPR measurements is inversely proportional to the polarization resistance. High values of polarization resistance usually yield low corrosion rates. A smaller ratio of solution resistance to polarization resistance often relates to a smaller error caused by solution resistance. The curves are seldom exactly linear. The accuracy of the linear fit to calculate polarization resistance value can therefore depend strongly on the potential range used. At the lowest overpotentials, the curve is most probably linear but may show the highest scatter especially when measuring low corrosion rates. If the corrosion rate is very low, the potential range must be large to obtain reliable current measurements. Then the curve may not be on the linear range. The linear range usually occurs at overpotentials below 30 mV.

Compared with the configuration used in laboratory experiments, field probes are simpler. A conventional reference electrode is not necessary since the measurement requires only a few minutes. One can assume that the corrosion potential of a corroding piece of metal acting as a pseudo reference electrode will remain constant during the measurement. The use of a corroding metal as a reference electrode is not sensible for corrosion potential monitoring purposes. Using a noble metal counter electrode is also not necessary, and any relatively corrosion resistant metal will suffice. The configuration is usually a flush-mounted probe with three concentric electrodes or a probe with three replaceable rod electrodes. Care is necessary to avoid using LPR in applications where the electrodes can contact oil or other viscous compounds or become covered with scale. These will prevent electrochemical measurements and can result in excessively low corrosion rates being monitored. In the presence of a biofilm, the LPR method can give an excessively high corrosion rate.

2.3 Galvanic current measurement

Galvanic current measurements are made between two dissimilar metals using a zero-resistance ammeter (ZRA). Equipment for galvanic current measurements is available from several manufacturers, but a common potentiostat is also suitable for this purpose. The idea behind zero-resistance amperometry is to set the dissimilar metals to the same potential and then measure the cell current flowing between the electrodes. Figure 5 shows the schematic coupling when using a potentiostat as a ZRA. One can often omit the resistor between counter and reference electrode connections.

The weight loss can be calculated from the measured current using Faraday's law. The calculated figure represents actual weight loss only if the sample connected as the anode has no local cells. No cathodic reactions should occur on the anode sample since the current consumed by these reactions will flow between the macroscopic samples. When local cells are present, the ZRA measurement shows only the increase in corrosion rate caused by the galvanic coupling.

Measurement of galvanic current is the only method to obtain information on actual reaction rates in a galvanic couple. Potential values given in an electrochemical series will not help in this purpose. Galvanic current measurement also has use to monitor the corrosivity of the environment by using two dissimilar metals such as carbon steel and copper or carbon steel and stainless steel. For example, oxygen entering a closed water system will increase the cathodic reaction rate. This will be evident as an increase of the current flowing between the samples. Thin strips of dissimilar metals could be installed on a metal surface under a thick coating or insulating material. When water or moisture diffuses through the coating, it will start corrosion between the dissimilar metals.

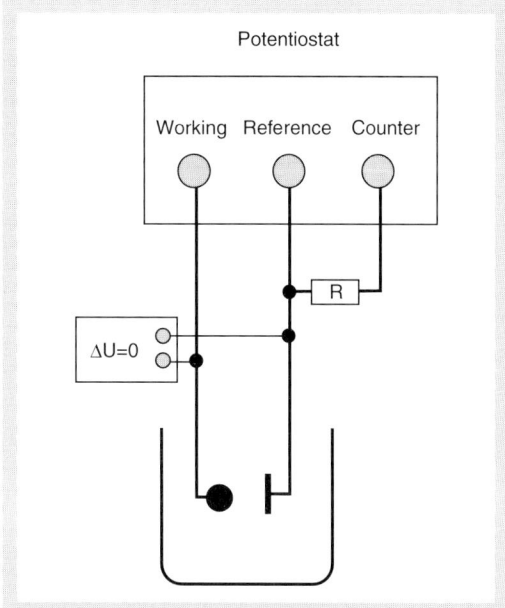

Figure 5. Using a potentiostat as a zero-resistance ammeter.

2.4 Tafel method

In the Tafel method, the sample is polarized from its corrosion potential to the anodic direction, cathodic direction, or both. At sufficiently high overpotential, the potential vs. logarithm of current density shows linear dependency. By extrapolating the linear ranges of anodic and cathodic polarization curves back to the corrosion potential, their intersections should meet and show the corrosion current density. Since the electrode reactions are never purely activation controlled in practice, linear ranges may be difficult to find, and the plot may not even show a clear linear component. Several charge transfer reactions, diffusion polarization, and ohmic polarization will distort the linearity. The measured current is converted to a corrosion rate by using Faraday's law. Using the Tafel method for corrosion monitoring purposes has many limitations. The curve fitting procedure to extrapolate the slopes should be automatic. The measured curves often do not obey Tafelian behavior. The fitting will then result in unpredictable and erroneous corrosion current densities. Every result of the fitting procedure requires visual validation.

Corrosion current densities measured with different electrochemical methods may give different results. As Fig. 6 shows, relatively large variations are possible even under controlled laboratory conditions. A polarization resistance measurement and a Tafel slope measurement conducted back-to-back have given corrosion current densities with a difference of a factor of three. In practice, only changes larger than a factor of ten in corrosion current density usually have significance.

CHAPTER 7

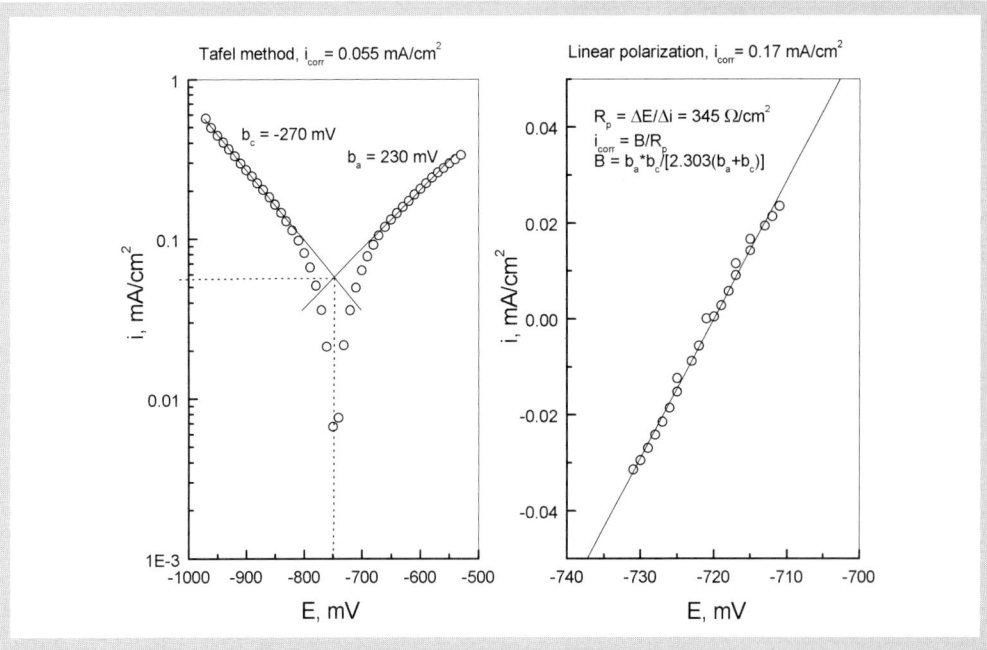

Figure 6. Corrosion current density measured by the Tafel method and linear polarization method.

2.5 Electrochemical impedance spectroscopy

Electrochemical impedance spectroscopy (EIS) is a new measuring technique that uses an alternating current signal to excite a corroding specimen. EIS monitors the electrical response of the corroding sample surface to the applied signal over a wide frequency spectrum usually from 10–100 Hz to 10–100 kHz. Lower measured frequencies equate to a longer measurement. Depending on the measurement method, measuring below the mHz range may easily require hours to complete one measurement. Depending on the frequency range, the measured impedance and phase angle describe different factors. At the highest frequencies, solution resistance is a determining factor. At the lowest frequencies, the charge transfer resistance is a determining factor. For the middle frequencies, capacitive behavior due to double layer and adsorbed species are determining. The analysis is complex compared with the commonly used ER or LPR techniques. One convenient way to use the EIS technique is to take the low frequency limit essentially the same as the polarization resistance determined by the LPR method. The EIS method also allows separation of the various components that are assumed to be part of the polarization resistance and can lead to errors in LPR corrosion rate determinations. The most important component is the solution resistance in low conductivity environments.

EIS techniques are also useful to examine coated or inhibited materials much more effectively than with dc techniques. The technique has frequent use in systems with a significant solution resistance or as a method to analyze the performance of a coating. EIS data has use in determining the properties of the surface layers such as pore resistance and film capacitance. Other areas of application for EIS are in the evaluation of corrosion of steel in concrete structures and of cathodic protection since both applications require significant compensation for resistive losses. The main limitations of these techniques are that the analysis of the data is complex and its interpretation is not fully developed for all applications. The techniques require application of a theoretical equivalent circuit to analyze and interpret the data. Curve fitting software is necessary for calculation of solution resistance, double layer capacitance, and polarization resistance. These techniques often require comparison with other more common corrosion monitoring techniques such as corrosion coupons to obtain meaningful data.

2.6 Electrochemical noise

Electrochemical noise is a passive electrochemical technique that requires no polarizing current. It measures the naturally occurring electrochemical potential and current variations that are due to corrosion. It can give accurate indications of general corrosion, pitting, and stress corrosion cracking when properly applied but requires expertise and computer power to be effective. Electrochemical noise measurement records the random fluctuations in the corrosion potential and current. Potential noise measures potential between a working electrode and a reference electrode. Current noise measures variations between two nominally similar electrodes. A further development is to measure noise resistance by measuring current noise simultaneously with a zero-resistance ammeter and potential noise with a reference electrode and a multimeter with respect to time for both. Noise resistance is calculated from these values, but the signals require correct synchronization with respect to time. Electrochemical noise has had use to identify localized corrosion and differentiate general and localized corrosion conditions. Electrochemical noise requires measurement of very small signals that are often prone to extraneous noise. This method has most common use with other electrochemical techniques.

The use of electrochemical noise measurements for corrosion monitoring has become a topic of increased interest. Some workers have proposed that the noise resistance relates directly to the polarization resistance. This is valid if the corrosion is uniform and samples are identical. Electrochemical noise is usually the result of nonuniform or localized dissolution. The analytical methods are under constant development. Fast Fourier transforms, maximum entropy, and Poisson analysis of peak maxima have had use. The electrochemical noise measurement may require so much expertise that it will not find acceptance as a monitoring tool by industry.

3 Other methods

Besides the physical measurements that register extent of corrosion damage and electrochemical measurements that register instantaneous corrosion rate, several other methods exist that have use for monitoring. Some use analysis of reaction products caused by corrosion reactions, e.g., hydrogen monitoring and metal ion analysis.

CHAPTER 7

3.1 Hydrogen flux monitoring

Atomic hydrogen is a common product of corrosion in acid environments resulting from a cathodic hydrogen evolution reaction. Atomic hydrogen can recombine to form molecular hydrogen and come off the metal surface or it may diffuse into the metal in atomic form. The presence of hydrogen sulfide and other compounds called "hydrogen poisons" prevents the recombination reaction and promotes hydrogen absorption. Hydrogen is an indication of corrosion rate but can also cause damage. Hydrogen flux monitoring uses measurement of the pressure increases caused by hydrogen entering the probe or electrochemical techniques. Probes are inserted into the vessel or pipe section. They consist of a steel sensor that has a hollow space inside connected to a pressure meter. As atomic hydrogen diffuses through the sensor, it recombines in the hollow space causing a pressure increase. The rate of pressure increase is assumed proportional to the hydrogen absorption rate. Hydrogen probes installed outside the equipment use an externally applied electrochemical cell to monitor the rate of hydrogen transfer.

The limitations of hydrogen probe monitoring are numerous. Hydrogen probe data may not correlate with weight loss corrosion since corrosion rate is only one factor involved in defining the severity of hydrogen absorption. The diffusivity of hydrogen in steel also changes rapidly with temperature, composition, and microstructure. This can produce large differences in hydrogen probe measurements. Such techniques can at least provide qualitative information used to monitor the severity of hydrogen charging and the potential for hydrogen cracking of exposed process equipment. For more quantitative monitoring applications, such considerations require inclusion in the analysis of the data.

3.2 Chemical analysis

Process water samples can also be a valuable source of indirect monitoring data. Information such as pH, dissolved metal content, and chloride, hydrogen sulfide, and ammonia content may give information on corrosivity of the environment and corrosion rates. For example, NACE standard RP0192 "Monitoring Corrosion in Oil and Gas Production with Iron Counts" describes an application for estimating corrosion rates in oil production equipment. The iron production rate given as kg/day reflects the effectiveness of corrosion prevention measures. Iron concentration expressed as mg/L will only be relevant if the flow rate is constant. Several samples taken from the same points in the same manner and analyzed by the same methods are necessary to provide baseline information.

Several solution properties allow monitoring of solution corrosivity. To link these values to corrosivity, the relationship between corrosion rate and pH, chloride content, etc., require determination beforehand. Samples of corrosion deposits can also be analyzed to determine the nature of the corrosion and the reactive species causing the attack.

4 Data collection and analysis

Corrosion monitoring with coupons gives a baseline for corrosion rate estimations. Corrosion coupons are sufficiently accurate to provide information on total corrosion rates but do not give details of why and when corrosion occurs. Corrosion coupons have limited use in active corrosion prevention. To relate corrosion rates to other process variables, information must be available quickly. More real-time data collection procedures equate to more active use. One should always consider corrosion data with other process variables. Only then will useful information result. Change in corrosion rate may relate directly to a change in a controllable parameter such as temperature, pressure, flow rate, or chemical composition. An industrial plant may have a high number of process chemicals. The complexity and dynamic nature of process environments makes estimating their effect on equipment corrosion difficult. Corrosion monitoring has use here to minimize corrosion damage.

Corrosion is seldom a steady-state process. Corrosion rates derived from weight loss coupons and presented as loss of thickness in a year can easily lead one to think of constant thinning of an equipment wall. Periods of low corrosion and high corrosion usually occur. Identifying the causes of high corrosion rates can give possibilities for effective corrosion prevention measures. Selection of appropriate instrumentation with balanced monitoring methods is a crucial element. Corrosion coupons will provide a basis for interpretation of data from corrosion probes and other on-line corrosion monitoring devices. They should be installed on retractable probes to allow installation and periodic removal during operation. Electrical resistance, corrosion potential, and linear polarization probes should be used with coupons to provide more continuous data. Corrosion coupons and ER probes are valuable when placed in a region of potential corrosion. They will provide useful information only if they actually corrode. ER probes are often the most common method in a monitoring program. Their use is a simple and inexpensive technique that produces a continuous and regular flow of data. A major advantage of the electrical resistance technique is the ability to work in almost any environment since it does not require an electrically conductive medium to function.

Several options for data collection are available. The simplest is a manual method using a hand-held instrument. Manual methods have low cost but provide less data than either continuous data logging or on-line instruments. Active monitoring requires more detailed data. Continuous data logging using field mounted units, periodic data collection, and on-line instruments linked to a control room give more continuous data than manual data collection. Separate data logging units are less expensive than real-time instruments that should preferably have use at critical points. For example, corrosion monitoring coupons and sensors should be located in sites where water will condense or impinge, chemicals will be added, or flow will change abruptly since corrosion is usually worst in these locations. Those parts of the system with the highest corrosion risk that are the most critical to the operation are proper places for corrosion monitoring.

References

1. Gartland, P. O., Horn, H., Wold, K. R., et al., FSM - Developments for monitoring of stress corrosion cracking in storage tanks", CORROSION 95, paper no. 545, NACE, Houston.

2. Grauer, R., Moreland, J. P., and Pini, G., A literature review of polarization resistance constant (B) values for the measurement of corrosion rate, NACE, Houston, 1982, 66 p.

Corrosion monitoring

CHAPTER 8

Corrosion management

1	**Design and material selection**	**349**
2	**Operation**	**351**
3	**Inspection**	**352**
4	**Maintenance**	**354**
	References	356

CHAPTER 8

Jari Aromaa and Anja Klarin

Corrosion management

Corrosion control problems are very complex. They include more than simple correction of a basic problem such as a leaking pipe. Corrosion control encompasses all the procedures used to design, build, and operate a plant without excessively severe corrosion problems. Effective corrosion control therefore requires cooperation of various groups including management. Proper corrosion control involves information distribution. Effective corrosion management uses a system engineering strategy to improve the performance of engineering systems by specifically including people[1]. Managing corrosion is a primary task of preventive maintenance programs.

Earlier workers have claimed that a significant amount of corrosion costs could be prevented by using existing technology and knowledge. For example, Battelle and the Specialty Steel Industry of North America estimated in 1996 that about one-third of the cost of corrosion is avoidable and could be saved by broader application of corrosion resistant materials and application of best anti-corrosive practice from design through maintenance.

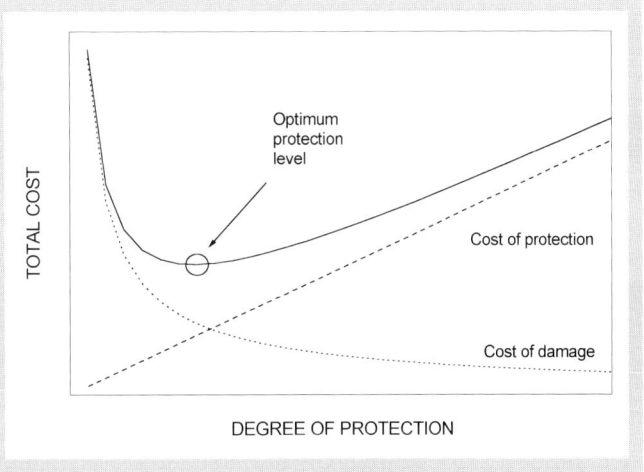

Figure 1. Schematic diagram for minimizing total cost of corrosion.

Corrosion control and treatment are important because corrosion of equipment and structures has considerable influence on operational and structural safety. Economics is another basic consideration since corrosion will eventually weaken structures to the point where replacement or reinforcement is necessary. Corrosion problems are solely economical problems when corrosion damage has no secondary occupational or environmental effects. Minimizing the total cost of linearly increasing protection costs and exponentially decreasing corrosion costs is then easy as Fig. 1 shows. With no protection the corrosion damages cause heavy costs. By applying some protection, a significant reduction in total costs is achieved. Above optimum protection level, the additional cost of protection is higher than the resulting decrease in cost of damage.

The relative economics in a chemical plant construction are extremely complex. The use of low-cost materials can cause high maintenance costs and higher production losses. Materials having better resistance to degradation are intrinsically expensive.

CHAPTER 8

Costs of inspection and maintenance are perhaps easier to estimate. As any dentist will advise, cleaning and periodic inspection will reveal minor damages. Repair costs then are not too high. Waiting until a system fails will always cause higher repair costs.

Corrosion management of an installation contains four different phases: design, construction, use, and remedial actions. Anything that occurs in one phase will have an influence on the other phases. Therefore, persons responsible for corrosion management must be appointed. For the final three stages, the people can belong to the maintenance force. The tasks in a corrosion management group should include process and equipment design, material selection, manufacturing and construction, operation, maintenance, and economics. Each task must develop and apply corrosion control procedures to satisfy particular requirements. Responsibilities must be assigned, and controls must be sufficiently stringent to ensure their fulfilment. The primary task of the entire system is to minimize the total cost of corrosion including initial investments, corrosion control measures, maintenance costs, additional safety measures, loss of production due to shutdowns, etc. Comparing costs of corrosion vs. production value using a suitable time interval is especially useful in an industrial facility. An example is cost of corrosion prevention measures vs. value of pulp produced in one year.

A properly managed corrosion control program will prevent and mitigate corrosion attack in its early stages. Continuing surveillance is necessary to detect corrosion attack. Various corrosion monitoring and inspection methods are available for this purpose. Time and available personnel permitting, all equipment requires inspection for signs of corrosion during scheduled and random shutdowns. These activities require organization and proper management to produce an effective corrosion control program. Minor maintenance can correct corrosion problems detected sufficiently early. Preventive maintenance is the most cost-effective method of controlling corrosion including problems caused by poor design. Without proper preventive maintenance, corrosion can seriously damage equipment. Material and equipment that require special treatment to protect them against corrosion are naturally those most vulnerable to corrosion attack and those that can lead to the most expensive damages. They require careful inspection and maintenance.

Preventive maintenance related to corrosion control could include the following specific tasks:

- An adequate cleaning program including periodic lubrication and periodic removal of accumulated water and other foreign matter

- Maintaining adequate drainage to prevent moisture entrapment

- Protection of sensitive equipment against water, dust, and chemicals by using covers or by storage in a protected environment (Systems used for process control are particularly vulnerable.)

- Detailed and scheduled inspection of equipment and systems to detect corrosion, damage, and failure of protective coating systems

- Appropriate remedial or repair action that is preferably preventive in nature

- Moving surfaces that must remain bare require frequent wiping to clean
- Regular inspection of absorbing materials such as thermal insulation that contact metals
- Inspecting electrical equipment often to prevent current leakage and stray-current corrosion that can be extremely rapid
- Diligent maintenance of electrical equipment providing corrosion protection such as anodic or cathodic protection.

1 Design and material selection

The main task in material selection is to have the right material for the specified service and to ensure its correct fabrication and maintenance. Persons involved in material selection must give careful attention to proper material specification using appropriate standards and dimensions, correct design and fabrication procedures, quality and inspection requirements, and instructions for in-service monitoring and maintenance. Material specifications should include known limitations and ensure that the composition and condition of the purchased material is identical to the material from which corrosion resistance data was obtained. Any special requirements on surface finish, hardness, etc., are also necessary.

Proper material selection involves five primary considerations:

- Mechanical properties
- Fabrication
- Corrosion resistance
- Availability
- Total cost.

The most important mechanical property is alloy tensile or yield strength. Hardness, fatigue resistance, impact strength, etc., often require consideration. The material must have sufficient corrosion resistance in the service environment or the mechanical properties will have no use. Poor corrosion resistance will cause early failure regardless of good structural or mechanical design or mechanical properties of the material. Ability to fabricate and availability are extremely important. Depending on the application, an ideal material should form easily and join to the required component shape. Alloys must keep their key properties with all fabrication methods. Availability also has considerable importance and often is the most crucial selection consideration. After studying all the relevant factors, making an estimation on the total cost is possible. Total cost can be estimated for different material options using initial price, installation costs, and costs of maintenance and corrosion prevention during the expected effective life.

Figure 2 shows the dependence of corrosion resistance on cost. For an environment with known corrosivity, evaluation of material performance is possible. When comparing similar materials such as stainless steels, they will often fall into two classes.

CHAPTER 8

One will have excessively high corrosion rates, and the other will have adequate corrosion rates. Between these groups is a transition area. Proper material selection is naturally the least expensive alloy with adequate corrosion resistance. If the system becomes more severe, the transition area moves to more expensive materials. To obtain some insurance against unexpected increase in corrosivity, a more resistant material than the primary candidate may be selected. Figure 2 is not useful for comparing totally different materials such as fiber reinforced plastic and titanium.

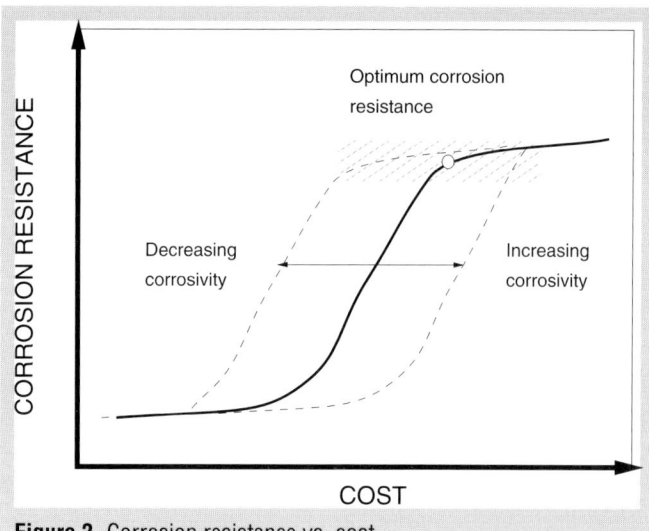

Figure 2. Corrosion resistance vs. cost.

Cooperation between designer, material expert, and corrosion expert is extremely important to minimize the risk of corrosion and avoid costly design and material selection mistakes. Pressurized vessels and components and especially those subjected to cyclic stresses are particularly challenging. The designer must know the physical and chemical factors affecting the materials. A corrosive environment influences the mechanical behavior of the system. The size of defects such as cracks on the surface becomes more critical when corrosive agents are present.

Corrosion is not an isolated phenomenon. It will affect production, economics, and safety of a plant. Corrosion problems require evaluation considering the other problems they may cause. Corrosion prevention costs should be justified by the costs they avoid. These costs can be due to production losses, replacement costs, impact on product quality, safety, etc. Corrosion is inevitable, so preventive measures are unavoidable. Careful design and material selection will reduce the risk of corrosion. Over-design or over-conservative selection of materials can add so much to capital cost that they make projects commercially unfeasible.

The design stage is very important since errors will not be evident until a unit has operated for some time. Anticipated corrosion rates for normal operation and starting and stopping periods require estimation. Construction details to include application of paint, joining techniques, accessibility for inspection and maintenance, and application of various corrosion prevention methods require careful planning in the design stage rather than addition as an afterthought. To assist during construction, one can use a check-list for various installation and construction stages. For example, the following details may apply during installation of valves:

- Install the valve with the valve spin facing upwards (Leaking of the seal may cause dripping on actuators and corrosion.)
- Minimize the risk of cavitation by installing valves on return lines (lines with lower temperature)
- Avoid erosion by not installing valves immediately after pipe bends or connections and close to pumps because of possible low pressure
- Prevent condensation outside the valve body in cooling applications by using an appropriate paint system
- Use saturated or superheated steam with steam applications (Removal of condensate must operate well or erosion will occur.).

Errors during the construction phase can result in corrosion problems at the construction site, the manufacturer's or subcontractor's site, or during storage and transit. Adherence to specifications, inspection of welded joints and coatings, and proper temporary corrosion prevention procedures will diminish the risk of fabrication mistakes. A major risk is confusing materials. Poor identification systems can result in material confusion, e.g., "stainless" vs. "acid-proof" steel. In large constructions, welding consumables and small items such as pumps and valves may have several material specifications. The purchaser may sometimes specify "stainless steel valves," but the manufacturer supplies valves with a carbon steel body and stainless steel valve spin. To avoid such misunderstandings, preliminary discussions and detailed specifications are necessary. Portable analytical equipment is extremely useful during major overhauls to avoid material confusion. Construction, inspection, and test reports must be archived and updated to contain correct information, i.e., "as-built." Proper inhibition and correct function of electrochemical protection systems must be ensured before system start-up and documented.

2 Operation

Effective corrosion management requires cooperation of operations and maintenance personnel. For the maintenance people to provide maximum equipment availability, operation personnel must accept the fact that some scheduled downtime is necessary for inspection and maintenance. An analogy is that the race car must go into the pit sometimes. The maintenance people must know about any changes in production methods and rates. They should be consulted about equipment overhauls, process repairs, improvements, and routine maintenance with which they can be directly involved. In the long run, equipment will not perform as intended or provide acceptable service life without this "cooperative" approach by operation and maintenance personnel.

Plant management can have a significant effect on corrosion problems that may appear during operation. Changes in production rates, production methods, or products will usually have some effect on the corrosivity of the process streams. Knowledge of the effect of these changes is necessary beforehand. A modern practice is to use process simulation or actual trials to define the altered environmental conditions. Small changes in operation can produce large effects. For example, a slight increase in temperature may initiate localized corrosion. Changing a chemical supplier may result in enough variation of the chemical purity to change corrosion rates. The corrosion management responsibility of plant management is to communicate all planned or actual changes in production methods to the persons responsible for materials, corrosion, maintenance, and inspection.

The people actually operating the equipment are extremely important when trying to reduce corrosion problems, improve management of corrosion, and obtain additional benefits in a better work environment for safety. Most operator training concentrates on running the process, and problems are rare. Most problems arise from errors in less frequent operations. For example, descaling with acid may occur every few weeks. This requires a temporary shutdown. A temptation might exist to use excessively concentrated acid or excessively high cleaning temperature to reduce downtime, but this could cause more corrosion.

Properly trained operators should be familiar with the operating and maintenance instructions provided by the equipment manufacturer covering automatic and manual operation of the equipment. Such personnel should also be familiar with how the process unit functions and the purpose of all controls. The controls should be tested periodically to ensure proper function. Written instructions should be available to handle emergency situations. A continuous training program with refresher training for emergency situations can save considerable trouble.

3 Inspection

Inspections are a critical part of any preventive maintenance program. Inspections define the system condition and identify maintenance requirements. Review of maintenance records and repairs, operating log sheets, flow charts, and reports of prior problems are important for guiding inspections. This will also assist in identifying any specific areas that may require additional inspections. Inspection procedures require modification under any of the following conditions:

- Material changes
- New failure mechanisms occurring
- New requirements occurring
- New analytical, testing, or detection tools becoming available.

A conscientious and organized inspection is an effective weapon against breakdowns, because inspections often detect faults that can be handled before the situation becomes critical. An inspection program includes two separate parts: inspection during operation and inspection after a failure. Inspection during operation includes subjective and objective inspection. The inspectors must know the acceptable and unacceptable conditions. Subjective inspections involve human senses of looking, listening, feeling, and smelling. They usually have nonspecific criteria such as a gearbox is more noisy or hotter. Objective inspections use measurement criteria to determine actual conditions or changes in condition. Training of inspectors and selection of inspection points is necessary in every corrosion management program.

Inspection programs can be optimum for cost and safety by combining process information, data from corrosion, and condition monitoring from inspections. The selection of inspection points is important. A list of potentially risky places is the following[2]:

- Abrupt changes in direction of flow such as elbows, tees, return bends, and changes in pipe size (These points can create turbulence.)
- Presence of "dead-ends," loops, crevices, obstructions, or other conditions that may produce turbulent flow and increase corrosion rate or cause stagnant flow allowing debris to settle or corroding materials to accumulate
- Junctions of dissimilar metals that might promote galvanic corrosion
- Stressed areas such as those at welds, rivets, threads, or areas that undergo cyclic temperature or pressure changes.

Most pulp and paper mill equipment is subject to vibrations, temperature changes, thermal and mechanical fatigue or shocks, and corrosion and erosion. Knowing whether a crack or leak presents a problem comes with experience. This experience encompasses an understanding of equipment, materials, welding, and heat treating with the ability to analyze conditions accurately by using meaningful techniques.

Scheduled inspection and preventive maintenance are essential to determine system status and provide early correction of weaknesses. Preventive maintenance reduces the total amount of labor used and the expense incurred and ensures corrosion will not prevent the particular system from performing its designed function.

Information from corrosion monitoring can be useful for corrosion management in several ways. It provides information for maintenance and inspection management, optimization purposes, etc. Corrosion monitoring is a vital part of any plant corrosion control program. It requires integration with other programs designed to optimize the process conditions.

Corrosion damages and other failures will occur. They require prompt action to minimize additional damage and lost production. An effective team with people from operation and management supported by experts from materials, corrosion, etc., can rapidly give suggestions on courses of action. Selection of personnel for such a team requires advance consideration.

When investigating a corrosion failure, an analyst must do the following:

- Estimate the extent of damage and the likelihood of additional and consequential failures
- Determine the failure mode
- Determine the failure cause
- Design and implement a corrective action
- Ensure implementation of the corrective action
- Ensure that the corrective action will prevent another failure.

CHAPTER 8

Table 1. Background information for a corrosion failure investigation.

Environment	Material	Construction	Form of corrosion
Chemical composition	Material type - steel - stainless steel - nickel alloy - coating - polymer	Crevices	Form of corrosion - uniform - pitting - crevice
pH and temperature	Chemical composition - standard designation - analysis	Deposits	Corrosion depth
Redox potential	Coating - paint type - paint specification - polymer coating	Mechanical stresses	Visible cracking
Dissolved gases	Construction - cast - welded	Abrasion	Erosion or cavitation
Immersion - always dry - continuous - intermittent	Heat treatment - annealing - prevention of sensitization	Liquid/gas interphases	
Pressure and temperature variations in gas phase	Manufacturer	Vibration	
Condensation from gas phase		Joining methods	
Flow rate - continuous - stagnant - variable			
Solids in flow			
Startup/shutdown period Maintenance/repair/revision			

The first task is to decide if production can continue. The next step is to determine what happened and why it happened. Finally, the problem requires solution with verification that corrective actions are in place to avoid reoccurrence. Correct failure analysis is an extremely important task when trying to solve corrosion problems. An expert with much experience can save considerable time and money merely by thinking rather than starting a large experimental program. Damage reports are an extremely valuable source of information. This experience should be available to everyone in design, material selection, maintenance, etc., to avoid making the same mistakes. Table 1 provides a list of factors to consider for background information.

4 Maintenance

Maintenance has two categories: corrective maintenance and preventive maintenance. The primary task of maintenance is keeping equipment efficient and extending its service life as cost-effectively as possible. Maintenance has become an increasingly vital business function through its goal of assuring maximum equipment uptime at peak performance with lowest total costs. The modern approach to preventive maintenance includes routine care, regular inspections and planned repairs, and overhauls and reconditioning to assure the greatest equipment service reliability. Corrosion prevention should be a major task in every preventive maintenance program. By using suitable tools of routine care, inspection, and planned repair work, mitigation of corrosion can give large benefits. The operating department and the maintenance department should cooperate to minimize process downtime by using preventive and predictive maintenance procedures and computerized maintenance management systems. These objectives require specific process and equipment knowledge that some maintenance contractors may not possess.

Predictive maintenance is actually an extension of preventive maintenance. Predictive maintenance is collecting and analyzing inspection data. Data collection and analysis can be accomplished while the equipment is running. Assuming that measurements of vibration, heat, tension, speed, corrosion rate, etc., are within acceptable levels, the equipment is operating efficiently. Wear causes these measurements to drift beyond established control limits. Then preventive maintenance is necessary to restore the equipment to optimum operating conditions. Predictive maintenance considers operating time or equipment condition. It gives tools to prevent production downtime (preventive maintenance) and more costly emergency repair (corrective maintenance).

Design of a system includes decision of its operational lifetime. The realization of this requires periodic inspection and maintenance. For example, measurement and monitoring of corrosion rates is necessary for preventive maintenance. By monitoring corrosion with an ER probe, one can estimate remaining wall thickness. When the wall thickness is less than a preset alarm level, starting planning for preventive maintenance is necessary. Preventive maintenance repair then begins when the wall thickness is less than a preset repair level as Fig. 4 shows.

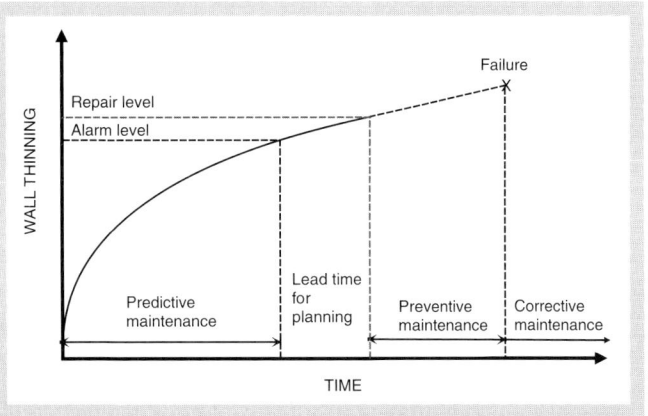

Figure 3. Monitoring of wall thickness and its relation to predictive, preventive, and corrective maintenance steps (not to scale).

A good monitoring program combined with selected inspections will give a base to confirm that the corrosion rates obtained from laboratory or pilot plant tests used in the design phase were relevant. A proper preventive maintenance program and a reliable corrosion monitoring program may even allow longer times between inspections required by government agencies or jurisdiction.

References

1. Trethewey, K. R. and Chamberlain, J., *Corrosion for Science and Engineering*, 2nd edn., Longman Scientific & Technical, Essex, 1995, Chap 11.
2. Abramchuk, J., *Materials Protection* 1(3):60(1962).

CHAPTER 9

Concepts of maintenance in the pulp and paper industry

1	**Industrial maintenance**	**359**
2	**Maintenance yesterday and today**	**360**
3	**Managing maintenance people**	**361**
3.1	Effective maintenance program	361
3.2	Maintenance manpower in different industries	362
3.3	Cost considerations of maintenance	364
4	**Reliability and maintainability**	**366**
4.1	Reliability and failures	367
4.2	Reliability and maintenance engineering	369
4.3	Condition assessment	372
5	**Equipment availability and efficiency**	**376**
5.1	Lost-time analysis	376
5.2	Production potential	378
6	**Evolution of team approach and multicraft**	**381**
6.1	Evolution of multicraft	381
6.2	Owner-operator	382
6.3	Costs and benefits of craft interference	382
7	**Trends in different decades**	**383**
7.1	Main trends in the pulp and paper industry	383
8	**Maintenance management trends by Idhammar**	**386**
8.1	Reliability centered maintenance	392
8.2	Equipment efficiency through operations and maintenance	393
8.3	Computerized maintenance management	394
8.4	Waste in maintenance	396
8.5	Autonomous maintenance	396
9	**Planning maintenance**	**398**
9.1	Selecting maintenance planners	399
10	**Examples of maintenance management**	**402**
10.1	Maintenance at International Paper mill	402
10.2	Preventive maintenance in a Norwegian mill	405
10.3	Maintenance training in Sweden	408
11	**International maintenance reviews**	**409**
11.1	Canadian reviews	409
11.2	North American surveys	413
12	**Maintenance costs in comparison with other costs**	**419**

CHAPTER 9

12.1	Maintenance costs compared with product costs	419
12.2	Maintenance costs compared with investment costs	420
12.3	Maintenance costs compared with breakdown costs	421
13	**Equipment vendor's role in maintenance**	**422**
13.1	Single equipment vendor	422
14	**Vendor of a turnkey project**	**423**
14.1	Action plan for a pulp and paper mill	423
15	**Conclusions**	**424**
	References	427

CHAPTER 9

Anja Klarin

Concepts of maintenance in the pulp and paper industry

1 Industrial maintenance

In today's competitive marketplace, pulp and paper companies must achieve high productivity while decreasing and controlling total costs. Maintenance must be a partner in this endeavor and not merely a "cost of doing business." The effectiveness of the maintenance function largely determines the availability and performance of production equipment. This has a large impact on efficiency. Maintenance therefore plays an increasingly significant role in determining the efficiency and perhaps survival of pulp and paper companies. An estimated US$ 250 billion is spent each year on maintenance and repair in the aggregate for all North American industrial plants[1].

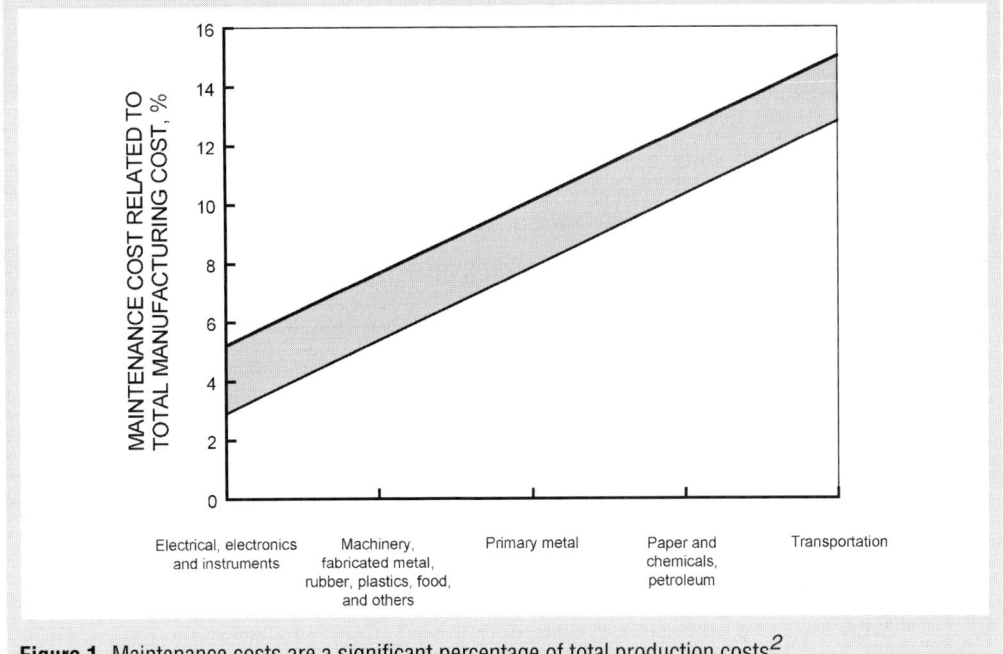

Figure 1. Maintenance costs are a significant percentage of total production costs[2].

Industrial maintenance is starting to make a transition from a "fix it" focus to that of a high-level business function. At many facilities, increasing maintenance costs constitute 4%–14% of production costs and are often greater than profits as Fig. 1 shows.

Even when maintenance productivity increases as evidenced by programs in certain Finnish pulp and paper mills, the costs of maintenance are still 5%–8% of the turnover of the mill depending on the type of paper produced[3]. A fact of industrial life is that machinery and equipment that receive proper care and maintenance will malfunction and eventually experience breakdown unless reconditioned or replaced.

2 Maintenance yesterday and today

The cost of maintenance per ton of product can be a gross measurement of maintenance productivity. In the 1950s, the average was 2 man hours/ton of paper[4]. Today, it can be as low as 0.5 man hours/ton. In the past, other measurements included the number of electric motor failures and machine availability. In today's industrial workplace of high labor costs, expensive machinery, and ever-threatening market competition, effective maintenance is not only important but also vital for profitable operation and economic survival.

The old and primitive approach of "fixing it only when it falls" results in the following:

- Costly interruptions to production
- Extensive and unnecessary stocking of costly spares
- Shortened equipment life
- Inefficient use of personnel
- Unsatisfactory equipment operation and difficulty with product quality control
- Constant "fire-fighting" to remain ahead of breakdowns
- Excessive, hidden maintenance-related costs.

The modern approach of preventive maintenance involves a commitment to prevent failures and breakdowns through the following:

- Routine care including cleaning, lubrication, and adjustments
- Regular inspections to detect faults before they progress to serious proportion allowing repairs on a planned basis with minimum interference with production schedules
- Planned repairs, overhauls, and reconditioning to assure the greatest equipment service reliability.

The overall results of preventive maintenance are as follows:

- Increased machine operation time
- Improved and consistent quality of pulp and paper products
- Decreased total costs for maintenance services.

Each dollar saved in maintenance can equate to as much as 10 dollars in product sales. To achieve the total maintenance goal (maximum uptime and peak equipment performance at minimum cost), the maintenance function must focus on using its resources in the most productive manner. These resources include instrumentation, equipment, materials and parts, budget monies, and most important of all the people in labor and administration.

3 Managing maintenance people

Management of maintenance requires three items:

- Organization: a structural plan for job functions and responsibilities ("doing" and "planning"), communications, work requests (a work order system), record keeping, and lines of supervision

- Strategy for action: policies and guidelines for work planning and scheduling, setting performance goals, training, commitment to preventive maintenance, and related improvements in reliability and maintainance

- Administration: leadership, assessment of productivity and work quality, monitoring workplace safety, facilitating needed changes and improvements, and cost control.

According to Baldwin[1], good maintenance performance is the result of a sound organizational structure designed to use people effectively. When people do the right actions at the right time, maintenance becomes a planned and rehearsed event rather than a chance occurrence. Maintenance programs designed to upgrade the maintenance effort can only be implemented and sustained within an organizational structure that is proper. A mill will pay a heavy price for a maintenance organization structure that does not perform the job correctly. The key point is that all maintenance personnel (supervision and labor) must be involved in managing maintenance and share in the associated responsibilities. The role of labor unions is of importance in North America and Nordic countries. In North America, not all maintenance labor belongs to a union as in the Nordic countries. Some unions especially in the Nordic countries have so much power that they obtain their demands with strikes if necessary.

Baldwin reported that only about 20% of maintenance work in North American industry occurs on a planned basis. Realizing significant gains in the areas of planning and scheduling requires that maintenance and production personnel treat maintenance as a mutual "investment." Maintenance people must emphasize continual effort and action, and production people must support and respect the maintenance strategy.

3.1 Effective maintenance program

Marks of an effective maintenance program include:

- Good organization and sharing of responsibility by everyone
- Using a "reliability-based" preventive maintenance approach to maintenance

CHAPTER 9

- Maximum work planning and scheduling
- Good communications and information flow between labor, supervisory staff, and administration
- Adequate training including technical and personnel interrelations
- Good record keeping with key information readily available for planning and action
- Constant commitment to improvement (equipment and practices)
- Close working relationship with production
- Constant effort to monitor maintenance effectiveness, work quality, and costs
- Effective provision of material and spare items on a cost-management basis
- Focus on people for a quality work environment and maximum support of the program by all personnel
- Strong and continual commitment to safety by all personnel in planning work, monitoring total conditions in the workplace, and performing regular daily activity.

3.2 Maintenance manpower in different industries

How many people would one expect in the maintenance function for a typical plant? Unfortunately, no single correct answer exists because a "typical" plant does not exist. Notable among the many factors affecting maintenance manpower requirements are the type of industry, size and age of the plant, the degree of automation implementation, extent of diversion of maintenance people to capital projects, caliber and productivity of the maintenance workforce, level of preventive maintenance practiced, and the standard applied to defining what constitutes a satisfactory degree of maintenance.

The Journal of Plant Engineering conducted a mail survey of 8424 of its readers in 20 major manufacturing industries. Table 1 shows a summary of the results of the survey of the 1637 responses presented by industry (type of products manufactured) and plant size (total number of employees)[5]. The overwhelming preponderance of problems concerned the inability to attract, train, and retain qualified people at all levels including crafts people, engineering support, and management. The most commonly cited reasons for the scarcity of qualified crafts people were improper preparation at the high school level, lack of time and money for training, rapidly advancing technology outpacing the skills of even the most competent crafts person, low pay scales, and a requirement to staff from within imposed by union contract. Other factors cited were the budgetary constraints of the department (raw headcount) and the department's perceived role as a stepchild of production. Numerous other comments dealt with the pros and cons of contract maintenance vs. in-house staff. These comments usually favored using contract maintenance personnel to supplement but not replace in-house staff.

The data in Table 1 regarding the pulp and paper industry are extremely low. In 1972, maintenance workers comprised about 25% of the workers in a typical Swedish pulp and paper mill. Today, that figure is closer to 31%[6].

Table 1. Maintenance staffing level by industry[5].

Total plant employment	Total maintenance employees					
	Under 10	0–25	26–50	51–100	101–200	Over 200
	Construction and material handling machinery					
Under 100 (3)	100					
100–499 (12)	41.7	58.3				
500–999 (3)	33.3	66.7				
1000 and over (2)			50			50
	Metalworking machinery					
Under 100 (0)						
100–499 (25)	56	28	16			
500–999 (4)		50	25	25		
1000 and over (5)		20	20	40	20	
	General machinery or equipment					
Under 100 (3)	100					
100–499 (21)	61.9	28.6	4.8	4.8		
500–999 (9)	11.1	44.4	33.3	11.1		
1000 and over (2)	50	50				
	Aircraft and aerospace equipment					
Under 100 (0)						
100–499 (16)	31.3	68.7				
500–999 (10)	20	40	40			
1000 and over (21)		11.5	15.4	15.4	46.2	11.5
	Paper and allied products					
Under 100 (4)	75					
100–499 (59)	39	37.3	18.6	3.4	1.7	
500–999 (18)		5	25	25	35	10
1000 and over (13)		15.4	7.7	7.7	15.4	53.8
	Chemicals, paints, and plastics					
Under 100 (7)	57.1					
100–499 (77)	19.5	39	16.9	19.5	5.2	
500–999 (18)		11.1	27.8	22.2	22.2	16.7
1000 and over (8)		12.5		12.5	25	50
	Rubber or plastic products					
Under 100 (2)	100					
100–499 (95)	24.7	43.2	20	2.1		
500–999 (18)	4.3	26.1	21.7	34.8	13	
1000 and over (15)			6.7	13.3	73.3	6.7
	Metal casting, rolling, and drawing					
Under 100 (0)						
100–499 (57)	28.1	33.3	31.6	7		
500–999 (16)	12.5	25	12.5	12.5	37.5	
1000 and over (14)			7.1	7.1	21.4	64.3

3.3 Cost considerations of maintenance

Cost management of maintenance activities should not be simply a matter of "spending the least money" but always to keep equipment in optimum economic operating condition. Money must be spent wisely and effectively. In more progressive companies, the maintenance department is a "cost center" where monies spent on the maintenance function determine to a great extent the total profitability of the company.

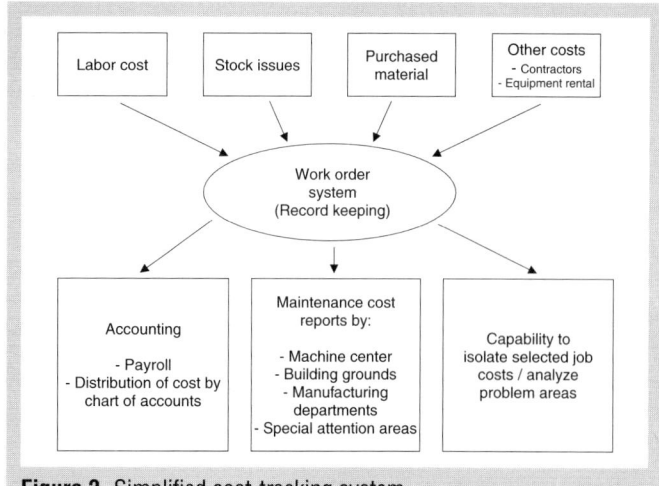

Figure 2. Simplified cost-tracking system.

Costs associated with maintenance include:

- Labor (trades, technical-support staff, and administration and management)
- Materials (losses due to erosion, corrosion, or both)
- Lubricants
- Replacement parts
- Stocking of spare tools, equipment, and instrumentation
- "Outside" services of contractors and shop training.

Figure 2 illustrates a simple cost-tracking system[1].

Industry-wide experience shows repeatedly that an effective preventive maintenance program reduces costs associated with breakdowns (lost production and repair costs). Figure 3 shows a general life span model for complex machines[2]. The figure shows that effective application of reliability-based maintenance significantly lowers the probability of failures particularly in the early and later life of a machine.

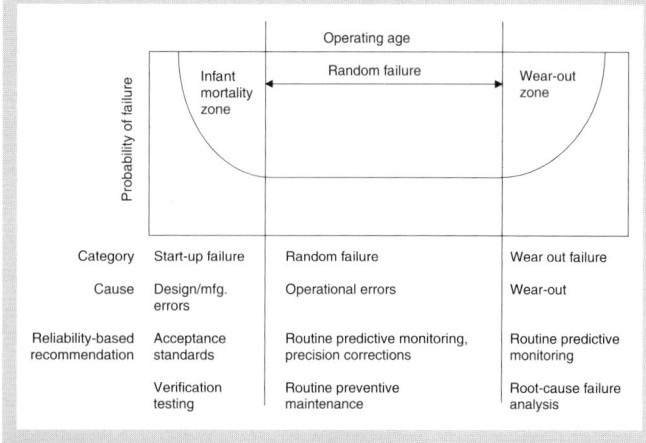

Figure 3. General life span model for complex machines.

The cost of preventive maintenance can increase with the extent of the preventive effort especially if inspections and preventive action require significant equipment downtime. As suggested by the diagram in Fig. 4, the objective is to establish an optimum preventive maintenance level for the lowest total cost. Beyond this limit, money spent on preventive maintenance results in little gain.

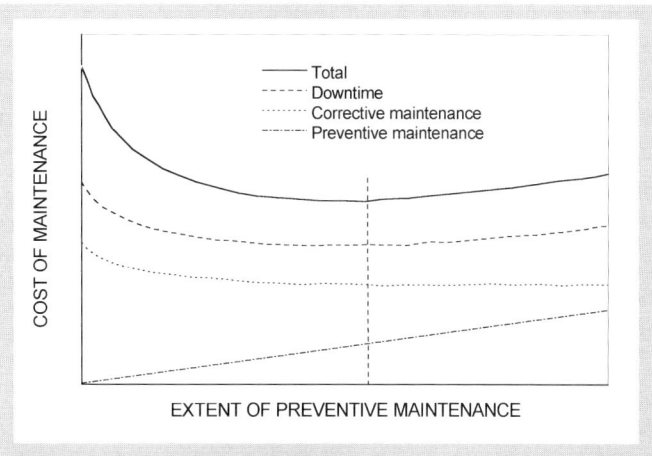

Figure 4. Total maintenance costs determined by preventive maintenance (Not to scale).

Total effective preventive maintenance means:

- Equipment condition determined using "on-line" measurements and monitoring as far as practical to minimize downtime

- Equipment shut down for maintenance only when this is the most cost-effective action

- Preventive maintenance requiring downtime has carefully planning with performance in the minimum time using all maintenance windows for equipment access

- Corrective, rebuild, and improvement maintenance strategically timed and planned for maximum effectiveness

- All condition monitoring "cost-justified."

The result is a significant reduction in total maintenance cost as Fig. 5 shows[7]. This approach might be second generation preventive maintenance that would be a possible strategy for the future.

CHAPTER 9

Major cost decisions (equipment modifications and replacement) require a systematic, analytical approach to assure short-term and long-term economies. Typically, situations involving expenses and returns occurring more than one year into the future must also assess the interest costs to reflect the true economic picture. Mill equipment performance or effectiveness therefore results from the following:

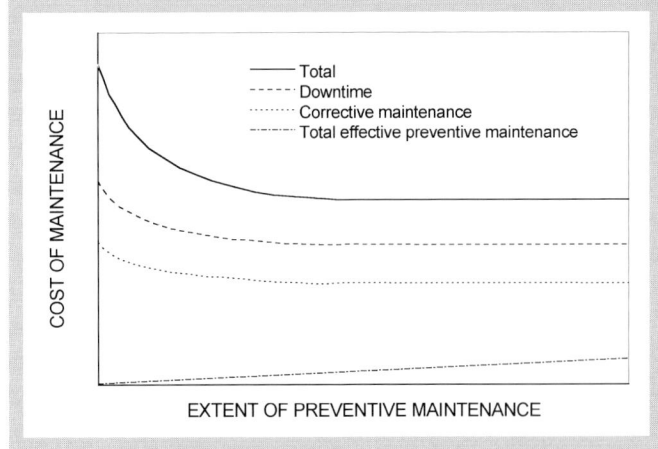

Figure 5. Reduction in total maintenance cost through total effective preventive maintenance (Not to scale).

- Design and fabrication of equipment
- Operation proficiency
- Efficacy of support with maintenance actions.

4 Reliability and maintainability

Although a maintenance department may have good organization with a staff of qualified and trained personnel whose action is managed well, reliability and maintainability still dictate the availability of properly installed equipment. Reliability is the probability that an equipment item or system can operate satisfactorily for a specified period of time without breakdown or other interruptive failure. A common expression for this is the mean time between failures (MTBF). Maintainability is the ease of providing preventive and repair maintenance often expressed as the mean time to repair (MTTR).

Availability is obviously the result of reliability and maintainability. Reliability is initially limited by the "quality" of the design for the equipment item or system (the quality of parts used in fabrication and details of assembly) and whether the item was properly installed or erected as Fig. 2 shows. Appropriate modifications that in turn can use the findings of failure analysis can improve reliability.

The following can also improve reliability:

- Timely rebuilds
- Effective schedule of inspections
- Special preventive maintenance
- Care in operation.

Notice that the first four causes of early life failures in Fig. 6 are the result of improper quality of equipment manufacturing and commissioning. Maintainability is therefore largely determined during the design stage. For instance, how accessible is the component for inspection and repair? It is important for time and cost reasons that all equipment be easily maintained. In the event of breakdown and interrupted production, the necessary corrective action can readily occur. Modifications to reduce maintenance requirements or improve maintainability (design-out maintenance) are economically justified in many cases.

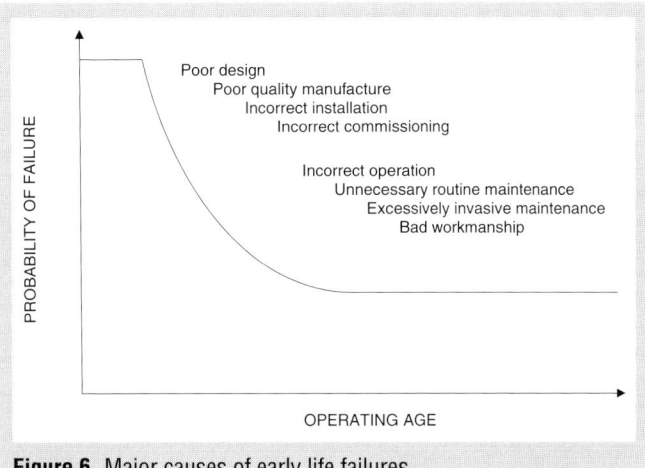

Figure 6. Major causes of early life failures.

Maintainability engineering can identify and plan ways of improving the reliability and maintainability of mill equipment and systems. To identify items warranting such attention, having accurate information about the maintenance and service history is essential. Computer-based maintenance management systems are proving effective in analyzing historical data and identifying equipment items that are candidates for improvements or special attention.

Maintenance personnel require good training. They need encouragement to identify equipment "weaknesses," monitor equipment for proper performance, and assure that all maintenance work has the highest quality.

4.1 Reliability and failures

Failure occurs when a machine, system, or component can no longer perform its intended function. This includes the following:

- Sudden breakdown causing mill stoppage for emergency repairs

- Serious deterioration in performance that effects mill production (quality and quantity)

- "Hidden" faults that eventually cause breakdown if uncorrected and secondary damage to related components (This could be a failed bearing that eventually seizes and seriously damages the mounting shaft.)

- Significant degradation in condition requiring premature and perhaps extensive repairs to restore satisfactory and reliable operation.

CHAPTER 9

Table 2 shows a survey in Pulp & Paper in 1981 that revealed the major causes of equipment failures[8].

Table 2. Major causes of equipment breakdown or failure (1 = most and 7 = least).

Causes	Southeastern United States	Northeastern United States	Midwestern United States	Eastern Canada	Western Canada	North American average
Normal wear	1	1	1	1	1	1
Operator error	3	3	4	2	2	2
Equipment design or manufacture	5	5	2	6	6	5
Corrosion	4	4	5	4	5	4
Insufficient maintenance time	2	2	3	5	3	3
Improper maintenance	6	6	6	3	4	6

Normal wear was the greatest cause of equipment breakdown, and improper maintenance was rarely a cause. Operator error and budget planning were the second and third causes, respectively. Corrosion and equipment design and manufacture were the next most common reasons for unexpected equipment failure. According to several correspondents, the machines do not actually break down. Instead, they are difficult to service, and design is not properly considered[8].

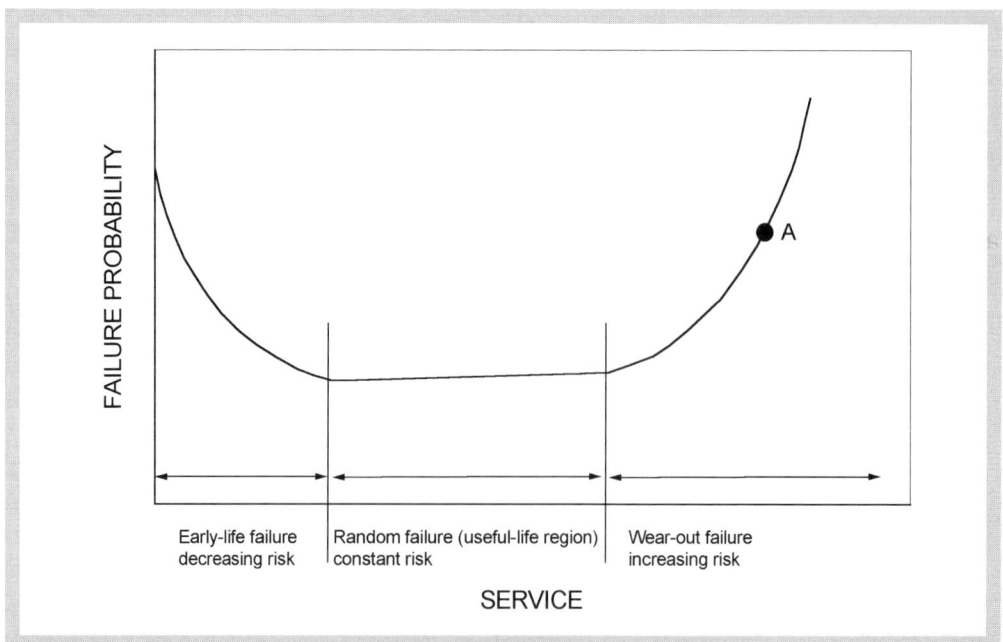

Figure 7. Typical failure rate diagram for service life of an item. Note that replacement or overhaul at point "A" will restore to nearly new condition with a total improvement in reliability.

The objective of preventive maintenance is to avoid equipment failure through regular care and action before the effects of failure occur. Effective preventive maintenance action depends on the item of equipment and on the stage in its service life. The "bathtub diagram" in Fig. 7 provides a simplified model of the three modes of failure:

- Early life failure (decreasing risk)
- Random failure (constant risk, useful life region)
- Wear failure (increasing risk).

A study of the failure diagram suggests the following preventive maintenance considerations for three failure modes:

- Mode 1: One should carry out operational trials before actual startup to detect manufacturing defects or installation deficiencies.
- Mode 2: No advantage in preventive parts replacement or overhauls since the probability of failure will not be reduced. It could even increase by the introduction of factory-defective parts or through poor maintenance workmanship. Failure in this stage is often the result of improper operation and abuse of equipment.
- Mode 3: A definite benefit from preventive replacement of parts or equipment or reconditioning before service time allows an item to reach a high risk state.

Note that most components or systems exhibit failures of the first mode ("burn-in" failures of electronic parts or mechanical wearing of gear reducers). Some items exhibit the life-probability trend that Fig. 7 illustrates. A notable example is rolling-element bearings. Failures occur in all three life stages.

A routine part of the total maintenance effort should be the analysis of equipment failures to determine mode of failure, "root" causes, and appropriate corrective action. A structured investigation sometimes aided by metallurgical or other tests will usually suffice. Maintenance records should include sufficient information to permit monitoring and accurate study of failures so equipment improvements can occur and effective future maintenance action is assured. Besides the three elements mentioned earlier, a complete preventive maintenance program has as its objective maintenance action for equipment improvement.

4.2 Reliability and maintenance engineering

Reliability engineering tries to prolong the life of components and assemblies[9] that results in fewer failures with more constant operating periods and conditions. The application of reliability engineering concentrates on the first phases of equipment life, i.e., from the conceptual phase to the specification phase and design phase as Fig. 8 shows. The same is true for maintainability engineering that aims at short repair periods. At the end of the design phase, note that only about 20% of the investment cost is spent, but more than 80% of the total future operating and maintenance costs are locked in.

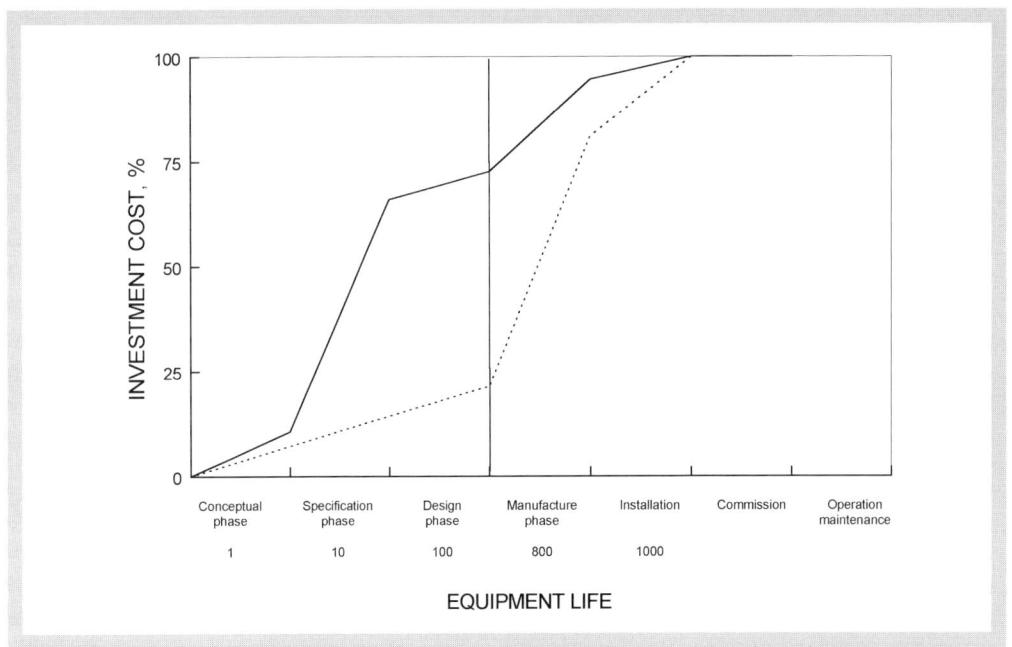

Figure 8. Investment development cost during the project life. The dotted line shows investment cost development during the equipment lifetime. The solid line shows the degree to which future operating and maintenance costs are established by decisions in that particular phase. The numbers 1 to 1000 indicate the cost index for modifications decided at the particular phase.

Modifications are less expensive during the early phases, and the cost increases rapidly as the project approaches the operating phase. If a modification is necessary, it should be introduced as soon as possible. If not, the modification cost might be so high that it cannot be economically justified. Note that more than 99% of the costs of this development project occur during periods when the equipment manufacturer is in charge.

Cooperation between an equipment manufacturer and the maintenance people in the pulp and paper mill is vital for good maintenance work at the mill. If this cooperation does not exist, a great risk exists that a mill with totally new equipment will not be easy to operate and maintain. The expected results are not likely to occur.

Careful consideration of reliability maintenance during a project involving new equipment or new processes will also give the following as by-products:

- Data for preventive maintenance
- Data for a spare part provision plan
- Data for the procurement of tools and other maintenance aids
- Availability data for production planning
- Framework for recruiting and training personnel.

These data form a sound basis for a complete maintenance plan.

Life cycle cost

Life cycle cost is a method of logisticians for their study of a system. In principle, it contains many calculations of the expected problems in operations and maintenance. It reveals costs and technical difficulties and provides a basis for the future support of a working system.

One important part of life cycle cost calculations is the work of reliability and maintenance engineers. These engineers work with selection of suitable components considering initial cost, expected component life, cost of replacement, cost of storing replacement parts, need for special tools and aids for maintenance, etc.

Just-in-time

Much capital can be tied up in storage of spare parts. This capital cost can be reduced if the operation of the mill can be guaranteed through a dependable maintenance procedure. The provisioning of spare parts is included in a well-designed maintenance plan. Parts can be ordered shortly before they are expected to be needed. This reduces the stores and frees the capital for use elsewhere. Table 3 gives the definition of critical spare parts.

If maintenance ensures high availability for all equipment in the production chain, the commonly used buffers such as work in process inventory between different steps in the process can be reduced or eliminated. Work in process inventory costs money. If these buffers can be eliminated without disturbing the process, the economy of the process can increase. This approach to preventive maintenance and connected repairs is an application of the "just-in-time" principle. In most cases, the maintenance organization costs much less than the benefits of production with no buffer.

Increased automation

Most modern plants are highly automated and contain electronic, pneumatic, and hydraulic devices with a high technology level. Even if the reliability of such devices is very high, they can still cause many problems for maintenance personnel without certain precautions.

In most cases, the manufacturer's service organization or specialists outside the user's plant perform maintenance of electronic equipment. Even so, the use of such equipment has a great influence on the user's maintenance organization. Consistent training and upgrading of personnel competence is necessary.

Table 3. Definition for critical spare parts according to Baldwin[1].

> A part or component that would cause a substantial shutdown of all or a significant part of a primary production line if the existing part were to become defective is a critical spare part.
>
> Critical spares use the following:
> - The concept that 20% of the machine components will create 80% of the downtime
> - Based on the above, the 20% that are truly critical have the following characteristics:
>   - Not easily repairable within an acceptable time
>   - A custom or proprietary part that requires long lead time
>   - Part may be available only at a distant location and is not usually stocked locally.
>   - Usually, a component for a critical machine that does not have a backup system.
>
> Critical spares are not the following:
> - Normal replacement parts that are readily available (These parts are handled as part of normal warehouse inventory)
> - Usually not components for equipment that can remain idle until the part is received without impacting mill operations.

4.3 Condition assessment

In 1957, Grimnes[10] explained that a conscientious and well-organized inspection service is the most effective weapon against breakdowns since inspections often detect faults that can be handled before the situation becomes critical. The work resulting from inspections is anti-breakdown maintenance rather than preventive maintenance because the latter term implies that one should never run into any such thing as a breakdown.

Inspectors should work closely with the operators in the various departments and should check operator reports on irregularities regularly. These reports must be examined so operators feel a part of the process and have motivation to keep the reports updated.

Condition-based maintenance has largely replaced the old practice of maintenance action on a fixed period of service. This approach sometimes has the term predictive maintenance. Predictive maintenance relies on information indicating the "health" of the equipment. It compares this information to some acceptable range, attempts to predict the progression toward failure, and most importantly dictates the preventive action necessary. Condition-based maintenance results in extended service life without sacrificing the reliability of the operation since maintenance uses actual indicated needs. Major methods used for condition assessment of mill equipment include:

- Vibration measurement involves measuring the back-and-forth motion of the machinery, structures, and mechanical components. The severity of vibration can assess mechanical condition since faults cause higher-than-normal vibration levels.

- Temperature measurement can reveal faulty components or abnormal operating behavior (poor electrical connection or gear box with insufficient thermal rating). Temperatures can be measured by contact thermometers or with non-contact infrared optical devices.

- Sound measurement involves measuring the sound intensity and patterns emitted by operating equipment in real-time or after recording. This indicates possible abnormal conditions or faults. Another method involves tracking ultrasonic sound waves through a material to detect flaws, cracks, and loss of thickness.

- Lubricant analysis involves periodic examination of a machine's lubricant for the presence of foreign particles that might indicate abnormal wear is occurring. This may suggest the need for improved lubrication, a change in service condition, or improvements to the equipment itself.

The most effective condition-assessment methods are those that monitor equipment "on-line" and determine the state of health under actual operating conditions. Condition-assessment methods not only indicate when a "problem" appears but provide information about its nature, severity, and actual cause. Increasingly more equipment items are being supplied with built-in condition monitoring features to assist the owners in achieving maximum reliability and productive uptime.

Basic care

In processes where a high "operator density" exists (operators in relation to equipment), operators should perform basic care and inspections. This is the most efficient way since the operators are close to the equipment. Little extra time is necessary to complete the tasks. In highly automated areas such as paper machines, pulp mill, and wood yard, the operators' work can also be organized so that they can be responsible for basic care and inspections[11].

Basic care includes the following:

- Detailed cleaning of components
- Lubrication
- Adjustments.

Inspection includes the following:

- Subjective inspection
- Objective inspection.

CHAPTER 9

Detailed cleaning

Who is responsible for the detailed cleaning of components such as electric motors, hydraulic and pneumatic units, gear boxes, etc.? In most mills, this is a "no-man's land." Detailed cleaning is important. Figure 9 shows that heavily contaminated motors will only achieve 2% (5 months) of their 100% (20 years) electric life[11].

Detailed cleaning gives as an additional effect the best inspection one can ever achieve. Cleaning reveals defects that might not be apparent to someone just walking past and looking. One needs competence to find the defects. Detailed cleaning should not be assigned to shifts but to certain individuals. Cleaning is an essential maintenance step. Because it is so basic, it is often overlooked.

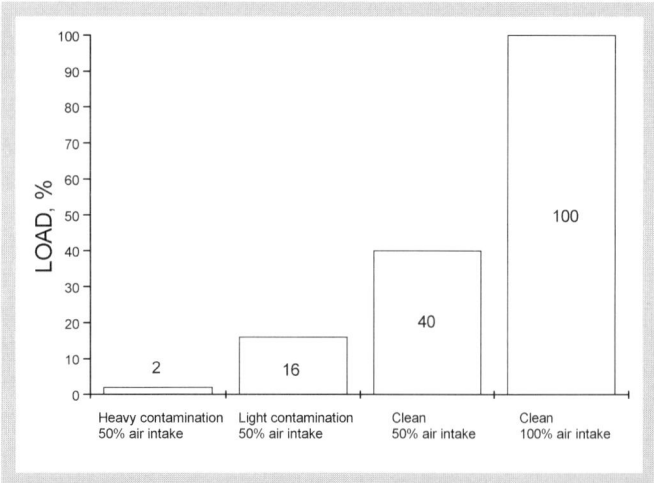

Figure 9. Maintaining 100% air flow and cleaning can increase electric motor life from 5 months to 20 years.

Lubrication

Operators should perform greasing, oiling, and maintenance of oil levels when distances and safety concerns make it practical. Maintenance people should be educated not to lubricate excessively. Blown-out seals are the beginnings of bearing failure. Excessive grease leads to overheating.

Adjustments

Operators must know how equipment works, where adjustments are necessary, and how to perform adjustments. Essential basic steps such as the correct procedure for tightening a leaking packing gland are important. One should not take for granted that everybody knows how to adjust or that a small water leakage is necessary to lubricate the packing.

Inspection

Subjective inspections use common sense. Look, listen, feel, and smell. The results are subjective because they do not use firm measurements. Objective inspections use measurements to determine actual conditions or changes in condition.

The first step in inspection training is to demonstrate what "mill blindness" is and educate inspectors about acceptable and unacceptable conditions.

Concepts of maintenance in the pulp and paper industry

Inspection techniques and condition monitoring techniques

Most pulp and paper mill equipment is subject to vibrations, temperature changes, thermal and mechanical fatigue or shocks, corrosion and erosion, or all these items. Realistic inspections and educated interpretations of situations can save mills significant money in maintenance costs. Knowing whether a crack presents a maintenance problem comes with experience. This experience encompasses an understanding of equipment materials, welding, and heat treating with the ability to analyze conditions accurately through the use of meaningful techniques.

Some inspection objects and techniques used in inspections are as follows:

- Cracks in equipment
    - Visual
    - Fluorescent penetrating liquid and magnetic powder
    - Ultrasonic detector
    - Radiography such as X ray
    - Eddy current
    - Acoustic emission
- Temperature
    - Contact thermometer
    - Infrared light
    - Thermography
    - Bearings
    - Misalignment
    - Thermal isolation
- Pumps
    - Flow
    - Repriming
    - Excessive noise
    - Frequent clogging
    - Temperature
    - Bearing temperature
    - Leakage
- Bolts and fasteners
- Heat exchangers
    - Mountings
    - Misapplication
    - Misalignment.

CHAPTER 9

"Knowing why" training is equally as important as "knowing how" training. Training should therefore include explanations of why inspections are being done and what a failure is. Understanding the difference between a failure and a breakdown is important.

5 Equipment availability and efficiency

The total mission of a mill maintenance program is to make the mill and its equipment available for a maximum production time with conditional operation at peak performance. A measure of equipment performance must reflect the total effectiveness of the equipment considering the combined effects of availability (%), pulp and paper quality rate (%), and production rate (%) as Table 4 shows.

Table 4. Definitions of availability, prime quality, operating speed, and total equipment efficiency by Idhammar[11].

1	% Availability or time up = $\dfrac{\text{Time up} \times 100\%}{\text{Time up} + \text{time down}}$ Time up + time down = 365 days per year	A
2	% Prime quality volume = $\dfrac{\text{Prime quality volume} \times 100\%}{\text{Prime quality volume} + \text{rejects}}$	Q
3	% Operating rate or speed = $\dfrac{\text{Target capacity} \times 100\%}{\text{Actual capacity} + \text{speed losses}}$	S
	Overall equipment efficiency (OEE) is OEE = %A x %Q x %S	

5.1 Lost-time analysis

According to Aurell and Isacson[12], Swedish paper companies appointed a task force to develop a standard method (SSG Standard 2000) for lost-time analysis. Several Swedish mills have now adopted and use this standard. Finland has adopted a similar standard. One company has found these methods applicable and of considerable value for most mills and recommends that mills adopt this type of lost-time analysis. This method uses time. It records not only production interruptions but also changes in the production rate through the department or production unit. Reasons for stoppages or changes in production rate are recorded in the production log that provides data for lost-time analysis. A high frequency of production rate adjustments is not desirable and may be revealed by the reports. The method identifies two key measures of production performance: availability and utilization.

Concepts of maintenance in the pulp and paper industry

Availability is a measure of reliability. It means the portion of the total operating time during which the department has been available for full production. Use gives the portion of the total operating time the department has been used for full production (within the limits set by the availability). The ratio between use and availability provides important information. This ratio is always less than 100%. A ratio close to 100% may indicate that the department is a bottleneck as Fig. 10 shows. A low ratio below 80%–90% indicates the production capacity of the department is not fully used. Size of the buffer storage ahead and after the department will influence results and require consideration when making the analysis.

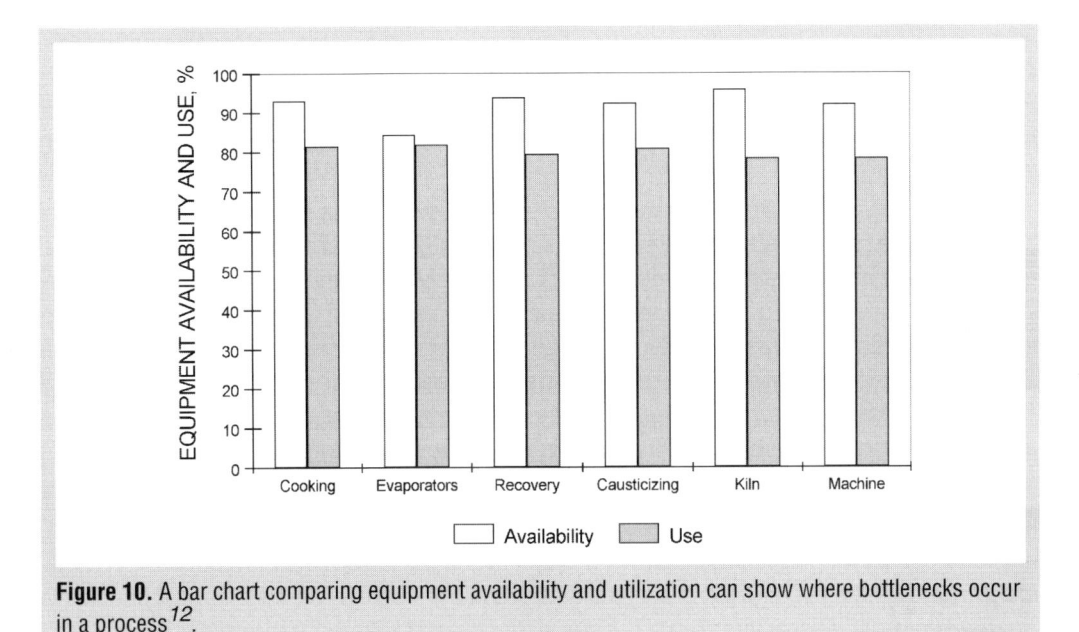

Figure 10. A bar chart comparing equipment availability and utilization can show where bottlenecks occur in a process[12].

The degree of sophistication of this method may be too high for some mills. It may also be difficult to use when analyzing existing operating records that do not follow the concept. In such cases, one may define the two key factors (availability and use) in a different manner that excludes the impact of running at or below capacity. Only production time and downtime are considered. Reasons for downtime should be split into external and internal causes.

With this simplified approach, availability has the following meaning:

$$A = \frac{T_p + T_{de}}{T} = \frac{T - T_{di}}{T} \tag{1}$$

where T_p is operating time
T_{de} external downtime
T total time available for production
T_{di} internal downtime.

CHAPTER 9

Use then means the following:

$$U = \frac{T - T_d}{T} = \frac{T_p}{T} \qquad (2)$$

where T_d is downtime.

Sanclemente[13] showed in 1992 that most mills track downtime and especially extended downtime. These mills react to the issues that contribute to the downtime.

A key element in effective tracking of lost production is to maintain simplicity when collecting, compiling, analyzing, and presenting data. Reasons for loss in production should be identified and their impact quantified in an easily understood manner. This should include not only the obvious downtime but also lost production due to slowness. Determining if the losses were due to the operating department or to external reasons is also important.

Aurell and Isacson[12] have described a simplified version of the Swedish SSG Standard 2000 for lost time analysis. They outlined the modified Swedish process and described a basic system for collecting data and analyzing the information. They showed the difficulty in defining maximum production potential (production at capacity) for a department.

5.2 Production potential

If the impact of individual departments on lost production is not important, arbitrarily assigning a value for production potential is all that is necessary. Lacking a good definition of production capacity for individual departments, most mills track downtime but do not track lost production associated with slowness. Most lost production capability is the result of slowness rather than downtime. Many issues associated with lost production are therefore not addressed. Table 5 gives estimations for these parameters[13].

Table 5. Slowness, availability, and use.

1. Slowness TRR x (MSR - AR) / MSR TRR is time at reduced rate, min. MSR is maximum sustainable rate AR is actual rate
2. Availability, % (TT - IDT - ISB) / TT IDT is internal downtime ISB is internal slowness TT = total time
3. Use, % (TT - IDT - ISB - EDT - ESB) / TT EDT is external downtime ESB is external slowness

Concepts of maintenance in the pulp and paper industry

Every major production unit in the mill should collect lost time specifying digester No.1, recovery boiler No.1, etc. Production personnel should note the time and reasons for lost production as Table 6 shows.

Table 6. Mill production variance report showing lost production by internal slowness (ISB), internal downtime (IDT), external slowness (ESB), and external downtime (EDT).

Date	Time from	Rate from	Rate to	I/E	Reason	Explanation of variance	ISB	IDT	ESB	EDT
35 795	700	800	15	2	3	Evaporators down for water wash	0	0	4	0
35 795	800	850	15	2	3	Ramping up after water wash	0	0	2	0
35 795	850	1 000	16	1	6		0	0	0	0
35 795	1 000	1 020	0	1	5	Kicked out	0	20	0	0
35 795	1 020	1 100	15	1	5	Ramping up after HPF KO´d	1	0	0	0
35 795	1 100	2 020	16	1	6		0	0	0	0
35 795	2 020	2 310	0	2	7	Power failure	0	0	0	170
35 795	2 310	110	14	2	7	Ramping up after power failure	0	0	8	0
35 795	110	700	16	1	6		0	0	0	0
35 796	799	100	16	1	6		0	0	0	0
35 796	100	220	16	2	2	Ramping down	0	0	5	0
35 796	220	700	14	2	2	Slowback	0	0	35	0

Pulp mill production variance report
Month: January

A Pareto chart can show the reasons for lost time and provide relative significance for each reason. A bar chart comparing availability and utilization defines the bottlenecks in the process as Fig. 11 shows.

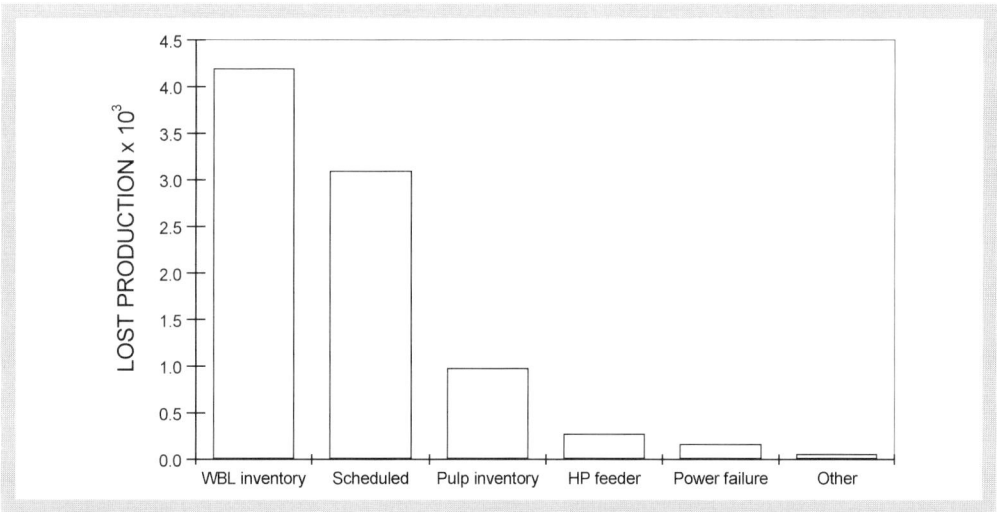

Figure 11. Pareto chart showing reasons for lost time and providing relative significance for each reason.

CHAPTER 9

To remove the ambiguity associated with defining departmental production potential, one should use actual operating data. The duration curve used by some companies removes this ambiguity. Production for a period of time such as the daily average for the previous year is sorted from the day of greatest production to the day of lowest production. To understand clearly how production has varied during the year, a curve showing sorted production vs. percentage of operating time is prepared according to Fig. 12[13].

The maximum sustainable production level is that level compiled and sorted from the day of greatest production to the day of lowest production. In Fig. 12, the average production of days 36 and 37 is the maximum sustainable rate. This technique is applicable to any operating department or any piece of equipment in the mill. Lost production in a pulp mill is equivalent to lost production on the paper machines or any other part of the mill if this method defines production potential. A close look at the duration curve clearly shows that the maximum sustainable rate for the pulp mill is 1000 bleached a.d. tons. The average rate for the period is 920. This average rate was exceeded 69% of the time.

The ratio between the maximum sustainable rate and the average rate is the capacity efficiency. This ratio can show how well the mill is running. When compared with similar mills, it can provide an indication of the production level one can realistically expect. After collecting and compiling the data, one must present it in an easily understandable form.

Distributing the information to those individuals in the mill who can best use it is obviously important. The distribution should include those actually doing the maintenance and members of management who determine allocation of time and capital.

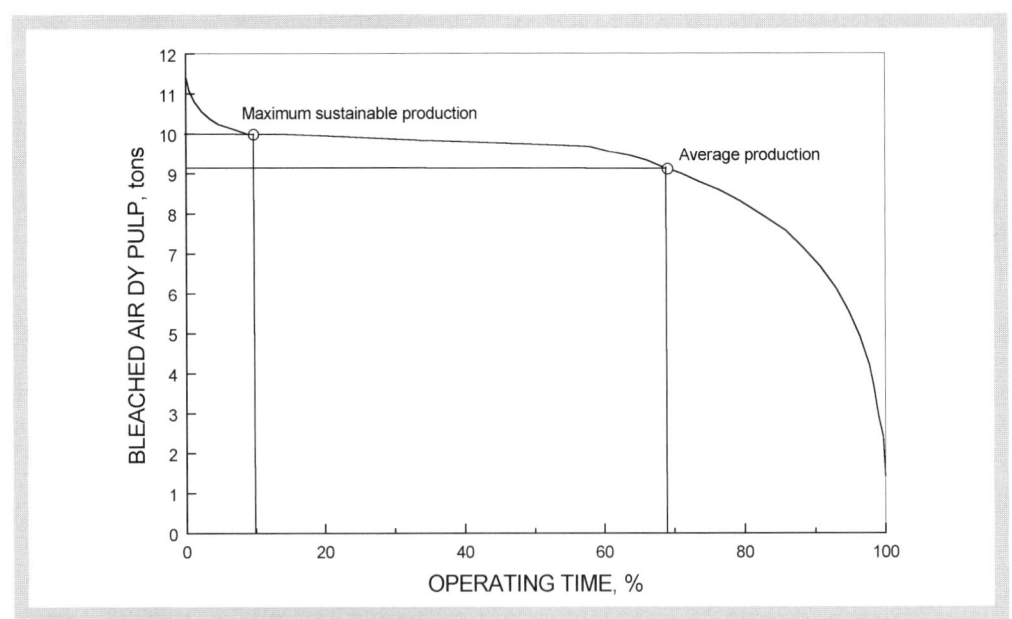

Figure 12. Duration curve showing sorted production vs. percentage of operating time can help understand production variation.

Concepts of maintenance in the pulp and paper industry

6 Evolution of team approach and multicraft

With the technically complex workplace and world market competition today, maintenance and production people must coordinate their efforts and share responsibility for profitable operation of the mill and its facilities. Proper operation and maintenance must be the common objective of both groups.

Maintenance must accept production's expectation of maximum equipment availability despite all the requirements for effective maintenance action. Production must simultaneously accept the fact that an adequate amount of appropriately scheduled downtime must be provided to allow for routine preventive maintenance, equipment overhauls, corrective work, and improvements. In the long run, equipment will not perform as intended nor provide acceptable service life without this "cooperative" approach to its maintenance.

Just as production must answer to the mill's customers regarding product quality, maintenance must practice the "customer" concept. In this case, the internal customer is the production group. The target must be quality work, efficient job execution, and a full scope of properly planned maintenance actions. Maintenance must simultaneously be treated respectfully as a "service" and not as a "servant" as is too often the case.

The "team approach" should extend to equipment manufacturers and suppliers of maintenance materials and services. Maintenance must establish and use a network of "key" industry people who can provide information, technical support, and quality materials on a continuing basis and at critical times of breakdown emergencies.

6.1 Evolution of multicraft

According to Frampton[14], craft interference is a problem or opportunity that has existed since the inception of the craft system. In the past, individuals had expertise in a particular trade. This system led to the establishment of craft unions and jurisdiction.
The trend in industry today is toward total productive maintenance. This is a system that takes advantage of the abilities and skills of all individuals in the organization. The concept is at odds with the traditional maintenance organization in which each craft person is responsible only for a particular area of expertise. Since development of the single craft system some years ago, expecting organizations to change overnight to the new concept is not realistic. The key is to remove the negative impact of craft interference defined as "the interval during which a trades person must wait for a trades person of a different craft to arrive and start work." This definition implies that the first trades person has worked up to the craft jurisdictional line and cannot complete any additional work without infringing on the "territory" of the second craft.

The next organizational design involves a multicraft supervisor who has responsibility for many different crafts. In pursuit of "total productive maintenance," some companies began creating "super" people or multicraft maintenance employees. These individuals are qualified in two or more trades and have the ability to complete jobs requiring numerous craft skills. Companies that undertook this approach often saw the number of maintenance craft groups decrease from 15–20 to less than six. The creation of the super craft people required that the company provide extensive training and

CHAPTER 9

cross-training to ensure that the individuals could safely and effectively perform their assigned tasks. The most important factor in the evolution of a multicraft organization mill is the stance of the labor unions. In cases where the unions do not accept the approach, the only result is continuing management problems and possibly strikes. In an operating multicraft environment, the burden of coordination decreases considerably especially for the supervisor.

6.2 Owner-operator

One approach to better maintenance is to begin transferring lower level maintenance tasks to the production organization and continuing to transfer responsibilities as production workers become trained and qualified. Typically, the transfer of responsibility begins with tasks such as equipment cleaning, simple visual inspections, and straightforward lubrication. As operators become qualified, they assume responsibility for routine preventive maintenance, common corrective maintenance, and limited troubleshooting and diagnostics.

In this environment, the maintenance department becomes a resource for this continuous improvement process. Clearly, maintenance will remain the repository for specialized techniques and sophisticated routines required to maintain the ever-increasing level of technology. Many people in the pulp and paper industry see the transition from traditional organizations with well-defined craft lines to more flexible structures as a desirable change, but they are unclear about the benefits and costs.

6.3 Costs and benefits of craft interference

A question that often arises is "What will it cost to remove craft boundaries and what will the gain be?" If every individual in the maintenance organization was totally skilled and capable of doing every job, no interference would exist. Two additional factors contributing to the problem of craft interference are inexperience or inadequately trained supervisors and ineffective planning and scheduling. When supervisors (many of whom rose through the ranks) have inadequate training, they often cannot coordinate the people, material, and equipment necessary for an effective maintenance organization.

Some people in the industry have negotiated multicraft or multiskill agreements with their units. Under these agreements, the company can expect maintenance personnel to work up to their level of qualification. In retrospect, this is an old maintenance tradition. For many years, farmers have owned, operated, and maintained their equipment. No self-respecting farmer would call an electrician to change a light bulb, a plumber to fix a leak, or a tractor mechanic to change spark plugs. Companies often view the benefits of a more flexible maintenance force in terms of reduced manpower. This is one benefit but not the major advantage to be gained. Increased machine availability and increased production are the true driving factors.

Since the pulp and paper industry is capital intensive, the increased return on investment gained from higher availability and better use of capital equipment far outweighs the reduction in payroll that may occur. According to Thornton and Frampton[15], the modification of a strict craft maintenance organization to include multicraft maintenance may be necessary to improve productivity in the pulp and paper industry. To meet competitive pressures, companies will need to analyze the cost of such organizations. Once the losses from a traditional craft organization are known, management will be responsible to implement the needed flexibility. While no canned solutions exists, several logical steps appear to exist.

First, a period of education is necessary during which labor and management become aware of the financial impact of the current maintenance organization on the company. This education should include the impact of competition and the need to reduce expenses. After this need is clearly understood, a joint effort by labor and management can arrive at an agreeable organization and implementation plan.

Multicraft work rules have applied in many mills in North America. In 1986, Canadian mills averaged 11 maintenance job classifications and United States mills in the South averaged only 5 maintenance job classifications. Operators could do some maintenance work at 60% of southern United States mills and 20% of the Canadian mills.

7 Trends in different decades

7.1 Main trends in the pulp and paper industry

Size and number of mills

According to Sward[9] and the Finnish Forest Industry Federation[16], the number of paper mills decreased, the average mill size increased, and the average machine size increased from 1960 to 1990. Table 7 shows these changes.

Table 7. Changes in paper and pulp mill sizes and numbers in Finland and Sweden.

	1960 Finland	1960 Sweden	1990 Finland	1990 Sweden
Pulp mills				
Number of mills	54	127	46	55
Total capacity, 1 000 tons	3 516	5 588	8 886	11 015
Average mill capacity, 1 000 tons	65	44	193	200
Paper mills				
Number of mills	42	76	44	54
Total capacity, 1 000 tons	1 970	2 280	8 967	8 940
Average mill capacity, 1 000 tons	47	30	204	166

The trend to form fewer pulp and paper corporations in Europe will result in bigger mills and differentiation in products.

Investment costs for a new greenfield pulp mill in Scandinavia have increased (in current prices) from SEK 1200 per ton/year in 1960 to SEK 7400 in 1986. This is an increase of more than six times. In 1990, the investment cost was about SEK 9000/ton.

In integrated paper mills in Sweden, the number of man hours/metric ton produced was reduced from 4 in 1970 to 2.8 in 1985 for operating personnel, but the number remained almost constant at 1.2 man hours/metric ton for maintenance personnel. Figure 13 shows the total number of man hours and maintenance man hours/ton in a modern Swedish newsprint mill[17]. When expressing the maintenance man hours/ton in a more results-oriented way, the Braviken mill is producing 2.33 prime quality tons/maintenance man hour[18]. These figures indicate the technical development in pulp and paper mills and the resulting changes in operational and maintenance needs. The increasing size of mills often combines with high automation and low labor costs in operations, but maintenance labor costs remain fairly constant. High investment costs also mean higher penalty costs for shutdowns. Downtime must therefore decrease to a minimum to ensure profitable operations. More complicated production equipment with electronic and hydraulic controls requires more competent maintenance personnel supported by a very effective maintenance organization.

Maintenance and operations connections

Twenty years ago, about 21% of all hourly employees in North American pulp mills were maintenance personnel. Today, that average is about 24%. In Sweden, this figure today is 32%. During the same period, production output has more than doubled. The fact that maintenance personnel have become a larger part of the payroll and therefore a larger part of total manufacturing costs is logical. In a highly-automated futuristic plant, maintenance people may comprise 70% or 80% of the total plant payroll. Productivity will depend upon reliable and maintainable equipment and efficient maintenance.

In 1985, Swedish mills spent an average of 27% of their maintenance budgets on outside contractors vs. 14% for North American mills. This fact can account for the high maintenance productivity figures of Swedish mills (about 0.6 maintenance man hours/ton) since the productivity figure accounts only for mill maintenance labor.

Supervisors per 100 mechanics at Swedish mills in 1985 averaged 21. In North American mills in 1980, they were 11.4 and in 1986 only 11.8 as Fig. 14 shows[18].

New equipment involves approximately 10% people and 90% technology. Improving maintenance might involve 90% people and 10% technology. Maintenance improvement means doing better with what one already has. Tools such as vibration analysis and computer systems are easily bought and changed, but people are not. Keys to improved maintenance and equipment efficiency include information and training. The former refers more to "knowing why" than "knowing how." This is part of training. People do not change unless they understand why the change is necessary and how they personally can benefit from the change. The effectiveness of maintenance personnel depends on many factors. Morale, organization, maintenance methods, procedures, and knowledge are among the most important factors[19].

The personnel "know-how" factor (often used but seldom understood) is an almost undefinable quantity that consists of four basic attributes: experience, judgement, skill, and knowledge. Each of these attributes directly relates to training of one type or another. Experience is a basic quantity that can be almost discounted in a discussion of formal training. As a rule, experience is something that "takes care of itself

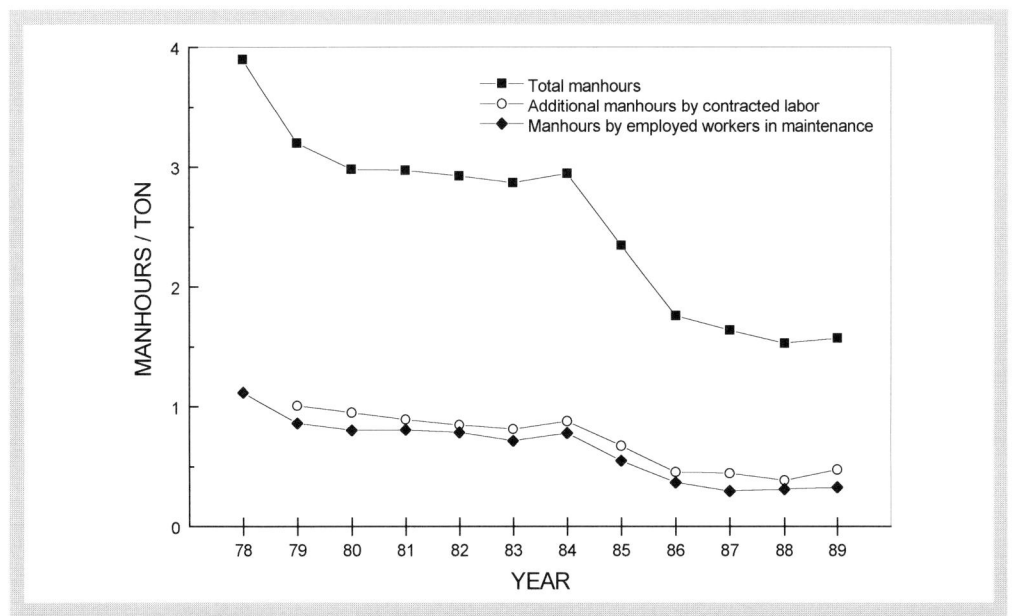

Concepts of maintenance in the pulp and paper industry

Figure 13. Total number of man hours and maintenance man hours/ton of newsprint at Holmen Paper's Braviken mill in Sweden.

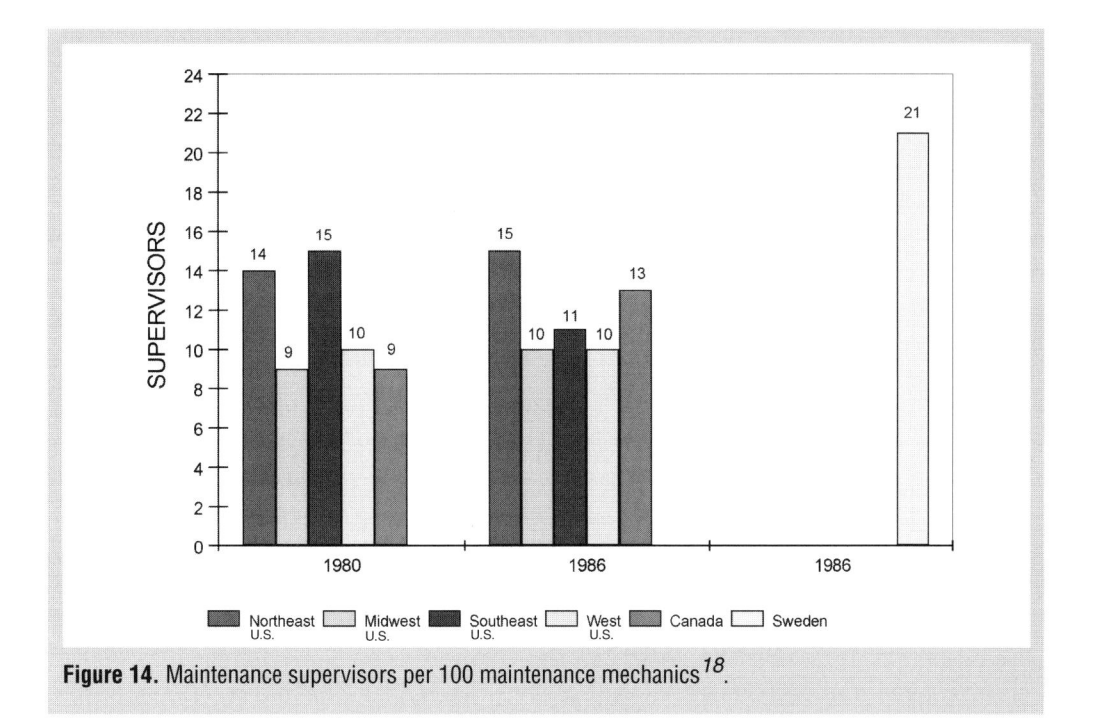

Figure 14. Maintenance supervisors per 100 maintenance mechanics[18].

over time." Judgment is usually a function of experience. It can be inspired very early during training in those people mentally capable of exercising it. The development of high morale and the encouragement of personal initiative contribute to good judgment. An employee's skill develops only through specific effort. Skill results partly by being shown how to do a job and partly by actually doing the job. Skill does not come through training alone because the aptitude factor enters into the picture. Finally, acquisition of knowledge is the simplest of the four factors. Many specific systems for acquisition of knowledge (teaching and learning) are available.

Those people charged with training maintenance personnel must know these four factors. They must also know that training must be a constant endeavor to reach its greatest potential. Training is ineffective when confined to intermittent courses, occasional lectures, or isolated correspondence courses taken by younger members of the maintenance department. A sustained training program is necessary for practical results.

Many papers are available on training maintenance people at pulp and paper mills. No single common solution is suitable for every mill. The trainer must realize the needs of the maintenance people and tailor the program as necessary.

Outsourcing of maintenance

Contracting has been used in maintenance for peak labor, specialist services, workshop, and project or shutdown work. These contracted services have been used to cover peaks in workload and optimize the number of on-site personnel to produce a cost-effective result.

In recent years, the concept of outsourcing has grown to an independent new business area. These new service companies are free from pulp and paper manufacturers. Is this hunting of cost reduction in maintenance a sign of despair of the business or not? How can an outsourced contractor possess the expertise systems, specialist skills, and a capacity to maintain and operate a customer´s plant in the long run without continuity in the contracting work? Who is paying the bills of the accidents to the labor of the outsourced contractors? In the short run, the insurance companies will cover this. What about the long run? Several open questions remain concerning outsourcing that still need answers.

8 Maintenance management trends by Idhammar

Christer Idhammar has been a major force in defining new ideas and trends in pulp and paper mill maintenance. Idhammar has been active almost 30 years in Sweden, Nordic countries, North America, and other areas of the world. According to Idhammar, plant management is more commonly concerned with the cost of planned maintenance than with its long-term benefits. Production management often develops a short-sighted focus on keeping the equipment running. The North American practice of running plants at or beyond design capacity to maximize short-term output aggravates the maintenance problem. The trend to reduce manpower when market demand is down aggravates this further.

Concepts of maintenance in the pulp and paper industry

Although having established approval procedures for capital appropriations is common, maintenance costs do not receive the same attention at budget approval time. This occurs despite the fact that return on investment and profitability will benefit substantially from effective and efficient maintenance practice. Part of the reasoning is that relating increased profit to improved maintenance is not always easy.

Maintenance efficiency

Most maintenance organizations spend considerable time on "fix it" maintenance often with short notice and without planning. This is expensive maintenance that slows results with much downtime. Planned and scheduled maintenance is safer, more efficient, and results in increased prime volume of output at a lower maintenance cost.

Planned work refers to a job that is planned and scheduled. Planned means that the job is prepared with tools, materials, spare parts, safety precautions, etc. Scheduled means a decision exists about who will do the job and when the job will be done. A planned job means planning and scheduling no later than the day before implementation. Most plants claiming high levels of planned maintenance are not planning but merely scheduling.

The quickest return on investment most often results from a focus on reduction of unplanned work. Measures having the most significant effect on reducing the amount of unplanned work are the following:

- Direct preventive maintenance cleaning and lubrication and fixed time maintenance where justified
- Indirect preventive maintenance condition monitoring including subjective inspections and objective inspections (predictive maintenance)
- Planning and scheduling procedures.

Information about improvement potential in Eq. 3 and goals should be available to all employees. Everyone concerned must have information about why and how changes are being made and what the results will be. Because this involves a large number of people, goals must be easy to understand and measure.

$$Improvement\ potential = U \cdot W = \frac{U}{W} factor \tag{3}$$

where U is unplanned work with lead time shorter than the day before it begins
W wasted time related to unplanned work: finding out what to do, finding the right people, finding the spare parts, finding tools, etc.

Multiplication of the estimated percentage of unplanned work (U) by the estimated percentage of wasted time related to unplanned work (W) gives the improvement potential. For example, if the percentage of unplanned work is 80% and the percentage of wasted

CHAPTER 9

time associated with this unplanned work is 60%, the maintenance efficiency improvement potential is 80% x 60% or 40%. Figure 15 shows the calculation in graphical format[20].

Accepting the premise that planned maintenance is safer, less expensive, and generally more efficiently performed than unplanned maintenance, one must only measure the U factor and use that as a goal.

Primary goals in a maintenance operation are to provide equipment efficiency and extend the technical life of equipment. A secondary goal is to deliver this as cost-effectively as possible. Unfortunately, many mills focus too much on cutting maintenance costs.

The PQV/M factor (prime quality volume per US$ 1000 invested in maintenance) must have two factors. One factor excludes capital maintenance jobs, and one factor includes capital maintenance jobs done by the maintenance department. One should not use these factors to compare plants with each other. The continuous improvement trend should measure progress in reaching goals. This approach is more positive and avoids many arguments that are often the only result of excessive comparison between plants.

Equipment efficiency

Common goals are to reduce downtime, speed losses, and run equipment at peak performance. These are joint ventures between maintenance and operation. The following is a summary of advice when measuring and analyzing equipment efficiency:

- Available time for production is 365 days per year

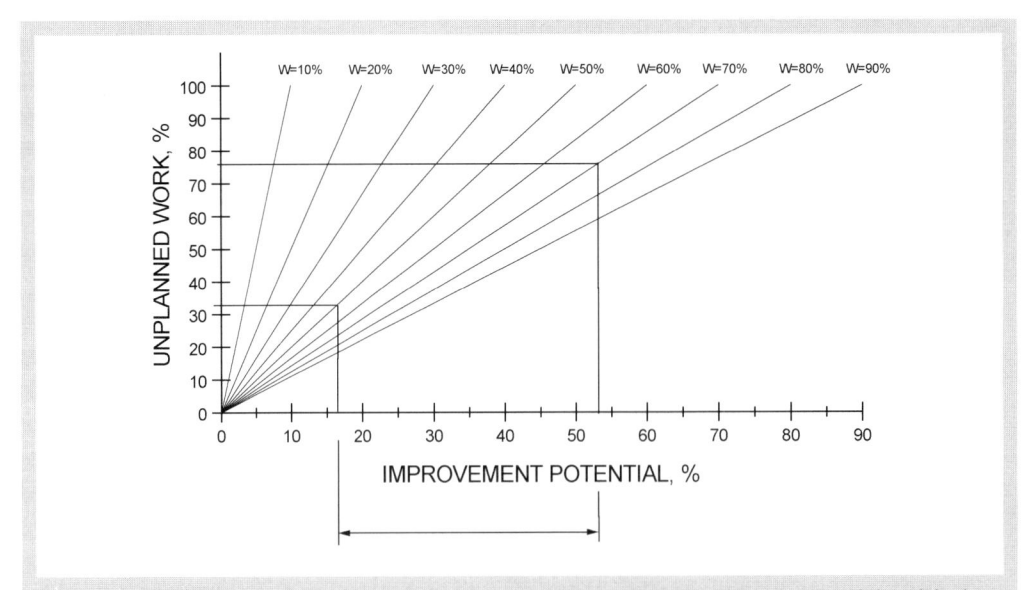

Figure 15. Improvement potential, E, calculated as the difference between the present and the wished unplanned work parameters. This means (78 x 70) - (32 x 50) = E = 38%.

- Record downtime, quality, and speed losses by symptom and action taken not by department
- Use the information to analyze the cause
- Design out the problem.

Planning and scheduling

Many mills believe they have planned maintenance because they have planners and planning systems including work orders. Often, planners are not planning because they have become "go-fers" for foremen. When they are planning, only 30%–40% of what is being planned and scheduled occurs according to the schedule. The U factor is 60%–70%. The inability to follow a schedule often happens because maintenance must work with break-in jobs added to the schedule. These are jobs requested or needed with a very short delivery time.

The most efficient way to accomplish longer lead times for maintenance is to implement preventive maintenance using condition monitoring. This should include basic measures (seeing the wear of a chain, pulley, or rotary joint carbon ring) to more sophisticated methods such as using infrared cameras, wear particle analyses, or vibration analyses. The major part of condition monitoring does not require sophisticated equipment and can be done by committed and trained operators and maintenance personnel practicing basic care.

High class maintenance resulting in improved equipment efficiency and reduced maintenance costs can occur when changing from a "fix it" mode of maintenance to a "prevent it, analyze it, and continuously improve it" mode of maintenance as Fig. 16 shows. A mill that truly adopts this philosophy has an opportunity to achieve excellence. In addition, employees will find their work more enjoyable. This will not happen solely with maintenance efforts but must be a joint venture with production in a flexible organization.

A world-class maintenance indicator is the distribution of total effort hours in maintenance as follows:

- Planned and scheduled work approximately 70%
- Continuous improvement work approximately 25%
- Break-in or unplanned work approximately 5%.

CHAPTER 9

Enhancing maintenance performance from an average level of about 30% planned maintenance to an elevated range of 90%–95% planned maintenance is not easy. A variety of obstacles require management in the process of changing to more planned maintenance. Benefits will come from increased manufacturing efficiency, lower costs, and increased safety among other things. Figures 16 and 17 describe the change from unplanned to planned maintenance.

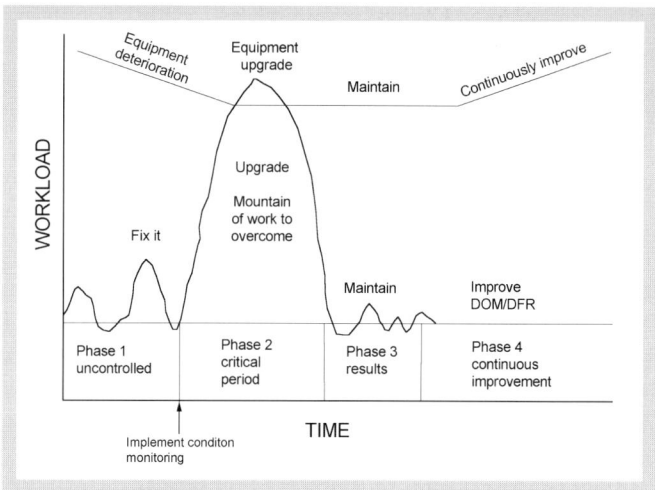

Figure 16. Impact of condition monitoring in changing from unplanned to planned maintenance[20].

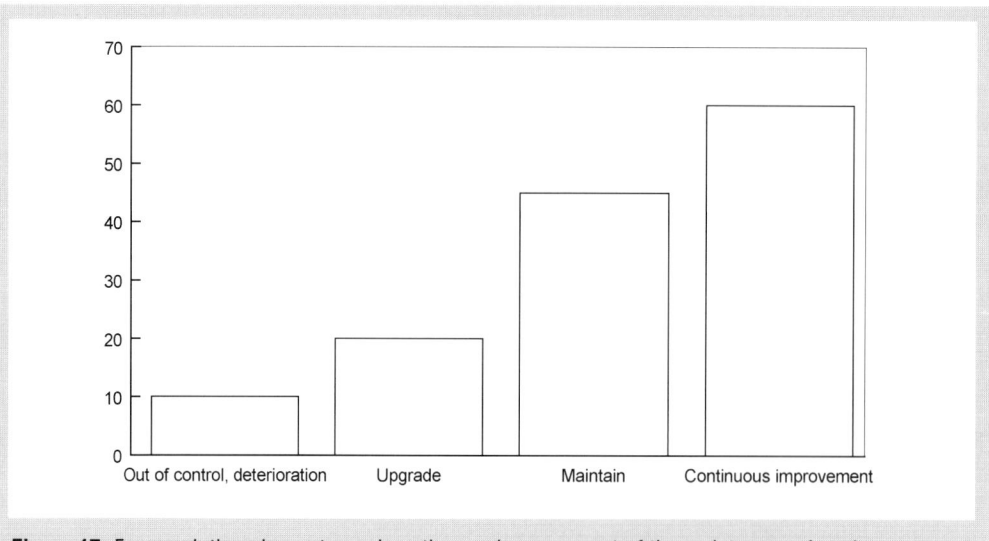

Figure 17. Four evolution phases toward continuous improvement of the maintenance function.

In Phase 1 of Fig. 16, equipment condition is poor and declining. The technical life of the equipment is declining rapidly. The ratio of planned to unplanned maintenance is typically more than 40%. Labor costs are high, and use is poor. Uptime is low, and maintenance costs are high.

When implementing a maintenance improvement project at the beginning of Phase 2, the workload decreases, and scheduled corrective maintenance action will occur early. Instead of waiting for a breakdown, a potential failure will be corrected earlier in a safe manner with a lower maintenance cost. Only little production losses will occur because much maintenance will happen when equipment undergoes tool, product, wire, or felt changes. The payoff period for these actions is usually very short. The cost increase is not really a cost but a cash-flow difference between paying US$ 1 now or US$ 10 later.

During Phase 3, planned maintenance can be 80%–95%. This is a realistic and attainable value, but it does differ between production areas and types of equipment.

During Phase 4, maintenance work has had such fine-tuning that a mill will have time to truly focus its efforts toward continuous improvements. The mill is also now in a position to optimize its maintenance effort hours. It will know how many people it needs. A sharp decrease in overtime to less than 4% will occur, and equipment efficiency will be extremely high.

Maintenance does not sell service. It sells equipment efficiency and delivers this through cost-effective service. The idea is that equipment will be busy making quality products between scheduled shutdowns and people will be busy working on planned and scheduled work and designing out problems (continuously improving).

A key indicator in results-oriented maintenance is the mix of work categories of Fig. 18. Results-oriented maintenance strongly opposes work measurement techniques and hands-on tool assessment. These methods might identify what needs improvement, but they cause friction between management and crafts people. Results-oriented maintenance believes that management must organize planning and scheduling so that it occurs in a professional manner. This results in the execution of maintenance work becoming more efficient as Fig. 18 shows.

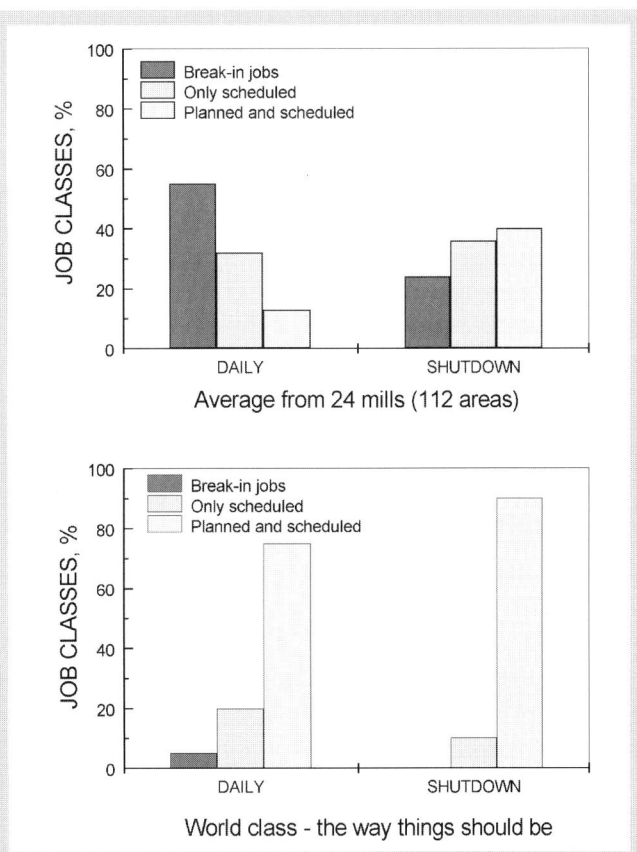

Figure 18. Planning and scheduling performance pattern in 24 North American pulp and paper mills (upper) and in a world-class maintenance mill (lower)[20].

Figure 18 shows the average planning and scheduling performance in 24 mills including about 70 maintenance areas. It also describes the status in 1987 in an example mill:

- 55% of daily maintenance activities happen as break-in jobs or unplanned and unscheduled ("Break-in" describes the fact that these jobs are breaking-in to the schedule that was done the day before the next day's job.)
- 32% of the jobs happen only as scheduled (No planning or preparation of the jobs occurred.)
- 13% of all jobs happen after they have been planned and scheduled.

Shutdown planning and scheduling performance is better but not very impressive. 24% of the jobs are break-in work, 36% of the jobs are only scheduled, and 40% are planned and scheduled.

Figure 18 also shows high class performance or how things could be: only 5% break-in work in daily schedules, 20% executed as scheduled because these jobs do not require planning, and 75% of all jobs executed as planned and scheduled. No break-in work should occur during a shutdown. About 10% of work might not need planning and occur as scheduled. 90% of jobs are planned and scheduled work.

The example mill made a flexibility agreement with its union in 1986. This agreement stated that operators would do maintenance work and eliminated craft lines in maintenance. All craft people had the right to do everything they were trained and capable to do safely. For training and compensation purposes, five crafts were recognized. People were trained in different crafts and achieved increased compensation when they mastered additional skills or crafts.

8.1 Reliability centered maintenance

The reliability centered maintenance concept uses failure development theory[20]. As with other maintenance concepts, reliability centered maintenance requires an up-to-date equipment record with clear identification of equipment and components. The record requires integration with spare part lists to give a bill of material per equipment unit.

Failure consequences in reliability centered maintenance can have the following classifications:

- Hidden failure consequences that might not lead to any direct consequences (For example, an alarm does not work.)
- Safety or environmental failure consequences that can result in personal injury or environmental damage
- Process failure consequences that inhibit production by causing losses of availability, quality, or speed performance
- Nonprocess failure consequences that have no effect on the process but do impact maintenance costs.

8.2 Equipment efficiency through operations and maintenance

Centralization of the maintenance function was common 20 years ago. The trend 10 years ago was to decentralize maintenance. Today, the goal is to decentralize maintenance into different areas with a centralized support function. Many pulp and paper mill companies are decentralizing the maintenance function into production areas to include integration between maintenance and operation workers. Operators are becoming more involved in maintenance work. Concepts such as team building, total productive maintenance, role expansion, multicraft, equipment owners, etc., are becoming more common. The question today is how and to what degree to decentralize[22]. The general opinion is that having a central maintenance function is still necessary to support the decentralized functional areas with services such as the following:

- Use of contractors
- Spare parts stores (Should report to maintenance)
- Skill level upgrade programs
- Support areas with specialists
- Keeping current with new materials and techniques
- Workshop facilities
- Reliability and maintainability analysis in projects and procurement of equipment.

Central maintenance functions become increasingly more specialized support functions. One element that supports decentralization is the increased amount of automation in a mill. Automation typically reduces the number of operators but increases the demand on skills of maintenance personnel. With increased automation, the time spent on maintenance work changes toward troubleshooting and fault finding. A longer time to find a failure often occurs in a complex control system compared with repairing it.

Integrating production and maintenance work in one function is a very big change that cannot occur without a plan, training, and the ability to change. The key to this integration is the process of changing people. This is difficult and can only occur with a plan using a common sense approach describing the objective, how to accomplish the objective, and a strategy not to dismiss people. Reduction in people should occur through attrition and changes in duties. Communication is critical to the process so people are informed, motivated, and trained. Figure 19 is a diagram showing a summary of acceptance and willingness to change.

The concept of change usually uses a team approach to organization. The supervision management style must therefore change. This can be very frustrating and difficult because everyone will not immediately endorse the team concept. To make it work, the supervisor needs to work as a coordinator, coach, and supporter instead of instructor or dictator. Training and information are key issues in organization development. Making change visible and ensuring that the change process is not too sluggish are

CHAPTER 9

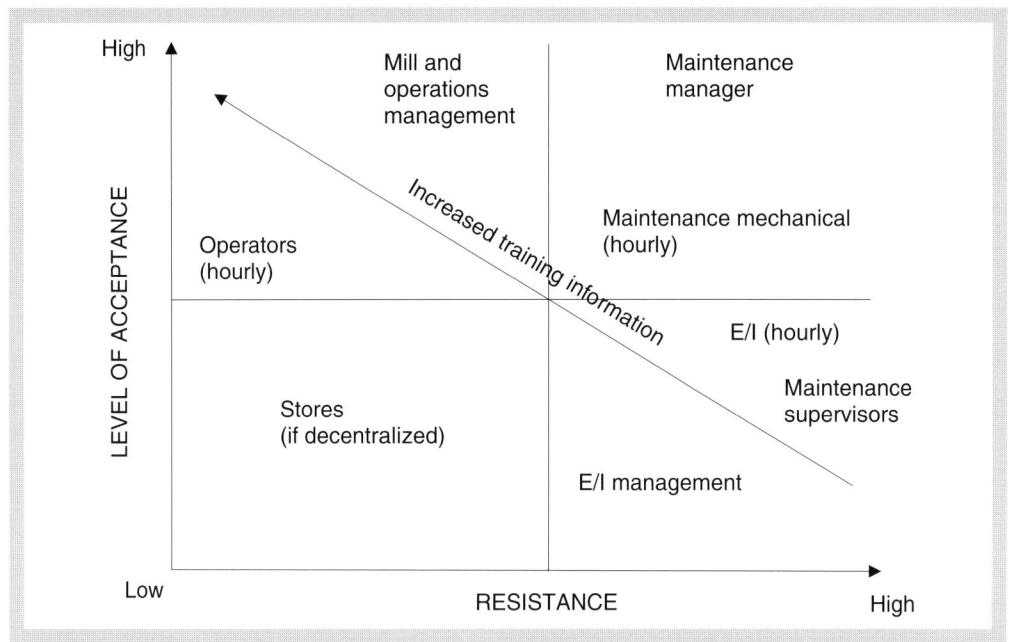

Figure 19. Mill and operations management most often favor an integrated organization. Many craft people have negative opinions. Increased training and information can therefore raise the acceptance of integration.

also necessary. An improvement effort often stops or slows due to excessive consideration given to one or two individuals while hundreds suffer.

8.3 Computerized maintenance management

During the 1970s and 1980s, a magic phrase in maintenance management was computerized maintenance management systems (CMMS). An estimated 300 CMMS are on the market today. Most claim to be the best with features no other systems have. Of all systems available, about 10 or perhaps 20 can truly claim to be complete maintenance management systems[23]. Studies have found that only 30% of the systems brought three years ago are still in use today. Of those in use, only 30% of the total capabilities find use.

Many studies have shown no correlation between effective maintenance and the existence of a CMMS. Correlations do exist between the use of CMMS and effective maintenance. Using these facts, Idhammer offers some explanations and advice.

What is a complete system?

No system can be totally complete. A complete system is one that covers the core functions most effectively. Those core functions include the following:

- Integration between equipment records and store items so a current bill of material is available at any time

- Complete storeroom management functions for single or multistorage functions
- Optional purchasing system module covering all necessary functions (Companies want the option of keeping their existing purchasing system or buying a separate module. Excessive integration might delay or even stop the selection process. In many cases, the selection and implementation process has been happening for years because purchasing, maintenance, and storeroom management cannot agree on the system's purchasing functions.)
- Work order planning and scheduling functions (Planning all work directly on-line must be possible including direct access to store items and a technical database for information about previous identical work. This ensures that a job planned once does not need planning again. The same applies to safety precautions.)
- Backlog management function that can sort all work orders according to status, e.g., waiting material, waiting approval, planned, ready to be scheduled, data to be done, etc.
- Preventive maintenance module capable of handling fixed time maintenance (time-based programmed replacements or overhauls) and condition monitoring and lubrication routes
- History per equipment including reports on equipment failures and repairs, downtime, and maintenance costs.

Purchase of a computer system from any of the top 10 system suppliers can lead to a tremendous success or a tragic failure. Success comes more from complete implementation and use of the system than from the system itself. Critical for success is that all users fully understand the capabilities of the system and make every effort to make the implementation successful.

Implementation includes changing mind sets and procedures and building databases. This does take time. Some advice follows:

- Do not develop your own system; development cannot be justified.
- Many functions can still be done by hand. If you do not have basic maintenance procedures well organized, you might have too much cultural shock if you try to computerize.
- Do not extend the selection process. People expect action soon.
- Buy a well-established system with many users and an active user group with a proven service record.
- Budget for investment in implementation and training. Training must not be limited to how the system works. Include "know why" training, what changes will be done, and what the benefits will be to everyone in the organization.

CHAPTER 9

8.4 Waste in maintenance

According to Idhammar[11], waste is the difference between the way items are now and the way they could and should be. The areas defined to have most waste or offer the greatest improvement are the following:

- Planning: What and how, need for parts, tools, drawings, etc.
- Scheduling: When and who should do it.

Very little planning exists. Most of what people perceive as "planned" activity is actually only "scheduled." In mills having planners, the planners are scheduling, buying spare parts, managing contractors, etc., instead of planning. Most planning and scheduling efforts focus on shutdowns while daily maintenance planning and scheduling is poor. A major opportunity exists in most mills to improve planning and scheduling and executing jobs according to schedule. This is schedule compliance.

The execution of work without planning and scheduling is obviously a symptom not a cause. Four major causes of poor planning and scheduling include:

- Wrong approach to maintenance efficiency
- Poor management
- Excessively short time between request and execution of maintenance work
- Insufficient systems to support planning.

All these causes represent huge opportunities to eliminate waste and improve productivity.

8.5 Autonomous maintenance

According to Idhammar, efficient maintenance can result from the following four-step plan:

1. Improve communications
2. Improve prevention
3. Improve planning
4. Improve scheduling.

Improve communications

Teach the entire maintenance and production organization what efficient maintenance is. A common joke is that many people in operations cannot count to more than one when it comes to priority for maintenance work. Operators commonly over-prioritize maintenance work. This is often due to a lack of trust in the ability of maintenance planning to complete the job without the operator pushing the job.

An excellent way to improve communication between operations and maintenance is to establish daily meetings to discuss and agree firmly on the next day's main-

Concepts of maintenance in the pulp and paper industry

tenance schedule. Most craft people like to know today where they are going to work tomorrow. Most supervisors do not want to inform craft people about their assignment for the next day. The meeting between supervisor and craft people should occur during the last 20–30 minutes of every day to discuss the next day's schedule. The meeting between operations and maintenance will reduce the number of emotional break-in jobs. The meeting between supervisors and crafts people will develop commitment to the schedule and understanding for planning and scheduling.

Improve prevention

Cleaning, lubrication, and condition monitoring are necessary. Cleaning and lubrication performed properly will reduce the need for corrective maintenance. Condition monitoring prolongs the lead time between work requests and execution to give time for professional planning and scheduling.

Improve planning

Most maintenance planners focus only on shutdown planning. Although scheduling is important, planning must include having parts, tools, and instructions available when the job begins.

Improve scheduling

An amazing international phenomenon is that all daily maintenance jobs are scheduled to take two people eight hours. One person can safely complete many daily jobs in two hours. Measurements of schedule compliance should include average time per scheduled and executed job. The goal is to improve schedule compliance and average time per job continually and eventually increase the total equipment efficiency.

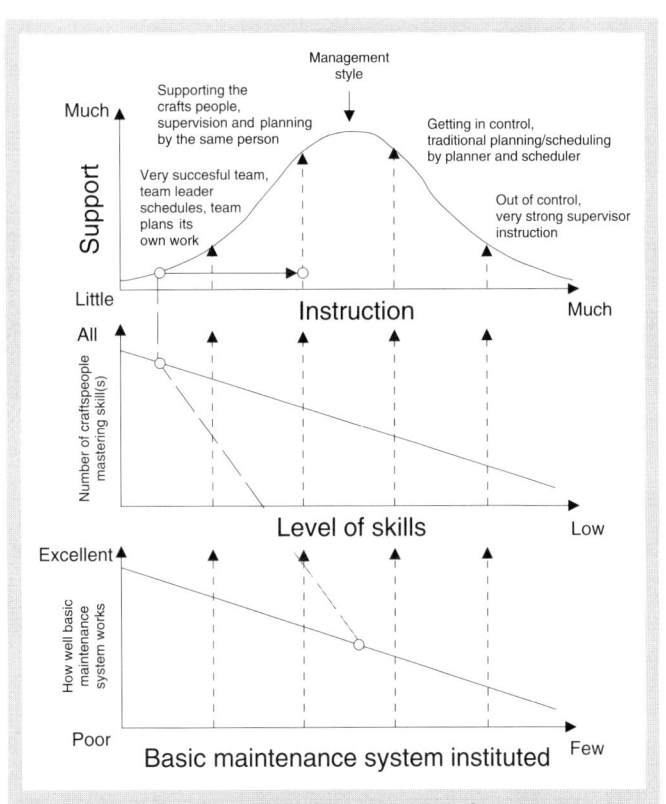

Figure 20. Changes needed in different levels before autonomous maintenance.

Before any organization can move toward autonomous maintenance, instituting the essential systems is

CHAPTER 9

necessary. Skills and motivation must be at a satisfactory level, and managers must change their management style with skill levels and system implementation as Figs. 20 and 21 show.

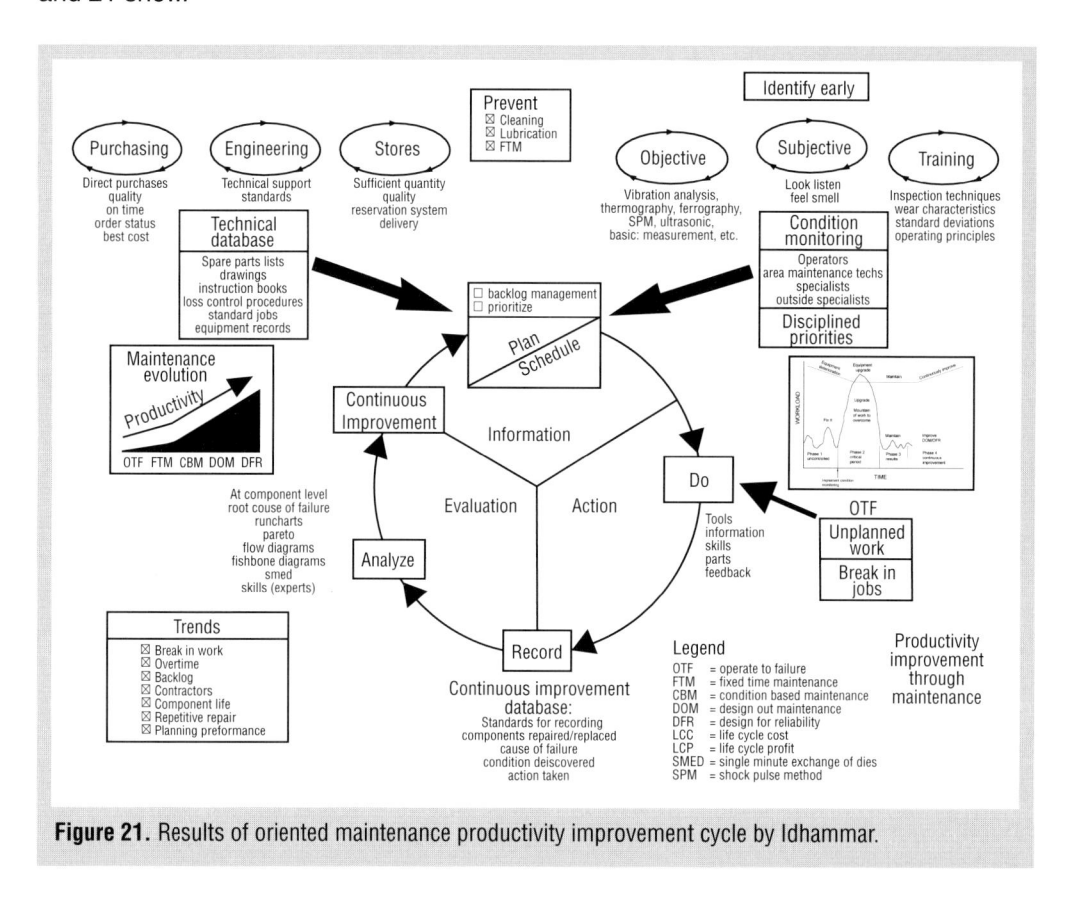

Figure 21. Results of oriented maintenance productivity improvement cycle by Idhammar.

9 Planning maintenance

Mills with undeveloped maintenance systems often have inadequate planning or no planning. Maintenance work happens when it is unavoidable often in the form of emergency repairs. In the long term, this form of maintenance is very expensive because deciding when to do the work is not possible. Expensive contract labor is very often employed, or mill personnel work on an overtime basis. The loss of revenue resulting from production downtime is also a consideration. Proper planning results in the correct maintenance work occurring at the right time with correct methods, appropriate trades people, available spare parts, and other resources being available when needed.

    Certain acute repairs must occur immediately. Gains will result even when time is pressing through planning and preparing the work before it starts. A repair can often be done better for lower cost when properly prepared with the correct tools, instructions, drawings, and labor estimates.

Any well-run maintenance organization should include a system for recording equipment history. The system should incorporate a file that is available to indicate:

- What work was achieved
- When it was completed
- Who performed the work
- How much time was used
- Observations in connection with the work that can have value for future maintenance planning.

This history of experience can ensure that maintenance will be as easy as possible. New experience will influence planning every time the cycle is completed. It allows maintenance work to be channeled toward the final objective of maintaining the optimum level of reliability and best human safety at minimal cost.

Most maintenance experts agree that maintenance planning should be centralized when warranted by size of the maintenance work force, the complexity of the facilities and equipment, or both. This central approach is even justifiable in a work force as small as 20–25 crafts people. A centralized planning function with one planner requires approximately 2.5–4 min of planning per crafts person. This planning cost is repaid if worker productivity can increase by only 5%. Additionally, the maintenance supervisor has more time for straight supervision without worrying about the planning responsibility. This in itself improves productivity. The priority matrix of Table 8 is an excellent tool.

9.1 Selecting maintenance planners

To avoid the "clerk" status of the planner, management should consider establishing entry level compensation for planners that would encourage competent trades people to apply for the position. By using this approach, trades people will accept the planner as a viable support resource.

After determining the compensation level, consideration should focus on identifying appropriate maintenance managers and supervisors to develop a selection criteria process that will define prerequisites necessary to master maintenance planning. The criteria should include:

- Communication capabilities
- Compatibility characteristics
- Trades person experience
- Formal educational achievements
- Positive attitude attributes.

CHAPTER 9

Table 8. Maintenance work request priority matrix[24].

Category	Safety	Production cost control Product quality Energy conservation	Morale Labor relations Public relations Intra relations	Environmental pollution control
	A	B	C	D
Emergency Immediate response of all necessary resources without limitation	Imminent danger to life or possibility of absolute destruction of property and facilities			
Breakdown Immediate response with appropriate resources A. Work is essential B. Automatic overtime to stabilize or prevent deterioration of condition C. Work may be "two phases" with permanent corrective action done under lower priority at later time	Breakdown that could result in injury or destruction of property and will probably become worse if not corrected	Breakdown that seriously limits production of essential facilities or will result in major product contamination, major shipment rejection, etc.	Breakdown that impairs production of an area other than that of the originators	Major breakdown could pollute or endanger the environment and would necessitate shutdown of essential facilities if not corrected. Required feedlines set by law.
Short range Necessary work to be investigated and scheduled immediately and scheduled to meet date and time required A. Work requested is necessary to continuing operation of facilities B. Work requested necessitated a shutdown of major equipment C. Routine shutdown work	Hazard to health or safety that is not expected to become worse before corrective action occurs	Repairs and projects to improve costs, increase production, improve quality, or conserve energy	Work necessary to meet contractual agreements or maintain morals and community relations	Problem that could be expected to become worse and would eventually require at least some equipment to be shutdown or reduced

Table 8. Maintenance work request priority matrix[24].

Category	Safety	Production cost control Product quality Energy conservation	Morale Labor relations Public relations Intra relations	Environmental pollution control
Long range Necessary work that can be planned and scheduled A. Overtime is not authorized without special approval B. Work to be planned and scheduled to optimize efficiency of effort and assure compilation by date required	Safety projects and preventive work to eliminate hazards or prevent them from developing or growing	Preventive work and long range projects related to production, cost, energy conservation, and quality improvements	Repairs and projects for employee facilities and public relations	Projects and preventive work on pollution control equipment and facilities
Competitive Routine work that can be planned and scheduled on a regular basis No overtime authorized	Preventative maintenance Housekeeping	Preventative maintenance Housekeeping	Preventative maintenance Housekeeping	Preventative maintenance Housekeeping
Contingency Desired work that may be rescheduled or deferred to suit availability of resources A. No overtime is authorized B. Resources will be used as available	Minor projects and work to continue program, e.g., signs, safety suggestion, bonus, etc.	Minor improvement projects and repairs	Minor repairs, yard cleanup, decorative painting, etc.	Repairs to auxiliary or backup equipment the operation of which is not imperative to continuing operations

The criteria selection development will provide an approved method for interviewing and selecting maintenance planners on an unbiased basis. The decision to select individuals to become maintenance planners should use a thorough understanding of the planning concept.

CHAPTER 9

When developing or using mill specific planner training programs, the following topics should receive consideration:

- Maintenance philosophy
- Computerized maintenance management
- Basic steps of planning and scheduling
- Work area organization
- Material procedures
- Effective communications
- Scheduling techniques
- Key factor performance monitoring.

These subjects should reflect current policies and procedures at the mill. Training materials should also include existing mill documentation. The decision to select and train qualified maintenance employees to become planners is a major step toward accomplishing a "planned environment" for improving equipment reliability and availability.

10 Examples of maintenance management

10.1 Maintenance at International Paper mill

International Paper's mill in Bastrop, Louisiana, was a winner of the "Peak Award." This award is for maintenance excellence was presented at the National Plant Engineering & Maintenance Conference in Chicago on February 25, 1992.

Mill manager Bob Janda stated, "First class organizations don't just happen; you make them happen." Mr. Janda cited several basic principles the Bastrop Mill follows with their personnel[25]:

- Help people understand the business: regular meetings explain economic conditions nationally and industry-wide
- Provide people with needed skills: training is available for technical, computer, safety, and interpersonal skills
- Define expectations and obligations: high standards are mutually set and followed with a policy of providing the best means available to help people excel
- Enable people to influence decisions that impact them: listen to what people have to say, act on it, and provide feedback so every employee working at the mill knows they can talk to and be heard by a supervisor concerning job-related issues.

A key ingredient in the outstanding performance of the mill was the adoption of a business unit concept in operations. Four business units exist: finished products, pulp, power, and operational services. A manager at staff level has total financial and cost responsibility for a segment of the mill including maintenance. Everyone in the business unit works together to achieve common goals for safety, production, and cost control. An important step in this concept is asking and accepting input from all members of the business unit in planning and decision making. Members take pride in operating their units. Several results show this:

- 2 million safe man hours worked in slightly less than one year
- Increased production
- Declining maintenance costs
- Ability to operate with fewer supervisors.

Planning

Corporate and plant engineering prepare a five-year facility plan using information supplied by the business unit teams. The teams identify needs and opportunities with input from maintenance, production, and engineering. Maintenance prepares a 1–3 year business plan to identify opportunities for improvement and performance goals.

Routine maintenance and production shutdown schedules coordinate on a mill-wide basis at least a week in advance. Major outages are planned one year in advance. Shutdown plans integrate with each area to support the business plan.

Operations are planned a week in advance. Maintenance manpower is then allocated according to the needs in each area. Outside contractors are used to reduce backlogs or provide better service at lower cost. Regular outside maintenance includes roof work, painting, insulating, and electric motor rewinding.

Quality of maintenance

Quality and maintenance service levels are measured through an audit and review procedure. Quality service is evaluated by examining the frequency of repeat failures, auditing work-in-progress, reviewing equipment availability, and monitoring downtime data.

Quality standards are administered and enforced through training sessions, crew meetings, on-the-spot corrections, and job reviews by crew managers. Quality contributions are rewarded with individual and crew or team recognition in mill publications and awards. Maintenance cost trends have shown a reduction, and equipment uptime has been stable.

Skills

Maintenance employees are multiskilled, and no job jurisdictional restrictions exist. Technical skill assessments are done biannually in one-on-one sessions and are used to plan training sessions. Persons with advanced skills and knowledge are given autonomy in jobs and provide leadership to peers and management. Predictive maintenance skills are developed and kept current by having mechanics learn maintenance routines. They are taught by maintenance engineering personnel and equipment manufacturer's representatives.

CHAPTER 9

Training

Each department identifies training requirements when new processes begin or processes change significantly. Training methods include classroom and on-the-job instruction and use of outside consultants, supervisors, and hourly personnel. Personnel showing aptitude and interest receive off-site training. Training is documented, certificates are issued, and continuing education units and college credits are awarded.

Individual training costs range from US$ 700–1600 for maintenance personnel. 100% participation in mandatory annual safety training exists and almost 90% exists in skills training. An emphasis on training is necessary. One manager said, "Training has a shelf life lower than bread."

Preventive maintenance

Emphasis is shifting from reactive to proactive maintenance. Past history and manufacturers' recommendations are useful to optimize preventive maintenance schedules. If experience dictates, intervals between preventive maintenance are shortened. If longer time intervals are necessary, they do not exceed manufacturers' recommendations.

Auditing and upgrading preventive maintenance procedures is a continuing process using mechanics' experiences. Area foremen and maintenance superintendents review mechanics' comments. The business teams regularly examine equipment downtime reports, production reports, and incidences of equipment failures. Such review ensures the effectiveness of predictive and preventive maintenance procedures. A primary objective is to eliminate repetitive failures. Equipment and process reliability is tracked through trend data stored in a computer. As a result, the mill has shown an upward trend in productivity.

Several technologies have use throughout the mill to observe equipment performance. The most common are continuous on-line vibration monitoring, monthly lubrication analysis, heat sensing on critical equipment, and infrared inspection of electrical equipment.

Computerized maintenance management systems

A computerized maintenance management system is useful for various maintenance functions throughout the mill including equipment specifications, bills of material, work schedules, preventive maintenance, equipment search, equipment cost, purchasing, and training. All maintenance personnel have access to a terminal linked with the CMMS. Approximately 95% are trained to a basic level. About 50%–60% of the maintenance mechanics are active daily users.

Safety

Promotion of employee awareness of environmental, health, and safety issues on and off the job takes several forms:

- The importance of complying with safety and health regulations is emphasized.

- Department safety committees actively encourage employee participation in safety and health issues.

- Safety and health incentives and award programs for employees are promoted. Morale is high and absenteeism is low. Supervisors are proactive on absenteeism and speak with employees immediately concerning any problems they are having.

10.2 Preventive maintenance in a Norwegian mill

M. Peterson & Son A/S has an integrated pulp and paper mill in Moss, Norway. The pulp mill produces 150 000 tons/year of sulfate pulp, and the paper mill produces 120 000 tons/year of unbleached kraft paper primarily for linerboard and sack paper. During 1980, an analysis was conducted regarding the availability of production equipment. Figure 22 shows the result[26].

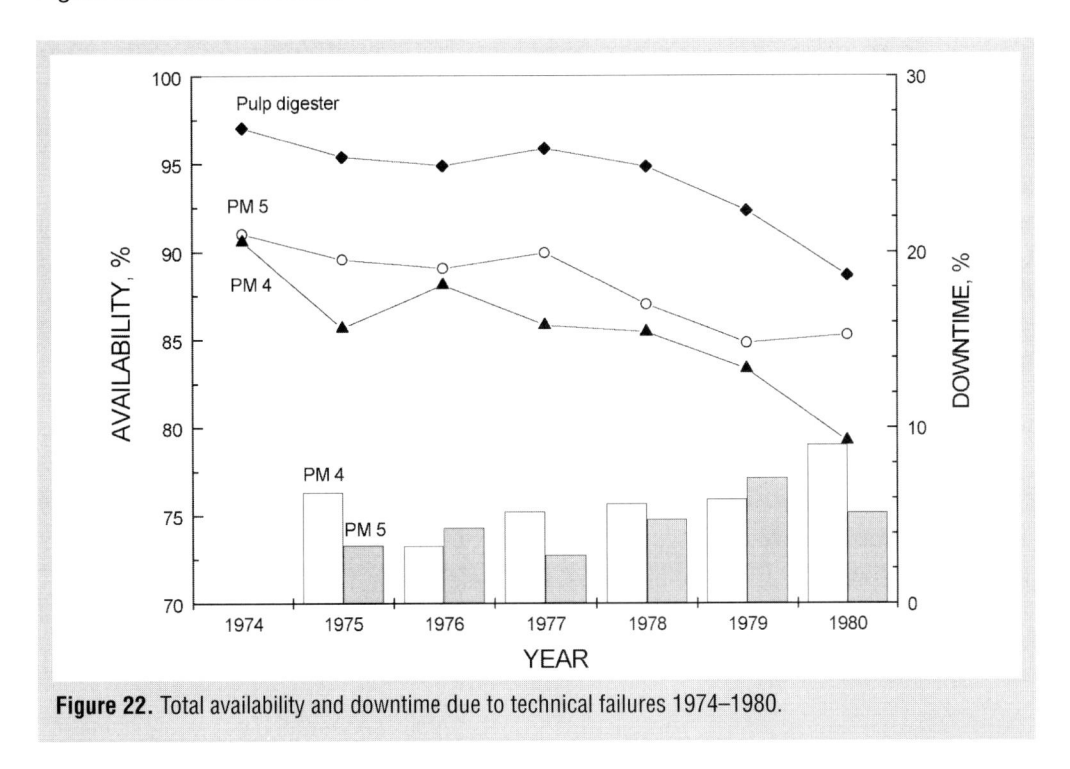

Figure 22. Total availability and downtime due to technical failures 1974–1980.

A comparison between the curves and the columns shows that a decrease in availability is almost equivalent to the increase in downtime due to mechanical failures. Almost one-half the total downtime of the production equipment was due to mechanical failures. Mechanical failures were therefore a dominating factor. Reduced availability

CHAPTER 9

meant lost production. At the Moss mill in 1986, a 1% reduction in availability meant a loss in production quantity of about US$ 200 000 per year.

The following actions occurred:

- The existing preventive maintenance system was improved through condition monitoring.
- A training program for all maintenance employees was implemented.
- A planning department was established to improve the administrative systems and routines for preparation, planning, technical documentation, purchasing, inventory control, and analysis.
- Relationships between production people and maintenance people were improved.

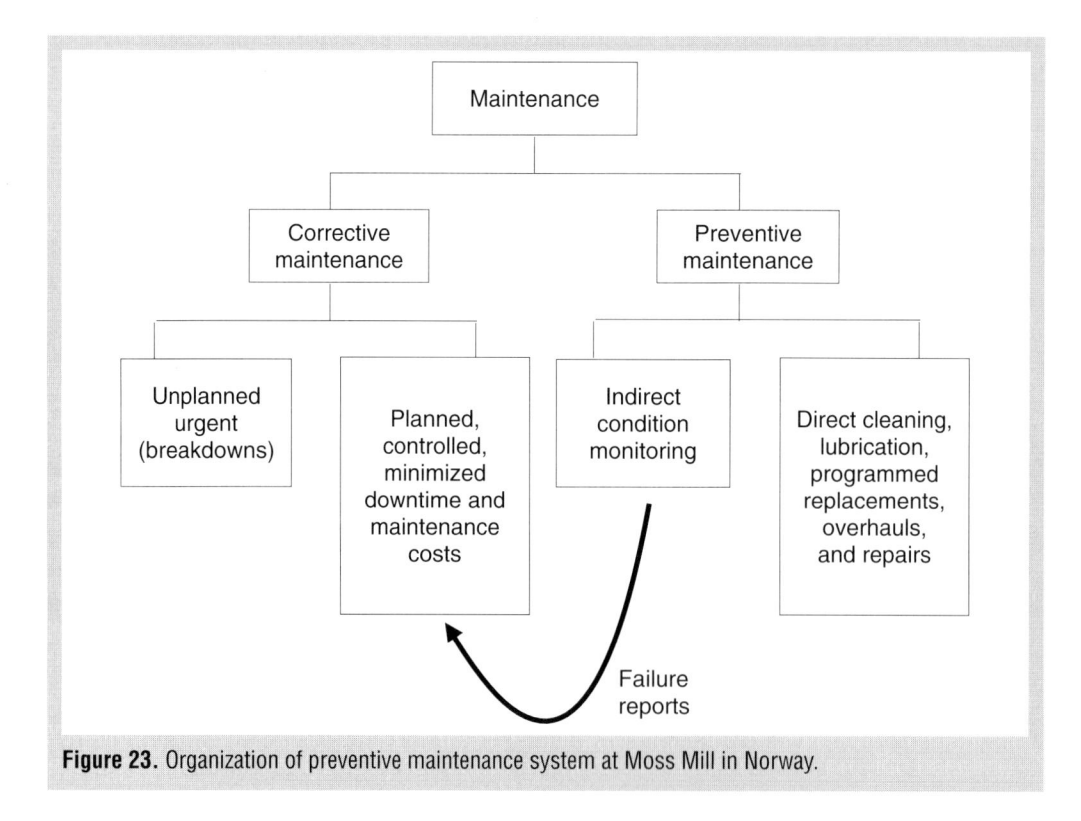

Figure 23. Organization of preventive maintenance system at Moss Mill in Norway.

Maintenance was divided into two categories: corrective maintenance and preventive maintenance as Fig. 23 shows. In most organizations, the major part of corrective maintenance is unplanned and very expensive. The planned part is very small. The reason for this is lack of preventive maintenance.

Concepts of maintenance in the pulp and paper industry

A project group of five members consisted of two from organized labor, one foreman, and two from the engineering or management staff. The project group visited pulp and paper mills in Sweden to study systems in use and results obtained. Most mills visited could show reductions in downtime of 40%–50% after the introduction of preventive maintenance. Breakdowns had been reduced to one-third over a period of three years. Overtime was reduced by more than 50%.

A preventive maintenance program was first implemented on paper machines Nos. 4 and 5. Later, the preventive maintenance system was extended to include the complete paper mill and the complete pulp mill. The system included all programmed maintenance activities and had four main parts:

- Main maintenance list
- Routine list
- Preventive maintenance cards
- Instructions.

In 1986, the Moss mill implemented the preventive maintenance system. Approximately 10 850 inspection points covered by nine inspectors were registered. These inspectors shown in Fig. 24 did inspection work only part of their time.

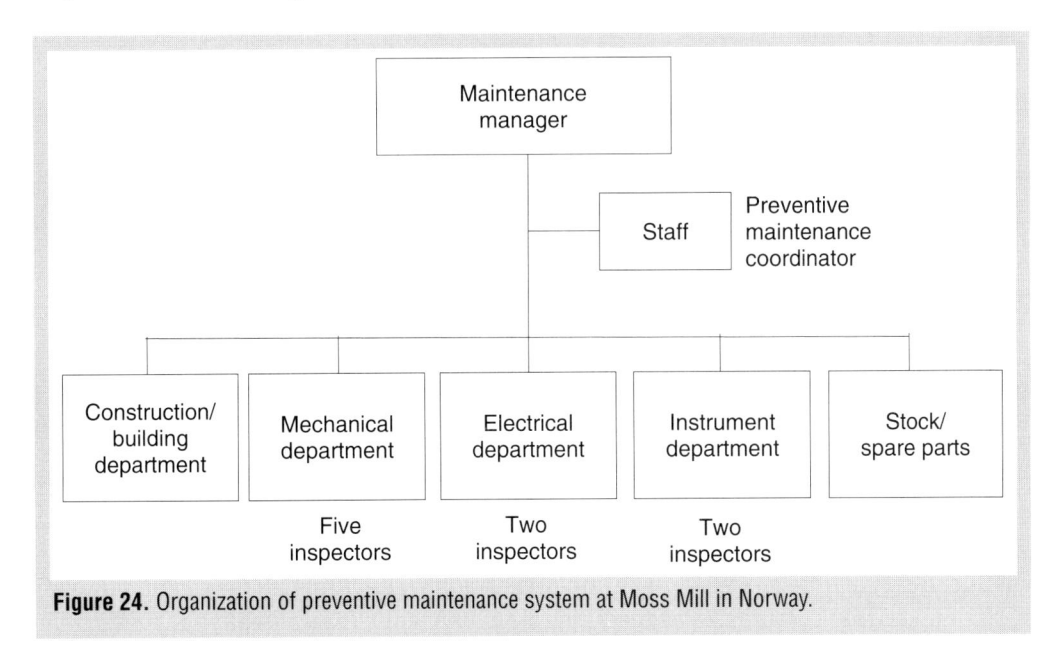

Figure 24. Organization of preventive maintenance system at Moss Mill in Norway.

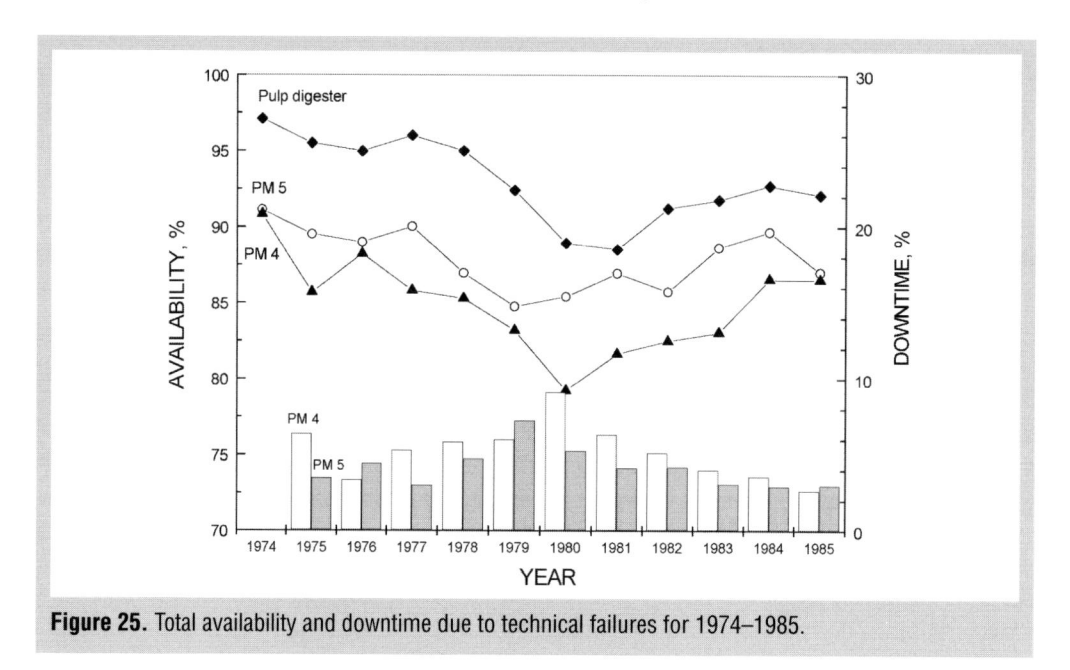

Figure 25 shows the trend curves for availability for 1981–1985.

Figure 25. Total availability and downtime due to technical failures for 1974–1985.

During 1981–1985, downtime due to technical failures for Nos. 4 and 5 paper machines decreased from 8.7% to 3.5% and from 5% to 3.1%, respectively. The reduction in downtime resulted in an increase in availability that gave a production increase of almost US$ 1.5 million. An interesting point is that downtime due to mechanical failures is now about 25% of total downtime. Planned corrective maintenance has increased considerably, and the amount of unplanned corrective maintenance has decreased.

10.3 Maintenance training in Sweden

A difference in social structure between Sweden or Finland and the United States allows Nordic mill employees added time and incentives for training. Working fewer hours and taking longer vacations and parental or sick leave, a Swedish mill might commonly run six shifts.

Swedish mills are more likely than U.S. mills to send hourly workers off-site for instruction. Braviken, Frövifors Bruk AB, and Skärbacka mechanics and electrical instrumentation personnel upgrade their skills with 1–3 weeks of training each year augmenting the 1.5–3 years of technical schooling they receive before employment.

Swedish mill supervisors receive supervisory training once or twice each year. According to Idhammar, U.S. supervisors receive training only once every five or six years. Individual supervisors in Swedish mills generally have fewer workers assigned to them than those in U.S. mills. According to Idhammar, Holmen's Braviken mill in Norrköping is an efficient mill where the maintenance ratio is 50 supervisors to 126 salaried employees. Maintenance man hours/ton are slightly under 0.5. This figure does not include maintenance by outside contractors.

Concepts of maintenance in the pulp and paper industry

11 International maintenance reviews

11.1 Canadian reviews

In 1976, 1984, and 1986, the Canadian Pulp and Paper Association, Technical Section, Mechanical Engineering & Maintenance Committee performed surveys on maintenance organizations, methods of operation, and costs for Canadian mills and some Scandinavian mills.

Maintenance organization

The maintenance work force in Canadian mills operates from a central workshop or an area supported by a central workshop. Most mills use shift workers. A survey revealed the results in Table 9.

Table 9. Workshop organization at Canadian pulp and paper mills in 1984 and 1986 with figures in parentheses giving average pulp production per day[27, 28].

	1984	1986
Central workshop	8 mills (330 tons/day)	14 mills (360 tons/day)
Area supported	3 mills (391 tons/day)	16 mills (773 tons/day)
Combination	63 mills (897 tons/day)	39 mills (799 tons/day)
Number of personnel	125 (54.6 tons/day)	112 (48.6 tons/day)

With a centralized maintenance shop near stores and technical information, better control and use of labor and material is possible. In a large mill, the distance between the work location and the centralized workshop and stores makes area support essential.

Maintenance workforce

Table 10 gives the average figures from the surveys.

Table 10. Maintenance workforce in Canada in 1976, 1984, and 1986 in tons/day/maintenance employee.

	1976	1984	1986
Tons of product/maintenance person	4.46	4.46	4.5
Maintenance person/supervisor	12.2	10.3	8.84

Figures 26 and 27 show the maintenance work force vs. tons of production per day and the number of supervisors vs. tons of production per day according to the surveys performed in 1984 and 1986.

CHAPTER 9

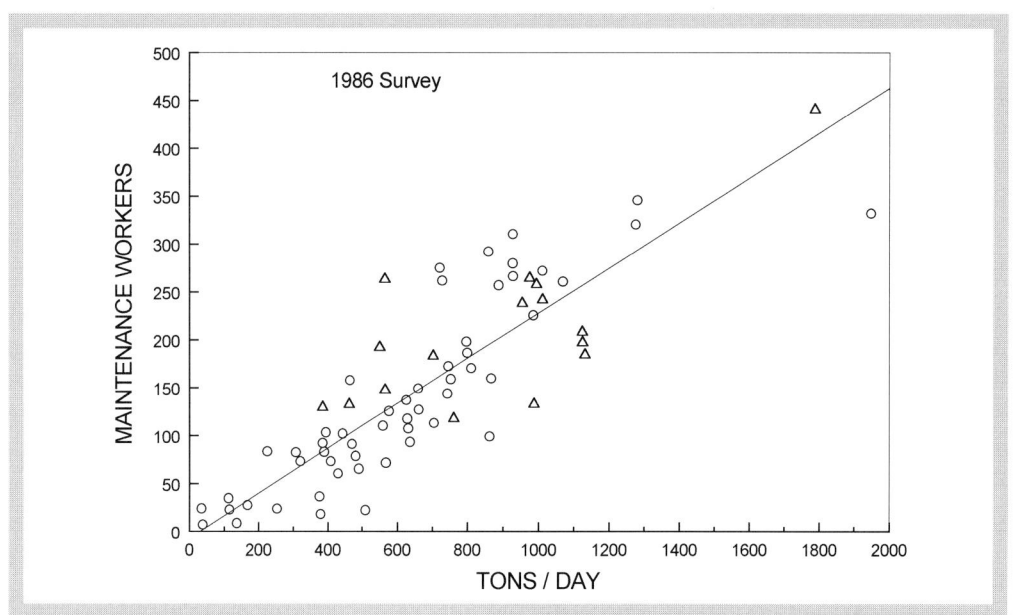

Figure 26. Maintenance work force vs. tons production/day in 1986 according to the Canadian questionnaires.

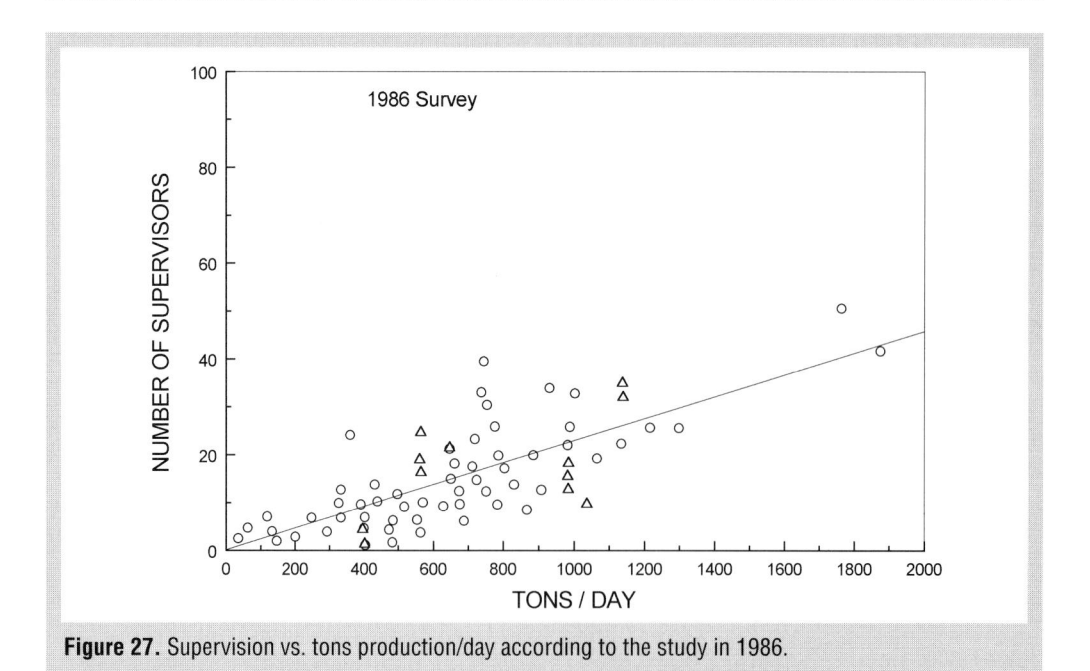

Figure 27. Supervision vs. tons production/day according to the study in 1986.

The supervision of the work force used trade specialization with welders, mill wrights, and pipe fitters having the same supervisor in several instances. The bar charts in Fig. 28 show the distribution of the labor force by trade specialty.

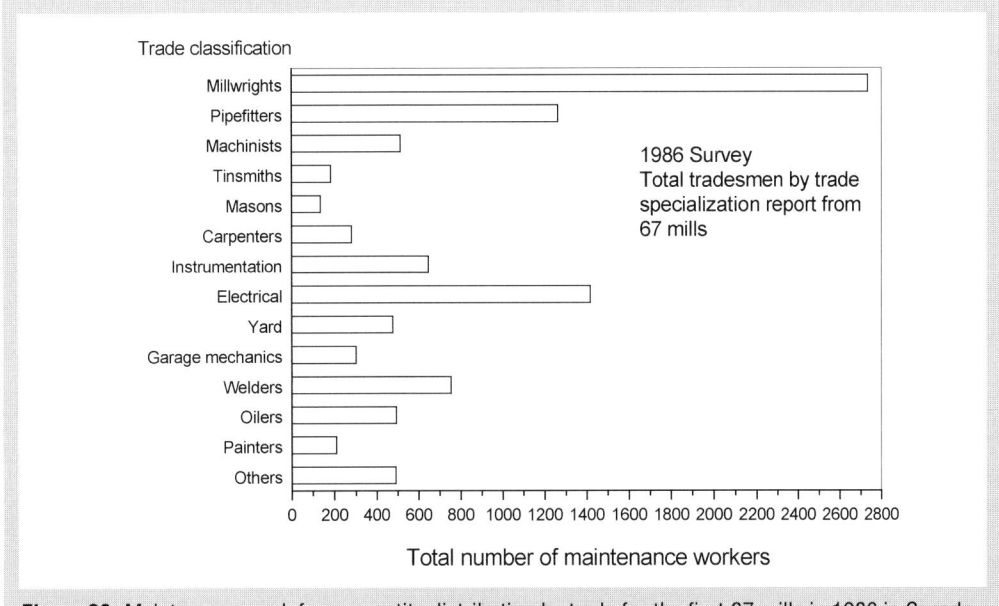

Figure 28. Maintenance work force quantity distribution by trade for the first 67 mills in 1986 in Canada replying to a CPPA maintenance survey questionnaire.

Material and cost control

The purchasing department supervised the stores operation at 39 mills in 1984 and 38 mills in 1986. At 29 mills in 1984 and four mills in 1986, maintenance services did that work. Other departments that supervised stores included services (none in 1984 and 22 mills in 1986), production (2 mills in 1984 and in 1986), engineering (2 in 1984 and none in 1986), and accounting (2 mills in 1984 and only one mill in 1986).

At least 22 mills in 1984 and 20 mills in 1986 were using a manual system to control inventory, and 43 mills in 1984 and 33 mills in 1986 used work order systems for material and cost control. Maintenance costs were compiled by department only at 45 mills in 1984 and in 1986, by equipment number at 4 mills in 1984 and 3 mills in 1986, and both department and equipment number at 25 mills in 1984 and at 21 mills in 1986.

Maintenance material cost

The 58 mills in 1984 and the 60 mills in 1986 completing this part of the questionnaire had a total material cost for 1983 of CAD$ 285.45 million and for 1986 of CAD$ 307.71 million. Figures 29 and 30 plot the individual mill returns using a mill identification number. Table 11 also shows this information for all the mills reporting.

CHAPTER 9

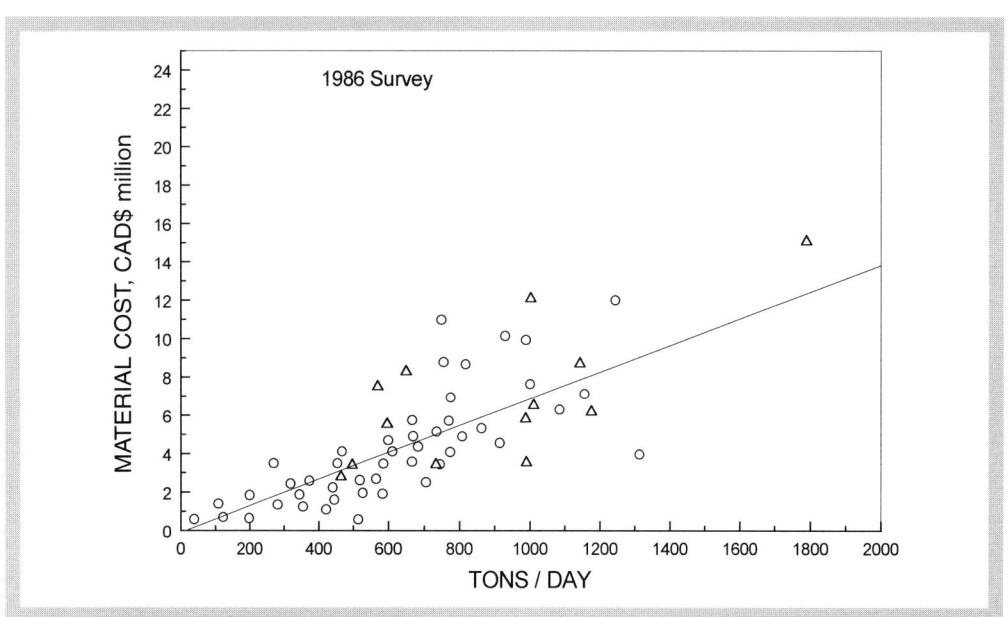

Figure 29. Material cost in millions of CAD$ vs. tons production/day in 1986 using data obtained from the first 67 replies received.

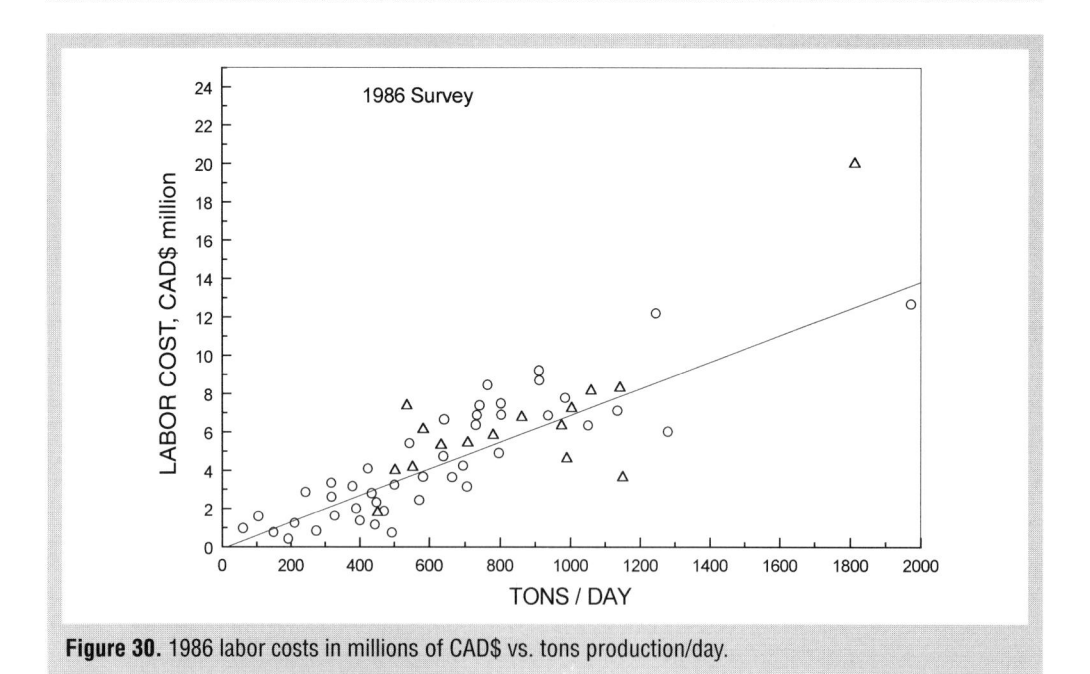

Figure 30. 1986 labor costs in millions of CAD$ vs. tons production/day.

Concepts of maintenance in the pulp and paper industry

The average maintenance material cost was CAD$ 19.5/ton in 1983 and CAD$ 21.78/ton in 1986. For a mill with 500 tons of daily production, these costs were CAD$ 3.36 million in 1983 and CAD$ 3.70 million in 1986.

The combined daily tonnage for the 58 mills in 1983 was 42 457 tons/day and 41 544 tons/day for the 60 mills in 1986. Since most mills also used outside contractors for small, major, and specialty maintenance, the total material costs included contracted maintenance in several returns. The real material costs were therefore higher than given here.

Maintenance labor cost

Maintenance labor costs reported by 56 mills in 1984 were CAD$ 301.54 million and by 60 mills in 1986 were CAD$ 308.14 million as Fig. 30 shows. These values do not include any contracted labor. The average maintenance labor costs per 500 tons of daily production were CAD$ 3.55 million in 1983 and CAD$3.71 million in 1986. This means CAD$ 20.4/ton in 1983 and CAD$ 21.81/ton in 1986. The increase in nominal labor costs is only 7%.

The surveys made no attempt to monitor maintenance effectiveness or identify specific high cost areas. A fair measurement of maintenance productivity can come from the surveys by comparing the tons of production for each maintenance person. This indicates an industry average of 4.5 tons/person in 1984 compared with a figure of 4.46 obtained in the 1983 and 1976 surveys. Assuming 8 h per day for each maintenance person, maintenance hours per ton of product in 1984 were 1.78 maintenance hours/ton. This figure is high compared with the figures from Idhammar where the Braviken Mill was 0.5 h/ton.

The average material costs included some contracted maintenance work charged to materials. While comparison of these factors is possible, factors such as age, type of production, use of contractors for certain maintenance jobs, and the present condition of the mill require consideration when considering the large variance between mills.

11.2 North American surveys

Pulp & Paper journal periodically conducts a survey of maintenance practices in the North American pulp and paper industry. Survey data are presented by region (western, southern, northeastern, and midwestern United States; Canada; and international) and paper grades (linerboard, printing and writing, and tissue and newsprint). Table 11 gives benchmark survey mill data by paper grade[29], and Table 12 gives some key figures[6,18].

Most maintenance costs came from labor. This is shown as labor and contract maintenance cost as a percentage of total maintenance costs. Canada had the highest combined labor and contract maintenance costs at 67.7% followed by northeastern mills at 58.5%.

CHAPTER 9

Table 11. Benchmark survey mill data summary by paper grade for 1992.

	Lineboard	Printing and writing	Tissue	Newsprint
Mill responses	22	23	3	9
Mill processes	14.0	19.0	8.0	7.0
Woodyard	15.0	16.0	2.0	2.0
Kraft pulp	10.0	22.0	4.0	4.0
Bleaching	3.0	7.0	0.0	8.0
TMP groundwood recovery	18.0	15.0	3.0	3.0
Boiler	21.0	23.0	4.0	7.0
Effluent treatment	20.0	19.0	3.0	5.0
Kiln causticizing	17.0	14.0	2.0	2.0
Pulp rnachine	6.0	12.0	3.0	0.0
Converting	2.0	13.0	3.0	1.0
Other	1.0	1.0	0.0	1.0
Operating efficiency				
Average salable tons/year	459 350.1	354 344.7	363 000.0	338 996.9
Average reject tons/year	29 995.8	30 209.0	19 750.0	21 568.0
Percent rejects	6.5	8.5	5.4	6.4
Average uptime	87.5	88.5	89.7	90.9
Average planned and scheduled downtime	3.8	3.9	4.7	2.5
Average operational downtime	5.7	5.6	2.9	4.2
Average unscheduled electrical downtime	1.2	1.3	0.9	1.1
Average unscheduled mechanical downtime	1.9	1.5	1.7	1.6
Maintenance costs				
Maintenance cost per ton	42.1	59.3	66.6	85.5
Maintenance labor hours per ton	1.0	13.6	1.3	9.6
Percent maintenance labor cost to total maintenance costs	36.8	39.4	47.3	35.4
Percent maintenance material cost to total maintenance costs	44.9	41.8	41.3	44.5
Percent contract maintenance. costs to total maintenance costs	14.4	16.8	12.5	19.6
Maintenance organization				
Central	2.0	5.0	1.0	1.0
Area	7.0	8.0	0.0	2.0
Team	3.0	3.0	0.0	0.0
Combination	9.0	10.0	6.0	6.0
Other	1.0	0.0	0.0	0.0
Maintenance work force				
Mills with union work force	20.0	23.0	5.0	7.0
Mills with nonunion work force	1.0	1.0	2.0	4.0
Mills with contract work force	4.0	1.0	0.0	2.0
Mills with flexi-craft or multicraft work force	16.0	17.0	6.0	6.0
Mills with strict craft line work rules	7.0	7.0	1.0	3.0
Average number of maintenance job classifications	6.5	7.0	6.2	5.0
Percent E/I to total skilled crafts personnel	24.7	26.8	30.4	27.5

Concepts of maintenance in the pulp and paper industry

Table 11. Benchmark survey mill data summary by paper grade for 1992.

	Lineboard	Printing and writing	Tissue	Newsprint
Maintenance labor support				
Average number of maintenance hourly people per first line supervisor	11.5	16.5	15.2	12.1
Average number maintenance hourly people per maintenance management	6.8	16.4	12.5	18.1
Average number maintenance hourly people per mill management	21.2	12.2	3.7	24.8
Average number mill hourly people per mill management	19.7	11.3	8.3	21.0
Average number maintenance hourly people to total mill hourly people	31.9	25.8	24.5	30.1
Percent mills where operators can perform maintenance	69.6	60.0	71.4	33.3
Percent of maintenance work performed by these operators	14.0	17.7	5.4	35.0
Maintenance hours worked				
Percent hourly maintenance overtime without shutdowns	9.8	9.6	9.1	9.4
Percent hourly maintenance overtime with shutdowns	14.3	13.5	17.8	14.8
Average total hours worked replacing electric motors	6.4	4.8	7.1	4.3
Average total hours worked replacing dryer bearings	5.0	3.4	8.1	3.0
Average total hours worked replacing rotary joints	10.0	16.6	42.9	3.7
Hours per year on education of maintenance hourly people	11 284.0	5 377.8	20 097.2	6026.0
Planning				
Percent work planned and scheduled weekly	58.7	54.0	33.8	53.7
Percent work planned and scheduled daily	65.0	66.6	54.4	64.7
Percent mills with formal planning	72.7	86.4	57.1	44.4
Percent labor hours spent in preventive maintenance	16.4	16.0	21.8	21.0
Average number of maintenance hourly people per planner	26.9	42.7	26.0	21.1
Storeroom support				
Percent mills where storeroom reports to maintenance	18.2	17.4	42.9	11.1
Average maintenance hourly people per storeroom personnel	18.3	18.1	27.2	25.4
Percent mills with storeroom pickup and delivery to job site	22.7	47.8	14.3	11.1
Number of storerooms with pickup and delivery during shutdown	4.0	8.0	0.0	1.0
Number of storerooms with pickup and delivery daily	4.0	9.0	1.0	1.0
Percent storeroom inventory costs to total maintenance costs	31.6	44.1	32.5	56.0
Computers				
Mills with computerized maintenance systems	21.0	22.0	6.0	7.0
Number of in-house computer systems	9.0	16.0	3.0	7.0
Number of package computer systems	7.0	3.0	1.0	0.0
Number of modified computer systems	6.0	5.0	2.0	1.0
How mills rate computer systems (1–10)	6.6	6.8	3.8	6.3

CHAPTER 9

Table 12. U.S. and Canada mill data in 1986, 1987, and 1992.

	Northeast	Midwest	South	West	Canada
Tons per year / maintenance employee 1986	1127	2532	2012	2199	1203
Maintenance employee/total employee 1986 1987	21% 18%	19% 18%	23% 24%	27% 26%	26 % 26 %
Job classification	7	9	6	8	12
Maintenance budget/total operating budget 1987	9%	18%	26%	15%	13%
Maintenance costs/tons per year 1992 Labor cost Material cost Contracting cost	57.3 39.1% 43.5% 19.4%	53.2 50.0% 43.5% 2.8%	64.1 36.2% 46.3% 15.1%	49.3 35.4% 40.6% 19.3%	68.1 34.8% 35.0% 32.9%
Contracting cost 1987	16.0%	14.0%	20.0%	15.0%	12.0%
Preventive maintenance 1987 1992	10.0% 20.0%	13.0% 15.0%	16.0% 17.0%	14.0% 17.0%	11.0% 17.0%

In the 1992 survey, southern mills reported that 100% use flexi-craft or multicraft work rules and 80% allowed operators to perform maintenance. Only half the Canadian mills in 1992 used flexi-craft or multicraft work rules, and 40% of the operators performed maintenance.

Canada had only 16.7% overtime maintenance hours (10.4% with shutdowns and 6.3% without), and northeastern mills had 27.7% (17.0% with shutdowns and 10.7% without). Canadian mills also spent the least amount of time replacing rotary joints using only 2% of the total maintenance hours for replacement. Midwestern mills used 28.8%, and international mills used 38.3% of the hours for replacement.

Linerboard mills had the highest production rates (459 350 tons/year) and the lowest maintenance costs per ton (US$ 42.1/ton). Mills producing linerboard had the largest amount of management support with 11.5 hourly maintenance personnel for every first-line supervisor and 6.8 hourly maintenance personnel for every maintenance management person. Linerboard mills spent substantial time on education with 11 284 h/year for hourly maintenance personnel (total maintenance hours 459 350 h/year). This means 2.45% of the maintenance hours were spent on education. Linerboard mills had a high rate of planning compared with other grades with 58.7% of the work planned weekly, 65% planned daily, and 72.7% of the mills having formal planning groups.

Printing and writing mills had the highest reject rate per annual salable ton (8.53%). They also had the lowest education (only 0.11% of the maintenance hours) and the lowest hours spent on preventive maintenance (16%). This group had the most mills with formal planning (86.4%) and a high percentage of work planned weekly (54%) and daily (66.6%).

Concepts of maintenance in the pulp and paper industry

Tissue mills had the lowest reject rate per annual salable ton (5.44%). Mills producing tissue had the highest education percentage (4.2%) of the total maintenance hours and spent the greatest percentage of hours on preventive maintenance (21.8%).

Newsprint mills had the highest maintenance cost per ton at US$ 85.5 and the highest percentage storeroom inventory cost to total maintenance cost (56%). The percentage of newsprint mills that allowed operators to perform maintenance work was low at only 33.3%.

Table 13 presents the survey data evaluation from the years 1986 and 1987 in the United States, and Table 14 presents data from four Swedish pulp and paper mills for comparison. These Swedish mills spent an average of 27% of their maintenance budgets on outside contractors vs. 14% for North American mills. This could account for the high maintenance productivity figures for these mills (0.59–1.28 labor hours per ton).

The objective of the surveys was not only to examine specific maintenance practices but also to assess the entire maintenance organization. The benchmark survey offers a process for measuring and monitoring various maintenance practices.

Table 13. U.S. and Canada mill data comparison.

	Northeast	Midwest	South	West	Canada
Production tons per year/maintenance employee					
1986	1127	2532	2012	21 992 547	12 031 372
1987	1386	1411	2899		
Maintenance employees/total employees					
1986	21%	19%	23%	27%	26%
1987	18%	18%	24%	26%	26%
Maintenance supervisors/100 mechanics					
1986	1514	1012	1111	1011	1310
1987					
Maintenance man hours/ton of production					
1986	137 154	69 118	94 071	76 062	158 110
1987					
Average electric motors replaced each year 1987	7%	8%	6%	8%	6%
Average dryer bearings replaced each year 1987	3%	4%	5%	3%	6%
Average rotary joints replaced each year 1987	34%	66%	31%	31%	31%

CHAPTER 9

Table 14. Swedish mill data summary.

	Korsnäs-Marma AB Gävle, Sweden	Billerud Uddeholm AB Gruvöns Bruk, Sweden	Svenska Cellulosa AB - Ortviken Sundsvall, Sweden	Svenska Cellulosa AB - Östrand Sundsvall, Sweden
Annual production (short tons)	545 000	597 000	594 000	402 000
Paper machine availability	88%	84%	86%	94%
Days/year of mill operations	342	355	349	353
Products	Linerboard Bleached board Bleached pulp	Medium Linerboard Bleached pulp	Newsprint	Bleached pulp CTMP pulp
Total mill employees	1330	1280	905	600
Total maintenance employees	440	390	249	250
Total maintenance labor hours/year	697 000	563 300	351 000	324 000
Maintenance labor hours/ton	1.28	0.94	0.59	0.81
Maintenance employees/total employees	33%	30%	28%	42%
Supervisors/100 maintenance mechanics	18	18	20	26
Maintenance budget spent on outside contractors	29%	39%	20%	20%
Total maintenance costs (US$/ton)	$29.24	$23.24	$18.32	$28.00

A confusing element in both surveys is showing integrated paper mills and mills using market pulp together. The number of unit processes included in production determines the total maintenance costs in US$/ton produced. The complexity of the processes also influences the final maintenance costs. Numbers and statistics can be misleading, and maintenance expenditures for the year may not correlate directly to the machine efficiency or mill profit for that year. One should therefore use a long-term method of determining the effectiveness of a mill maintenance program.

Information on maintenance at Finnish pulp and paper mills is very limited. At the Veitsiluoto Oy's Kemi pulp and paper mill, maintenance organization uses area business units. A central workshop includes electric repairs, warehouse, repairs of engines and small equipment, painting, and normal in-house maintenance[30]. According to the maintenance supervisor, "The quality of the paper is good if the maintenance is well managed and organized." Maintenance teamwork is important at the Kemi mill. The estimated total maintenance costs in 1993 were about FIM 75 million. The production in 1992 was 566 178 tons of paper. This means an average maintenance cost of FIM 132/ton of paper produced. The maintenance cost of a paper machine at the Kemi paper mill is FIM 75–120 depending on the complexity of the process, i.e., the paper

Concepts of maintenance in the pulp and paper industry

grade. The maintenance cost of 132 FIM/ton of paper includes only the production of paper not the total costs of an integrated pulp and paper mill.

According to maintenance specialists, maintenance costs at a Finnish bleached pulp mill are approximately FIM 100/ton of product. Cost of materials is about 40% of the total maintenance cost. The cost of maintenance labor is about 45%, and the cost of outside contractors is approximately 15% of the total maintenance cost.

12 Maintenance costs in comparison with other costs

The maintenance portion of the North American pulp and paper industry's total operating budget is 9%–26%.

12.1 Maintenance costs compared with product costs

The price of pulp[31] has varied during the years as Fig. 31 shows. The price of pulp increased dramatically in the late 1980s. In 1985, the price of bleached pulp was US$ 400/ton, and in 1989 the price doubled. A similar increase in price occurred with printing and writing papers, but the change was only from US$ 1000/ton to US$ 1400/ton.

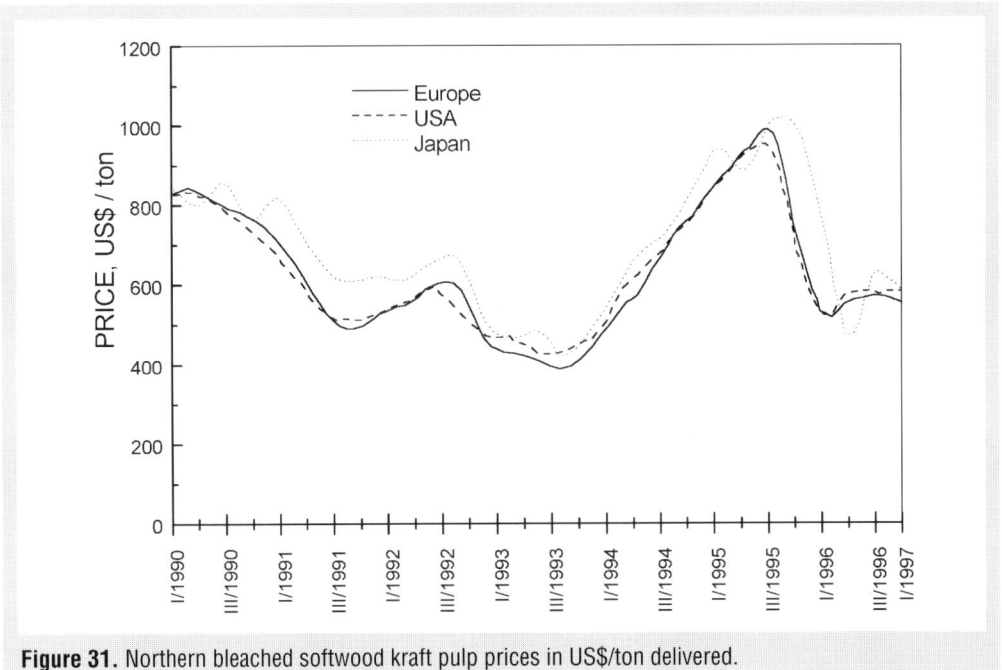

Figure 31. Northern bleached softwood kraft pulp prices in US$/ton delivered.

The average maintenance costs in 1992 were between US$ 42/ton for tissue and US$ 85.5/ton for newsprint. The maintenance costs for printing and writing paper were US$ 59.3/ton. The prices of these paper grades were approximately US$ 420/ton and US$ 1020/ton for newsprint and printing and writing, respectively. This means that in 1992 the maintenance costs for newsprint in the U.S. market were 20% of the sales

CHAPTER 9

price. For printing and writing papers, the maintenance costs were about 6% of the sales price. These figures are significant compared with the profits of the mills. The price of pulp and paper in 1992 was extremely low due to simultaneous oversupply, economic recession, and environmental pressures.

In 1992, the total production of pulp and paper board in the world was 245 million tons. Assuming variation of the maintenance costs between US$ 42–85/ton, the pulp and paper industry in 1992 consumed US$ 10–20 billion in maintenance globally.

In Finland, the production of pulp and paper in 1992 was 9.2 million tons[32]. Assuming that the maintenance costs in Finland are approximately at the same level as in the United States, the total maintenance expenditures for the Finnish pulp and paper industry were US$ 386–786 million in 1992.

The production of pulp in the world in 1992 was 155 million tons. Assuming that the maintenance costs for pulp are FIM 100/ton, the maintenance expenditure in the pulp industry globally in 1992 reached FIM 15 billion (about US$ 3 billion). The production of pulp in Finland in 1992 was 8.75 million tons. This leads to a maintenance cost estimation of FIM 875 million in 1992 in the Finnish pulp industry.

12.2 Maintenance costs compared with investment costs

Figure 32 shows that the capacity expansion boom that drove capital spending to record levels in the late 1980s played itself out in 1991 leaving environmental improvement as the main engine of capital spending.

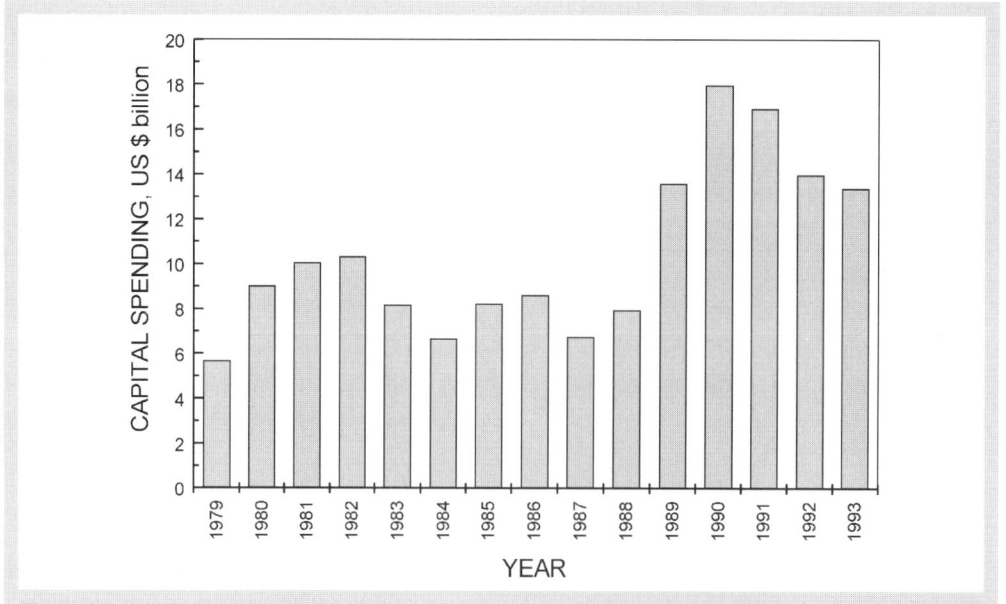

Figure 32. U.S. three-year capital spending plans reported by companies for projects completed the previous years and planned for the next years in the pulp and paper industry[33].

Environmental projects in 1993 accounted for 14.8% of total spending plans. This was an increase from 13.8% in 1992 and 11.1% in 1991. Mills had to work harder to comply with stricter regulations covering dioxin formation, odor control, and recycled fiber content.

The boom years of the late 1980s saw the paper companies (many reporting record profits) investing in new capacity. This oversupply then led to falling prices and collapsing profits, and the borrowing to implement new projects left some companies looking distinctly unhealthy. To make matters worse, this overcapacity coincided with a pronounced and severe slowdown in the general economy of the industrialized world for many years. As a result, many plans for further expansion were delayed to await the next round of tight market conditions[34].

Projects completed in the pulp and paper industry in 1992 accounted for US$ 3.4 billion of the total U.S. spending. Reported capital expenditures for 1992–1994 for pulp production related equipment was US$ 5.3 billion and for rebuilds was US$ 1.34 billion. On average, this meant that U.S. companies in the early 1990s would invest US$ 1.77 billion annually for pulping equipment and US$ 0.22 billion for rebuilds. Half the rebuilds would be for paper machines. In the near future, the U.S. pulp industry will invest US$ 1.99 billion for pulping equipment.

One may assume that 500 000 tons of pulp is produced by machinery valued at US$ 0.5 billion. This means that 59 million tons of pulp produced in the United States annually will require machinery valued at US$ 59 billion. Assuming that the expected life of a pulp mill is 30 years and the salvage value is nil, depreciation without interest will be 59/30 = US$ 1.97 billion annually. To keep production at a level of 59 million tons, investments for machinery and rebuilds should be more than US$ 1.97 billion. The planned capital expenditure of US$ 1.99 billion is insufficient because the capacity of pulp production in the United States will increase. The annual maintenance costs for the U.S. pulp industry that today are US$ 20/ton x 59 million tons = US$ 1.18 billion will increase significantly if capital spending does not resume soon.

Understanding the low investment activity is easy because the pulp price during the last ten years has been US$ 420–800. This means that the 59 million tons of pulp produced in the United States during the high price years will give a net cash flow compared to the low price years of at least US$ 800–420/ton x 59 million tons = US$ 22.42 billion. The price of pulp is therefore the primary determinant in the pulp industry. Maintenance costs compared with net annual cash flow during a high pulp price boom is only a marginal figure. It is only 5.3% of the net "good time" cash inflow.

12.3 Maintenance costs compared with breakdown costs

Experience has shown that sudden breakdowns occur occasionally in a pulp and paper mill even if the mill follows a preventive maintenance program. The sales value of the lost production for one breakdown day at a pulp mill producing 1100 tons/day of bleached pulp can be FIM 2.3–3.8 million depending on the market price of the pulp. The average maintenance cost for one day at the same mill is estimated to be about FIM 0.110 million. This means that an investment of 1 FIM today in incorrect maintenance procedures can cost

CHAPTER 9

FIM 20–40 tomorrow if the procedures result in a breakdown.

The price of a breakdown at a paper mill producing 1100 tons/day can be as high as FIM 10 million/day. The maintenance costs of a paper mill would be only FIM 0.145 million/day. Here the ratio between daily maintenance cost and breakdown cost is almost 70. The high cost of a breakdown compared to maintenance costs results in the conservatism seen at pulp and paper mills. This conservatism is contrary to any changes that increase the risk of a breakdown. The conservatism has a strong influence on equipment manufacturers and all outside contractors doing jobs in the pulp and paper industry. Reliability of an equipment vendor or a contractor is the primary element of importance.

13 Equipment vendor's role in maintenance

The role of the original equipment manufacturer has traditionally been important in the training and education of maintenance people. In most cases, the manufacturer who designed the equipment and the processes understands the methods to optimize results and maximize the benefits of the equipment. That manufacturer can also provide "hands-on" support with the latest techniques. During training, personnel from the manufacturer can analyze existing methods vs. improved ones and discuss the equipment in detail.

13.1 Single equipment vendor

Suppliers of pulping systems today must be more competitive than ever. To remain competitive, they must give the best possible service and support. Service must include sale and stocking of spare parts; training seminars; technical expertise to support corrective maintenance; and programs to develop, install, and help implement preventive maintenance systems. Figure 33 shows the interface between an equipment manufacturer and a customer.

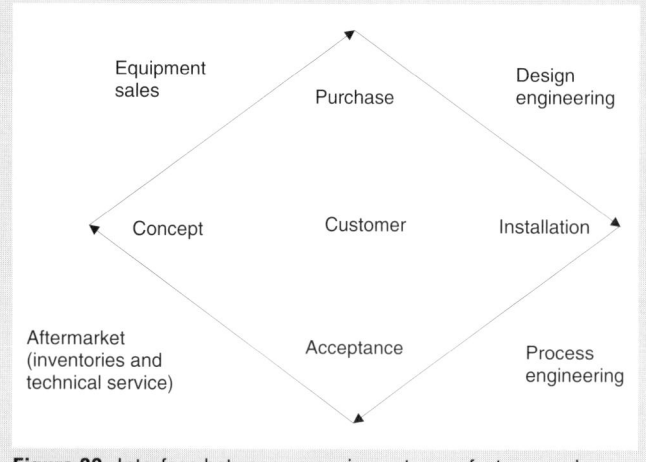

Figure 33. Interface between an equipment manufacturer and a customer[35].

Sales people have responsibility from concept to purchase. Design engineers then assume responsibility through installation and pass it to the process engineering people. It remains there until contract requirements are met and the equipment is turned over to the mill personnel. At that point, the service people assume the total responsibility for continued performance of the equipment.

During the guarantee period, the quality of the technical support needed can be crucial for the continuation of the future cooperation work between these two companies. Figure 33 shows that a different vendor will likely be chosen if any link breaks.

14 Vendor of a turnkey project

Approximately 30 years ago, the Nordic pulp and paper industry began to require turnkey projects from the manufacturers of recovery boilers and wood-fired boilers. In the late 1980s in the United States, a stronger trend toward vendor-led turnkey projects grew. Some owners perceived a cost advantage and a more valid performance warranty as benefits from this approach. This caused dramatic changes to the manner in which owners today implement projects and to the traditional roles of owners, engineers, constructors, and equipment vendors. Equipment vendors from Scandinavia and the remainder of Europe where vendors have led projects for several years in this way have spearheaded the approach.

Under a single contract, the turnkey vendor can provide total project management, design engineering, procurement of equipment and materials, construction, operations and maintenance manuals, operator training, flush-out, testing and checking services, startup assistance, and supervision. The turnkey vendor will remain on the site during operations and conduct quality and operational performance tests according to the turnkey vendor's warranties and guarantees.

In summary, the turnkey vendor will do everything short of actually operating the facility. In some cases, financing of the project is part of the turnkey package. Most vendor turnkey projects are for areas for which they have developed specific technical packages suitable for a variety of projects.

A threat exists that continuation of this trend may lead to less plant-wide process expertise available to pulp and paper companies from outside sources other than equipment vendors. Large engineering companies are certainly suspicious. Today, the very existence of engineering companies specialized in the pulp and paper industry is under challenge. In a turnkey project, the reliability of the vendor is even more important and critical than with single equipment deliveries. In a project that consists of many processes, the systems can possibly be incompatible leading to serious production problems and low reliability during commissioning of the project.

14.1 Action plan for a pulp and paper mill

Results of North American reviews have revealed that measurements of maintenance efficiency vary at different mills. No single technique can evaluate maintenance effectiveness. Each company or mill should therefore agree on its own measurements.

The first action to improve maintenance of a pulp and paper mill is to gather the history of the maintenance facts measured normally. Methods for evaluating lost time and identifying bottlenecks in the process then require consideration. Equipment that has a high risk of sudden breakdown requires identification for maintenance and production people. All improvements to decrease the risk should be planned and implemented if economically possible.

The trend in the pulp and paper industry is toward a maintenance system that takes advantage of the abilities and skills of all individuals in an organization. Since the maintenance craft system was developed over a number of years, changing it quickly is difficult. An evaluation of how labor unions will react to the demand for multicraft employees is important.

One approach to achieving better maintenance is to introduce lower level maintenance tasks to the production organization and continue to transfer responsibilities as production workers become trained and qualified. Typically, the transfer of responsibility begins with equipment cleaning, simple visual inspections, and straightforward lubrication. As operators become more qualified, they can take responsibility for routine preventive maintenance, common corrective maintenance, and limited troubleshooting and diagnostics.

The maintenance department therefore becomes a resource for a continuous improvement process. Clearly, the maintenance department will remain the keeper of the specialized techniques and highly sophisticated routines required to maintain the increasing level of technology found in modern mills.

Even if the instruments for on-line measurements and monitoring of the parameters affecting maintenance are top quality and a well-functioning maintenance management program exists, one parameter still requires handling with care. This concerns the people at the mill. It includes not only maintenance people but also production labor and staff. A living and rewarding cooperation between production and maintenance is vital for good maintenance performance. Methods for establishing and measuring this cooperation are necessary. Top management should give rewards to the teams that promote cooperation through their actions.

The example at an International Paper mill shows the proper principles for implementation of a good maintenance environment at a mill. These are the following:

- Help people understand the business
- Provide people with needed skills
- Explain expectations and obligations
- Enable people to influence decisions that impact them.

15 Conclusions

In 1992, the total production of pulp and paperboard in the world was 245 million tons. Assuming the variation of maintenance costs at US$ 42–85/ton, the pulp and paper industry in 1992 consumed US$ 10–20 billion in maintenance globally. The total industrial maintenance cost in the world is about US$ 250 billion. Maintenance costs for the pulp and paper industry is therefore almost 10% of global industrial maintenance costs.

In Finland, the production of pulp and paper in 1992 was 9.2 million tons. Assuming that the maintenance costs in Finland are approximately at the same level as in the United States, the total maintenance expenditures in the Finnish pulp and paper industry were US$ 386–786 million for 1992.

The production of pulp in the world in 1992 was 155 million tons, and maintenance expenditures in the pulp industry globally in 1992 reached FIM 15 billion (about US$ 3 billion). The production of pulp in Finland in 1992 was 8.75 million tons leading to a maintenance cost estimate of FIM 875 million in the Finnish pulp industry.

The capacity expansion boom that drove capital spending plans to record levels in the late 1980s played itself out in 1991. In the near future, the U.S. pulp industry will invest annually only US$ 1.99 billion in pulping equipment. The 59 million tons of pulp produced in the United States annually will need machinery valued at US$ 59 billion. Assuming that the expected life of a pulp mill is 30 years and the salvage value is nil, the depreciation without any interest will be US$ 1.97 billion annually. Because the capacity of the U.S. pulp production will increase in the near future, a capital investment of US$ 1.99 billion annually is not sufficient. This means that the annual maintenance costs for the U.S. pulp industry that today is US$ 1.18 billion will increase drastically if capital spending does not resume soon.

Pulp prices during the last ten years have fluctuated at US$ 420–800. This means that the 59 million tons of pulp produced in the United States during the high price years give a net cash flow compared with the low price years of at least US$ 22.42 billion. This net cash inflow gives a payback period of 2.6 years. The price of pulp is therefore the primary factor in the pulp industry. Maintenance costs compared with this net annual cash inflow during a high pulp price boom is only a marginal figure. It is only 5.3% of that net "good time" cash inflow.

The sales value of the lost production for one breakdown day at a pulp mill producing 1100 tons of bleached pulp daily is FIM 2.3–3.8 million depending on the market price of pulp. The average maintenance costs for one day at the same mill is approximately FIM 0.110 million. This means that an improper investment of 1 FIM in maintenance can cost FIM 20 or even FIM 40 tomorrow if this wrong investment results in a breakdown of the mill.

The price of a breakdown at a paper mill producing 1100 tons/day can be as high as FIM 10 million/day. The maintenance costs of a paper mill would be only FIM 0.145 million/day. Here the ratio between daily maintenance cost and breakdown cost is almost 70. The high cost of a breakdown compared with maintenance costs means that breakdowns caused by equipment manufacturers should not exist. This is why reliability of an equipment vendor or a contractor is the most important parameter to check before a cooperative venture begins. An equipment vendor should therefore focus on improving the reliability and maintainability of its products. Such actions are the following:

- Run quality projects
- Improve project management skills
- Organize a collection of feedback from mill maintenance and production people after each project
- Improve communications inside the company and thereby decrease blindness due to isolation and arrogance

CHAPTER 9

- Make parameters affecting reliability and maintainability the first priory during the design phase of a project.

Many pulp mills will encourage a closer working relationship between maintenance and production with a philosophy of joint responsibility for profitable operation. Progressive mills will discover the benefits of using the "team" approach toward the planning and scheduling of maintenance action and problem solving. Lost-time analysis will focus on cause and correction and replace the traditional preoccupation with who (maintenance, production, or other group) should receive blame for the downtime.

Equipment will become more automated and reliable by design but will require an increasingly skilled and trained maintenance force to assure intended availability and performance. Training will assume a role of greater importance since maintenance personnel must be aware of the latest technology and technical practices.

The main trends affecting maintenance at a pulp and paper mill will probably include the following:

- Increasing emphasis and reliance on preventive rather than repair maintenance

- Greater dependence on condition-assessment monitoring that could indicate the need for action to prevent breakdown (More machinery will be factory-equipped with monitoring sensors and instrumentation. This will mean closer cooperation with equipment manufacturers.)

- Operating a mill for longer periods and scheduling downtime strategically for routine operator activities and planned maintenance (The dependance on outside contractors should not be too strong.)

- Increasing acceptance of the benefits of using computerized systems for record keeping, planning, scheduling, and documentation

- Team approach between maintenance and production to provide enhanced technical support

- Increasing "flexibility" in trades manpower for maximum productivity and effectiveness

- Greater emphasis on people at all levels of the maintenance organization (More flexible organizational structures to allow the greatest effectiveness and functioning of personnel as partners in the maintenance mission.).

Maintenance will emerge as an increasingly vital business function through its goal of assuring maximum equipment uptime at peak performance with lowest overall costs.

References

1. Baldwin, R. F., Managing Mill Maintenance; The emerging realities, Miller Freeman, Boston, 1990, Ch. 1.

2. Pardue, F., Piety, K., and Moore, R., Elements of reliability-based maintenance; Future vision for industrial management, 1992 Conference on Pulp and Paper Maintenance, Pulp & Paper, Chicago, p. 1.

3. Plattonen, J., Results from a practical case in the pulp and paper maintenance development, 1990 Proceedings of 24th EUCEPA Conference, EUCEPA, Stockholm, p. 339.

4. Allison, A. W. and Shoudy, C. A., in Maintenance in Pulp and Paper Mill Equipment, 1961 Paper Trade Journal, New York, 1961, p. 180.

5. Palko, E., Plant Engineering 43(8):55(1989).

6. Smith, K. E., Pulp & Paper 61(9):97(1987).

7. Chute, J. R., Pulp and paper maintenance for today and tomorrow: What you really need to know, 1991 Maintenance Conference, CPPA, Montreal, p. 29.

8. Coleman, M., in Maintenance Practices in Today's Paper Industry (K. L. Patrick, ed.), Miller Freeman, San Francisco, 1986, p. 11.

9. Sward, K., in New Maintenance Strategies; Organizing, Implementing, and Managing Effective Mill Programs (K. L. Patrick, ed.), Miller Freeman, San Francisco, 1992, p. 36.

10. Grimnes, S. H., in Pulp, Paper and Board Mill Maintenance (J. F. W Evans, ed.), Paper Trade Journal, New York, 1957, p. 7.

11. Idhammar, C., Results Oriented Maintenance in Pulp and Paper Manufacturing, TAPPI 1992 Engineering Conference Proceedings, TAPPI PRESS, Atlanta, Book 2, p. 701.

12. Aurell, R. and Isacson, C.-I., Pulp & Paper 56(8):156(1982).

13. Sanclemente, M. R., in New Main Strategies, Organizing, Implementing, and Managing Effective Mill Programs (K. L. Patrick, ed.), Miller Freeman, San Francisco, 1992, p. 36.

14. Frampton, W. C., in New Main Strategies, Organizing, Implementing, and Managing Effective Mill Programs (K. L. Patrick, ed.), Miller Freeman, San Francisco, 1992, p. 20.

CHAPTER 9

15. Thornton, R. T. and Framton, W. C., Craft Interference in Maintenance - Causes, Costs, and Solutions, TAPPI 1992 Engineering Conference Proceedings, TAPPI PRESS, Atlanta, Book 2, p. 713.

16. Molkentin-Matilainen, P., Finnish Forest Industry Federation. Personal information, 1997.

17. Young, J., in: New Main Strategies, Organizing, Implementing, and Managing Effective Mill Programs (K. L. Patrick, ed.), Miller Freeman, San Francisco, 1992, p. 206–209.

18. Smith, K. E. and Carpenter, B. A., Pulp & Paper 60(9):60(1986).

19. Barker, E. F., in Pulp, Paper and Board Mill Maintenance (J. F. W. Evans, ed.), Paper Trade Journal, New York, 1957, p. 112.

20. Idhammar, C., in New Main Strategies, Organizing, Implementing, and Managing Effective Mill Programs (K. L. Patrick, ed.), Miller Freeman, San Francisco, 1992, p. 3.

21. Idhammar, C., Reliability Centered Maintenance - RCM, 1992 Conference on Pulp and Paper Maintenance, Pulp & Paper, Chicago, p. 1.

22. Idhammar, C., Equipment efficiency through operations and maintenance management, 1990 Proceedings of 24th EUCEPA CONFERENCE, EUCEPA, Stockholm, p. 222.

23. Idhammar, C., Pulp & Paper 66(11):35(1992).

24. Pierce, R., Maintenance productivity - it can be achieved, 1980 Maintenance Conference, CPPA, Montreal, p. 117.

25. Foszcz, J. L., Plant Engineering 46(11):50(1992).

26. Larsen, A. S., Pulp & Paper 60(9):74(1986).

27. Pounds, D. P. W., Wazny, G. M., and Reithaug, H., Summary of returns from maintenance survey questionnaire, 1984 Maintenance Conference, CPPA, Montreal, p. 17.

28. Pounds, D. P. W., Wazny, G. M., and Reithaug, H., Summary of returns from maintenance survey questionnaire, 1984 Maintenance Conference, CPPA, Montreal, p. 1.

29. Harrison, A., Pulp & Paper 67(2):43(1993).

30. Lauermaa, K., Paperi ja Puu 75(4):195(1993).

31. Anon., Pulp & Paper Week, Price Watch, (3); 2(1993)

32. Anon., Pulp & Paper International 35(1):33(1993).

33. Espe, C., Pulp & Paper 67(1):75(1993).

34. O'Brian, H. and Pearson, J., Pulp & Paper International 35(1):42(1993).

35. Hall, D., Pulp & Paper 61(9):111(1987).

Conversion factors

To convert numerical values found in this book in the RECOMMENDED FORM, divide by the indicated number to obtain the values in CUSTOMARY UNITS. This table is an excerpt from TIS 0800-01 "Units of measurement and conversion factors." The complete document containing additional conversion factors and references to appropriate TAPPI Test Methods is available at no charge from TAPPI, Technology Park/Atlanta, P. O. Box 105113, Atlanta GA 30348-5113 (Telephone: +1 770 209-7303, 1-800-332-8686 in the United States, or 1-800-446-9431 in Canada).

Property	To convert values expressed in RECOMMENDED FORM	Divide by	To obtain values expressed In CUSTOMARY UNITS
Area	square centimeters [cm^2]	6.4516	square inches [in^2]
	square meters [m^2]	0.0929030	square feet [ft^2]
	square meters [m^2]	0.8361274	square yards [yd^2]
Density	kilograms per cubic meter [kg/m^3]	16.01846	pounds per cubic foot [lb/ft^3]
	kilograms per cubic meter [kg/m^3]	1000	grams per cubic centimeter [g/cm^3]
Electrical conductivity	millisiemens per meter [mS/m]	0.1	microsiemens per centimeter [μS/cm]
	millisiemens per meter [mS/m]	0.1	micromhos per centimeter [$\mu\Omega^{-1}$/cm]
Energy	joules [J]	1.35582	foot pounds-force [ft • lbf]
	joules [J]	9.80665	meter kilograms-force [m • kgf]
	millijoules [mJ]	0.0980665	centimeter grams-force [cm • gf]
	kilojoules [kJ]	1.05506	British thermal units, Int. [Btu]
	megajoules [MJ]	2.68452	horsepower hours [hp • h]
	megajoules [MJ]	3.600	kilowatt hours [kW • h or kWh]
	kilojoules [kJ]	4.1868	kilocalories, Int. Table [kcal]
	joules [J]	1	meter newtons [m • N]
Frequency	hertz [Hz]	1	cycles per second [s^{-1}]
Length	nanometers [nm]	0.1	angstroms [Å]
	micrometers [mm]	1	microns
	millimeters [mm]	0.0254	mils [mil or 0.001 in]
	millimeters [mm]	25.4	inches [in]
	meters [m]	0.3048	feet [ft]
	kilometers [km]	1.609	miles [mi]
Mass	grams [g]	28.3495	ounces [oz]
	kilograms [kg]	0.453592	pounds [lb]
	metric tons (tonne) [t] (= 1000 kg)	0.907185	tons (= 2000 lb)

Property	To convert values expressed in RECOMMENDED FORM	Divide by	To obtain values expressed In CUSTOMARY UNITS
Pressure, stress, force per unit area	kilopascals [kPa]	6.89477	pounds-force per square inch [lbf/in^2 or psi]
	Pascals [Pa]	47.8803	pounds-force per square foot [lbf/ft^2]
	kilopascals [kPa]	2.98898	feet of water (39.2°F) [ft H2O]
	kilopascals [kPa]	0.24884	inches of water (60°F) [in H2O]
	kilopascals [kPa]	3.38638	inches of mercury (32°F) [in Hg]
	kilopascals [kPa]	3.37685	inches of mercury (60°F) [in Hg]
	kilopascals [kPa]	0.133322	millimeters of mercury (0°C) [mm Hg]
	megapascals [Mpa]	0.101325	atmospheres [atm]
	Pascals [Pa]	98.0665	grams-force per square centimeter [gf/cm^2]
	Pascals [Pa]	1	newtons per square meter [N/m^2]
	kilopascals [kPa]	100	bars [bar]
Speed	meters per second [m/s]	0.30480	feet per second [ft/s]
	millimeters per second [mm/s]	5.080	feet per minute [ft/min or fpm]
Volume, fluid	milliliters [mL]	29.5735	ounces [oz]
	liters [L]	3.785412	gallons [gal]
Volume, solid	cubic centimeters [cm^3]	16.38706	cubic inches [in^3]
	cubic meters [m^3]	0.0283169	cubic feet [ft^3]
	cubic meters [m^3]	0.764555	cubic yards [yd^3]
	cubic millimeters [mm^3]	1	microliters [μL]
	cubic centimeters [cm^3]	1	milliliters [mL]
	cubic decimeters [dm^3]	1	liters [L]
	cubic meters [m^3]	0.001	liters [L]

Index

A
abrasion 59–60, 290, 313–314
absorption 59–62, 163, 342
acid 81–82, 141, 239–241
acrylics...288
activation...................................73–76, 80–81
active alkali...195, 214
active metal 128–130, 304
activity..69–70
activity coefficient..70
additives.......................54, 156, 279, 287–288
adsorption 77, 140, 227, 278, 283
aeration 176, 227–228, 273
aerobic 213, 225–229, 231
alkali...........168, 178, 180, 267, 274–275, 299
alkaline extraction........................63, 203, 210
alloying................24, 31–40, 43–46, 270–271
amphoteric metal 304
anaerobic.............................213, 225–229, 231
anion 139–141, 178, 185
annealing..26–27, 30
anode 68–69, 77–82, 302, 312, 317–321, 323–326
anodic protection 315–319
anodic reactions 68, 79, 140–141, 282
anolyte 77
AOD process 35
arc strike 53
arc wire spray...308
arrest temperature....................................138
atmospheric corrosion ..91–92, 166, 221–224, 259–260, 303
austenite...............................29, 34–35, 38–39
austenitic stainless steel52, 158–159, 201, 217
austenitizing 30
auxiliary electrode 109

B
Bacillus..225
bacteria 213, 219, 225–231, 283–286
bainite 30

ball valve 259
barrier coatings 75, 287, 296, 299–301
base...199–200
bimetallic corrosion....................................128
binder........................219, 287–288, 296–297
biocide...283–285
biofilm 225, 227, 229–231, 283–284
biofouling 225, 283
black liquor...............182, 194–195, 214–218, 237–239
black oxide 315
blackening 315
blast furnace..29
blistering 161–163, 234, 293
body-centered cubic..................21–22, 30, 45
boilers 167–170, 217–218, 237–242, 423
brightening 204–205, 208, 210
brittle fracture....................21–24, 30, 38, 153
brittleness 24, 61
brush plating 312
Butler-Volmer equation.......73–74, 76, 97, 102

C
calomel electrode 112, 324
carbon equivalent............................31, 48, 52
carbon steels...31–35, 48, 230, 240, 267–269
carburization 236
cast materials..26
cathode 68–69, 75–82, 229, 318–319
cathodic inhibitor 282
cathodic protection............................320–326
cathodic reactions 68, 79–82, 86, 176, 186, 296, 338
catholyte..77
cation 179, 199
caustic...............................179, 264, 274, 317
caustic black...315
caustic embrittlement................................154
caustic soda 60–61, 203, 207
cavitation............133, 135–137, 257–258, 351
cell potential 69, 79, 82

cellular glass 220
cementation 82, 253
cementite 30, 32
ceramics.................... 264, 307–309, 312–313
chalking.. 293
chemical recovery 193–194, 198–199, 214–215
chill zone 26
chlorates 47, 278
chlorides 143, 185, 188, 205–207, 219–220, 270, 275, 277–278
chlorine 181–183, 187–191, 202–207, 275, 277–280, 283–285
chlorine dioxide......... 183, 188–189, 205–207, 209–210, 278–280
chromium equivalent.. 35
cladding 33, 42, 237–238, 304–306
cleaning............. 210–211, 283, 293, 348, 360, 373–374, 382, 387, 397, 424
cleavage 162
cleavage fracture 22, 30, 47
coal tar 290
coating 129, 286–288, 339, 341, 348
Coccus 225
cold work 30, 154
composites.................... 54, 57, 59–60, 63, 309
concentration polarization 82, 103
concrete 33, 205, 259, 341
condensate . 133, 168–169, 214, 231, 253, 281
conditioning film 227
conductivity.......... 81, 176, 184, 279, 335–337
conversion coating 314
cooling water..................... 148, 230–231, 285
corrosion allowance 125–126, 258, 332
corrosion cell 78–79, 249, 320
corrosion current density 80–81, 320, 323
corrosion effect 123
corrosion fatigue 153, 155–157, 159, 203
corrosion monitoring 119, 331–332, 355
corrosion potential . 80–81, 266, 279, 336–339
corrosion product........................... 83, 86, 131
corrosion rate........................ 78–79, 106–109
corrosion resistance........................ 19–21, 86
corrosion system 67, 123
corrosiveness................................... 202, 331
corrosivity 86, 175–176

counter electrode 111–113, 338
coupon tests.. 333
cracking 89–91, 117–118, 153–159, 198, 263–265, 268–271, 293
creep 34, 163–164
creep resistance 34, 170, 234
crevice corrosion................... 89–90, 116–117, 137–138, 140–141, 158, 203, 253, 258, 270–272
critical current density .. 85, 134, 196, 317–319
critical humidity............................... 221–222
crystal structure 21, 28
current density 72–76, 80–81, 94–95, 107–108
cyclic polarization curve 97–98
cyclic stress 153, 156

D

deaeration 241, 280
dealloying...160
degree of reduction 237, 239
delignification............. 183, 193, 195, 203–210
delta ferrite 153
dendritic structure 21
depolarization...76, 229
design 125–126, 131–132, 134–137, 249–253, 257–260, 317–318, 366–369, 382, 386, 389, 422–423, 426
detonation coating..309
differential aeration cell 145
diffusion layer..77, 82
diluent 287
dip plating 312
disk refining 201–202
dislocation... 21–22
dispersion strengthening..............................32
dissociation pressure 232
dithionite 209
double layer..77–78
ductile fracture 22
ductility 24, 27, 158, 162–163
duplex stainless steel 41, 159, 270
dynamic equilibrium............................. 69, 71

E

elastomer 54, 60, 259, 314
electrochemical cell 68–69, 310, 342

electrochemical corrosion 67, 137, 181–182
electrochemical equivalent 72, 106–107, 114, 119
electrochemical impedance 103, 340
electrochemical potential............ 129, 336, 341
electrochemical protection 267, 315–316, 336–337
electrochemical series 70, 86, 93, 132, 339
electrochemical tests..................................... 86
electrode 68–73, 109–113, 116, 318–319, 337–339
electrode potential 68–73, 94, 109–113, 319
electrodeposition 302, 312
electrolysis 69, 71, 189–190, 320
electrolyte.............. 72–73, 143–144, 176–177, 191–192
electroplating...................... 162, 301, 310, 312
Ellingham diagram 232
embrittlement........... 31, 52, 62, 154, 161–163
enamel coatings 313
endurance limit... 156
epoxy 54, 221, 287–290, 296–297
epoxy esters 288, 290
equilibrium diagram 24
equilibrium potential........... 69–72, 84–85, 182
equivalent circuit 103–106, 115, 341
erosion 133, 135–137, 198, 257–258, 287, 309, 364, 375
erosion corrosion 133, 135–137, 176, 198, 203, 216, 319, 364
eutectic temperature..................................... 25
Evans diagram... 80
exchange current................ 71–72, 74–75, 102
exfoliation.. 149

F
face-centered cubic 21, 30
fatigue limit 156
fatigue resistance 24, 59, 156, 349
ferrite 29, 34–35, 153, 158, 160–161
ferritic... 31, 34–35
filiform corrosion 129
flame spraying.. 308
flow-induced corrosion 132–133, 135
flue gas 60, 239, 275, 301, 306
formic acid 47, 126, 179
fouling 225, 239
fracture........................... 21–24, 135, 156–157
fracture toughness 24, 135
fretting............................... 46, 133, 309–310
furan... 59, 314
fusion line... 49, 52–53

G
galvanic cell............................ 69, 78, 131, 154
galvanic corrosion 128–132, 258, 353
galvanic couple........... 128, 131, 160–161, 339
galvanic series...................... 93, 129, 132, 251
galvanizing 127, 132, 221, 260–261, 302–304
galvanostatic test 98
general corrosion 123, 198, 231, 319, 335
Gibbs free energy 69–70, 232
grain 21–22, 149–152, 251
grain boundary 98, 160
green liquor 126, 136, 187, 216–217
groundwood 201–202

H
Haber-Luggin capillary 112–113
halogens................................... 178, 185, 236
hard water 185, 304
hardness 27, 135, 269, 309–310
headbox... 211, 213
heat tint 28, 53, 131
heat treatment 26–28, 150, 153–156, 306
heat-affected zone 26, 49, 52, 151, 163
Helmholtz double layer 77
hexagonal structure................................ 21, 45
high-temperature oxidation 334
hot corrosion................ 34, 166, 170, 235, 275
hot cracking 155
hot dip coatings.. 302
hot dipping 301–304
hot flocking 297
hot working 21, 27, 153
Huey test 90
humidity test 92
hydrochloric acid........ 159, 178, 185, 218, 273
hydrodynamic layer 77
hydrogen attack.. 161
hydrogen blistering 161–162
hydrogen damage.............. 161–163, 167, 242
hydrogen embrittlement 131, 161–163, 172

hydrogen evolution.......... 75, 81–83, 162–163, 180–182, 229, 280, 318, 342
hydrogen overvoltage.................................75
hydrolysis........... 140–141, 143, 187, 199, 220
hydrosulfite 209
hypochlorites................ 47, 189, 275, 278, 284

I

immersion plating 312
immunity 84, 160
impact strength 24, 310, 349
impingement attack 48, 133, 257, 259
impressed current cathodic protection....... 320
inclusion...................................... 123, 162, 342
incubation time 90
inert anode 326
inhibitor 185, 250, 282–283
inorganic coatings 312
insulation 60, 218–221, 253
intergranular corrosion.............. 36, 38, 52–53, 149–155
intergranular cracking 158
intermetallic phases 39, 146
internal oxidation.. 236
interstitial atom 21
iron-based alloys 35
isocorrosion diagram 264, 274–275

J

joining 49, 253–254, 350
joints .. 132–133, 138, 154, 253–261, 335, 416

K

knife-line attack 49, 53
kraft process 193, 195, 199, 214, 314

L

Langelier index 186–187, 243
lattice defects 21–22
lime mud 216
limiting current density 75–76, 81, 94, 97
linear oxidation... 235
linear polarization 74, 102–104, 337–338
linings 56, 59–60, 288, 301, 312–314
localized corrosion 82, 137–138, 185, 225, 253, 263, 317, 322, 326, 332
low alloy steels . 28, 31–34, 167, 169, 264, 269
Luggin capillary 112

M

magnetite 28, 163, 180–181, 240–242, 318
martensite 27, 30, 32
mass loss 106–107, 196, 332–333
mass transfer ... 75, 80–83, 112, 180, 266, 336
materials selection 149, 249, 261, 263, 326
mechanical pulping 41, 192, 201–202
metallic coatings 131, 301, 307
microbiological corrosion................... 224–225
microstructure....... 24–27, 35–36, 38–40, 153, 270–271
mild steel................................. 126, 169, 269
mill scale 28, 131, 250, 286, 293, 302
mineral acids 62, 64, 178, 273
mineral wool..................................... 219–220
mixed potential theory....... 67, 79–80, 119, 321
molten salts..277
molybdenum 31, 34, 36–45, 143, 146, 191, 200–201, 270–271
mortar 313–314

N

Nernst equation 70
nickel............. 31, 34–38, 40, 42–45, 143, 146, 180–181, 200, 270–271, 273–275
nickel equivalent..35
nickel-base alloys 302
noble metals................180–181, 263, 276, 302
normalizing 27, 52
nucleation...........................137–138, 164, 257

O

ohmic drop...................... 76, 80–81, 113–114
orange peeling..293
organic coatings 255, 297, 299
overlay welding 41, 268, 305–306
overpotential 72–76, 229
overprotection 163, 322–323
overvoltage.. 75
oxidation 67–70, 164–166, 180–182, 189–191, 232–236, 288, 309
oxidation resistance 169–170, 234, 254
oxide film 28, 133, 163–164, 197, 232–236
oxide scale 164, 166, 233
oxidizer................ 70, 181, 187, 208–209, 228, 236, 279
oxidizing agent 89, 285

Index

oxygen 75–76, 81–83, 140–141, 164, 180–183, 221–222, 238, 280–281
oxygen corrosion 167, 241
oxygen evolution 83, 138
ozone 181–183, 203, 205, 301

P

painting 221, 287, 292–293, 403, 419
paints 286–290, 296–297, 299
parting 160
passivation 82–85, 192–193, 273, 279, 317–319
passive 28, 84–85, 137–138, 196, 282, 336–337
passivity 67, 83–85, 278, 317
pearlite 30
pH 176, 178–180, 230, 240–243, 278–280, 342
phase diagram 24–25, 29–30
phase stability diagram 233
phenolic resin 299, 314
pickling 28, 162–163, 286, 302
pig iron 29
pigment 287, 297
Pilling-Bedworth ratio 234
pitting 126, 128, 136–140, 157–158, 266–267, 317, 336, 341
pitting corrosion 98, 112, 137–140, 191, 271–272
pitting factor 145, 333
pitting resistance equivalent ... 38–39, 146, 270
plasma spraying 297, 308
plastic deformation 22–24, 32, 256
plating 42, 256, 310, 312
plug lining 306
point defects 21–22
polarization 74–76, 79–82, 317–318, 325–326, 332
polarization curve 76, 84–85, 92, 94–95, 97–99, 114
polarization resistance 102–104, 106, 338–341
polyester 58–59, 219, 288, 299–300
polyethylene 54, 57, 63
polyisocyanurate 219–220
polymer 54–57, 297–299
polysulfide 187, 197, 245

polyurethane 219–220, 314
polyvinyl chloride 58, 63, 290
porosity 26, 52, 164, 304
potential 67–70, 182–183, 266, 279, 317–326, 378, 380, 386–388, 391
potentiodynamic polarization 97, 102, 338
potentiostat 109, 113, 338–339
potentiostatic test 99
Pourbaix diagram 83–84, 177, 180
powder flame spray 308
Prandtl layer 77, 82
precipitate 128, 151, 153
primer 288, 292–293, 298, 313
probes 332, 334–336, 338, 342–343
protection current density 323–324
protection potential 138, 317, 322–325, 336
pulping 187, 192–195, 198–202, 319, 421–422, 425
quenching 26–27, 30, 32, 46, 153

R

rate law 165–166, 234
reaction rate 71–73, 77–78, 94, 164, 191–192, 228–229, 282, 339
recausticizing 216–217
recovery boiler 167, 198, 214–217, 236–239, 310, 379
recrystallization 27, 30
redox potential 68, 89–90, 101, 144, 180–183, 206, 208, 210, 273
reducing agent 34, 189, 207, 281
reduction 68–70, 81–82, 141, 178, 189, 228–229, 237–239
reference electrode 109–113, 318–319, 337–338, 341
reinforced plastic 57, 62, 206, 300, 314, 325, 350
repassivation potential 97, 138, 146
residual stress 155, 256
resins 57–60, 287–288, 290, 299–300, 314
resistivity 113, 130, 143, 323, 334
Ryznar Stability Index 187

S

sacrificial anodes 286, 315, 320, 326
salt bridge 112
scaling 164, 240–241, 280

SCC 118, 153, 155–159, 268–269, 325
Schaeffler diagram 35–36
Schikorr reaction 240
selective leaching 128, 159–161
semichemical pulping 148, 201–202
sensitization 36, 52–53, 98, 149–153
sentinel holes 332
sentry holes.. 332
sheet lining.. 305, 329
sodium chloride 184–185, 189, 224
sodium hydroxide...... 126, 179, 193–197, 216, 242, 274–275
solid solution................................... 24, 30, 43
solvent 176, 287–288, 296–300
Spirillum... 225
stabilization 34, 151, 208, 306
stainless steel......... 33–38, 131, 146–149,152, 158–159, 266–268, 270–271, 273–274, 304–306, 322
standard electrode potential 69–71
steam 126, 167–170, 215–216, 237–242, 351
steel 28–34, 107–108, 126–127, 167–170, 180, 263–271
stray-current corrosion 349
strengthening 28, 31–32, 40, 43
stress corrosion cracking.. 117–118, 153–159, 268–271, 273–275, 325
stress relieving 27, 52, 251, 304
striations 157
strip lining.. 306
sulfate-reducing bacteria............ 225, 228–231
sulfide stress cracking................................. 158
sulfidity 195–196, 216–217, 237
sulfite process............ 193, 199, 201, 215, 217
sulfuric acid....... 178, 188–189, 200–201, 224, 228–229, 239, 268, 273

T
Tafel method........................... 74, 97, 339–340
tarnishing 164, 224
tempering 27, 30, 32
tensile strength........................ 22–24, 162–163
test cell.. 109, 112
thermal insulation......... 92, 218–219, 221, 349
thermal spraying 304, 306–307
thermomechanical pulping 192, 201

titanium 45–48, 146, 206, 208–209, 263–264, 276–279, 315
top coating 293
toughness 24, 51, 59, 135, 254
under-deposit corrosion 145, 200, 218, 257
uniform corrosion 92, 123–129, 157, 263–264, 288, 332–333

V
vacancy 21, 234
vinyl............................... 57, 287–290, 299–300

W
Wagner number.. 130
wallpapering 306
Warburg impedance 104–105
waterline corrosion.............................. 33, 129
weathering steel 267
weight loss 62, 86–90, 106–109, 145, 333
weld decay 53, 151
weld overlay 193
welding.................. 48–53, 149–155, 255–256, 304–306, 375
white liquor 181, 195–197, 214, 216–217
white water................. 147, 210–214, 230–231
wire flame spray 308
work hardening 32, 43
working electrode........ 109, 111–112, 116, 341

Y
yield strength 23–24, 155, 157–158, 162–163, 349

Z
zinc.............. 75, 180–181, 221–224, 287, 292, 296–297
zinc coating 254, 260, 302